Dialogues and Games of Logic

Volume 5

The Dialogue between Sciences, Philosophy and Engineering.

New Historical and Epistemological Insights

Homage to Gottfried W. Leibniz 1646-1716

Volume 1
How to Play Dialogues: An Introduction to Dialogical Logic
Juan Redmond and Matthieu Fontaine

Volume 2
Dialogues as a Dynamic Framework for Logic
Helge Rückert

Volume 3
Logic of Knowledge. Theory and Applictions
Cristina Barés Gómez, Sébastien Magnier, and Francisco J. Salguero, eds.

Volume 4
Hintikka's Take on Realism and the Constructivist Challenge
Radmila Jovanović

Volume 5
The Dialogue between Sciences, Philosophy and Engineering. New Historical and Epistemological Insights. Homage to Gottfried W. Leibniz 1646-1716
Raffaele Pisano, Michel Fichant, Paolo Bussotti and Agamenon R. E. Oliveira, eds., with a Foreword by Eberhard Knobloch

Dialogues and Games of Logic Series Editors
Shahid Rahman   shahid.rahman@univ-lille3.fr
Nicolas Clerbout
Matthieu Fontaine

The Dialogue between Sciences, Philosophy and Engineering.

New Historical and Epistemological Insights

Homage to Gottfried W. Leibniz 1646-1716

Edited by

Raffaele Pisano,

Michel Fichant,

Paolo Bussotti,

Agamenon R. E. Oliveira,

with a Foreword by

Eberhard Knobloch

© Individual author and College Publications 2017.
All rights reserved.

ISBN 978-1-84890-227-5

College Publications
Scientific Director: Dov Gabbay
Managing Director: Jane Spurr
Department of Informatics
King's College London, Strand, London WC2R 2LS, UK

www.collegepublications.co.uk

Original cover design by Laraine Welch
Printed by Lightning Source, Milton Keynes, UK

All rights reserved. No part of this publication may be reproduced, stored in a retrieval system or transmitted in any form, or by any means, electronic, mechanical, photocopying, recording or otherwise without prior permission, in writing, from the publisher.

CONTENTS

Foreword — vii
Eberhard Knobloch (Germany)

Introduction — xi
1646-1716. An Interdisciplinary Tribute to Gottfried Wilhelm von Leibniz' Anniversary
Raffaele Pisano (France) and Paolo Bussotti (Italy)

Acknowledgements — xvii

Remarks for the reader — xix

List of Contributors — xxi

Leibnizian Intelligibility — 1
Jacob Archambault (USA)

Leibniz's Foundation of Arithmetic and its Influence on Frege — 21
Mattia Brancato (Italy)

Historical and Philosophical Details on Leibniz's Planetary Movements as *Physical–Structural Model* — 49
Paolo Bussotti (Italy) and Raffaele Pisano (France)

Leibniz and the Impossibility of Squaring the Circle — 93
Davide Crippa (Czech Republic)

A Scientific Re-assessment of Leibniz's Principle of Sufficient Reason — 121
Antonino Drago (Italy)

Leibniz's Theory of Series — 141
Giovanni Ferraro (Italy)

Prospects for an Idealist Interpretation of Leibniz's Dynamics — 169
Glenn A. Hartz (USA)

Asymmetric, Pointless and Relational Space: Leibniz's Legacy Today — 179
Joseph Kouneiher (France)

We Live in the Best of Possible Worlds: Leibniz's Insight Helps to Derive Equations of Modern Physics Vladik Kreinovich (USA) and Guoqing Liu (China)	207
A Background Condition for Analysis Montgomery Link (USA)	227
From Leibniz to the Information Age. Leibniz's deep Footprints in Wiener's Scientific Path Leone Montagnini (Italy)	255
Leibniz and the Sciences of Engineering Agamenon Rodrigues Eufrásio Oliveira (Brazil)	287
***Logica Mathematica*: Mathematics as Logic in Leibniz** Anne Michel-Pajus (France) and David Rabouin (France)	309
Describing Reality: Bernoulli's Challenge of the Catenary Curve and its Mathematical Description by Leibniz and Huygens Miguel Palomo (Spain)	331
The Dialectics of Recognition in Leibniz's *Theodicy* and a Possible Route to Modern Political Thought Ana-Maria Pascal (UK)	341
The Equivalence of Hypotheses and Leibnizian *Vires* Tzuchien Tho (Italy)	363
Leibniz's Defence of Heliocentrism Friedel Weinert (UK)	381
Index	403

Foreword

The universal genius and outstanding mathematician Gottfried Wilhelm Leibniz died in Hannover on November 14, 1716. The maxim of life of this restless scholar is to be found in a mathematical manuscript presumably written in October 1673 (A VII,3, p. 537): *Malo enim bis idem agere, quam semel nihil.* (For I prefer to do the same twice instead of doing nothing once.) That he really did. We know for example thirty (vain) attempts to solve the so-called "six squares problem" in number theory (A VII, 1). Many books will appear in order to celebrate the 300th anniversary of his death including the present volume, and that for good reasons.

The richness of Leibniz's scientific and philosophical legacy cannot be overestimated. His literary estate comprehends about 100 000 sheets of paper. The definitive edition of his *Complete Works and Letters* appears since 1923, originally thanks to the Prussian Academy of Sciences, today thanks to the Berlin-Brandenburg Academy of Sciences and Humanities and the Göttingen Academy of Sciences. Up to now (March 2017) fifty-seven volumes have been published, but more than hundred-thirty are expected. In other words, we know less than one half of his estate.

This is especially true of his mathematical (series VII) and scientific, medical, and technical writings (series VIII). I established these series in 1976 and 2001, respectively. The more volumes have appeared, the better we can evaluate the development of his ideas, the value of his insights and methods, the correctness of his own assertions and affirmations.

Let us consider three examples.

1 Leibniz's Mathematical Career, the Influence of other Mathematicians, especially on his Invention of the Calculus

In 1920 James Marc Child believed to be able to demonstrate the so-called "Barrow hypothesis", already brought forward by Jakob Bernoulli and the Marquis de l'Hospital (Child 1920): Leibniz allegedly depended on Isaac Barrow's *Lectiones geometricae*. In 1926 Dietrich Mahnke published a fundamental paper that relied on a painstaking, thorough analysis of Leibniz's original mathematical manuscripts (Mahnke 1926). These manuscripts verified Leibniz's affirmation that Blaise Pascal, Honoré Fabri, Grégoire de St. Vincent, Christiaan Huygens were the most influential authors for him with regard to the invention of the calculus, not Isaac Barrow. Since 2008 the manuscripts are published and available for everybody (A VII, 4 and 5). It is useless to continue a fruitless discussion after these publications.

Eberhard Knobloch (2017) Foreword. In: Pisano R, Fichant M, Bussotti P, Oliveira ARE (eds.), *The Dialogue between Sciences, Philosophy and Engineering. New Historical and Epistemological insights. Homage to Gottfried W. Leibniz 1646-1716.* College Publications, London, pp. vii-ix
©2017 College Publications Ltd. ISBN: 978-1-84890-227-5 www.collegepublications.co.uk

2 Financial and Insurance Mathematics

In 1683 Leibniz published his only paper on financial mathematics. There he explained how to calculate the current value of a certain amount of money, that is, how to calculate the "interest accruing in the meantime" (Leibniz 1683). At the very end of his paper he promised another paper on the estimation and purchase prices of life annuities. Such an article never appeared. Only in 2000 a volume containing the fifty most important Leibnizean writings on insurance and financial mathematics revealed Leibniz's impressing results and far-reaching ideas in this respect (Leibniz 2000). Half of these studies were reprinted in 2001 in the academy edition (A IV, 4).

Leibniz turned out to be a pioneer of mathematical modelling of human life. He pleaded for solidarity by establishing public insurance companies in order to help those who suffered from great misfortunes. But he emphasized at the same time that everybody is responsible for himself so that he has not to burden his community. For him solidarity and justice were closely connected with each other.

3 The use of Infinitesimals, Foundational Questions

Already in 1926 Dietrich Mahnke had pointed out that Leibniz must not be blamed for having left out an exact foundation of his new calculus (Mahnke 1926, p. 61 note 1). He repeated this admonition in 1931 (Mahnke 1931, p. 596). Yet, still in 1972 Morris Kline maintained (Kline 1972, p. 384, 387): *Neither Newton nor Leibniz clearly understood nor rigorously defined his fundamental concepts [...] They relied on the coherence of the results and the fecundity of the methods to push ahead without rigor.*

Kline did not take notice of Lucie Scholtz's partial edition of Leibniz's *Quadratura arithmetica circuli* etc. (Scholtz 1934) though already the title of her PhD dissertation *The exact foundation of the infinitesimal calculus in Leibniz* should have warned him. Only in 1993 the first complete edition of this Leibnizean treatise appeared (Leibniz 1993) and was published in the academy edition of Leibniz's Complete writings and letters in 2012 (A VII, 6). It is the longest mathematical treatise he ever wrote. Its importance cannot be overestimated. It provides an exact foundation of infinitesimal geometry and integration theory based on a well-defined notion of infinitely small: infinitely small means smaller than any given quantity. Such a quantity must be a variable and can be replaced by Weierstrassean epsilon technique. In 2004 a French translation appeared (Leibniz 2004), in 2016 a bilingual Latin-German edition (Leibniz 2016). Since then a long series of papers has been written about this treatise by various authors among them having a paper in the present volume. It enabled historians of science to make a big step forward. We are waiting for still twenty-four volumes of series 7 and ten volumes of series 8: Its second volume has appeared in 2016. They will tremendously enrich our knowledge about Leibniz as mathematician and scientist. For the time being we know only a rather limited part of his literary estate. Very much has still to be done.

References

A = Gottfried Wilhelm Leibniz, *Sämtliche Schriften und Briefe*, edited by the Berlin-Brandenburgische Akademie der Wissenschaften and the Akademie der Wissenschaften zu Göttingen. Leipzig-Berlin since 1923.
A VII, 3 means: Academy edition series VII, volume 3.

Child, J. M. 1920. *The early mathematical manuscripts of Leibniz*. Translated from the Latin texts published by Carl Immanuel Gerhardt with critical and historical notes. Chicago-London.

Kline, M. 1972. *Mathematical thought from Ancient to Modern Times*. New York.

Leibniz, G. W. 1683. Meditatio Juridico-mathematica de interusurio simplice. *Acta Eruditorum* October: 425-432.

Leibniz, G. W. 1993. *De quadratura arithmetica circuli ellipseos et hyperbolae cujus corollarium est trigonometria sine tabulis*, kritisch herausgegeben und kommentiert von Eberhard Knobloch. Göttingen. (*Abhandlungen der Akademie der Wissenschaften in Göttingen* Mathematisch-physikalische Klasse 3. Folge Nr. 43).

Leibniz, G. W. 2000. *Hauptschriften zur Versicherungs- und Finanzmathematik*, hrsg. von Eberhard Knobloch und J.-Matthias Graf von der Schulenburg. Berlin.

Leibniz, G. W. 2004. *Quadrature arithmétique du cercle, de l'ellipse et de l'hyperbole et la trigonométrie sans tables trigonométriques qui en est le corollaire*. Introduction, traduction et notes par Marc Parmentier, texte latin édité par Eberhard Knobloch. Paris.

Leibniz, G. W. 2016. *De quadratura arithmetica circuli ellipseos et hyperbolae cujus corollarium est trigonometria sine tabulis*. Herausgegeben und mit einem Nachwort versehen von Eberhard Knobloch. Aus dem Lateinischen übersetzt von Otto Hamborg. Berlin-Heidelberg.

Mahnke, D. 1926. Neue Einblicke in die Entdeckungsgeschichte der höheren Analysis. *Abhandlungen der Preussischen Akademie der Wissenschaften* Jahrgang 1926 Physikalisch-mathematische Klasse Nr. 1. Berlin.

Mahnke, D. 1931. Zusätze zu den ungedruckten Handschriften, in: Joseph Ehrenfried Hofmann, Heinrich Wieleitner, Dietrich Mahnke: Die Differenzenrechnung bei Leibniz. *Sitzungsberichte der Preußischen Akademie der Wissenschaften* Jahrgang 1931 Physikalisch-mathematische Klasse. Berlin: 562-600. Here: 590-600.

Scholtz, L. 1934. *Die exakte Grundlegung der Infinitesimalrechnung bei Leibniz*. Marburg. (Partial edition of the PhD dissertation).

2017, March, Berlin, Germany
Eberhard Knobloch
eberhard.knobloch@tu-berlin.de

Introduction

1646-1716. An Interdisciplinary Tribute to Gottfried Wilhelm von Leibniz' Anniversary

Science and history of science are two topics, which have been studied for a long period both in the Eastern and in the Western cultural milieu. The results obtained in the course of the centuries depend on the theoretical and the practical discoveries of several categories of people: scientists, artisans, architects, engineers, and so on. From an anthropological and individual point of view, the ambitions has been an important element in such a complex picture.

Within this context, Leibniz was a particular important and polyhedral figure. Generally speaking, Leibniz shares with Descartes and Spinoza the deep conviction that human reason is sufficient to discover the profound nature of reality, and, because of this, he is traditionally classified with as a rationalist philosopher. Unlike Descartes, Leibniz particularly during his earlier career sought to reconcile the new mechanical philosophy of the Scientific Revolution of the 17th century with the Aristotelian-Scholastic traditions, which Descartes and Spinoza largely rejected. Besides the fundamental contributions not only to philosophy, the foundation with Newton, but independently, of the infinitesimal calculus, as well as an original theory of motion, some important aspects of Leibniz's work are until now not completely known.

Mainly during 1679 up to 1695 Leibniz worked as engineer-surveyor at silver mines (Harz Mountains). He was busy with designing and wind-powered pumps. In this period, he produced many designs for mining technology manufactures; and in the same period (1670s-1680s) he wrote essays on Logics and Universal language and concerning his conceptualization of the truth. Particularly, the correspondence with Denis Papin and Huygens (1680–1686) needs a new deep exploration. These technological manuscripts refer to Leibniz activities in up cited Harz where also economic aspects of his projects are shown in details besides Leibniz's work as engineer. One year later, *Discourse on Metaphysics* (1686) was written and Leibniz–Arnaud correspondence began. Further, Leibniz provided fundamental contributions to theoretical science, too. n Leibniz's dynamics, and in particular on his concepts of "vis viva", "vis mortua" and "inertia", there is a huge literature and these aspects have been explored in depth, even if, since their interpretation is far from being easy and universally accepted, new interesting contributions are possible. However, Leibniz also supplied contributions to astronomy. This part of Leibniz's scientific production is extremely interesting because he tried to construct a system of the world different from Newton's. The fact that his system was basically coherent from a mathematical point of view, but did not provide the correct explanation of the celestial bodies' movements is an interesting chapter in the complex history of the relation between physics and mathematics. Furthermore, important publications exist on Leibniz's astronomy, but they are few in number and in general date back to the period 1960-1990. Hence, new contributions to this section of Leibniz's thought are necessary.

In order to homage Leibniz's anniversary this collection volume presents new historical and philosophical insights taking into account science faced by Leibniz. The contributions composing this book are very well written by international and interdisciplinary scholars, who are expert on the proposed subjects. The contributors explain the various ways in which the Leibnitian sciences allowed to construct advanced modelling on the one hand, and to develop new scientific and philosophical ideas on the other hand. The scholars from different traditions have discussed in this book the emergency style thinking in a methodology and in a theoretical perspective, aiming to summarize the already known picture on Leibniz's and to offer new interpretations and insights. The book contains chapters on other dimensions that have their place in any rounded history and philosophy of science, and deep scientific aspects.

Jacob Archambault (USA) in "Leibnizian Intelligibility" starts from a classical subject of Leibniz *Forschung*: the critics addressed by Leibniz to Newton's conception of gravity as an unintelligible and unacceptable concept. Archambault analyses the correspondence Leibniz and Clarke – which is fundamental for this theme – as well as the *Antibarbarus Physicus* and other works by Leibniz. The inquire concerning the specific argumentations used by Leibniz against Newton's gravitation leads the author to profoundly explore the concept of Leibniz "intelligibility", reaching the conclusions: a) Leibniz connects his critics to Newton with those against the occasionalists; b) this idea can be fully justified within Leibniz's general physical conceptions; c) Newton's conception of gravity is not consistent with Leibniz's idea of the "best of all possible worlds".

Mattia Brancato (Italy) in his "Leibniz's Foundations of Arithmetic and its Influence on Frege" points out first of all the novelties of Leibniz's approaches in several fields and specifically in the *analysis situs*. This is due to his qualitative approach and the resort to relations such as homogeneity and similarity. Despite this, Brancato stresses Leibniz's influence on the philosophy of mathematics of the 19th century is almost exclusively limited to the birth of mathematical logic with Frege (this is also probably because the manuscripts on the *analysis situs* were unknown in the 19th century). Brancato refers to a long and interesting series of references, which show the specific influence of some works by Leibnis on Frege's *Grundlagen der Arithmetik*.

Paolo Bussotti (Italy) and Raffaele Pisano (France/Australia) in their "Historical and Philosophical details on Leibniz's Planetary Movements as

Physical-Structural Model" face two problems of Leibniz as a scientist: a) planetary theory; b) gravitational theory. Bussotti and Pisano describe and explain the two subjects. Afterwards they introduce the notion of physical-structural model: a model in which a mechanical explanation of the *actions* of the "forces" (this exists in Newton, too) and of their *origin* (this does not exist in Newton) is offered. They show that, with his planetary theory and his ideas on gravity – rethought in the course of the years – Leibniz tried to overcome Newton as a physicist by offering a more complete theory than Newton's and a theory free form the action at a distance.

Davide Crippa (Czech Republic) in "Leibniz and the Impossibility of Squaring the Circle" investigates the influence exerted on Leibniz by James Gregory's researches concerning the quadrature of the central conics. Thence the protagonists of these papers are Leibniz and Gregory to whom Huygens is added. For, Crippa argues that one of the reasons why Leibniz dealt with the arithmetic quadrature is connected to Huygens' convictions that the circle can be squared algebraically. In contrast to Huygens' idea, Leibniz proved that the circle cannot be algebraically squared.

Antonino Drago (Italy) in his "A Scientific Re-Assessment of Leibniz's Principle of Sufficient Reason" analyses the theories, which are not explained in a deductive manner. Their common denominator is the resort to the intuitionistic logic. Nonetheless, within these theories, the principle of sufficient reason is often implicitly used. Drago shows that the resort to this principle makes the logic of a theory to converge towards a classical logic. In particular, Markov's theory of computable numbers is analysed where Drago identifies two constrains which makes it impossible to apply the principle of sufficient reason. In the light of a more general reflection, Drago argues that, under such constrains, the principle of sufficient reason cannot be applied to metaphysical problems.

Giovanni Ferraro (Italy) in "Leibniz's Theory of Series" analyses the specific problem concerning Leibniz's work on the numerical series starting from the fundamental *Quadratura Arithmetica*, written between 1675 and 1676. Leibniz did not publish this work, but it was the basis of many of Leibniz's ideas on series published in several papers during the following years. Ferraro is basically interested in the internal motivations which lead Leibniz to design his theory. He focuses on an interesting aspect: the connections between the convergence of a series and the formal manipulations used by Leibniz. His thesis is that, albeit Leibniz used the sequence of the partial sums to calculate the sum of a series, his conviction is the limit-process does not actually arrives at providing the sum of the series. Rather it is only a method to operate upon the series.

Glenn A. Hartz (USA) in "Prospect for an Idealist Interpretation of Leibniz's Dynamics" starts analysing two idealistic interpretations of Leibniz: those offered by Robert M. Adams and Donald Rutherford. These two scholars try to insert Leibniz's dynamics within the general picture of his monadology. Hartz thinks that these interpretations are inadequate. In his opinion, Leibniz considers dynamics and monadology mutually autonomous disciplines. It is, hence, a mistake to force dynamics within monadology. Dynamics has proper concepts and methods, which cannot be reduced to those of the other discipline and which, on some occasions, can be useful to catch some aspects of body's metaphysics. To be clear, Hartz claims idealism cannot be exported to every part of Leibniz's work. As to dynamics, idealistic interpretation does not work.

Joseph Kouneiher (France) in his "Asymmetric, Pointless and Relational Space: Leibniz's Legacy Today" starts from an interpretations of Leibniz's monadology based on various issues. The most relevant are two: 1) the individuals are distinguished by differentiation, which subtends a relational geometric calculus tool; 2) the monads are interpreted as indivisible dynamical modes, similar to the dynamic modes defined by the modern quantum field theory. Relying upon this interpretation, Kouneiher enucleates: 1) a profound link between Leibniz and Grassmann's geometrical calculus as well as with some aspect of Cartan's work; 2) the Leibnizian root of modern physics until reaching general relativity.

Vladik Kreinovich (USA) and Guoqing Liu (China) in "We Live in the Best of Possible Worlds: Leibniz's Insight Helps to Derive Equations of Modern Physics" show that, although the idea of the best of possible worlds seems far from physics, in fact, it inspired significant development in modern physics. Kreinovich and Liu refer in particular, to the fact that, for the modern dynamics of the physical fields, the functional action satisfies certain conditions of optimality. The actions can be expressed by a series of expressions, the actual expression of the action, within this series, fulfils a criterion of optimality. This idea, which is designed and explained with all the necessary mathematical details by the authors is used to deduce the general equations of relativity, quantum mechanics, etc. The comparison is best possible worlds – Leibniz/criteria of optimality-modern physics.

Montgomery Link (USA) in "A Background Condition for Analysis" starts reminding the reader the classification of the totality of the doctrinal truths developed by Leibniz in *New Essays on Human Understanding*: 1) synthesis; 2) analysis; 3) classification by terms. Link points out the differences connoting the three approaches with the theoretical and practical advantages and problems connected to each of them. The author considers the logic and metaphysical aspect as well as the mathematical one, focusing, in particular on the analytical aspect. He reaches the conclusion that Leibniz's theoretical framework is not grounded in a definitive way. To overcome this difficulty he claims it is maybe possible to refer to the third form of classification. He then draws the conclusions from such hypothesis.

Leone Montagnini (Italy) in his "From Leibniz to the Information Age. Leibniz's deep Footprints in Wiener's Scientific Path" highlights a profound similarity between Wiener and Leibniz: both of them were multidisciplinary geniuses. Montagnini clarifies that Wiener appreciated Leibniz from his youth and considered Leibniz the first one to open the road towards the modern computers. From here, a long line of reasoning begins which connects Wiener with some threads departing from Leibniz and reaching Wiener himself: basically the threads Leibniz-Babbage and Leibniz-Maxwell-Gibbs. These threads are implemented with several considerations concerning Boolean algebra and mathematical logic. At the end of this paper, the reader will get a clear idea as Leibniz's influence on Wiener and modern cybernetics.

Agamenon Rodrigues Eufrásio Oliveira (Brazil) in "Leibniz and the Sciences of Engineering" points out that many aspects of Leibniz's scientific thought have been profoundly analysed by philosophers and historians of science, in particular his mathematics, dynamics, logic and the relations between these disciplines and metaphysics. Thence, one could reach the conclusion that Leibniz's entire scientific production has been examined. This is not the case because there is a discipline, to which Leibniz gave significant contributions, and which is rarely studied. This discipline is engineering. Oliveira after having analysed some aspects of Leibniz's scientific thought (in particular dynamics, differential and integral calculus) explains the main features of Leibniz's production as an engineering and the connections with such aspects of his scientific work.

Anne Michel-Pajus (France) and David Rabouin (France) in "*Logica Mathematica*: Mathematics as Logic in Leibniz" claim that, although Leibniz is credited to be one of the precursors of modern mathematical logic, it is not easy to find precise evidences in his texts, which confirm this interpretation. Starting from such considerations the authors try to find a precise characterization of Leibnizian concepts as *logica mathematica*, *mathesis universalis* and *ars combinatoria*, relying upon textual evidences. They touch, thence, a series of interesting questions connected to the way in which Leibniz developed his mathematics (for example the idea that also a qualitative mathematics could exist) and how he saw the relations logic-mathematics. Basically, the authors are interested in how Leibniz saw the calculation as a logic rather than in the subject "logic considered as a form of computation".

Miguel Palomo (Spain) in "Describing Reality: Bernoulli's Challenge of the Catenary Curve and its Mathematical Description by Leibniz and Huygens" deals with the description of the catenary curve by Leibniz in a particular perspective: Palomo analyses the connections between this specific mathematical problem and its influence on the general theory of knowledge by Leibniz as well as on his metaphysics. The picture is designed by following the epistolary Leibniz-Huygens and the further debate, in which Bernoulli was involved, too. Within this itinerary, the superiority of the analytical methods by Leibniz compared with those geometrical by Huygens is discussed as well as the relations between the description of catenary and the quadrature of hyperbola.

Ana-Maria Pascal (UK) in "The Dialectics of Recognition in Leibniz's *Theodicy* and a Possible Route to Modern Political Thought" deals with the dialectic of reason in Leibniz. Pascal thinks that this is particularly significant for the history of socio-political ideas. She analyses a quite important theme: "the recognition of the other", which became the cornerstone of Hegel's *Phenomenology of Spirit*, but which was also profoundly considered in the period between Leibniz and Hegel. Within this context, Leibniz's *Theodicy* plays an important role. Pascal shows how the dialectic faith-reason is connected to the theme of the recognition of the other; by a story, she herself defines as an epic, a story in which Leibniz is a fundamental reference point.

Tzuchien Tho (Italy) in "The Equivalence of Hypotheses and Leibnizian *Vires*" analyses some important aspects of Leibniz's dynamics, in particular that of *vis* under the perspective of the principle called "equivalence of hypotheses", according to which motion and rest are relative in a physical system. On the other hand, on several occasions – for example in his famous correspondence with Clarke – Leibniz spoke of "true motion". This seems a contradiction with the equivalence of hypotheses and many authors tried to find a conciliation between these two assertions. Nonetheless, this aspect has always been problematic in Leibniz *Forschung*. Tho offers a new perspective by means of a reinterpretation of the causal nature of Leibniz's *vires*.

Friedel Weinert (UK) in "Leibniz's Defence of Heliocentrism" claims that one could think that Leibniz, due to his mechanistic conception, was a strong defender of the Copernican system as a physical reality. In contrast to this, Leibniz often referred to the heliocentric system as a hypothesis. To understand what kind a Copernican was Leibniz, Weinert points out that it is necessary to go beyond kinematical-philosophical statement as the relativity of motion and the equivalence of hypotheses – which, obviously played a role – and to reach his dynamical conceptions as far as the planetary motions are concerned. This approach guides Weinert to analyse works as the *Tentamen motuum coelestium causis*, the *Specimen dynamicum* and the correspondence with Huygens.

The volume brings together cutting-edge writing by leading authorities on the history and philosophy of science and around Leibniz: historical, scientific aspects. It provides authoritative insights to scholarly contributions that have tended to be too much randomly spread in journals and books, sometime not easily accessible to the young researcher or reader, as well. The book should prove to be absorbing reading for historians, philosophers and scientists.

2017, January, Lille, France

Raffaele Pisano and Paolo Bussotti
raffaele.pisano@univ-lille3.fr
paolobussotti66@gmail.com

Acknowledgments

The genesis of such a lengthy book as this collection of articles needed to be found in deep roots. The book has been a long time in the making and has fulfilled its promise. Therefore, we heartily express our appreciation to all contributing Authors for their efforts to produce papers of interest and of elevated quality worthy of this College Publications volume. The result is outstanding.

We express my warm and pleasant gratitude to Eberhard Knobloch for his friendly and leading *Foreword*, and to all our distinguished Authors.

In addition, finally yet importantly, our acknowledgments one more time are addressed to Shahid Rahman, Jane Spurr and College Publications for their good job and positive reception of our project in the *Leibniz's Anniversary Science Research Book*.

The Editors
January 2017

Remarks for the Reader

All of the papers in this international volume have been independently blended-peer reviewed. However, the Editors has respected different individual ideas, historical philosophical, epistemological and scientific accounts from each of the eminent Authors. The book contains introductory material, Table of Contents and Authors' contributions appear in alphabetical order. Following editorial guides for authors, the numbers of pages of the documents cited in the running text appear both with "p." or "pp" and without. Each of the Authors is responsible for his or her own opinions, which should be regarded as personal scientific and experienced background.

Pisano R, Fichant M, Bussotti P, Oliveira ARE (eds.), *The Dialogue between Sciences, Philosophy and Engineering. New Historical and Epistemological insights. Homage to Gottfried W. Leibniz 1646-1716*. College Publications, London, p. xix
© 2017 College Publications Ltd | ISBN: 978-1-84890-227-5 www.collegepublications.co.uk

Contributors

Jacob Archambault Fordham University USA

Mattia Brancato Milano University, Italy

Paolo Bussotti Udine University, Italy

Davide Crippa Research Fellowship Academy of Sciences, Praha, Czech Republic

Antonino Drago Napoli Federico II University, Italy

Giovanni Ferraro Molise University, Italy

Michel Fichant Sorbonne University, Paris , France

Glenn A. Hartz Ohio State University, USA

Eberhard Knobloch Berlin University of Technology, Germany

Joseph Kouneiher Nice and Sophia Antipolice University, France

Vladik Kreinovich Texas University at El Paso, USA

Montgomery Link Suffolk University Boston, USA

Guoqing Liu Nanjing Technology University, China

Leone Montagnini Biblioteche di Roma, Italy

Anne Michel-Pajus Paris Diderot University, France

Agamenon Rodrigues Eufrásio Oliveira Rio de Janeiro Federal University, Brazil

Miguel Palomo Sevilla University, Spain

Ana-Maria Pascal Regent's University London, UK

Raffaele Pisano Lille University, France / Lorraine University, France / Sydney University, Australia

David Rabouin Paris Diderot University–CNRS, France

Tzuchien Tho Milano University, Italy

Friedel Weinert Bradford University, UK

Pisano R, Fichant M, Bussotti P, Oliveira ARE (eds.), *The Dialogue between Sciences, Philosophy and Engineering. New Historical and Epistemological insights. Homage to Gottfried W. Leibniz 1646-1716*. College Publications, London, p. xxi
© 2017 College Publications Ltd | ISBN: 978-1-84890-227-5 www.collegepublications.co.uk

Leibnizian Intelligibility

Jacob Archambault

Abstract. In the Leibniz-Clarke correspondence, the *Antibarbarus Physicus*, and elsewhere, Leibniz rejects various Newtonian accounts of gravitation as unintelligible. The immediate intention of this paper is to explain this rejection. However, doing so requires a broader understanding of Leibniz's notion of intelligibility as such.

After reviewing current literature on Leibniz's disputes with the Newtonians, I challenge the received analysis of Leibniz's rejection of Newtonian gravitation, which holds Leibniz rejected Newtonian gravitation on mechanistic grounds. Next I survey various theses Leibniz rejects as *un*intelligible. From here, I move to Leibniz's positive account, as explicated in his correspondence with Wolff, of intelligibility in general. Last, I show how this account applies to Newtonian gravitation in particular. We find that: i) Leibniz polemically links the intelligibility of Newtonian accounts of attraction to that of the occasionalist hypothesis; ii) as a criticism of Newton's own understanding of attraction, this is justifiable; and iii) each of the three main accounts of attraction offered by Newton and his followers straightforwardly undermines the Leibnizian theological tenet that this is the best of all possible worlds.

Keywords: Intelligibility, Leibniz-Clarke correspondence, Isaac Newton, Gravitation, Dynamics, Force.

1 Introduction

> That means of communication, says he, is invisible, intangible, not mechanical. He might as well have added inexplicable, unintelligible, precarious, groundless, unprecedented (AG 345 = G VII. 418).

Thus, in his final letter of their exchange, does Leibniz describe the account of gravitation given by Samuel Clarke. It is well-known that Leibniz rejected the accounts of gravitation offered by Newton and his followers. But literature on this topic up to present has been content either i) to explain this rejection as a consequence of Leibniz's rejection of the intelligibility of action at a distance; or ii) to regard Leibniz's rejection, given his acceptance of his *own* notion of force, as unjustified.[1] This article provides a fuller account of *why* Leibniz regarded Newtonian gravitation as unintelligible.

The argument of the paper is as follows. After a brief review of recent literature on Leibniz's disputes with the Newtonians, I show that neither of the above-mentioned accounts explains Leibniz's rejection of Newtonian gravitation. From here, I explain the notion of intelligibility standing behind Leibniz's unintelligibility charge. Cataloguing a list of physical hypotheses that Leibniz takes to be *un*intelligible yields three necessary conditions on the *Leibniz-Intelligibility* of a hypothesis in physics, each of which I explicate in turn: *distinguishability*, *conceivability*, and *reducibility*. I then move to Leibniz's positive account of intelligibility. Thereby, I show each of the three main accounts of gravitation given by the Newtonians straightforwardly undermines confidence in the Leibnizian tenet that this is the best of all possible worlds.

2 Recent Work on Leibniz's Dispute with the Newtonians

The main monographs on Leibniz's disputes with Newton and his followers remain Hall (Hall 1980), and Bertoloni Meli (Bertoloni Meli 1993), with the former focusing primarily on the priority dispute surrounding the discovery of the calculus, and the latter devoting greater attention to disputes in natural philosophy. More recent work includes Bertoloni Meli (Bertoloni Meli 1999, 2002, 2006), Hall (Hall 2002), Attfield (Attfield 2005), Brown (Brown 2007), and Janiak (Janiak 2007). Of these, Hall and Bertoloni Meli's contributions are more historical, while those of Attfield, Brown, and Janiak take up a more evaluative stance.

2.1 The Historical Backdrop

Leibniz's disputes with the Newtonians arise mainly in connection with two issues: i) the nature of force generally and gravitation in particular, and ii) the discovery of the calculus.[2] These conceptually distinct disputes were sometimes intertwined.

2.1.1 The Earlier Exchanges

Newton's earliest attempted interaction with Leibniz is in a letter from June 1676. The letter, however, did not reach Leibniz until quite long after this date. Newton then sent a second letter (GM I, pp. 122-147), primarily concerned with the calculus and cordial in its tone, in October of the same year, which Leibniz received and replied to in June of 1677. In 1684, Leibniz published his first paper on the calculus in the *Acta Eruditorum*, which, as Hall (Hall 2002) suggests, prompted

---

[1] Cf. Garber 1995; Brown 2007.

[2] This section draws from Hall (Hall 2002) and Bertoloni Meli (Bertoloni Meli 2002).

Newton to put his own method in print in the 1ˢᵗ edition of the *Principia* in 1687, along with an admission that he had received a letter from Leibniz on his method of the calculus in 1677. As shown by Bertoloni Meli (Bertoloni Meli 1993), Leibniz read and took notes on the *Principia* in Vienna prior to writing the *Tentamen de Motuum Coelestium Causis* in Italy in 1689. In 1690, Leibniz published the *De Causa gravitatis, et defensio sententiae Autoris de veris Naturae Legibus contra Cartesianos* in the *Acta Eruditorum*, a work betraying the degree to which Leibniz had already begun to associate Newton's work on gravitation with "the Cartesians" – a name usually used by Leibniz to refer to the occasionalists.[3]

In 1693, Leibniz reestablishes contact with Newton, leading to a single exchange of brief letters that year. During the 1690's Leibniz is also working tirelessly on his own, massive *Dynamics*, a work he never publishes, and the only public hint of which is his *Specimen Dynamicum*, published in the *Acta Eruditorum* in 1695.

These earlier exchanges, both private and public, are marked by a high degree of cordiality among all parties. Leibniz's publications, correspondence, and other writings in natural philosophy from this period form part of an attempt to *win over* atomists, occasionalists, and others to his own physico-philosophical system, typically introducing them to it via his demonstration that Cartesian quantity of motion is not conserved.[4]

2.1.2 Later Interactions

The only other direct correspondence between Leibniz and Newton is in 1712, when Leibniz sends a lengthy philosophical letter to which Newton gives no response. In 1713, spurred on by Roger Cotes, Newton publishes the second edition of the *Principia*, with Cotes providing the preface to the work. Leibniz's primary engagements with the Newtonians in this later period are in the *Antibarbarus Physicus* and in the correspondence with Samuel Clarke.

The later interactions differ from the earlier in several respects. First, Leibniz is not attempting at the time to introduce his own physics to the public; and so these works are more focused on what is wrong with Newtonian commitments, rather than how they might be corrected by Leibnizian ones. In both the *Antibarbarus Physicus* and the correspondence with Clarke, Leibniz attacks Newton's views on force and gravitation by assimilating them to a discredited or unpopular model: in the former, to scholasticism; in the latter, to occasionalism. Second, Leibniz becomes involved at this time in the priority dispute surrounding the invention of the calculus, having sent a letter to the secretary of the Royal Society disputing accusations of plagiarism, only to have those accusations reaffirmed. Thirdly, the vantage point of these later works is affected both by the negative reception of the dynamics in the late 1690s, as well as by the comparative success of Leibniz's recently published *Theodicy*, especially the favor that work received from Caroline, Princess of Wales,

---

[3] G IV, pp. 509, 520; G VII. p. 356. This identification seems to have been commonplace. Cf. G IV, pp. 488-89.

[4] See, for instance, GM VI, pp. 123-128.

who initiated and was an important third party to the Leibniz-Clarke correspondence (Bertoloni Meli 1999, 2002). As a result, much of the polemics on both sides of this correspondence attempts to show how the natural philosophy of the opposing side implies an attenuated natural theology.

2.2 Historical Understandings

The best-known view surrounding Leibniz's rejection of Newtonian gravitation, put forward by Garber (Garber 1995) and provisionally settled on by Brown (Brown 2007), is that Leibniz rejected it because it would have to involve action at a distance. Given that Newton himself rejects this option, this view is usually tied to the claim that Leibniz failed to consider Newton's own position on gravitation, and fails to recognize the way that Newton's position depends on his own distinction between mathematical deductions and metaphysical hypotheses (*Ibidem.* Cf. Attfield 2007). According to this position, Leibniz fails to see that for Newton, "gravitation' was just the name of an observable phenomenon, the accurate description of which allowed the motions of bodies to be understood' (Attfield 2007, p. 239. Cf. Garber 2012). Given this failure, Leibniz's rejection was unjustified.

Before settling on the above view, Brown (Brown 2007) suggests two alternative reasons why Leibniz might have thought Newtonian attraction unintelligible: first, because it is non-mechanical; second, because it violates Leibniz's principled wall of partition between metaphysics and physics (*Ivi*, p. 150). After rejecting the first explanation on the grounds that mechanical explanations are *not* more informative than non-mechanical ones,[5] and the second on the grounds that 'Newton himself had constructed a methodological wall very similar to Leibniz's own' between "hypotheses' and 'what can be deduced from the phenomena" (*Ivi*, p. 151), he settles on the conclusion that Leibniz rejected Newtonian attraction because it involved action at a distance. Brown then suggests Leibniz could have accommodated Newtonian gravitation by including it in God's act of creation, but that accepting Newtonian attraction would have forced Leibniz to give up too many of his own philosophical commitments. Leibniz's rejection of Newtonian gravitation is entirely accounted for by his rejection of action at a distance; and Leibniz's rejection of action at a distance is, in Russell's words, a 'mere prejudice' (Russell 1900 [1937], § 47).

In short, according to the dominant historiographical narrative, Leibniz's dispute with Newton serves as a classic case of adherents of different Kuhnian paradigms talking past each other. 'Leibniz is a heritor of the natural philosophical tradition of Descartes, and Newton is a heritor of the mathematical tradition that Galileo followed. The very different ways in which Leibniz and Newton treat the notion of force are […] reflections of that fundamental difference' (Garber 2012, p. 47).[6]

---

[5] At least in the sense of 'informative' captured by having predictive power.

[6] An indirect threat to this received narrative comes from Janiak (Janiak 2007), who shows one aspect of this story – the assimilation of Newton's distinction between the mathematical and real treatment of force to that between the

As explanations of Leibniz's rejection of Newtonian attraction as *unintelligible*, both the received explanation and Brown's provisionally suggested alternatives remain incomplete. Leibniz's insistence on the unintelligibility of action at a distance does not provide a fundamental reason for his rejection of Newtonian gravitation, but only pushes the question back to that of why Leibniz regarded action at a distance as intelligible. Furthermore, while positing gravitation as a primitive quality of bodies was an option taken by some,[7] this was not the only option available to Newtonians – indeed, it was rejected by Newton himself – nor was it the only one Leibniz considered. Leibniz also addresses the view according to which gravitation is effected directly by a law or direct act of God, advanced in the second edition of the *Principia* by Newton himself;[8] as well as the non-mechanical ether account offered by Newton in the queries to the *Optics*.[9]

Copernican model of planetary motion as a heuristic tool and a physical hypothesis – is in important respects mistaken. Janiak writes,
> If Newton intended to treat gravity as purely mathematical, as a mere calculating device, we would not expect him to contend that it 'really exists.' And he does not dodge the implication that, as a real force, gravity bears causal relations [...]. So although Newton is agnostic in the *Principia* on the underlying *cause* of gravity [...] his agnosticism does not hinder him from claiming that gravity prevents the moon from following the inertial trajectory along the tangent to its orbit (*Ivi*, p. 130).

According to Janiak, 'mathematical' and 'physical' do not name modifications of the type of entity under examination – for instance, gravity; rather, they are modifications of the *way* of treating that type, which is antecedently regarded as really existing (*Ivi*, p. 132).

[7] Notably, by Roger Cotes in his preface to the second edition to the *Principia* (PM. p. 392). Cf. Locke, *Works* III, pp. 467-68:
> The gravitation of matter towards matter, by ways inconceivable to me, is not only a demonstration that God can, if he pleases, put into bodies powers and ways of operation, above what can be derived from our idea of body or can be explained by what we know of matter, but also an unquestionable and everywhere visible instance, that he has actually done so.

[8] 'Thus, the ancients and the moderns, who own that gravity is an *occult quality*, are in the right, if they mean by it that there is a certain mechanism unknown to them, whereby all bodies tend towards the center of the earth. But if they mean that the thing is performed without any mechanism by a simple primitive quality, *or by a law of God* [emphasis mine], who produces that effect without using any intelligible means, it is an unreasonable occult quality, and so very occult, that it is impossible it should ever be clear, though an angel, or God himself, should undertake to explain it' (G III, p. 519 = AJ, p.112).

'I objected that an attraction properly so called, or in the Scholastic sense, would be an operation at a distance without any means of intervening. The author answers here that an attraction without any means of intervening would indeed be a contradiction. Very well. But then, what does he mean when he will have the sun to

Part of my contention in what follows will be that though the concern with action at a distance – and the corresponding assimilation of Newtonianism to the abuses of scholasticism – is central to the polemic of the *Antibarbarus Physicus*, the concern with action at a distance is secondary to Leibniz's polemic against Newton and his followers. Increasingly, Leibniz's main objection, and the one on which he presses Samuel Clarke the hardest, is that Newton's physical explanation of gravitation turns gravitation into a perpetual miracle; and the main *strategy* Leibniz follows to discredit the Newtonians will be to assimilate their position on gravity to a kind of 'localized occasionalism' of the sort associated with Malebranche.

3 Leibnizian Intelligibility

3.1 Leibnizian Unintelligibility

A good starting point for our enterprise might be to look at examples of what Leibniz considers *unintelligible*, and his reasons for doing so. Among other things, Leibniz calls the hypotheses of an influx of soul into body (G II, p. 275 = AG, p. 181), action at a distance (G V, p. 54 = AG, p. 301), that matter can think (ibid.), Newtonian attraction (G VII, p. 342 = AG, p. 318), and that God could have created the world sooner (G VII, p. 405 = AG, p. 341) variously 'unintelligible' or 'not intelligible.'

3.1.1 Distinguishability

Regarding the last mentioned of these, Leibniz writes the following in his fifth letter to Clarke:

> Since I have demonstrated that time, without things, is nothing else but a mere ideal possibility, it is manifest that if anyone should say that this same world which has been actually created might have been created sooner without any other change, he would say nothing that is intelligible. *For there is no mark or difference whereby it would be possible to know that this world was created sooner* (G VII, p. 405 = AG, p. 341; emphasis mine).

---

attract the globe of the earth through an empty space? Is it God himself that performs it? But this would be a miracle if ever there was any' (G VII, p. 418 = AG, p. 345).

At first sight, positing gravity as a *law* of God, and positing it as an *action* would appear to be two different things. However, it was common among occasionalist philosophers to identify God's laws with God's very act of willing. This identity is likely assumed by Leibniz's Newtonian interlocutors. See McCracken 1983, p. 91.

[9] See G VII, p. 340 = AG, p. 318.

Here, Leibniz indicates one criterion on the intelligibility of a hypothesis—let us call it the *Distinguishability criterion*. This criterion states that a class of hypotheses $\mathcal{H}$ is intelligible only if there is some possible way to distinguish between the state of affairs posited by a member h and that which would obtain under a contrary hypothesis h' in $\mathcal{H}$.[10] In the above case, Leibniz states that there would be no noticeable difference between the present world and a world identical in all respects except that of having been created earlier (or later) than the present world. Therefore, the hypothesis that God could have created the world earlier than he had is unintelligible.[11]

3.1.2 Conceivability

A second determinant Leibniz places on intelligibility is *conceivability*. In his preface to the *New Essays*, he writes 'everything in conformity with the natural order can be conceived or understood by some creature' (G V, p. 58 = AG, p. 304). This does not mean Leibniz rejects everything that he does not understand; it merely entails that Leibniz rejects explanations he views as *incapable* of being understood. For instance, defending the view that matter is mechanically neither capable of sensation nor of reasoning, Leibniz writes, 'I recognize that we are not allowed to deny what we do not understand, though I add that we have the right to deny (at least in the order of nature) what is absolutely unintelligible and inexplicable'

---

[10] Note that this is not Leibniz's principle of the Identity of Indiscernibles, but it *is* a close cousin of it. The identity of indiscernibles is a principle on *objects*: objects which cannot be distinguished from each other are identical. A straightforward extension of the principle holds that *states of affairs* which are indistinguishable are the same state. Assuming, then, that hypotheses are distinct to the degree that they posit distinct states of affairs, it follows that hypotheses are distinct when they posit states of affairs that are distinguishable from each other. A special case of this, used in the above passage, holds that a hypothesis positing a state of affairs as distinct *from itself* is unintelligible.

[11] Leibniz appeals to this same criterion in his argument against absolute motion. Responding to Clarke's claim that a finite material universe might move through some absolute space, Leibniz writes:
> It does not appear reasonable that the material universe [...] should have any motion otherwise than as its parts change their situation among themselves, *because such a motion would produce no change that could be observed,* and would be without design. [...] motion indeed does not depend upon being observed, *but it does depend upon being possible to be observed* (G VII, p. 403 = AG, p. 340; emphases mine).

Here again, the intelligibility of a hypothesis depends upon the distinguishability of the state of affairs consequent upon it from other possible states of affairs where the hypothesis does not obtain.

(*Ibidem*). Here, Leibniz insists that the natural order is fundamentally able to be understood by intelligent creatures like ourselves.[12]

Leibniz's use of this criterion implicitly engages in a debate with the occasionalists. Consider the following quote from Malebranche:

> As I understand it, a true cause is one in which the mind perceives a necessary connection between the cause and its effect. Now, it is only in an infinitely perfect being that one perceives a necessary connection between its will and its effects. Thus God is the only true cause, and only he truly has the power to move bodies. I further say that *it is not conceivable* that God would communicate to men or angels the power he has to move bodies (Malebranche, *Oeuvres*, vol. I, p. 649 = Lennon and Olscamp, p. 450. Emphasis mine).

Here, Malebranche argues for occasionalism on the grounds of the inconceivability of the opposing hypothesis. Part of Leibniz's strategy against the occasionalists is to show that it is inconceivable that the occasionalist thesis itself would hold.[13]

It is in relation to such a conceivability test that Leibniz sets forth his views about the necessity of mechanical explanations in the physical realm. Shortly after the passage quoted above from the preface to the *New Essays*, Leibniz writes:

> Thus, we can judge that matter does not naturally have the attraction mentioned above, and does not of itself move on a curved path, because it is not possible to conceive how this takes place, that is to say, it is not possible to explain it mechanically, whereas that which is natural should be capable of becoming distinctly conceivable, if we were admitted into the secrets of things (G V, p. 59 = AG, p. 304).

We should glean several points from the above passage.

First, Leibniz calls the hypothesis of material attraction *conceptually impossible*. For Leibniz, the unintelligibility of Newtonian gravitation is thus not surmountable by claiming, as Brown (Brown 2007) claims, that matter is essentially attractive.[14]

---

[12] One might object that conceivability would obtain even if only god or angels, or souls after death, were able to understand the nature of the universe. But the context of the quote militates against this interpretation. Leibniz thinks that he himself – a terrestrial human being – *positively* understands *that* Locke's view is unintelligible, and not merely that he does not understand it.

[13] An important part of this critique will be that the inconceivability of the occasionalist hypothesis is not grounded in an internal contradiction in the *idea* of occasionalism; rather, it is inconceivable that the occasionalist hypothesis would be *true*, in accordance with the principle of sufficient reason.

[14] Leibniz disagrees with Hobbes precisely on this point, stating 'Hobbes, who claimed that truths are arbitrary [...failed to consider] the fact that the reality of a definition is not a matter of decision and that not just any notions can be joined to one another' (G IV, p. 425 = AG, p. 26).

Second, in equating conceivable explanations in physics with mechanical explanations, Leibniz claims that mechanical explanations are the *only* conceivable explanations of physical phenomena.

Third, in making the above claims, Leibniz sets the dependency relation between intelligibility and the mechanist hypothesis in one direction: it is not that mechanical explanations are intelligible *because* they are mechanical; rather, mechanical explanations are intelligible because they meet a conceivability standard that hypotheses like universal attraction fail to meet.

3.1.3 Reducibility

Leibniz's third determinant on intelligibility might be variously described as a simplicity, reducibility, or anti-primitiveness criterion. While it might seem unlikely that a philosopher committed to the existence of the actual infinite – indeed, to an infinite of actual infinites – would care much for parsimony, Leibniz's philosophy is deceiving on this point. Consider the following passage:

> [W]e should criticize those who hold these subordinate principles as primitive and inexplicable, as, for example, those who fabricated miracles, or those who fabricated incorporeal ideas that produce, regulate, and govern bodies, those who put forward the four elements [...] as if they contain the ultimate explanation of things, or those who, uninterested in understanding the particular force by which we evacuate with pumps, [...] set up in nature which abhors, as it were, the vacuum a primitive, essential, and insuperable quality. And whoever isn't, with us, eager to know qualities hitherto hidden [...] has invented qualities of eternal obscurity, [...] which not even the greatest genius can know or render intelligible (G VII, p. 341-42 = AG, p. 317).

The common thread binding together the above examples is, according to Leibniz, that they all treat subordinate principles as if they were elementary, a procedure which Leibniz identifies as inventing 'qualities of eternal obscurity' – qualities that are not merely unintelligible, but necessarily so.[15]

There is a *prima facie* conflict between what I have referred to as Leibniz's commitment to simplicity and his commitment to an actual infinity of substances. But Leibniz's is not a commitment to simplicity in terms of tokens, but *types*. The

---

[15] The same reductionism is present in the *Antibarbarus Physicus*:
> [I]n nature, things must proceed by steps, and one cannot go immediately to first causes. [...] But if certain people, abusing this beautiful discovery [i.e. gravity], think the explanation given is so satisfactory that there is nothing left to explain [...] then they slip back into [...] the *occult qualities of the Scholastics* (G VII, p. 338-39 = AG, p. 314).

point may be made by comparing his own commitment to simplicity with that of Newton.

In a letter from 1693, Newton writes to Leibniz that 'since nature is very simple, I have myself concluded that all other causes are to be rejected and that the heavens are to be stripped as far as may be of all matter' (GM I, p. 171 = AJ, p. 109). Here, Newton takes his belief in the simplicity of nature as license to empty space of unnecessary ontological tokens (i.e. he commits himself to the hypothesis of a vacuum), and posit an additional ontological type – Newtonian attraction.

By contrast, Leibniz commits himself on the grounds of simplicity to an abundance of ontological tokens and – strictly speaking – only *one* ontological type, to which all other phenomena are to be reduced: the monad or simple substance.[16] In a letter to De Volder he writes, 'the phenomena of aggregates come from the reality of monads' (G II, p. 250 = AG, p. 176), and in another letter to the same he writes that '[C]orporeal mass, which, is thought to have something over and above simple substances, is not a substance, but a phenomenon resulting from simple substances, which alone have unity and absolute reality' (G II, p. 275 = AG, p. 181). Thus, Leibniz is convinced that Newtonian attraction will ultimately reduce to simple substance or a phenomenon of simple substance, and that there is no need to posit gravity as an *additional* primitive type, or to have recourse to the perpetual miracle according to which God himself causes gravity directly.

3.2 Intelligibility

3.2.1 In General

From here, we transition to Leibniz's positive views on intelligibility. First, we address Leibniz's identification of intelligibility with truth; second, his association of intelligibility with perfection.

In a 1689 paper addressing Copernican planetary theory, Leibniz writes that, 'the truth of a hypothesis is nothing but its intelligibility' (AG, p. 91), and again that 'the truth of a hypothesis should be taken to be nothing but its greater intelligibility' (AG, p. 92).[17] In the second of these quotes, Leibniz identifies the truth of a hypothesis with *greater* intelligibility, thereby implying hypotheses admit *degrees*

---

[16] That Leibniz identifies these two notions is apparent, for instance, when he writes in the *Principles of Nature and Grace* that, 'A *composite substance* is a collection of simple substances, or *monads*' (G.VI, p. 598 = AG, p. 207), and in a letter to Des Bosses that, 'Simple substances or monads are either intelligent or without reason' (G II, p. 438 = AG, p. 200).

[17] While the context of the above is Leibniz's engagement in a political endeavor (viz. attempting to persuade the Catholic Church to lift the ban on Copernicanism), we ought not therefore to take his words as insincere. Leibniz's attempt in the essay is grounded in a belief that we have every indication that he held wholeheartedly – namely, the relatively of all motion.

of intelligibility.[18] In the paper, the Copernican hypothesis is advocated as *more intelligible* than the Ptolemaic, without wholly denying the intelligibility of Ptolemaic models.

Regarding the second point, Leibniz writes in a 1715 letter to Wolff:

> The perfection about which you ask is the degree of positive reality, or what comes to the same thing, the degree of affirmative intelligibility, so that something more perfect is something in which more things worthy of observation are found (GLW, p. 161 = AG, p. 230).

Here, Leibniz identifies intelligibility with both perfection and positive reality.

Now the following questions, among others, may be raised about the above identifications. First, what does Leibniz mean by truth? Second, what does Leibniz mean by perfection? Third, are the identities in question necessary or contingent?

Leibniz answers the first question by stating that truth is 'always an implicit or explicit identity' (AG, p. 31), and that truth is 'the containment of the predicate in the subject' (AG, p. 98).[19] For Leibniz, primary truths are explicit identities, e.g. A = A (AG. 30). Secondary truths, however, are *implicit* identities, arrived at via a conceptual analysis wherein the concepts in question are reduced back to an identity, i.e. where the predicate is shown to be contained in the subject.[20] Furthermore, Leibniz distinguishes between necessary and contingent truths by holding that necessary truths admit of a finite proof, whereas contingent truth 'obtains only by an argument of infinite length' (Rescher 2001, p. 149). All truths are analytic, but contingent truths are incapable of finite resolution.

Thus, Leibniz's statement that 'the truth of a hypothesis is nothing but its intelligibility' (AG, p. 91) gives us several clues about his views on intelligibility. First, a hypothesis is made more intelligible when it is more explicitly able to be reduced via conceptual analysis to an identity. Second there is a sense in which, via

---

[18] This is further corroborated by some of Leibniz's other comments. For instance, in the *Antibarbarus Physicus* he calls the hypothesis of gravitation by means of a thread "more intelligible" than the primitive attractive force advocated by many Newtonians (G VII. 342 = AG. 318). Leibniz also describes himself as rehabilitating substantial forms "in a way that would render them intelligible" (G IV. 479 = AG. 139), thus implying not only that intelligibility admits of degrees, but also that something formerly obscure can, by an apt explication, *become* more intelligible.

[19] 'For Leibniz, it inheres in the very definition of truth that all truths are analytically true' (Rescher 2001, p. 149). '[Truth] is shown by giving a reason [for the truth] through the analysis of both terms into common notions' (AG, p. 98).

[20] On this construal, Leibniz need not hold that something is only an identity when each term can *salva veritate* be reciprocally substituted for the other. For instance, Leibniz claiming that all truths are reducible to identities need not entail that the truth 'Whales are mammals' is such that the term 'whales' is always able to be replaced by the term 'mammals' in a truth-preserving way (for then, for instance, it would be true that all mammals live in the ocean in their natural habitat).

this process of analysis, a hypothesis that was once obscure can *become* more intelligible.[21] Leibniz describes his rehabilitation of Aristotelian forms in just this way:

> Just as our age has already saved from scorn Democritus' corpuscles, Plato's ideas, and the Stoics' tranquillity in light of the most perfect interconnection of things, so now we shall make intelligible the teachings of the Peripatetics concerning forms or entelechies, notions which seemed enigmatic for good reason, and were scarcely perceived by their own authors in the proper way (GM VI, p. 235 = AG, p. 118).

Leibniz analyses perfection – the second concept with which he identifies intelligibility – into two components: variety and order.[22] Thus a given state of affairs x is maximally intelligible if and only if it is brought about by the most simple laws by which said state is able to be obtained; and one state of affairs is more intelligible than another to the degree that it is both simpler in its laws and more varied in its phenomena. Hence, Leibniz's claim that brute Newtonian attraction is unintelligible would by transitivity entail that it – be it miraculous, an occult quality of bodies, or a non-mechanical ether – would contribute to a less than optimal world.[23]

Unfortunately, Leibniz never to my knowledge directly addresses the question of whether the above identities are necessary or contingent. But considerations internal to Leibniz's philosophy and correspondence can provide us with a relatively certain answer.

Leibniz's identification of intelligibility and perfection occurs within a correspondence with Wolff, who begins this thread of enquiry by requesting 'to know how Your Excellency [i.e.Leibniz] usually defines perfection' (GLW, p. 160 = AG, p. 230). After hearing Leibniz's answer, Wolff writes 'I have […] found that your definition of perfection answers my needs in many ways' (*Ibidem*). Responding to Wolff's letter, Leibniz writes, 'I am gratified to know that you are not displeased with my very general definition of perfection' (GLW, p. 161 = AG, p. 231). In the correspondence, Wolff and Leibniz are looking for a definition, or the essence or meaning of a given concept—in this case, perfection. Thus, insofar as Leibniz is defining perfection, he sees himself as uncovering the essential structure

---

[21] Not *per se* more intelligible, but more so to the person who is given the explanation.

[22] 'God has chosen the most perfect world, that is, the one which is at the same time the simplest in hypotheses and the richest in phenomena' (G IV, p. 431 = AG, p. 39). 'It follows from the supreme perfection of God that he chose the best possible plan in producing the universe, a plan in which there is the greatest variety together with the greatest order' (G VI, p. 603 = AG, p. 210). 'And this is the way of obtaining as much variety as possible, but with the greatest order possible, that is it is the way of obtaining as much perfection as possible' (G VI, p. 616 = AG, p. 220).

[23] I.e. because the same state of affairs could conceivably have been brought about by simpler means – without primitive attraction.

of the idea thereof. So, the identity between intelligibility and perfection is necessary.

Leibniz's identification of intelligibility and truth, however, would have to be contingent. For if the most perfect state of affairs is necessarily the most intelligible, and the most intelligible is necessarily true, then it would follow that the most perfect state of affairs—i.e. this best of all possible worlds—would necessarily be the true state of affairs, i.e. that which actually obtained. In other words, the best of all possible worlds would *necessarily* be the actual world, or to put it another way, this world would necessarily exist. But Leibniz was zealous to deny this Spinozist conclusion.[24] Thus, while the essential identification of intelligibility and perfection is necessary and finitely analytic, the kind of necessity associated with the identification between intelligibility and truth would have to be that associated with the principle of sufficient reason, and therefore not finitely resolvable.[25]

3.2.2 With Respect to Newtonian Gravitation

3.2.2.1 Leibniz's Theological Polemic

In the letter to Caroline beginning his correspondence with Clarke, Leibniz wrote:

> Sir Isaac Newton and his followers also have a very odd opinion concerning the work of God. According to their doctrine, God Almighty needs to wind up his watch from time to time, otherwise it would cease to move. He did not, it seems, have sufficient foresight to make it a perpetual motion. No, the machine of God's making is so imperfect, according to these gentlemen, that he is obliged to clean it

---

[24] 'It was always one of [Leibniz's] paramount aims to avert a Spinozistic necessitarianism, and he regarded the contingency of the world's constituents and processes as an indispensable requisite towards this end, one in whose absence the idea of divine benevolence would be inapplicable' (Rescher 2001, p. 145).

[25] Leibniz offers evidence for this claim when he states, regarding the identity of indiscernibles (ID), that 'When I deny that there are two drops of water perfectly alike, or any two other bodies indiscernible from each other, I don't say it is absolutely impossible to suppose them, but that it is a thing contrary to the divine wisdom, and which consequently does not exist' (G VII, p. 394-95 = AG, p. 334). A world in which ID did not obtain would be less rich in phenomena, therefore less perfect, therefore less intelligible. Leibniz seems to imply that a world in which bodies attract each other is possible when he writes that 'on the strength of what God can do, we would grant too much license to bad philosophers, allowing them those *centripetal virtues* or those *immediate attractions* at a distance, without it being possible to make them intelligible' (G V, p. 54 = AG, pp. 300-301). In the above passage, Leibniz implies God *could* have created a world with centripetal virtues. In other words, such a state of affairs is possible. Cf. Rescher 1996.

> now and then by an extraordinary concourse, and even to mend it, as a clockmaker mends his work. (G VII, p. 352 = *LC* § 4)

In the above, Leibniz is alluding to an analogy used to compare his own system of pre-established harmony to the Cartesian system of occasional causes. Here is the version of the analogy Leibniz gives in his clarifications of Bayle's difficulties with Leibniz's system:

> One can imagine three systems for explicating the communication one finds between the soul and the body, namely: (1) the system of influence of the one on the other, which is that of the schools [...] (2) that of a perpetual supervisor, who represents in the one what occurs in the other, a little as if a man were charged to make two bad clocks, which were incapable of agreeing with each other themselves, always to agree, and this is the system of occasional causes, and (3) that of a natural agreement of two substances, such as there would be between two clocks perfectly exact (G IV, p. 520, translation mine).

Leibniz uses this example on multiple occasions in a number of different ways: sometimes to explicate the union of soul and body, at other times to explain the harmony of substances with each other; at others, to explain the unity of perceptions in a single substance.[26] The analogy seems to have originally been inspired by an experiment of Huygens (G IV, p. 498). But despite its varied uses, it is always used to compare Leibniz's system to occasionalism. In the passage in the letter to Caroline, the allusion is to the fact that the Newtonians, following the Cartesians, took the measure of force preserved in the universe to be quantity of motion, or mass multiplied by velocity. As Malebranche put it,

> God chose the simplest laws based on the single principle that the stronger must prevail over the weaker; and on the condition that there would always be the same quantity of motion in the world from the same direction, I claim that the center of gravity of bodies before and after their impact always remains the same, whether that center is at rest or in motion. (*Dialogues* X.XVI = JS, p. 190).

Leibniz, by contrast, had taken the correct measure to be mv^2, in accordance with his *Brevis Demonstratio* of 1686 (GM VI, p. 119-23). Newton and his followers, however, were aware that quantity of motion is not naturally conserved, and had used this to explain the need for 'active principles [...] or the dictates of a will' to intervene in the system of the world to amend it (*LC* § 107; cf. G VII, pp. 354-55 = *LC* § 6). Throughout the Leibniz-Clarke correspondence, Leibniz repeatedly presses this analogy to tie Newton to the occasionalists, and in particular to suggest that gravitational attraction is miraculous (G VII, p. 357 = *LC* § 9; G VII, pp. 366-7 = *LC* § 17-18; G VII, p. 376 = *LC* § 26-27).

---

[26] G IV. 498; 520; 522; G VI. 540-41.

The second, more surprising way Leibniz ties Newtonian philosophy to the occasionalists is through his attempts to characterize Newton's God as the soul of the world. According to Leibniz, Malebranche's occasionalism, since it does not preserve the distinction between the natural and supernatural, collapses into Spinozism. In a reply to Bayle, he writes:

> It does not seem necessary to me to remove action or force from creatures, under the pretext that they would create if they produced modes. For it is God who conserves and continuously creates the forces [of creatures] [...][W]ithout this, I find [...] that God would produce nothing, and there would be no substance apart from his own, which takes us back to all the absurdities of Spinoza. It also seems that the error of this author comes from nothing besides that he puts forth the consequences of the doctrine that remove force and action from creatures (G IV, pp. 567-68, translation mine; cf. G IV, pp. 509, 515).

In ascribing actions immediately to God, occasionalists make God the very nature of the things otherwise thought to perform the action. While the Newtonians don't appeal to God to account for *all* bodily interactions, they nevertheless posit what Leibniz views as the root of the occasionalist view, that 'Natural and Supernatural are nothing at all different with regard to God, but distinctions merely in Our Conceptions of things.' (Clarke, G VII, p. 362).

Furthermore, some Newtonians fully embraced the language of occasionalism in their explication of Newton's *Principles*. Here, for instance, is Cotton Mather's description of Newtonian gravitation:

> You will see [if you study physics] that the Influences of one thing upon another in the Course of Nature are purely from the Omnipotent and Omnipresent God, actually forever at Work, according to His own Laws, and putting His Laws in Execution, and as the Universal Cause producing those Effects, whereof the Creatures are but what One may call, The Occasional Cause. You will also be often and quickly carried up into those Immechanical Principles, from whence, The next step is unto God! The Gravitation of Bodies is One of them; for which No cause can be assigned, but the Will of the Glorious God, who is the First Cause of all (*Manductio ad Ministerium*, 50. in McCracken 1983, p. 319).

3.2.2.2 Occasionalism and the Newtonians

Leibniz's attempt to assimilate Newtonianism to occasionalism, and through occasionalism to Spinozism, was justified to some extent by Newton's own philosophical tendencies. Newton does hold absolute motion is proportionate to quantity of motion (*Principia*, Bk. 1, scholium to definitions = Cohen 412). Along with Clarke (G VII, p. 354), Newton is suspicious of granting objects powers and actions, lest God be deprived of his glory: 'Indeed, however we cast about we find almost no other reason for atheism than this notion of bodies having, as it were, a complete, absolute, and independent reality in themselves' (AJ, p. 32). God's presence in the world is secured by depriving bodies of superfluous principles that would make them more independent. We see this, for instance, in Newton's reasons for denying the existence of substantial forms in the *De Gravitatione*:

> [F]or the existence of these beings it is not necessary that we suppose some unintelligible substance to exist in which as subject there may be an inherent substantial form; extension and an act of the divine will are enough. Extension takes the place of the substantial subject in which the form of the body is conserved by the divine will; and that product of the divine will is the form or formal reason of the body denoting every dimension of space in which the body is to be produced (AJ, p. 29).

Lastly, Newton is explicit in assuming not merely that God produces gravity directly, but 1) that he does so by being directly present to the objects attracting each other, and that 2) this is possible because of the way space is an effect of God. According to Newton,

> [S]pace is an emanative effect of the first existing being, for if any being whatsoever is posited, space is posited. [...] So the quantity of the existence of God is [...] infinite in relation to the space in which he is present (AJ, p. 25).

Newton goes on to say

> [S]pace is eternal in duration and immutable in nature because it is the emanative effect of an eternal and immutable being. If ever space had not existed, God at that time would have been nowhere; and hence he either created space later (where he was not present himself), or else, which is no less repugnant to reason, he created his own ubiquity (AJ, p. 26).

Thus, we have it that space is uncreated.

We further have it that Newton himself identifies space with extension: after making the above remarks on the nature of space, he continues, 'Now that *extension* has been described, it remains to give an explanation of the nature of body' (AJ, p. 27). The context is one wherein Newton is distinguishing space/extension from body, and arguing against the commonly held view that extension is a property of bodies. And so, not only is space uncreated, space is uncreated extension, which is an emanative effect of God. If, then 'emanative effect' could be identified with either 'attribute' or 'affection', Newton would agree straightforwardly with the first half of the following corollary of Spinoza:

It follows that *res extensa* and *res cogitans* are either attributes of God or (by axiom 1) affections of God's attributes (*Ethica*, Bk. I, proposition XIV, corollary ii, my translation).

If this is right, we can see in Newton a kind of piety that seeks to ensure the presence of God in the world by requiring His immediate presence for otherwise ordinary operations: gravitation, material composition, etc.; and a corresponding tendency to remove active principles from created substances as far as possible to ensure creaturely dependence.

3.2.2.3 Recapitulation and Summary

With the above analysis complete, it is now possible to give a more unified account of why Leibniz rejected Newtonian attraction as unintelligible.

First, neither Newton nor any of his followers offer Leibniz a satisfying explanation of what Newtonian attraction *is*. Leibniz states that gravity is, according to the Newtonians, either an occult quality, a miracle, or unintelligible. So the best scenario for the Newtonians would be one in which Newtonian attraction *was* intelligible *per se*, but they had failed to make their case to Leibniz.

But this last option is barred by the approach of Newton, Clarke, Cotes, and others, who insist in various ways that gravity is primitive. Furthermore, the world in which gravity has some sort of irreducible ontological status is, by Leibnizian lights, less simple than the optimal world, because it would contain more ontological types than necessary, and therefore less intelligible.[27]

But while earlier I suggested Leibniz would have thought such primitive positing rash, Leibniz's reasons behind his objection to Newtonian attraction go deeper than this. Since Leibniz identifies intelligibility and perfection, and since Leibniz holds Newtonian attraction to be unintelligible especially on account of its alleged primitiveness, it follows that the world in which primitive Newtonian attraction obtains could not be the best of all possible worlds. The God who made such a world would be, by Leibnizian lights, arbitrary, and the principle of sufficient reason – according to which God acts – would not obtain.[28]

---

[27] On this point, consider the following passage from the Theodicy: 'If the effect were assumed to be greater, but the process less simple, I think one might say, when all is said and done, that the effect itself would be less great, taking into account not only the final effect but also the mediate effect' (G VI, p. 241 = H § 208).

[28] The principle of sufficient reason plays a more important role than is generally recognized in Leibniz's physics. Leibniz uses the principle to deduce certain consequences which he takes to be certain – for instance, the existence of the plenum and the non-existence of atoms. Take, for instance, his comment that the existence of two identical bodies is 'a thing contrary to the divine wisdom, *and which consequently does not exist*' (G VII, pp. 394-95 = AG, p. 334; emphasis mine). For Leibniz, metaphysics 'becomes real and demonstrative by means of these principles [i.e. of sufficient reason and the identity of indiscernibles], whereas before it did generally consist in empty words' (G VII, p. 372 = AG, p. 328).

Leibniz indicates this essential conflict between primitive Newtonian gravitation and the principle of sufficient reason when he writes in his final letter to Clarke:

> Has not everybody made use of this principle upon a thousand occasions? It is true, it has been neglected out of carelessness on many occasions, but *that neglect has been the true cause* of chimeras such as are, for instance, an absolute real time or space, a void, atoms, attraction in the Scholastic sense, a physical influence of the soul over the body, and a thousand other fictions, either derived from erroneous opinions of the ancients, or lately invented by modern philosophers (G VII, pp. 419-20 = AG, p. 346; emphasis mine).

As Leibniz sees it, to accept any of the above Newtonian hypotheses is to deny the principle of sufficient reason, and to deny that principle is to be reduced to absurdity.

4   Conclusion

Leibniz's ultimate reason for rejecting Newtonian attraction is not that it involves action at a distance, but that the existence of such a primitive attraction would lead to a less intelligible and less perfect world. Nor would Leibniz be able to make Newtonian gravitation intelligible by mere fiat: the very method by which it is posited is at odds with Leibnizian criteria of simplicity.

Consequently, Leibniz's rejection of (and Newton's acceptance of) attraction involves him in a graver question than the received analysis implies: namely, the question of whether this *actual* world is fundamentally intelligible. Leibniz, based on theological considerations of God's benevolence and goodness, inflexibly insists that it is; Newton, based on theological considerations of God's omnipotence and lordship, holds – at least by Leibniz's lights – that it is not. Thus, the debate between Leibniz and Newton over Newtonian attraction echoes a larger debate of enduring meaning in which philosophers, scientists, and historians of both these disciplines might partake.

Abbreviations

AJ = *Newton: Philosophical Writings*. Ed. Andrew Janiak. Cambridge: Cambridge University Press, 2004

AG = *G. W. Leibniz: Philosophical Essays*. Ed. and trans. Roger Ariew and Daniel Garber. Indianapolis: Hackett, 1989.

G = *Die philosophischen Schriften von G. W. Leibniz*. Ed. C. I. Gerhardt. Berlin: Weidman, 1875-1890. Reprint, Hildesheim: Georg Olms, 1978.

GLW = *Briefwechsel zwischen Leibniz und Christian Wolff*. Ed. C. I. Gerhardt. Hildesheim: Georg Olms, 1963.

GM = *Leibnizens Mathematische Schriften.* Ed. C. I. Gerhardt. Berlin: A. Asher, and Halle: H. W. Schmidt, 1849-1863. Reprint, Hildesheim: Georg Olms, 1971.
H = *Theodicy.* Trans. E. M. Huggard. New Haven: Yale University Press, 1952. Reprint, La Salle, Illinois: Open Court, 1985.
JS = *Nicolas Malebranche: Dialogues on Metaphysics and on Religion.* Ed. Nicholas Jolley and David Scott. Cambridge: Cambridge University Press, 1997.
LC = *G. W. Leibniz and Samuel Clarke: Correspondence.* Edited, with Introduction, by Roger Ariew. Indianapolis: Hackett, 2000.
LO = *Nicolas Malebranche: The Search after Truth.* Trans. T. M. Lennon and P. J. Olscamp. Columbus: Ohio State University Press, 1980; republished with a new introduction by T. Lennon, in Cambridge Texts in the History of Philosophy, 1997.
OC = *Oeuvres complètes de Malebranche.* Ed. A. Robinet. 20 vols. Paris, Vrin, 1958-1976.
PM = *The 'Principia'. A new translation by I. Bernard Cohen and Anne Whitman assisted by Julia Budenz. Preceded by A Guide to Newton's Principia, by I. Bernard Cohen.* Berkeley: University of California Press, 1999.

References

Attfield, R. 2005. Leibniz, the Cause of Gravity and Physical Theology. *Studia Leibnitiana* 37:238-244.
Bertoloni Meli, D. 2006. Inherent and Centrifugal Forces in Newton. *Archive for History of Exact Sciences* 60:319-335.
Bertoloni Meli, D. 2002. Newton and the Leibniz-Clarke correspondence. In Cohen and Smith, 455-464.
Bertoloni Meli, D. 1999. Caroline, Leibniz, and Clarke. *Journal of the History of Ideas* 60:469-486.
Bertoloni Meli, D. 1993. *Equivalence and Priority: Newton versus Leibniz.* Oxford: The Oxford University Press.
Brown, Gregory. 2007. Is the Logic in London Different from the Logic in Hanover? In *Leibniz and the English Speaking World*, ed. Pauline Phemister and Stuart Brown. Dordrecht: Springer, pp. 145-162.
Cohen, I. B, and G. E. Smith (Eds.) 2002. *The Cambridge Companion to Newton.* Cambridge: The Cambridge University Press.
Garber, D. 2012. Leibniz, Newton, and force. In *Interpreting Newton: Critical Essays*, ed. Andrew Janiak and Eric Schliesser. Cambridge: The Cambridge University Press, pp. 33-47.
Garber, D. 1995. Leibniz: physics and philosophy. In *The Cambridge Companion to Leibniz*, ed. Nicholas Jolley, pp. 270-352. Cambridge: Cambridge University Press.
Hall, A. R. 2002. Newton versus Leibniz: from geometry to metaphysics. In Cohen and Smith 2002, pp. 431-454.
Hall, A. R. 1980. *Philosophers at War: The Quarrel between Newton and Leibniz.* Cambridge: The Cambridge University Press.

Janiak, A. 2007. Newton and the Reality of Force. *Journal of the History of Philosophy* 45:127-147.
McCracken, C. J. 1983. *Malebranche and British Philosophy*. Oxford: The Oxford University Press.
Rescher, N. 2001. *Contingentia Mundi*: Leibniz on the World's Contingency. *Studia Leibnitiana* 33:145-162.
Rescher, N. 1996. Leibniz on Possible Worlds. *Studia Leibnitiana* 28: 129-162.
Russell, B. 1900 [1937]. *A Critical Exposition of the Philosophy of Leibniz*. London.
Stevenson, G. P. 1997. Miracles, Force, and Leibnizian Laws of Nature. *Studia Leibnitiana* 29:167-188.

Jacob Archambault, Fordham University, USA
jacobarchambault@gmail.com

Leibniz's Foundation of Arithmetic and its Influence on Frege

Mattia Brancato

Abstract. The *Initia rerum mathematicarum metaphysica* is one of Leibniz's most important works on the foundations of arithmetic: given its qualitative approach, the references to the *analysis situs* and the use of relations such as homogeneity and similarity, its innovation rivals that of much later works in the same field. However, Leibniz's influence on the late 19th century's philosophy of mathematics is usually related to the idea of what Frege called a *lingua characteristica* or *calculus ratiocinator*, mainly because the *Initia rerum* and other important writings were published after 1863, casting doubt on whether the greatest logicians and mathematicians of that time were really influenced by them, or if they just found in these works a later confirmation of their independently developed theories. Yet, Frege's *Grundlagen der Arithmetik* show an approach to the foundations too similar to that of Leibniz to be easily dismissed. I will argue that in the works written by Leibniz and known by Frege all the elements that constitute the novelty of the *Initia rerum mathematicarum metaphysica* are present and that they greatly influenced Frege's draft of the *Grundlagen*. Leibniz's fundamental intuitions proved to be in Frege's eyes the only ones in need of improvement, rather than a complete rejection.

Keywords: Algebra, *analysis situs*, definition of number, foundation of arithmetic, Frege, geometry, homogeneity, identity, Leibniz, *mathesis universalis*.

1 Introduction

In recent years, several studies have been made on Leibniz's influence in the late 19th and 20th century. The reason of such interest is understandable: the scientific community of that time witnessed a peculiar situation in which, on one hand, fundamental discoveries were made in fields such as logic or mathematics and, on the other hand, Leibniz's works on topics very similar to those of such discoveries were published for the first time. Having earned the attention of some of the most important philosophers and scientists of those years, questioning about the extent and effectiveness of Leibniz's influence is no doubt necessary.

An in-depth study on Leibniz's reception then seems legitimate, especially concerning Gottlob Frege: in his works, he frequently expressed his debt to

Leibniz's intuitions and, at the same time, he is considered one of the fathers of modern logic. However, Frege's confrontation with Leibniz poses several problems, the first one being a mere problem of primary sources and their availability: beyond doubt, the last volume of Gerhardt's edition of Leibniz's *Philosophische Schriften* was published in 1890 (Leibniz 1875-1890), a year in which Frege was far from being retired, but at the same time, references to Leibniz were made by Frege in works that were published or written before this year, such as the *Begriffsschrift* (1879) and the *Grundlagen der Arithmetik* (1884). Since the publication of the Gerhardt's edition started in 1875, it could be that Frege had at least a partial knowledge of those works, but it seems peculiar that Frege, having Gerhardt's edition at his disposal, felt the need of quoting Leibniz from the older and incomplete edition prepared by Erdmann, available since 1840 (Leibniz 1840).

The problem of determining what was the edition of Leibniz's works that Frege actually read is connected with the problem of the extent of Leibniz's influence, because the works and fragments presented in the Gerhardt's and Couturat's editions are more akin to those of Frege on the development of a new language used in logic. As a starting point on this topic, the following passage from *Begriffsschrift* is often quoted:

> Leibniz too, recognised - and perhaps overrated - the advantages of an adequate system of notation. His idea of a universal characteristic, of a *calculus philosophicus* or *ratiocinator*, was so gigantic that the attempt to realise it could not go beyond the bare preliminaries[1].

Frege considered in some way his *Begriffsschrift* a natural development of what he called Leibniz's universal characteristic:

> It is possible to view the signs of arithmetic, geometry, and chemistry as realisations, for specific fields, of Leibniz's idea. The ideography proposed here adds a new one to these fields, indeed the central one, which borders on all the others. If we take our departure from there, we can with the greatest expectation of success proceed to fill the gaps in the existing formula languages[2].

---

[1] Frege 1981, p. 6. "Auch Leibniz hat die Vortheile einer angemessenen Bezeichnungsweise erkannt, vielleicht überschätzt. Sein Gedanke einer allgemeinen Charakteristik, eines *calculus philosophicus* oder *ratiocinator* war zu riesenhaft, als dass der Versuch ihn zu verwirklichen über die blossen Vorbereitungen hätte hinausgelangen können" (Frege 1977, p. XI).

[2] Frege 1981, p. 7. "Man kann in den arithmetischen, geometrischen, chemischen Zeichen Verwirklichungen des Leibnizischen Gedankens für einzelne Gebiete sehen. Die hier vorgeschlagene Begriffsschrift fügt diesen ein neues hinzu und zwar das in der Mitte gelegene, welches allen andern benachbart ist. Von hier aus lässt sich daher mit der grössten Aussicht auf Erfolg eine Ausfüllung der Lücken der bestehenden Formelsprachen, eine Verbindung ihrer bisher getrennten Gebiete zu dem Bereiche einer einzigen und eine Ausdehnung auf Gebiete ins Werk setzen, die

Frege was therefore aware of Leibniz's project. Is this a real influence or a mere tribute to an eminent ancestor?

The secondary literature, both in accepting and denying Leibniz's direct influence, seems to align to the interpretation of Leibniz's philosophy that was developed in the early years of the 20th century: works such those of Russell (Russell 1900) and Couturat (Couturat 1901) focus on Leibniz's logic as a fundamental key in order to understand his entire philosophy. Although advancing different interpretations, it became soon clear that they were claiming their authority from the availability of primary sources never published before, the former recognising later their consistency with his interpretation and the latter being personally responsible for their publication[3]. As a direct consequence of this approach, these primary sources became a sort of *corpus logicum*, considered mandatory in order to achieve a correct interpretation of Leibniz's philosophy. Moving to the secondary literature that focuses on the relationship between Leibniz and Frege, it seems that this approach has been respected: in Lenzen (Lenzen 2004) for example, we read that the best exposition of Leibniz's logic is contained in the *Generales inquisitiones de analysi notionum et veritatum*, a work first published in Couturat's edition. Understanding Leibniz's influence becomes then a matter of understanding Frege's access to these specific writings on logic. In other words, the fact that some fundamental intuitions were already present in the Erdmann's edition of Leibniz's works and in the other sources already available at that time has never been considered a key factor, avoiding an in-depth analysis of their content in this regard[4]: if Frege was not aware of Couturat's edition, then the secondary literature tends to ascribe the similarities with Leibniz to a happy accident.

The problem of what defines Leibniz's logic and, consequently, if the fundamental principles are already accessible in the earlier editions is a matter perhaps too big to master, but in this paper, I will focus on a topic that I consider

---

bisher einer solchen ermangelten" (Frege 1977, p. XII). This passage continues following Leibniz's idea: the ideography is meant to be useful not only because it formalises the existing theories, but also because it creates a coherent language for new and undiscovered fields, as a sort of 19th century's *ars inveniendi*. As for Leibniz, on the notion of *ars inveniendi* see for example Leibniz (1923, VI-4A, p. 509) and the following texts.

[3] See the preface to the second edition of Russell's *A Critical Exposition of the Philosophy of Leibniz* (Russell 1900, p. V).

[4] A different approach is that of Peckhaus (Krömer et al. 2012): he recognizes in Frege's writings a direct influence of the Erdmann's edition and Tredelenburg's studies on Leibniz. A lecture that Tredelenburg gave in 1856, entitled *Über Leibnizens Entwurf einer allgemeinen Charakteristik* and reprinted in his *Historische Beiträge zur Philosophie* in 1867 (Tredelenburg 1867, pp. 1-47) focuses on Leibniz's conceptual language, using specific terms that we could find also in Frege. This reference is a decisive proof of the relation between ideography and Leibniz's universal characteristic. However, Frege's interest in Leibniz was beyond his access to secondary literature and the mere logical calculus: as I will show, the knowledge he exhibits in the *Grundlagen* is related to specific mathematical concepts and it was possible only after a thorough study of the Erdmann's edition.

one of the most important in order to understand Leibniz's impact on Frege: the foundation of arithmetic. Again, it's probably due to Russell's influence that this topic is often neglected. The fact that Frege's theory on the foundation of arithmetic is wrong derives from Russell's well-known decisive observation and it is nowadays a truth universally acknowledge. Frege's contribution to logic is considered a side effect of that effort and perhaps no other side effect has ever gained so much attention, in spite of the original theory, as this one. However, if this side effect has been generated following a fundamental intuition, it could be that the novelty introduced by Frege's logic is present in the intuition itself: the foundations of arithmetic are in Frege's eyes the best example of the effectiveness of his logic, as much as Leibniz's derivation of the principles of mathematics is one of the greatest achievement of his universal characteristic. During their development, Leibniz and Frege were constantly dealing with mathematical concepts, so much that these concepts influenced the very nature of their logic.

Following Kluge[5] (Kluge 1980a, p. 235), we can safely assume that Frege read not only Erdmann's edition, but also the *Mathematische Schriften* published by Gerhardt between 1849 and 1863[6], before the *Philosophische Schriften* and before the draft of Frege's *Begriffsschrift* and *Grundlagen der Arithmetik*. I will argue that these writings, together with some fundamental writings of the Erdmann's edition such as the *Dialogus*[7], the *Nouveaux essais sur l'entendement humain* and the Leibniz-Clarke correspondence, made a huge impact on Frege's foundational theory. However, in order to understand this influence, a detailed explanation of Leibniz's foundation of arithmetic and its peculiar approach is needed. Hence, after briefly discussing the metaphysical status of Leibniz's foundations, I will move to an in-depth analysis of the *Initia rerum mathematicarum metaphysica*, one of the most important work on this topic, written by Leibniz and published in Gerhardt's *Mathematische Schriften*. This analysis will show a partial consistency in Leibniz's writings on such matters. Finally, a confrontation with Frege's *Grundlagen der Arithmetik* will prove that Leibniz's influence affected the foundations of arithmetic and beyond, reaching the realm of pure logic[8].

---

[5] Kluge quotes a letter by Leibniz to Gallois, published in Gerhardt's *Mathematische Schriften* (Leibniz 1849, p. 181). A reference to this letter was found in Frege, in a posthumously published work dated 1880-1881 (Frege 1969, pp. 9-52). Kluge's aim is that of proving Frege's knowledge of Gerherdt's *Philosophische Schriften*, but for our purpose, this reference is enough to prove that he read at least the *Mathematische Schriften*.

[6] See Leibniz (Leibniz 1849-63) for the exact reference.

[7] In the Erdmann's edition, it was entitled *Dialogus de conexione inter res et verba* (Leibniz 1840).

[8] For these reasons, in order to show a correspondence between Leibniz and Frege, quotes from Leibniz's texts already available in the Akademie's edition are presented as they were edited in the Erdmann and Gerhardt's editions.

Homage to Gottfried W. Leibniz 1646-1716

2.1 The *Initia Rerum* as a Starting Point on Leibniz's Philosophy of Mathematics

Before providing an analysis of the *Initia rerum*, I would like to point out several problems that choosing this work as a starting point poses, both from Leibniz's and Frege's point of view. First of all, we are not sure if Frege actually read the *Initia rerum*: we can only prove on a general basis that he read Gerhardt's *Mathematische Schriften*, where this work is contained. Furthermore, I believe that understanding exactly how much Frege spent on Leibniz's writings is hard, if we don't turn our attention to the similarities in their approaches, provided they can be found in passages in which Frege refers directly to Leibniz. However, in this regard, the *Initia rerum* are worth analysing, because they establish a difference between geometry and arithmetic that Frege, as it will be shown, ascribes directly to Leibniz. This difference is only sketched in other writings, hence the use of this work in order to understand what Leibniz meant. What I would like to prove is not that Frege read Leibniz in the light of the *Initia rerum*, rather, that the light of the *Initia rerum*, i.e. the precise distinctions that Leibniz establishes here between logic, metaphysics and mathematics, is reflected in several works of Leibniz that Frege could have read.

If our aim is that of understanding the difference between arithmetic and geometry, another problem arises: in the *Initia rerum*, Leibniz's definition of number is given in a context heavily contaminated by the use of geometry and by its terminology, so much that it appears as if the foundation of arithmetic itself is completely based on geometry. As it will be shown in detail in the next section in fact, Leibniz defines number as "that which is homogenous to Unity, in other words that which is to Unity as a straight line to another straight line" (Leibniz 1863, p. 24). This assumption is somehow contradictory if confronted with what Leibniz said in a letter to Clarke, dated 1715:

> The great foundation of mathematics is the principle of contradiction or identity, that is, that a proposition cannot be true and false at the same time and that therefore A is A and cannot be non-A. This single principle is sufficient to demonstrate every part of arithmetic and geometry, that is, all mathematical principles[9].

Here the foundation of arithmetic, together with the foundation of mathematics in general, seems to be based on a purely logical principle. The principle of contradiction is in fact described, at least in this case[10], as an assumption concerning

---

[9] Leibniz 1976, p. 677. "Le grand fondement des Mathématiques, est le principe de la contradiction, ou de l'identité, c'est-à-dire, qu'une énonciation ne sauroit être vraie et fausse en même tems; et qu'ainsi A est A, et ne sauroit être non A. Et ce seul Principe suffit pour démonter toute l'Arithmétique et toute la Géométrie, c'est-à-dire, tous les Principes Mathématiques" (Leibniz 1840, p. 748).

[10] During his life, Leibniz used the principle of contradiction in several different contexts. For example, in modal logic the principle is used to define possibility and necessity, as it was since the *Confessio philosophi* (Leibniz 1923, VI-3), but it is

the truth value of something, i.e. 'A'. What is interesting about this approach is that Leibniz identifies the principle of contradiction with identity. This means that stating something about A in the form of 'A is A' means also expressing something about what A is not. Both these aspects lend mathematics to a foundation on logical terms. The role of logic here is not different from that which Frege ascribes to his *Begriffsschrift*, as it was previously mentioned: it has a place that is somehow above arithmetic and geometry from a metaphysical point of view, whereas these two disciplines seem to stand on the same ground, both deriving their principles from that of contradiction.

As is known, since the *Initia rerum* were written by Leibniz after 1714, we should find at least a kind of consistency with the idea expressed in the letter to Clarke in the same years. In the *Initia rerum*, the closest correspondence with this claim is the following:

> It is also to be noted that the whole of algebra is an application to quantities of the art of combinations, or of the science of abstracted forms, which is the universal characteristic and belongs to metaphysics[11].

It is no coincidence that this statement appears just after Leibniz's definition of number. What emerges from this passage is that algebra derives from an application of something, which is the science of abstracted forms that belongs to metaphysics, to something else, which is described as quantity. In this sense, we are not far from the relationships between the different areas of mathematics and logic that Frege derives from Leibniz. Our first task then would be that of understanding the true role of geometry in Leibniz's foundation of arithmetic and why he describes this process as a metaphysical foundation.

On a side note, the fact that Leibniz describes this foundation as one that belongs to metaphysics is particularly relevant, given the evolution of Leibniz's philosophy in the last years of his life. Leibniz was writing the *Initia rerum* while composing works such as the *Monadologie* and the *Principes de la nature et de la grâce fondé en raison*, works that are considered essential from a metaphysical point of view. This observation leads to the problem of finding a coherent 'metaphysical space' to things such as the principle of contradiction and the foundations of arithmetic in a context which was dramatically influenced by Leibniz's discoveries

---

also used in the proof of God's existence as a key element in the completion of Descartes and Spinoza's proofs. See for example *De probanda divina existentia* (Leibniz 1923, VI-4B, p. 1390) or *Si possibile est ens necessarium, sequitur quod existat* (Leibniz 1923, VI-4B, p. 1636).

[11] Leibniz 1976, p. 670. "Notandum est etiam, totam doctrinam Algebraicam esse applicationem ad quantitates Artis Combinatoriae, seu doctrinae de Formis abstractae animo, quae est Characteristica in universum, et ad Metaphysicam pertinet" (Leibniz 1863, p. 24).

in physics at the end of the 17th century[12]. What I would like to suggest here on this topic is that Leibniz saw this work as at least a last effort on the realisation of a *Mathesis Universalis*, the dream of a superior science that describes itself as an ideal bridge between geometry and arithmetic. Being an application of the universal characteristic to something else, there is no identification between *Characteristica Universalis* and *Mathesis Universalis*, since the former has a wider range of use in Leibniz's view[13]. As such, the *Mathesis Universalis* is not a purely formal theory, however, it contains, as it will be shown, a decisive implementation of formal properties. A proof of this interpretation is contained in the *Matheseos Universalis Pars Prior*, a work that Frege could have found in the same volume of Gerhardt's *Mathematische Schriften*:

> It follows the until now ignored and neglected subordination of Algebra to the art of Combination, that is that of the Specious (Literal) Algebra to the General Specious, of the science of the formulas denoting quantity to that of formulas in general, such as order, similarity, relation etc.; that is that of the universal science of quantity to the universal science of quality, so that our Specious Mathematics is nothing more than an eminent example of the Art of Combination[14].

Since this passage revolves around the idea of the application of quality to quantity and the subordination of algebra to the science of forms, we can safely assume that the *Initia rerum* are indeed Leibniz's last attempt on the foundational theory of mathematics sketched here, following the idea of a *Mathesis Universalis*. This last work on the metaphysical origin of mathematics is an ideal starting point in order to

---

[12] For a coherent reconstruction of these times, related to the draft of the *Initia rerum*, see De Risi 2007, p. 93.

[13] That of Crapulli (Crapulli 1969) is still an excellent book on the origin of the term, especially because it highlights the importance not only of the words used, but also of the amount of work done around Proclus' commentary on Euclid's Elements. It shows that a reconstruction of the influence of this term in Leibniz can't be focused only on the obvious reference to Descartes' *Regulae ad directionem ingenii*: the debate on a *Mathesis Universalis* was advanced and widely known, so much that a reference to Proclus' commentary is found in Suárez (Suárez 1965, I.4.15). I believe that this reference is fundamental to sketch the influence on Leibniz, because Suárez here establish a relation between metaphysics and mathematics in the context of disputations I-III, dedicated to the principle of contradiction, as the most important metaphysical principle. Suárez's reasoning on these topics is very similar to that of Leibniz and Leibniz knew it since his first writings.

[14] "Hinc etiam prodit ignorata hactenus vel neglecta sub-ordinatio Algebrae ad artem Combinatoriam, seu Algebrae Speciosae as Speciosam generalem, seu scientiae de formulis quantitatem significantibus ad doctrinam de formulis, seu ordinis, similitudinis, relationis etc. expressionibus in universum, vel scientiae generalis de quantitate ad scientiam generalem de qualitate, ut adeo speciosa nostra Mathematica nihil aliud sit quam specimen illustre Artis Combinatoriae" (Leibniz 1863, p. 61).

understand Leibniz's mature theory: here in fact, concepts like quality, quantity, order, similarity and relation are thoroughly defined. Therefore, it's now time to move to a detailed analysis of these concepts and their relation.

2.2 Leibniz's Definition of Number

In the *Initia rerum*, before the definition of number, Leibniz introduces several concepts and principles. Following Leibniz's approach, undoubtedly rigorous, is difficult, because it requires a deep knowledge of all the concepts involved and their relation. Therefore, in this exposition of Leibniz's theory, I will start from the definition of number and then gradually introduce the other concepts involved, following a logical scheme, more than the order in which they are presented.

Leibniz's definition of number, quoted in the previous section, is here presented in its entirety:

> Number in general, integer, fractional, rational, surd, ordinal, transcendental, can be defined through a general notion, as that which is homogenous to Unity, in other words that which is to Unity as a straight line to another straight line[15].

As a preliminary remark, I would like to point out the basic consequence of this definition: every number is something which bears some kind of relation to unity. If every number must be defined through a confrontation with unity, unity becomes the cornerstone of Leibniz's foundation of arithmetic. As it will be shown, Frege understood very well this aspect of Leibniz's theory. Other than unity however, this notion relies on a peculiar relation between unity and numbers and on a reference to geometrical objects that is worth analysing.

Homogeneity In the next section I will discuss the use of the term 'straight line' and the problems related to this choice, but for now, I would like to focus on the peculiar relation that a number must bear in order to be defined: homogeneity. This relation is essential to understand if the foundation is based on a purely geometrical definition or not. At first glance in fact, the definition of number seems to rely completely on the geometrical notion of a straight line, but the relation of homogeneity adds a new level of complexity, which shouldn't be underestimated. Leibniz defines homogeneity in this way:

---

[15] "Numerum in genere integrum, fractum, rationalem, surdum, odinalium, trascendentem generali notione definiri posse, ut sit id quod homogeneum est Unitati, seu quod se habet ad Unitatem, ut recta ad rectam" (Leibniz 1863, p. 24). It's quite unfortunate that this passage of the *Initia rerum* was cut from Loemker's edition. By doing so, the passage already quoted on Algebra as an application of the universal characteristic to quality looks incoherent, whereas it is connected with the definition of number.

Two entities are homogeneous to which two other entities can be assigned which are equal to them and similar to each other. Given A and B; if L is taken equal to A, and M equal to B, and L and M are similar, we call A and B homogeneous. Hence, I usually say also that homogeneous entities are those which can be made similar to each other by means of transformations, like curves and straight lines. That is, if A is transformed into its equal L, it can be made similar to B, or to its equal M into which B is assumed to have been transformed (Leibniz 1976, p. 667)[16].

Here, two definitions of homogeneity are hidden: the first one is what I would call the general definition of homogeneity, which involves the introduction of two new fundamental relation, equality and similarity, applied to four different entities — A, B, L and M — following a certain scheme or rule. Equality is applied to A and L, and to B and M, whereas similarity is applied to L and M. The other definition is what I would call the purely geometrical definition, which derives from the geometrical exemplification expressed by Leibniz between curves and straight lines. In this case, the scheme of the general definition is not completely respected, because the fourth entity, that is M, is taken as unnecessary. A bears the same equality relation to L, but in this case Leibniz admits the possibility of a direct connection between L and B through similarity. This is legitimate, given the geometrical exemplification, because the transformation happens in a geometrical context in which B could be already quantified and suitable for the similarity relation. I believe that this is a relevant distinction, because it means that in Leibniz's geometrical example B and L are ontologically closer than the same entities described in the general definition. Understanding which homogeneity relation is applied to the definition of number is therefore necessary, because it will also show the ontological connection between the concept of number and pure geometry.

If we now apply the general definition of homogeneity to the definition of number, the result is the following: a number (B) is something homogeneous to unity (A), i.e. unity (A) is equal to a straight line (L), the given number (B) is equal to another straight line (M), while these straight lines (L and M) are similar to each other. It is possible to apply the general definition because in the quoted passage Leibniz openly uses four entities, two numbers and two straight lines. However, it doesn't necessarily mean that in this case the purely geometrical definition can't be applied: since this definition is a special case of the general one, it could still be that the straight line L, related to unity, is also similar to the given number B. Even if this new definition of number seems more detailed than the original one then, it is

---

[16] "Homogenea sunt quibus dari possunt aequalia similia inter se. Sunto A et B, et possit sumi L aequale ipsi A, et M aequale ipsi B sic ut L et M sint similia, tunc A et B appellabuntur Homogenea. Hinc etiam dicere soleo, Homogenea esse quae per transformationem sibi reddi possunt similia, ut curva rectae. Nempe si A transformetur in aequale sibi L, potest fieri simile ipsi B vel ipsi M, in quod transformari ponitur B" (Leibniz 1863, p. 19).

now evident that the definitions of equality and similarity are needed for a complete understanding.

Leibniz defines equality as the relation between things that have the same quantity. Quantity then is defined in this way:

> *Quantity or magnitude is that in things which can be known only through their simultaneous compresence - or by their simultaneous perception. Thus, it is impossible for us to know what a foot or a yard is unless we actually have something to serve as a measure which can be applied to successive objects after each other*[17].

Until now, we have considered Leibniz's foundation of arithmetic without questioning its coherence, but after introducing this definition, it is possible to explain a passage of the original definition of number that could be considered incoherent at first glance: if Leibniz's aim is that of defining a number and his first step in this direction is conceiving it as something equal to something else, that is to say, something that has the same quantity of something else, then a contradiction arises, because the concept of quantity, related to numbers, is contained in the definition of number itself. However, Leibniz's definition of quantity is not that of a simple numerical description, but it is based on the idea of a simultaneous compresence. In this regard, stating something as 'A measures 6 meters' is completely meaningless, until A is introduced in a peculiar set of relations, in which an order between different objects is established. The introduction of similarity, in a context where different things are compared, is then necessary, otherwise equality would be a mere statement on the identity of two things, in the form of 'A is B':

> A foot or a yard can therefore not be explained adequately by a definition; that is, by one which does not include something similar to the thing defined. For though we may say that a foot is twelve inches, the same question arises concerning the inch, and we gain no greater insight, for we cannot say whether the notion of the inch or of the foot is prior in nature, since it is entirely arbitrary which we wish to assume as basic[18].

The difference between a numerical definition of quantity and Leibniz's definition is essential, because it shows that Leibniz had a significant idea of a foundation of

---

[17] Leibniz 1976, p. 667. "Quantitas seu Magnitudo est, quod in rebus sola compraesentia (seu perceptione simultanea) cognosci potest. Sic non potest cognosci, quid sit pes, quid ulna, nisi actu habeamus aliquid tanquam mensuram, quod deinde aliis applicari possit" (Leibniz 1863, p. 18).

[18] Leibniz 1976, p. 667. "Neque adeo pes ulla definitione satis explicari potest, nempe quae non rursus aliquid tale involvat. Nam etsi pedem dicamus esse duodecim pollicum, eadem est de pollice quaestio, nec majorem inde lucem acquirimus, nec dici potest, pollicis an pedis notio sit natura prior, cum in arbitrio existat utrum pro basi sumere velimus" (Leibniz 1863, p. 19).

arithmetic and not a mere exposition of the concept of number in geometrical terms: the problem of defining a number, without relying to a numerical reference or to the idea of counting, is in fact very frequent in writings concerning this topic, as it is for example in Frege. This passage also confirms that an adequate definition is one that involves similar things. Similarity is defined through quality, instead of quantity:

> Quality, on the other hand, is what can be known in things when they are observed singly, without requiring any compresence. Such are the attributes which can be explained by a definition or through the various modes which they involve [...] Similar are things having the same quality. Hence, if two similar things are different, they can be distinguished only when they are co-present[19].

Quality is then something that pertains to an object, without requiring compresence. It could be described as a set of properties that intrinsically pertain to something. In the case of the straight line for example a quality would be that of proceeding interminably in both directions, or the idea of a length with no width. These properties pertain to a straight line in a way that would make us unable to distinguish another straight line from the one considered, if they are taken as the only object existing in our space:

> The straight line is uniform between its extremities. For there is no determining factor given from which we can infer a reason for variation. It follows necessarily, therefore, that one locus of a point moving along it can be distinguished from another only in relation to its extremities. Hence any part of a straight line is also straight, and a line is everywhere internally similar to itself, nor can two parts be distinguished from each other if they cannot be distinguished by means of their extremities[20].

In other words, similar objects are those that cannot be distinguished if not considered in a reference system. It also follows that some properties instead derive from compresence: talking about the slope of a straight line for example has no meaning if only that line is conceived, without establishing the coordinate system.

It is possible now to apply these distinctions again to the original definition of number, obtaining in this way the most detailed description of Leibniz's original

---

[19] Leibniz 1976, p. 667. "Qualitas autem est, quod in rebus cognosci potest cum singulatim observantur, neque opus est compraesentia. Talia sunt attributa quae explicantur definitione aut per varias modificationes quas involvunt [...] Similia sunt ejusdem qualitatis. Hinc si duo similia sunt diversa, non nisi per compraesentiam distingui possunt" (Leibniz 1863, p. 19).

[20] Leibniz 1976, p. 671. "Rectam esse inter sua extrema aequabilem. Neque enim aliquid assumitur, unde reddi possit ratio varietatis. (3) Itaque oportet, ut unus locus puncti in ea moti ab altero discerni non possit seposito respectu ad extrema. Hinc et pars rectae recta est, itaque intus ubique sibi similis est, nec duae partes discerni possunt inter se, cum suis extremis discerni non possint" (Leibniz 1863, p. 26).

passage, considering the relations involved: a number is something that can be known through a relation of homogeneity to unity, that is to say, conceiving both unity and the given number in a simultaneous compresence with straight lines. These lines have the same intrinsic properties and they are co-present with each other. Having exposed every relation involved, it's finally evident that the homogeneity between a given number and unity is not the purely geometrical homogeneity, as it was defined before. If this was the case, L would be similar to B: the straight line equal to unity would be similar to the number defined. However, if being similar means having the same quality, by no means the intrinsic properties of a straight line are the same of those of a mere number. In Leibniz's terms, this means that a number and a straight line would be completely undistinguishable if taken as single entities, which is obviously wrong. It's clear now that Leibniz's foundation is not a purely geometrical foundation: unity and the other numbers are *treated* as geometrical entities but they are not *defined* as geometrical entities. In other words, geometry is used as a tool, an extremely useful tool indeed, but it doesn't express the essence of numbers: in Leibniz's definition, unity and the other numbers stand untouched by the geometrical qualities, they are just assigned to objects having geometrical properties, through the quantitative relation.

At first glance, the difference between a geometrical foundation and a foundation that uses geometry as a tool could be seen as a trivial distinction, but it bears an important consequence: ideally, equivalence and similarity can be used in the definition of number without a direct reference to geometry. It would be as admitting that quality and quantity can be found in non-geometrical contexts. According to Leibniz, this consequence seems at least legitimate: the exemplification used in the definition of quantity is consistent with this assumption, because it highlights how the notion of a measure is arbitrary, if taken alone, while the definition of quality relies on the idea of intrinsic proprieties, i.e. something legitimately conceivable in other contexts where other proprieties are given. Describing these new contexts as 'every context in which the existence of qualities is admitted' is perhaps a metaphysical assumption too big to master, especially in Leibniz's mature philosophy, where some important differences between what we would call reality and the realm of mathematics are stressed[21]. However, if this extension is limited to mathematics, there are contexts other than geometry where the definition and identification of qualities is possible. A step in this direction is done by Leibniz in this passage of the *Initia rerum*, which is worth quoting in its

---

[21] See Mugnai (Mugnai 1992) for Leibniz's difference between abstraction and reality. However, I would like to point out that this distinction is not always as rigid as it would seem, even in Leibniz's mature writings. This could be true especially for Leibniz's works concerning binary arithmetic: in Leibniz (Leibniz 1863, p. 227), dated 1703, for example he compares the use of the binary system to his characteristic, but in another writing on the same topic (Leibniz 1863, p. 239) he compares the same method to the creation of our world through God and nothingness, taken as 1 and 0. We could say that, although in the end rejecting the reality of certain mathematical entities, Leibniz was at least tempted to give a metaphysical meaning to some of them.

entirety because it directly follows the quote on algebra as an application to quantity of the science of abstracted forms discussed in the previous chapter:

> The product of the multiplication of $a + b + c +$ etc. by $l + m + n +$ etc. is nothing but the sum of all the binary combinations which can be built out of the letters of the two series, and the product of the three series $a + b + c +$ etc., $l + m + n +$ etc., and $s + t + v +$ etc. is the sum of all the ternary combinations which can be built from the three series of letters. Other forms will be produced from other operations. Thus, our calculation usefully obeys not only the law of homogeneous entities but also the law of justice [*lex justitiae*], according to which the relations between entities in our conclusions or results must correspond to the similar relations in the data, and must therefore be treated in the same way in mathematical operations, insofar as this is practical. The proposition is true in general that when the data proceed in a certain order, the conclusions proceed in a corresponding order[22].

Here, the science of the abstracted forms of calculations is applied to similar relations, established between specific numbers. A part from the law of justice, Leibniz admits that also the law of homogeneous entities is respected in this case, which means that quantity and quality are used in a non-geometrical context, hence the possibility of finding correspondences in "*similar* relations".

Finally, it is possible to understand why in several statements about the connection between logical-metaphysical principles and principles belonging to mathematics, Leibniz considers universal characteristics above arithmetic and geometry: in this framework, a general notion is applied to two specific field with a certain degree of autonomy. It follows that the real essence of Leibniz's foundational method is the distinction between quantity and quality and the way in which these concepts are applied in order to achieve a set of coherently related entities. If this is possible in non-geometrical terms, as shown, it means that the choice of the straight lines in the definition of number is in a sense arbitrary. The next step then would be proving that this choice is the best one, or at least one of the best, with respect to Leibniz's foundational purpose. To prove this, a thorough analysis of the geometrical entities chosen by Leibniz is needed.

---

[22] Leibniz 1976, p. 670. "Productum multiplicatione $a + b + c +$ etc. per $l + m + n +$ etc. nihil aliud est quam summa omnium binionum ex diversi ordinis literis, et productum ex tribus ordinibus invicem ductis, $a + b + c +$ etc. in $l + m + n +$ etc. in $s + t + v +$ etc. fore summam omnium ternionum ex diversi ordinis literis; et ex aliis operationibus aliae prodeunt formae. Hinc in calculo non tentum lex homogeneorum, sed et justitiae utiliter observatur, ut quae eodem modo se habent in datis vel assumtis, etiam eodem modo se habeant in quaesitis vel provenientibus, et qua commode licet inter operandum eodem modo tractentur; et generaliter judicandum est, datis ordinate procedentibus etiam quaesita procedere ordinate" (Leibniz 1863, pp. 24-25).

Straight Lines and Line Segments. In Leibniz's definition of number, the qualitatively similar entities chosen to establish a relation between numbers and unity are straight lines. Until now, I deliberately postponed the analysis of these entities. However, the notion of a straight line poses a problem about its very nature: is Leibniz referring to the notion of a straight line as something that proceeds interminably in both directions, or is he referring to the notion of a line segment, i.e. something that is conceived as a finite part of a straight line? The origin of this problem lies in an incoherent use of this term by Leibniz in the *Initia rerum*: in the definition of number, Leibniz uses the word *recta* and, according to another passage[23], he makes a distinction between a straight line, called *recta* as well, and a general line, called *linea*. From this passage, it seems that Leibniz is referring to a straight line, but in a third passage[24], very close to the one previously quoted, Leibniz writes about *rectae* that have some kind of finite quantity related to them, as in a comparison between line segments. It would be wise not to underestimate this problem, because straight lines and line segments could have, in Leibniz's terms, different qualities, so that, ideally, one could be more adequate than the other as a geometrical tool used in the definition of number.

In Leibniz's definition of number, a specific term is used to define the relation between the given lines, that is *Ratio*[25]. Now, this term is also used in another section of the *Initia rerum*:

> The simplest of all relations is that called ratio or proportion, and this is the relation between two homogeneous quantities which arises from themselves alone without assuming a third homogeneous quantity. So if y is to x as any number is to unity or $y = nx$, x being taken as abscissa, and y as ordinate, the locus, or the line in which the ordinates end, is a straight line[26].

Leibniz uses here the equation of a straight line in the form $y = nx$, where the y-intercept is equal to 0, to describe the same relation used in the definition of number, that is *Ratio*. This is also the only passage in which a reference to the discussed definition of number is given, because "y is to x as any number is to unity", in the

---

[23] "Recta, quae est linea intus sibi similis" (Leibniz 1863, p. 21), which can be adapted both to a straight line and a line segment. However: "Ex duobus punctis prosultat aliquid novi, nempe punctum quodvis sui ad ea situs unicum, horumque omnium locus, id est recta quae per duo puncta proposita transit" (Leibniz 1863, *Ibidem*). The verb '*transit*' used here suggests the use of straight lines.

[24] "Sint datae duae rectae, quae inter se comparentur utcunque. Verb. Gr. detrahatur minor ex majore" (Leibniz 1863, p. 23).

[25] See Leibniz 1863, p. 24.

[26] Leibniz 1976, p. 670. "Sed omnium Relationum simplicissima est, quae dicitur Ratio vel Proportio, eaque est Relatio duarum quantitatum homogenearum, quae ex ipsis solis oritur sine tertio homogeneo assumto. Veluti si sit y ad x ut numerus ad unitatem seu $y = nx$, quo casu x positis abscissis, y ordinatis, locus est recta, locus inquam seu Linea quam ordinatae terminantur" (Leibniz 1863, p. 23).

same way in which a number is to unity as a straight line to a straight line. However, in this case, if we replace x and y with unity and a given number, then n, i.e. slope, would be equal to the ratio between these numbers. In other words, it would seem appropriate to identify numbers with line segments more than straight lines, because they are equivalent to scalars, that is to say, the line segments generated by the projection of a point belonging to a straight line to the x axis and the y axis. Yet, in the definition of number, Leibniz adds a new depth to this idea, because he writes that even the ratio itself is homogeneous to unity[27]. Being a quotient, even a ratio is a number and every number can be expressed as a ratio, but in this context a ratio is also the slope of a straight line with a y-intercept equal to 0. As much as every number can be seen as a ratio, every line segment can be seen as a straight line having a slope equal to that ratio and a y-intercept equal to 0. This interpretation is confirmed by Leibniz himself:

> It follows that similar lines (*lineas*) have the same ratio of the homologous straight lines (*rectarum*)[28].

Line segments are indeed similar to each other. In the end, the incoherent use of the term *recta* to identify both straight lines and line segments hides Leibniz's belief that they can be seen as homologous. I believe that he was aware of this problem and I would like to suggest that perhaps Leibniz left this terminological incoherence because he didn't want to openly admit that a finite geometrical entity can be expressed through an infinite one, in the context in which the *Initia rerum* were meant to be published[29].

Putting aside the mere terminological problem however, considering line segments as slopes of straight lines having the y-intercept equal to 0 highlights some important features of Leibniz's foundation of arithmetic. If the succession of different numbers is given through a description of the relation they have with unity and the other numbers, comparing line segments to see which one is greater than the other could lead to the false opinion that a comparison between numerical quantities is still involved. In the representation of every number, as a straight line that differs only in having a different slope instead, it is clear that this is not the case: every number is, so to speak, a unity, because every infinite straight line is similar to that of unity, undistinguishable if not taken in a context in which the different slopes are perceivable, that is to say by means of compresence. At the same time, every number is unique, because it is defined through the infinite relations with the other numbers. The infinite relations are however expressed following a specific order that gives the idea of succession to the definition. In this sense, any number can be

---

[27] "Manifestum est etiam, si Ratio a ad b consideretur ut numerus qui sit ad Unitatem, ut recta a ad rectam b, fore Rationem ipsam homogeneam Unitati" (Leibniz 1863, p. 24).

[28] "Ex his sequitur, lineas similes esse in ratione rectarum Homologarum" (Leibniz 1863, p. 24).

[29] The *Initia rerum* were probably meant to be published in the *Acta Eruditorum*: as the opening suggests with Leibniz's reference to Wolff, this work was not conceived by Leibniz only for a personal use.

conceived following this order step by step. This way of defining objects is no other than an application of what Leibniz called *Situs*[30]:

Situs is a mode of coexistence. Therefore, it involves not only quantity but also quality[31].

> *Situs* is a certain relationship of coexistence between a plurality of entities; it is known by going back to other coexisting things which serve as intermediaries, that is, which have a simpler relation of coexistence to the original entities[32].

From the geometrical representation of numbers through specific straight lines it also follows the most important consequence: the straight line that represents unity has the form $y = x$, that is identity: it means that every numerical identity — $3 = 3$, $8 = 8$, and so on — can be conceived as a point that belongs to the straight line which represents unity, or as the ratio of this line[33]. In Leibniz's eyes this should have been

---

[30] It is worth noting that Leibniz's concept of *Situs* was heavily influenced by that of Erhard Weigel, one of his teachers during the early years of his education. Weigel's influence on Leibniz needs a separate and thorough research, even more since it was completely neglected by the secondary literature, but for the specific purpose of this article it is worth pointing out mainly for one reason: Leibniz's studies on the foundations were not as original as Frege thought they were, but they should be analysed as the final outcome of a tradition developed in Germany at the beginning of the 17th century. The influence of Weigel, known for his attempt to restore some of Aristotle's fundamental intuitions in the light of Euclid, also explains why the *Initia rerum* were meant to be published in the *Acta Eruditorum*, a place considered by Leibniz appropriate for these topics. The easiest way to show Leibniz's commitment to Weigel's tradition is showing Weigel's definition of *ratio*, contained in his *Idea matheseos universae*: "*Ratio* est *quantitas primo respective concepta*, h.e. valor unius termini in mensura alterius terminis similis & homogenei, priusque cogniti, vel saltem suppositi & assumpti. Ut si quaeratur, quam longa sit linea AB? Respondeaturque per linea pedalem CD tanquam per mensuram, dicaturque, linea AB esse v.g. *quadrupedalem*. Ubi quantitas respective concepta est quadruplicitas, seu quadruplum, ratio nempe sive valor quem habet linea AB ad lineam CD. *Proportio* vero est *quantitas bis respective concepta* h.e. ratio per similem rationem, quam habent alii duo termini, expressa: Ut si ad quaestionem, quam longa sit linea AB? supposita linea CD tanquam ejus mensura, assumptisque simul duobus aliis terminis similiter se habentibus v.g. semuncia & drachma respondeatur: lineam AB ad CD praecise tantam esse, quanta est semuncia ad drachmam" (Weigel 1669, pp. 14-15).

[31] Leibniz 1976, p. 667. "Situs est coexistentiae modus. Itaque non tantum quantitatem, sed et qualitatem involvit" (Leibniz 1863, p. 18).

[32] Leibniz 1976, p. 671. "Situs quaedam coexistendi relatio est inter plura, eaque cognoscitur per alia coexistentia, intermedia, id est quae ad priora simpliciorem habent coexistendi relationem" (Leibniz 1863, p. 25).

[33] "Unitatem autem repraesentare Rationem aequalitatis" (Leibniz 1863, p. 24).

considered a great achievement, and it is perhaps one of the most convincing explanation, at least in a geometrical form, of the idea expressed in the correspondence with Clark about the possibility of deriving both arithmetic and geometry from identity, i.e. the principle of contradiction. Another proof of this theory rests in the analysis of the key aspects of Leibniz's foundation of arithmetic: unity and coexistence are the cornerstones of Leibniz's theory and they are both founded using the principle of identity or contradiction. As it is for unity in fact, coexistence is contradictory if it doesn't follow a specific rule: *the whole is greater than a part*, otherwise comparing similar coexisting objects wouldn't be enough to establish a precise order. Again, in the demonstration of this principle, taken from Hobbes since the early years but also significantly reworked to a certain extent, identity has a key role[34]. In the end, straight lines are perfect as geometrical tools used to express relations, but only if they are supported by a distinction between quality and quantity, considered above geometry, and by a principle that belongs to logic and metaphysics: it's in this sense that Leibniz's *initia* are in fact metaphysical.

Leibniz's approach to the foundation of arithmetic has proven to be complex and deep. A final remark should be added on the choice of identifying the *Initia rerum* with Leibniz's most coherent and concrete effort in the foundation of arithmetic, among other purposes. Despite the evidence given by the title itself in fact, one could argue that other hints given by Leibniz in previous works, for example in the *Nouveaux essais*, should be interpreted as different approaches, rather than steps in the direction culminated with the *Initia rerum*. However, if the complete tools of the *analysis situs* were developed by Leibniz only after 1679[35], it is clear that, since that time at least, his intention was that of defining numbers through the concept of homogeneity, as many passages show[36]. I believe that this is sufficient to highlight a 'logicistic' and mathematical path that pervades Leibniz's passages on the foundation of arithmetic and that leads to this last work.

Now that most of the aspects involved have been described, a direct confrontation with Frege's foundations is possible, in order to understand if Leibniz's theory influenced the history of logic and mathematics.

---

[34] In Leibniz 1863, p. 20 the demonstration is composed by a definition and a identical preposition: "B est aequale parti ipsius A". It's thanks to the use of this identity that the proof is possible. Leibniz then specifies that he is using the principle of identity, or contradiction: "Unde videmus demonstrationes ultimum resolvi in duo indemonstrabilia: Definitiones seu ideas, et propositiones primitivas, nempe identicas, qualis haec est B est B, unumquodque sibi ipsi aequale est, aliaequale hujusmodi infinitae" (*Ibidem*).

[35] See De Risi (De Risi 2007) and Echeverría's work in Leibniz (Leibniz 1995).

[36] See Leibniz 1923, VI-4A, p.278, 310, 381, 383, 390, 392, 418, 421, and Leibniz 1923, II-2, p. 251.

3 Frege's Reception in the *Grundlagen der Arithmetik*

The analysis of Leibniz's definition of number in the *Initia rerum* showed several proprieties of his foundation of arithmetic that can be summarized as follows:

- Unity is the cornerstone of Leibniz's theory: every other number is defined through a comparison with unity. The order generated in this way can express, ideally, all the relations involved, so that it is possible to conceive any given number following the path that leads to unity.
- The comparison is done using qualities and quantities, that is to say, using entities that have the same intrinsic proprieties. Such entities are undistinguishable, if not taken as coexisting, and are assigned to numbers through a relation of identity.

- The qualitatively similar entities chosen for this scope are straight lines, or line segments, because they can successfully express all the possible numbers[37] in a coherent reference system where ratios are examined. However, the use of geometrical qualities is not mandatory, if another qualitatively and quantitatively consistent system is possible.

- Even if a number is defined through quantity, it doesn't mean that quantity is defined using numbers: the difference between quantities rests in their interaction with quality, so that, through coexistence, it can be established if a given quantity coincides with another one, or if it is only similar to it. The order is generated following the universal principle that the whole is greater than a part. In a specific reference system then, identity expresses not only the propriety of being itself, but also the propriety of not being what it isn't: identity is the expression in positive terms of the principle of contradiction.

On a purely theoretical level, the similarities with Frege's theory are already evident. Frege's attempt to define a number challenges the same problems, as it is shown in the *Grundlagen der Arithmetik*. This work is the most suited to judge Leibniz's influence, because it was published in-between the *Begriffsschrift* and the *Grundesetze*, in a time in which the new editions of Leibniz's writings were far from being published, meaning that if a real influence occurred, it should have been based solely on the Erdmann's edition and Gerhardt's *Mathematische Schriften*. It follows that there should be some sort of consistency between Leibniz's works appeared in these editions and the *Initia rerum*, a consistency that Frege noticed.

---

[37] Another great achievement of Leibniz's theory, at least in his eyes, was the possibility of defining even Surds in this way. It's not the possibility of the geometrical construction that counts in this case, but the fact that every Surd has a unique place, related to the other numbers and unity.

The *Grundlagen* is a work that constantly refers to Leibniz and to the philosophical issues and the historical context in which Frege thought himself. As it was explained for Leibniz, the definition of number poses an important paradox in Frege:

> The expression '*n* is a Number' is to mean the same as the expression 'there exists a concept such that *n* is the Number which belongs to it'. Thus the concept of Number receives its definition, apparently, indeed, in terms of itself, but actually without any fallacy, since 'the Number which belongs to the concept F' has already been defined[38].

The only way in which this definition can be accepted without contradictions is to conceive 'the Number which belongs to the concept F' as a definition that doesn't rely on quantity, taken in the naïve sense. Frege does so relying on the concept of equinumerosity. This means that it is possible to establish a one-to-one correspondence between two sets. As it is known, Frege applies this intuition to determine identical numbers, using the so-called Hume's principle: bijection is the tool used by Frege to determine equality and inequality through compresence or coexistence. It's clear that we are not too far from Leibniz's approach, so much that even if the tool used is different — equinumerosity instead of *situs* in a geometrical context — the exemplification of this intuition contained in the *Grundlagen* uses a spatial reference:

> 'Equal' we defined in terms of one-one correlation, and what must now be laid down is how this latter expression is to be understood, since it might easily be supposed that it had something to do with intuition. We will consider the following example. If a waiter wishes to be certain of laying exactly as many knives on a table as plates, he has no need to count either of them [...] Plates and knives are thus correlated one to one, and that by the identical spatial relationship[39].

---

[38] Frege 1980, p. 85. "Der Ausdruck: '*n* ist eine Anzahl', sei gleichbedeutend mit dem Ausdrucke: 'es giebt einen Begriff der Art, dass n die Anzahl ist, welche ihm zukommt'. So ist der Begriff der Anzahl erklärt, scheinbar freilich durch sich selbst, aber dennoch ohne Fehler, weil 'die An- zahl, welche dem Begriffe F zukommt' schon erklärt ist" (Frege 1884, p. 85).

[39] Frege 1980, p. 82. "Dazu ist es nöthig, die Gleichzahligkeit noch etwas genauer zu fassen. Wir erklärten sie mittels der beiderseits eindeutigen Zuordnung, und wie ich diesen Ausdruck verstehen will, ist jetzt darzulegen, weil man leicht etwas Anschauliches darin vermuthen könnte. Betrachten wir folgendes Beispiel! Wenn ein Kellner sicher sein will, dass er ebensoviele Messer als Teller auf den Tisch legt, braucht er weder diese noch jene zu zählen, wenn er nur rechts neben jeden Teller ein Messer legt, sodass jedes Messer auf dem Tische sich rechts neben einem Teller befindet. Die Teller und Messer sind so beiderseits eindeutig einander zugeordnet und zwar durch das gleiche Lagenverhältniss" (Frege 1884, pp. 81-82).

Arguing that Leibniz's theory served alone as inspiration for that of Frege would be overestimating Leibniz's influence, because Frege himself admits his debt to Hume, and he was also aware of the works of Cantor in this direction. However, that of Hume is considered in Frege's eyes as a mere intuition, although essential. In the history of the foundations of arithmetic offered in the *Grundlagen* instead, Leibniz has a major role in several occasions. While philosophers such as Kant or Mill are openly criticized, Leibniz is the only one whose definition of number is considered something that needs to be perfected, more than abandoned. Frege's belief in the correctness of Leibniz's approach is so strong that in many cases it is considered a touchstone for entire theories:

> How probability theory could possibly be developed without presupposing arithmetical laws is beyond comprehension. Leibniz holds the opposite view, that the necessary truths, such as are found in arithmetic, must have principles whose proof does not depend on examples and therefore not on the evidence of the senses[40].

In the footnotes, Frege refers directly to a passage from the *Nouveaux essais sur l'entendement humain*, taken from the Erdmann's edition. It is an interesting passage indeed, because it connects the principle of contradiction to the foundations of arithmetic and geometry on the topic of innate knowledge[41]. It shows that Frege's interest on Leibniz was not only based on the general distinction between truths of fact and truths of reason, as it would appear without referring to Frege's footnotes, but it was focused on the specific topic of the foundation of mathematics. Frege then was aware not only of the general idea of a universal characteristic, but also of the specific application of logic and its relation with arithmetic and geometry.

Even the idea of unity as a cornerstone of the foundational theory and the definition of a path that leads to unity from any given number is derived from a

---

[40] Frege 1980, pp. 16-17. "Wie diese Lehre aber ohne Voraussetzung arithmetischer Gesetze entwickelt werden könne, ist nicht abzusehen. §11. Leibniz meint dagegen, dass die nothwendigen Wahrheiten, wie man solche in der Arithmetik findet. Principien haben müssen, deren Beweis nicht von den Beispielen und also nicht von dem Zeugnisse der Sinne abhangt, wiewohl ohne die Sinne sich niemand hätte einfallen lassen, daran zu denken, ‚Die ganze Arithmetik ist uns eingeboren und in uns auf virtuelle Weise'. Wie er den Ausdruck ‚eingeboren' meint, verdeutlicht eine andere Stelle: 'Es ist nicht wahr, dass alles, was man lernt, nicht eingeboren sei; — die Wahrheiten der Zahlen sind in uns, und nichtsdestoweniger lernt man sie, sei es, indem man sie aus ihrer Quelle zieht, wenn man sie auf beweisende Art lernt (was eben zeigt, dass sie eingeboren sind), sei es…'" (Frege 1884, pp. 16-17).

[41] "Dans ce sens on doit dire que toute l'Arithmétique et toute la Géométrie sont innées et sont en nous d'une manière virtuelle, en sorte qu'on les y peut trouver en considérant attentivement et rangeant ce qu'on a déja dans l'esprit, sans se servir d'aucune vérité apprise par l'expérience ou par la tradition d'autrui" (Leibniz 1840, p. 208).

passage of the *Nouveaux essais*: the famous proof of the numerical formula *2 + 2 = 4*. Once again, Frege proves to be an excellent interpreter of Leibniz's intuitions. At first, he recognizes that there is a gap in the proof, because Leibniz avoids the use of brackets to highlight the different relations involved[42], but a part from this, the analysis of the proof is made in a Leibnizian spirit: *2 + (1 + 1) = (2 + 1) + 1* is considered a special case of *a + (b + c) = (a + b) + c*. Algebra is used here in a way very similar to that seen in the *Initia rerum*: it highlights a formal relation that is valid for every given number. The idea that a general relation expresses a truth that is universally maintained in the specific cases is also consistent with what Leibniz writes in the *Dialogus de conexione inter res et verba*, available in the Erdmann's edition[43]. Frege believes that assuming the law hidden in Leibniz's proof, every formula of addition can be proved. The result of this observation is decisive:

> If we assume this law, it is easy to see that a similar proof can be given for every formula of addition. Every number, that means, is to be defined in terms of its predecessor. And actually, I do not see how a number like 437986 could be given to us more aptly than in the way Leibniz does it. Even without having an idea of it, we get it by this means at our disposal none the less. Through such definitions, we reduce the whole infinite set of numbers to the number one and increase by one, and every one of the infinitely many numerical formulae can be proved from a few general propositions[44].

The infinite set of numbers is compared to unity. Also, the definition allows the existence of a path that leads to unity, as it was shown in Leibniz's account. This solution was adopted by Frege to criticise Kant's awkward distinction between small and large numbers[45], related to their intuition.

---

[42] Cfr. Frege 1840, p. 7.

[43] In the *Dialogus* (Leibniz 1840, pp. 77-78) the exemplification of the connection between truth and formal-relational proprieties is based on mathematics, on the square of a binomial in particular. Leibniz's reasoning is similar to the passage of the *Initia rerum* on the derivation of algebra from the application of universal characteristic on quantity and similar to Frege's reasoning in this passage. Numbers are substituted with letters in order to express universal rules or relations: "Vides utcunque pro arbitrio sumantur characteres, modo tamen in eorum usu certus ordo et modus servetur, semper omnia consentire" (*Ibidem*).

[44] Frege 1980, pp. 7-8. "Setzt man dies Gesetz voraus, so sieht man leicht, dass jede Formel des Einsundeins so bewiesen werden kann. Es ist dann jede Zahl aus der vorhergehenden zu definiren. In der That sehe ich nicht, wie uns etwa die Zahl 437986 angemessener gegeben werden könnte als in der leibnizischen Weise. Wir bekommen sie so, auch ohne eine Vorstellung von ihr zu haben, doch in unsere Gewalt. Die unendliche Menge der Zahlen wird durch solche Definitionen auf die Eins und die Vermehrung um eins zurückgeführt, und jede der unendlich vielen Zahlformeln kann aus einigen allgemeinen Sätzen bewiesen werden." (Frege 1884, pp. 7-8).

[45] Frege 1980, p. 6.

In general terms then, Frege's foundation follows that of Leibniz, but what about the essential idea of identifying numbers through coexistence? Answering this question means finding a connection between Leibniz's coexistence and the concept of bijection used in Hume's principle. A possible proof of this connection is Frege's reference, in the footnotes of the passage just quoted, to Leibniz's *Non inelegans specimen demonstrandi in abstractis*, appeared in the Erdmann's edition. This work starts with the famous definition '*eadem sunt quorum unum potest substitui alteri salva veritate*', adopted by Frege[46]. However, this definition shouldn't be taken alone: it's used in a specific context to determine what does not follow from it and, at a later stage, to determine relations between parts and the whole[47]. The same use of identity is found in the *Initia rerum*: in this prospective, if we assert $a = 3$ we do not, strictly speaking, *assign* the value *3* to *a*, rather, we state that in every occasion in which *a* is substituted with *3*, our system remains consistent and, consequently, we also state that, for every other value, the consistent system will show us a contradiction after the substitution. This is possible because every other relation in our reference system is rationally determined, maintaining both a nominalist approach in the definition of identities and yet a connection with truth, similar to that exposed in Leibniz's *Dialogus*[48]. Bijection is no doubt different on a specific level, but both are tools developed to avoid the idea of counting.

---

[46] See Frege 1980, p. 76. Angelelli (Angelelli 1967) observed that the notion of substitution is not valid for Leibniz in every occasion. The phenomenon of *reduplicatio* would be a decisive proof in this direction: when considered in themselves, things maintain a specific nature that would be lost in identity, so that, for example, a triangle, considered in relation to its propriety of having three angles is different from a triangle considered as having three sides, even if the substitution is always possible. However, it should be pointed out that every time Leibniz introduces the problem of *reduplicatio* in these terms, he introduces the concept of identity as well. *Reduplicatio* then should be interpreted as a specification of the rule and not as a violation of it. Again, the concept of quality is essential to understand Leibniz's definition: every intrinsic propriety has its peculiar description and Leibniz, more than trying to rehabilitate 'opaque contexts', establish a precise use of identity in order to maintain the richness of the qualitative approach. By no means however, identity is based solely on quality: coexistence is essential as well, so that the trilateral and the triangular are still conceivable through their different qualities, but through coexistence they always 'happen' to be the same.

[47] "Diversa sunt, quae sunt non eadem, seu in quibus substitutio aliquando non succedit [...] si A et B simul sumta coincidant ipsi L, A ut et B dicetur inexistens vel contentum, at L dicetur continens" (Leibniz 1840, p. 94).

[48] I won't discuss here the difference between reality and mathematics from Leibniz's point of view alone, because the *Dialogus* was written about thirty years before the *Nouveaux essais* and showing a consistency throughout such a large part of Leibniz's life would be hard. However, The *Nouveaux essais* were the starting point of Frege's interpretation of Leibniz. This premise shouldn't be underestimated, because Leibniz here establishes a connection between logical principles — mainly identity and contradiction — and his gnoseology. Understanding the relation

It is also legitimate to question about Frege's knowledge of Leibniz's tool, i.e. the use of straight lines geometrically defined and compared. At first, Leibniz's influence seems to end here, because this way of determining numbers is openly rejected:

> I should like straight away to oppose the attempt to think of number geometrically, as a ratio between lengths or surfaces [...] Newton proposes to understand by number not so much a set of units as the relation in the abstract between any given magnitude and another magnitude of the same kind which is taken as unity[49].

It is true that Newton uses a definition similar to that of Leibniz in the *Arithmetica Universalis*, as quoted by Frege[50], but some key difference must be pointed out: the relation attributed to Newton is considered 'in the abstract'. A number then is considered in a strictly qualitative sense. Compared to the *Initia rerum*, it would be as saying that a number *is* a straight line, but for Leibniz straight lines are not similar to numbers. Considered alone, straight lines are indistinguishable, whereas numbers are defined through straight lines in such a way that every straight line occupies a different place. Every number then has a peculiar definition that makes it different from the other ones, unlike straight lines. If the purely geometrical approach is correct instead, homogeneity would be useless, because there is no need for a transformation of an entity into another one. In other words, the distinction between quantity and quality, and the fact that they must be combined for a successful foundation of arithmetic, is the true novelty of Leibniz's theory. Quality alone in fact is not sufficient and, even if it's clear that Frege refuses the geometrical

---

between Leibniz's logical coherence about reality, despite nominalism, and Frege's famous notions of *Sinn* and *Bedeutung* would be another important outcome on Leibniz's influence on Frege. Leibniz's solution is expressed in the *Dialogus* and in the *Nouveaux essais*, both available in the Erdmann's edition. In the *Dialogus* we read: "Sed hoc tamen animadverto, si characteres ad ratiocinandum adhiberi possint, in illis aliquem esse situm complexum ordinem, qui rebus convenit, si non in singulis vocibus (quamquam et hoc melius foret) saltem in earum conjunctione et flexu, et hunc ordinem, variatum quidem in omnibus linguis, quodammodo respondere [...] etsi characteres sint arbitrarii, eorum tamen usus et connexio habet quiddam, quod non est arbitrarium" (Leibniz 1840, p. 77). In the *Nouveaux essais*, the same idea is expressed, with an interesting exemplification taken from geometry, using similar lines (Leibniz 1840).

[49] Frege 1980, pp. 25-26. "Hier möchte ich mich nun gleich gegen den Versuch wenden, die Zahl geometrisch als Verhältnisszahl von Längen oder Flächen zu fassen [...] Newton will unter Zahl nicht so sehr eine Menge von Einheiten als das abstracte Verhältniss einer jeden Grösse zu einer andern derselben Art verstehen, die als Einheit genommen wird" (Frege 1884, p. 25).

[50] Frege quotes the second chapter of Newton's *Arithmetica Universalis* (Netwon 1732, pp. 36-41). Here the notion of a fraction is indeed introduced in comparison with unity.

exemplification chosen, it doesn't mean that he is not aware of Leibniz's distinction and its usefulness:

> We shall do well in general not to overestimate the extent to which arithmetic is akin to geometry. I have already quoted a warning to this effect from Leibniz. One geometrical point, considered by itself, cannot be distinguished in any way from any other; the same applies to lines and planes. Only when several points, or lines or planes, are included together in a single intuition, do we distinguish them [...] points or lines or planes [...] stand as representatives of the whole of their kind. But with the numbers it is different; each number has its own peculiarities[51].

Leibniz's quote by Frege is again taken from the *Nouveaux essais*[52], showing a consistency in Leibniz's foundation of mathematics, at least in the later years. If Frege was able to understand this consistency in such a detailed way, it means that his reference to Leibniz was far from a mere homage: the frequent quotes and the fact that Leibniz is present in every essential part of Frege's theory show the extent of this influence.

---

[51] Frege 1980, p. 19. This passage is worth quoting in its entirety: "Ueberhaupt wird es gut sein, die Verwandtschaft mit der Geometrie nicht zu überschätzen. Ich habe schon eine leibnizische Stelle dagegen angeführt. Ein geometrischer Punkt für sich betrachtet, ist von irgendeinem andern gar nicht zu unterscheiden; dasselbe gilt von Geraden und Ebenen. Erst wenn mehre Punkte, Gerade, Ebenen in einer Anschauung gleichzeitig aufgefasst werden, unterscheidet man sie. Wenn in der Geometrie allgemeine Sätze aus der Anschauung gewonnen werden, so ist das daraus erklärlich, dass die angeschauten Punkte, Geraden, Ebenen eigentlich gar keine besondern sind und daher als Vertreter ihrer ganzen Gattung gelten können. Anders liegt die Sache bei den Zahlen: jede hat ihre Eigenthümlichkeit. Inwiefern eine bestimmte Zahl alle andern vertreten kann, und wo ihre Besonderheit sich geltend macht, ist ohne Weiteres nicht zu sagen" (Frege 1884, pp. 19-20). In this last passage, when Frege refers to the specific nature of a number that cannot be defined in advance, we see how numbers are conceived as a set of relations.

[52] Frege, quoting Leibniz, writes: "Leibniz recognized this already: for to his Philalète who had asserted that 'the several modes of number are not capable of any other difference but more or less; which is why they are simple modes, like those of space,' he returns the answer: 'That can be said of time and of a straight line, but certainly not for figures and still less of the numbers, which are not merely different in magnitude, but also dissimilar [...] for two unequal figures can be perfectly similar to each other, but never two numbers" (Frege 1980, p. 14). On a side note, this passage also proves that for Leibniz numbers are dissimilar, hence the use of the general law of homogeneity in his definition.

4 Concluding Remarks

Frege's studies on Leibniz have proven to be decisive for his foundational theory. In the framework of the *Grundlagen* this debt is meaningful, especially if we highlight that in this work Frege's attempt on the foundations of arithmetic is based on Hume's principle and not on the unpopular Basic Law V. The inconsistency of the latter overshadowed the great accomplishment of Hume's principle: if taken as an axiom in fact, it can prove Peano's axioms[53]. Leibniz contributed to this task supplying Frege with a specific idea of the relation between logic, arithmetic and geometry, a focus on the definition of numbers through coexistence and a distinction between quantity and quality that shows the limitations of a purely geometrical approach. The use of coexistence through quality and quantity is in fact very reminiscent of Frege's use of bijection in his foundation of arithmetic. While the use of straight lines together with the concept of *Ratio* in order to define numbers is considered a common approach in the 17th century, shared by Newtonians and Cartesians alike, these innovative ideas belong instead only to Leibniz's unique take on the foundations: a formal approach that tries to overcome the difficulties in defining numbers without using the idea of counting and relying on logical relations is in fact present for the first time in the history of science in Leibniz, but perhaps only perceptive minds such that of Frege were truly able to understand the difference.

The fact that Leibniz's take on the foundations is reported in such details in Frege is especially remarkable, considering the availability of Leibniz's works at the time of the *Grundlagen*. It follows that some of Leibniz's greatest intuitions on these topics are present in works that are not usually considered as belonging to logic or mathematics and, consequently, that Frege's studies on Leibniz must have been intensive and thorough. In this regard, we could say that Leibniz's renaissance was possible even before the seminal works published by Couturat and Russell and, perhaps, it was more coherent in defining the distinctions between logic, metaphysics and mathematics: the similarities between Leibniz and Frege force us to see how modern logic was developed with mathematics in mind as its main field of application, making it hard to give a coherent account of logic completely separated by its mathematical background. Even if it's true that the universal characteristic is by no means used for the sole purpose of defining mathematical objects, this neglected possibility is in Leibniz's eyes a *specimen illustre*, as he himself defined it. Similarly, Frege's ideography is developed in order to be applied to a specific mathematical content, that of the foundations. Ideography has a wider range of application indeed, but it can be applied both to arithmetic and geometry, obtaining in this way a kind of *analysis situs*, as Frege defines it in the *Begriffsschrift*[54]. By the time of this work then, Leibniz's influence on Frege is evident and it is the result of

---

[53] On this topic see Hintikka et al. 1995.

[54] "It seems to me to be easier still to extend the domain of this formula language to include geometry. We would only have to add a few signs for the intuitive relations that occur there. In this way we would obtain a kind of *analysis situs*" (Frege 1981, p. XI).

the circulation of Leibniz's brilliant ideas, already available at the end of the 19th century, before Couturat's edition.

References

Angelelli, I. 1967. On identity and interchangeability in Leibniz and Frege. *Notre Dame Journal of Formal Logic* 8:94-100.
Arthur, R. 2001. Leibniz and Cantor on the actual infinite. In In *Nihil sine ratione. Mensch, Natur und Technik im Wirken von G. W. Leibniz, Akten des VII. Internationalen Leibniz-Kongress*: 41-46. Berlin: Leibniz-Gesellschaft.
Couturat, L. 1901. *La logique de Leibniz d'aprés des documents inédits*. Paris: Alcan.
Crapulli, G. 1969. *Mathesis universalis: genesi di un'idea nel 16° secolo*. Roma: Edizioni dell'Ateneo.
Dascal, M. et al. 2008. *Leibniz: what kind of rationalist?*. Berlin: Springer.
De Risi, V. 2007. *Geometry and Monadology. Leibniz's Analysis Situs and Philosophy of Space*. Basel: Birkhäuser.
Di Bella, S. 2005. *The science of the individual: Leibniz's ontology of individual substance*. Dordrecht: Springer.
Fichant, M. 1994. Les axiomes de l'identité et la démonstration des formules arithmétiques: "2 + 2 = 4". *Revue internationale de philosophie* 48:173-211.
Frege, G. 1884. *Die Grundlagen der Arithmetik. Eine logisch mathematische Untersuchung über den Begriff der Zahl*. Breslau: Verlag von Wilhelm Koebner.
Frege, G. 1969. *Nachgelassene Schriften*. Hamburg: Meiner.
Frege, G. 1977. Begriffsschrift, eine der arithmetischen nachgebildete Formelsprache des reinen Denkens. In *Begriffsschrift und andere Aufsätze*, ed. Ignacio Angelelli. Darmstadt: Wissenschaftliche Buchgesellschaft.
Frege, G. 1980. *The foundations of arithmetic: a logico-mathematical enquiry into the concept of number*. Oxford: Blackwell.
Frege, G. 1981. Begriffsschrift, a formula language, modeled upon that of arithmetic, for pure thought. In *From Frege to Gödel: a source book in mathematical logic*, ed. Jan van Heijenoort. Cambridge: Harvard University Press.
Frege, G. 1996. *Grundgesetze der Arithmetik: Begriffsschrift abgeleitet von G. Frege*. Hildesheim: Olms.
Goethe, N., Beeley, D., Rabouin, D. (ed.). 2015. *G.W. Leibniz, Interrelations Between Mathematics and Philosophy*. Dordrecht: Springer Netherlands.
Hintikka, J. (ed.). 1992. *From Dedekind to Godel: essays on the development of the foundations of mathematics*. Dordrecht: Kluwer.
Ishiguro, H. 2005. Leibniz et la distinction frégéen entre "sens" et "référence". In *Leibniz et les puissances du langage*: 201-210. Paris: Vrin.
Kluge, E. H. 1980a. *The Metaphysics of Gottlob Frege. An Essay in Ontological Reconstruction*. Dordrecht: Springer-Science+Business Media.
Kluge, E. H. 1980b. Frege, Leibniz and the notion of an ideal language. *Studia Leibnitiana* 12:140-154.

Korte, T. 2010. Frege's Begriffsschrift as a Lingua Characteristica. *Synthese* 174: 183-94.
Krömer, R. Chin-Drian, Y. (ed.). 2012. *New essays on Leibniz reception*. Basel: Birkhäuser.
Leibniz, G. W. 1840. *Opera philosophica quae exstant latina Gallica Germanica omnia*, ed. Johann Eduard Erdmann. Berlin: G. Eichler (Printer).
Leibniz, G. W. 1849. Leibnizens Mathematische Schriften, Band I, ed. Carl Immanuel Gerhardt. In *Leibnizens Gesammelte Werke aus den Handschriften der Königlichen Bibliothek zu Hannover*. Halle: Schmidt.
Leibniz, G. W. 1863. Leibnizens Mathematische Schriften, Band VII, ed. Carl Immanuel Gerhardt. In *Leibnizens Gesammelte Werke aus den Handschriften der Königlichen Bibliothek zu Hannover*. Halle: Schmidt.
Leibniz, G. W. 1923-. *Sämtliche Schriften und Briefe*, series I-VII. Darmstadt (Leipzig, Berlin): Akademie der Wissenschaften (Akademie-Ausgabe).
Leibniz, G. W. 1976. *Philosophical papers and letters*, ed. Leroy E. Loemker. Dordrecht (Boston): D. Reidel Publishing Company.
Leibniz, G. W. 1995. *La caractéristique géométrique*, ed. J. Echeverría, M. Parmentier. Paris: Vrin.
Lenzen, W. 2004. Leibniz und die Entwicklung der modernen Logik. In Wolfgang Lenzen, *Calculus Universalis. Studien zur Logik von G. W. Leibniz* Paderborn: Mentis.
Look, B. 2011. Leibniz, Kant and Frege on the existence predicate. In *Natur und Subjekt: IX. Internationaler Leibniz-Kongress unter der Schirmherrschaft des Bundespräsidenten, Hannover, 26. September bis 1. Oktober 2011*:616-624. Hannover: Leibniz-Gesellschaft.
Luciano, E. 2006. The influence of Leibnizian ideas on Giuseppe Peano's work. In *Einheit in der Vielheit: Vortäge; VIII. Internationaler Leibniz-Kongress, Hannover, 24. Bis 29. Juli 2006*. Hannover: Leibniz-Gesellschaft.
Mugnai, M. 1992. *Leibniz's theory of relations*. Stuttgart: Steiner.
Newton, I. 1732. *Arithmetica universalis sive De compositione et resolutione arithmetica*, 2 volumes. Lugduni Batavorum (Leiden): apud Joh. Et Herm.
Pasini, E. 1997. Arcanum Artis Inveniendi: Leibniz and Analysis. In *Analysis and Syntesis in Mathematics*: 35-46 Dordrecht: Kluwer.
Pasini, E. 2001. La philosophie des mathématiques chez Leibniz. Lignes d'investigation. In *Nihil sine ratione. Mensch, Natur und Technik im Wirken von G. W. Leibniz, Akten des VII. Internationalen Leibniz-Kongress*: Berlin: Leibniz-Gesellschaft, pp. 954-963.
Peckhaus, V. 1997. *Logik, Mathesis universalis und allgemeine Wissenschaft: Leibniz und Die Wiederentdeckung der formalen Logik im 19. Jahrhundert*. Berlin: Akademie Verlag.
Rabouin, D. 2009. *Mathesis universalis: l'idée de «mathématique universelle» d'Aristote à Descartes*. Paris: Presses Universitaires de France.
Russell, B. 1900. *A Critical Exposition of the Philosophy of Leibniz*. Cambridge: The University Press.
Suárez, F. 1965. *Disputationes metaphysicae*, 2 volumes, reprinted. Hildesheim: G. Olms.
Trendelenburg, F. A. 1867. *Historische Beiträge zur Philosophie*. Berlin: Bethge.

Wallwitz, G. 1991. Strukturelle Probleme in Leibniz' *Analysis Situs*. In *Studia Leibnitiana*, XXII:111-118.
Weigel, E. 1669. *Idea matheseos universae*. Jena: Typis & sumptibus Johannis Jacobi Bauhoferi.

Mattia Brancato, University of Milan, Italy
mattia.brancato@outlook.com

Historical and Philosophical Details on Leibniz's Planetary Movements as *Physical–Structural Model*

Paolo Bussotti and Raffaele Pisano

Abstract. In February 1689 Gottfried Wilhelm von Leibniz (1646–1716) published the *Tentamen de motuum coelestium causis* in *Acta Eruditorum*. This work deals with a planetary model. This essay is not a mere formal–or–virtuosic mathematical exercise only. It represents a significant conceptualization within a general physical theory that Leibniz was going to construct. It is well known that Leibniz rejected the action at a distance. He was convinced that each interaction between bodies should be explained by means of mechanical causes only. However, we remark: 1) he was a contemporary of Huygens and Newton and – obviously – he had to take into account Newton's results; 2) he thought that also the *final causes* should play a role in the physical explanations. In fact, his introduction of the *vis viva* is connected to the idea that the final causes could play a role within physics. This fact offers an interesting and stimulating picture, which has fertile consequences both in his planetary model and in the role played by planetary model in his entire theoretical system. In order to inquiry historically and philosophically Leibniz's planetary model, we introduce the epistemological key–notion of the *physical–structural model*. This is connected with Leibniz's attempt of edifying a physical system: observation, physical quantities and mathematical interpretations are involved in such model. Within this context both the effects of the most important actions (specifically gravity) and physical causes were dealt with by Leibniz. In this paper, we present and discuss, by means of the above key–concept, Leibniz's planetary model. This paper is part of a larger research programme (RP) concerning the relationship between physics and mathematics in the history and philosophy of science.

Keywords: Leibniz, Acta Eruditorum, Tentamen, Phyisca nova, Tentaminis, Planetary movements, Modelling, Geometry, Kinematics, Dynamics, Physical–structural model, Relationship Physics–Mathematics.

1   Introduction

In this introduction, we will clarify the different kinds of models characterizing mainly the astronomical tradition. This is useful for a complete explanation of the concept of *physical–structural model* which is pivotal for our research and whose features will be expounded in the course of our paper. We worked on a PDF copy of the original *Acta Eruditorium*[1] of 1689 (Leibniz 1689, pp. 82-96), and sometime discussed figures as exposed by Gerhardt in his edition (Leibniz 1689 [1860] [1962] VI, pp. 161-187) and by Aiton (Aiton 1960).

1.1 Kinematical and Dynamical Approaches

The long scientific tradition dedicated to the analysis of the planetary movements by means of geometrical models relies, basically, upon a kinematical approach. This means that a series of movements described by means of geometrical figures (almost exclusively circles) were conceived, from the composition of which the apparent motion of a planet with respect to the earth was explained. Such explanation was only *kinematic* because neither an inquire on the causes of these movements nor an interest in understanding if the model described the real planetary movements existed. The capability of the kinematical–mathematical model (Pisano 2011, 2013, 2016) to predict (nowadays we say mechanically) the positions of the planets was required: *to save the phenomena*[2], if one uses the famous expression popularized by Pierre Duhem (1861-1916). Ptolemy's (ca. 100 – ca. 170) *Almagest* is the first systematic treatise we have, where this approach was designed. Obviously, as well known, this does not mean that Ptolemy was the first who developed geometrical–kinematical models for planetary motion. Though not all the scholars completely agree, Copernican theory, in the form originally presented in the *De revolutionibus orbium coelestium*, was still based on kinematical models[3].

A completely different approach is the one, in which one does not only restrict to structure a functioning geometrical–kinematical model, but he also tries to determine the *physical causes* (dynamics) of the planetary movements and to explain the movements based on the features connoting such causes (e.g., the forces). We precise that both the words *physical* and *causes* were – as well known – so polysemous to be ambiguous[4].

---

[1] https://books.google.it/books/about/Acta_eruditorum.html?id=Q2NCmqygGiwC&redir_esc=y

[2] *Sauver les Phénomènes. Essai sur la Notion de Théorie Physique de Platon à Galilée* (Duhem P. M. 1969 – posthumous translation in English – backing to Plato's *moto* [1908 in original Greek language: σώζειν τὰ φαινόμενα, sozein ta phainomena]). The *Aim and Structure of Physical Theory* also appeared two years before in French language. See also Goldstein 1997.

[3] As to Copernicus, there are two problems, which have to be kept separated: 1) *Did Copernicus really believe in heliocentrism or – as Osiander wrote – was heliocentrism only a mathematical hypothesis?* 2) *Did Copernicus offer a physical basis to his system?* The two problems are different. In general, the most shared answer to 1) is "yes", while to 2) is "no". We adhere to this vision. An interesting and different perspective on 2) is offered by De Pace 2009.

[4] In our case study, we take into account a (nowadays called) *dynamical system* as composed of bodies and planetary movements: mass, position, time and correlated derivatives, fixed rules–deterministic law (or simply finite degrees of freedom).

Although Leibniz already wrote about *dynamica*[5] and first used this term, a scientific dynamic theory, inside which the planetary movements were explained, was structured by Isaac Newton[6] (1642-1727) such as universal gravitation. We have no enough room to enter into the minefield whether Kepler (1571-1630) had already reached an outline of dynamical theory or if – in the period between Kepler (Radelet de Grave 1996, 2007) and Newton–some scientists developed a dynamical approach[7]. In this context, our aim is not to trace the history of the initial phases of dynamical astronomy, but to clarify a conceptual difference between a *kinematical* and a *dynamical* point of view. Newton's gravity theory is a suitable example of the latter approach. Precisely, this paper is essentially divided into two main sections:

1) In the former, which is by far the longer and more important one, we will present kinematical and dynamical approaches and mainly Leibniz's planetary model; we also explain why it is physical-structural (§ 1). We present a general picture of: a) the aspects of Leibniz's planetary model useful to develop our thesis on his *physical–structural model* and b) the aspects of Leibniz's gravity theory usable in our argumentation (§ 2).
2) In the latter, which are essentially our concluding remarks, we will provide the readers with an explanation why Leibniz felt the need to construct such a theory (§ 3).

In addition, this research concerns a particular aspect of the general Leibniz's planetary theory. We refer to Leibnizian *scientific model* to set its fundamental quantities and their relations; so we do refer directly to it as a theory. It is interesting to remark that nowadays (generally speaking) a theory is considered a scientific and logical synthesis of elements (well-tested and verified hypotheses about particular natural phenomena). Whereas, we will see that in Leibniz's planetary theory, some of these elements are lacking – for example the well tested hypotheses: it is enough to think of the vortices Leibniz introduced to explain gravity –, while there are other elements, which, nowadays would not be considered scientific. For example: the

---

Below we also propose a table to explain significances of some key Leibnizian words with respect to our historical and epistemological analysis.

[5] Probably the two most significant works by Leibniz on dynamics were *Essai de dynamique sur les lois du mouvement* (Leibniz 1691, [1860] [1962]); and *Specimen Dynamicum*, part 1 and 2 (Leibniz 1695, [1860], [1962]). During Leibniz's lifetime, only the first part of the *Specimen* was published. Obviously, there are many other works in which Leibniz dealt with dynamics, but the two mentioned seem to us the most relevant. On Leibniz's new dynamics see Fichant 1994; see also Knobloch 1994, 1995; Fichant 2001.

[6] Bussotti and Pisano 2014a, 2014b; *Id.*, *Philosophiae Naturalis Principia Mathematica. Full Transcription and Translation from Le Seur and Jacquier Jesuit Edition*, 5 Vols. The Oxford University Press, pre-print.

[7] In this paper we deal with Leibniz, not with Kepler. Nonetheless, as to the physical aspects of Kepler's astronomy, we refer to the fundamental Stephenson 1987, [1994] and to Pisano and Bussotti 2012, 2013, 2017.

idea that the planetary orbits have the same *vis viva* (section two below). This induces us to think that in different epochs and authors, some – not all of them, obviously – of the elements which identify a scientific theory can change. We discuss it as *scientific model* such as mental structured representation of a phenomenon: planets and their movement.

For example, in the taught planetary model of the atom scientists observe the atom as a nucleus with electrons orbiting around it in a manner analogous to the way that planets revolve around the Sun. This is useful (even if in modern physics it is replaced by a mathematical function of an electron etc.) for a very-simplified description of an atom; but it cannot, i.e., predict all of its attributes. Therefore Leibniz's planetary model simplified and mathematically well–approximated his known Universe, but, like we will show below a precise explanation of the celestial bodies' movements lacked; not all quantities, force, attractions, and related properties/attributes were fully described/predicted by his model. Without entering into the discussion (because we are not interested on that) we just historically remark it such as an evident difference with respect to Newtonian universal gravitation theory.

1.2 A Physical Astronomy Tradition

Ptolemy's *Hypotheses Planetarum* is a beautiful example of what we can call *physical astronomy* tradition. Ptolemy tried to give a physical reality to the models[8] he had conceived to determine the planetary movements in the *Almagest*. He thought that the universe was composed of a series of nested spherical cells with appropriate movements. The planets were rigidly connected to the spheres, whose composed movements determined those of the planets. The tradition of the celestial spheres was well consolidated in the Western science, before and after Ptolemy. It is enough to remind the reader Aristotle's (383-322 B.C.) *De Coelo* (Aristotle 1955) or Georg von Peuerbach's (1423-1461) *Theoricae Novae Planetarum* (Peuerbach 1473 [posthumous]) only to mention two famous works. The celestial spheres and correlated planetary movements were physical models which were largely independent from the kinematical theory as expounded, in its most general and precise form, in Ptolemy's *Almagest*. Thus they could not be considered as part of a unique dynamical approach. In fact, any discussion about the causes of the spheres' movements was lacking. After all, such movements remained as mysterious as the causes of the planetary movements. Around the end of the 16th century – beginning of the 17th century – these physical models were progressively abandoned: the *dissolution of celestial spheres*[9]. Therefore these models are cosmological within the natural philosophy at that time –, but neither kinematical nor dynamical.

---

[8] We also refer to spherical triangles to solve specific astronomical problems (Pisano and Bussotti 2015; see also Pisano and Casolaro 2011).

[9] *The Dissolution of the Celestial Spheres 1595-1650* is the title of a fundamental book by William H. Donahue on this subject (Donahue 1981). See also: Rosen 1985, Goldstein and Parker.

Newton's physical–mathematical model was a dynamical model, but – as well known – it was completely independent from the possible origin of gravity, which was assumed as a given force. This does not mean that Newton had no idea on the origin of gravity, but these ideas did not enter into the structure of the theory expounded in the *Principia*[10].

At the end of the 17th century there was a quite interesting attempt to pass from a modelling of planetary motion to a *complete* theory of planetary motion. This attempt has the result to offer models which are both dynamical and physical–cosmological, namely models in which the actions are considered both as to their *effects* (dynamical) and to their *origin in the material composition of the universe* (cosmological). We call these models *physical-structural* (Bussotti, 2015). In the following we list some peculiar aspects of approaches concerning planetary movements:

1) The *origin of planetary movement* (a sort of a *general structure of the universe*) is presented. However, within this structure there is no reference to the actions determining the movements and to the origins of such actions. The theory of the celestial spheres of Aristotle, Ptolemy and Peuerbach are typical examples of cosmological models.	Cosmological approach
2) The *movement of a planet* was described by geometrical constructions. This is a kinematical point of view.	Kinematical approach
3) The movement of the planets are explained according to the action of a force. Therefore, the cause of the movement is looked for, but the origin of the cause of the	Dynamical approach

---

[10] We have no room to deal with the complex problem concerning the origin of gravity in Newton, on which an abundant literature exists. We simply refer to the queries 21 and 22 of the second English edition of Newton's *Opticks* (Newton 1717 [1721]). These two queries are fundamental sources for Newton's ideas on the origin of gravity, which, however, had no influence on the physics developed in the *Principia*.

movement, that is a sort of a superior entity, is not looked for. This dynamical approach was typical by Newtonian mechanics.	
4) From a theoretical point of view, this is the most complete kind of model because the causes of the movements (forces), the origin of the forces and the structure of the universe are looked for.	Physical approach
5) Specifically, in Leibniz the model (see 4)) is connected to a *mechanical explanation* (jointly *kinematical* and *dynamical*) relying on the *physical structure* of the universe. This structure was mechanically (in the sense of kinematical and dynamical) provided by the *vortices*. By the vortices, the *mechanisms* pretended also to explain the origin of the forces[11] themselves as responsible for the whole planetary movements.	Mechanical structural approach

As told, for our aims, we focus on the fourth–and–fifth items, which, jointly, represent what we call a *physical–mechanical structural model*. These five terms encompass kinematical–philosophical–physical backgrounds as above proposed. Specifically, for Leibniz's planetary model/view, we historical-epistemologically mean:

---

[11] By term "force" we do not refer to the Newtonian forces as explained in *Principia*, but we mean any possible–general cause of a movement or change of a movement. We are aware of the danger behind the use of this term, but in the contexts we deal with, no ambiguity will appear.

Physical–Structure Model

Physical
Referred to mechanical (kinematical and dynamical) aspects of the movements where quantities, causes, movements and their interactions were geometrically–mathematically described.

Structural
Referred to a precise set of mechanical relations and structures (i.e, vortices).

Model
Referred to a conceptual simplified representation in order to produce hypotheses of explanations (i.e, by means of vortices) both about philosophical origins of the basic actions (i.e, gravity) and mechanical interactions (kinematical and then dynamical).

In addition, the distinction among *Physical cause*[12] and *Source of physical causes* (see also our resume of keywords Leibnizian concepts below) what follows is very important for our aim:

- *Physical causes*: forces such as dynamics interactions between planets or planets– and–Sun and their movements provided by mechanical structures (vortices).

- *Source of physical causes*: it is known that Leibniz's phenomenological world is not a mere consequence of physical cause–and–effect process, and he often proceeds looking for a final physical source (cause). Thus, the research on the origin of physical causes depends on the specific case displayed and discussed by the author.

Leibniz conceived a *Physical–structural model* (Figs. 2) of the planetary movements, many of whose fundamental elements were drawn from Descartes'

---

[12] We precise that in this paper we do not aim to discuss Leibnizian *final causes*; also because secondary literature is sufficiently wide and, in any case, Leibnizian causation would require by us more room. Just for completeness of argument and because the concept itself of *vis viva* has, as well known, connections to the *final causes*, in the following we only report its main significance. Generally speaking, the principle of *final causes* consists in finding the final causes, that is, the laws of natural phenomena through metaphysical reasoning, as well. Thus, final causes can be, i.e., forces or related laws which include forces. Leibniz faced that in *Unicum Opticae, Catoptricae, & Dioptricae principium* (*Acta eruditorum* 1682) proposing his laws of light. In addition, on the action, as well known, the discussion on Leibniz' *efficient causes* and *final causes* is an important and complex subject.

(1596-1650) *physical works* (Bussotti and Pisano 2013; Dhombres 1978, 1998) and Huygens' (1629-1695) vortex theory, but many were original. If Leibniz's ideas on planetary movements had succeeded, his theory would have been more complete than Newton's.

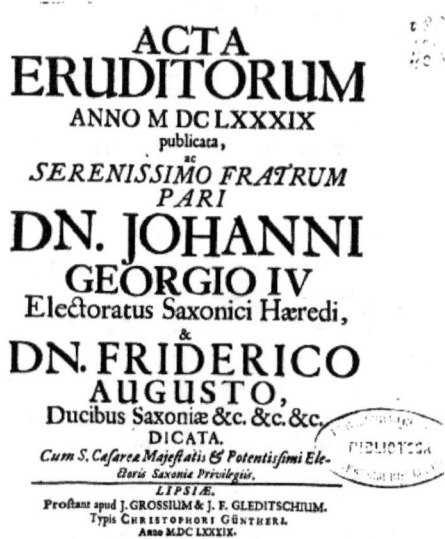

Fig. 1. *Acta Eruditorum*'s frontispiece and first page of *Tentamen de Motuum Coelestium Causis* (Leibniz 1689, pp. 82-96; see also Leibniz 1689, [1860], [1962], VI, pp. 144-161; see also posthumous *Zweite Bearbeitung* (Leibniz 1690, [1860], [1962]). Images source: Google books – Public domain.

Finally the difference between *kinematical* and *dynamical theories* is well explored in the literature, but the *physical-structural model,* as *key–notion,* is not common.

1.3 Building a Planetary Model

In 1689 Leibniz published the *Tentamen de motuum coelestium causis* (Leibniz 1689, pp. 82-96; hereafter *Tentamen*) in *Acta Eruditorum* (Leibniz 1689 [1860], [1962], VI, pp. 144-161). These crucial fourteen pages paper offered a new model of planetary motion based upon assertions which were, at least in part, different from Newton's laws[13]. The model's foundations relied on a variant of the vortex theory. Nonetheless, the origin of the interactions responsible for the planetary movements

---

[13] It is worth underlying that Leibniz claimed to have developed the theory expounded in the *Tentamen* before having read Newton's *Principia*. On this question, a well pondered and documented answer is given in Bertoloni Meli 1993, chapter 5, 95-125.

appear only at the beginning of the *Tentamen* (Leibniz 1689, pp. 82-83). During his whole contribution Leibniz tried to present a functioning *dynamical* model of the solar system, independently from the physical origin of the *dynameis*. On that, much material concerning Leibniz's ideas on the planetary movements is available in his correspondence[14]. For our aims, the most significant letters are those to Huygens[15]. However, the *Tentamen* remained the only published work on this subject. Leibniz's planetary model was, in general, not well accepted. There were several kind of criticisms:

1) Leibniz introduced some completely useless notions (as that of *circulatio harmonica*), to reach, after all, the same results Newton had already obtained in the *Principia*. This was Huygens' critics (Huygens to Leibniz, 11 July 1992, LSB, III, 5, 335-342).
2) More radical critics were addressed by Newton (Newton, in Edleston 1850, pp. 311-313) and the Newtonians and supporters of the controversy Newton–Leibniz, such as for example and especially, David Gregory (1659–1708) and John Keill (1671-1721). Among these critics, the most important were: a) Leibniz used fictitious entities as the centrifugal forces in an incorrect manner without understanding that the real force is that centripetal; b) there are mathematical mistakes in the treatment; c) there is a serious physical mistake as to the concept of velocity; d) Leibniz considers the vortices as stable entities, while Newton had proved they are not; e) the third Kepler law cannot be deduced inside Leibniz's theory (Bussotti 2015, Chapter 2).

Leibniz wrote further important manuscripts. Probably around 1690, he wrote a *Second Rework* (*Zweite Bearbeitung* Leibniz 1690, [1860] [1962], pp. 161-187) of the *Tentamen de Motuum Coelestium* and, probably around 1706, he wrote the *Illustratio Tentaminis de Motuum Coelestium Causis* (Leibniz 1706, [1860] [1962], pp. 254-276). But he did not publish these two works. The former adds a series of considerations on gravity and a deeper mathematical treatment in respect to the published *Tentamen de Motuum Coelestium* whereas the latter is a systematic

---

[14] We refer to the Leibniz Edition directed by Eberhard Knobloch at the Academy Edition of the Scientific, Medical and Technical Writings of G. W. Leibniz (Complete Writings and Letters, Series 8 - and previous ones).

[15] In the epistolary Leibniz-Huygens, we remind the reader three particularly significant letters: Leibniz to Huygens 1690, (Leibniz [1860] 1962], 187-193). The letter was not sent. On this see Aiton 1964, p. 114, note 16; Leibniz to Huygens, April 1992, LSB III, 5, 287-291; Huygens to Leibniz, 11 July 1992, LSB, III, 5, 335-342. The letters to Varignon, De la Hire, Papin, de Volder, de Duillier are relevant studies, as well. However, for our purposes, it is enough the reference to the mentioned correspondence with Huygens.

attempt to answer Newtonians' – and in particular Gregory's – criticisms. Until the middle of 1960s, the literature dedicated a scarce interest to Leibniz's planetary model. Things changed with the fundamental works by Eric J. Aiton, who, also as an answer to the ideas expressed by Alexandre Koyré (1892–1964) in his *Newtonian Studies*, proved that many supposed mistakes in Leibniz's theory are only *imaginary*. In fact, the theory is correct, at least as far as the fundamental elements are concerned. The other unavoidable reference point is represented by Bertoloni Meli's studies, whose profound, both conceptual and philological, analyses are among the most beautiful results with regard to our subject[16].

Notwithstanding, all these studies have the tendency to treat Leibniz's planetary theory as an isolated part of his production. In some cases, they connect it to some aspects of Leibniz's mathematics[17], whereas we are convinced that many of the basic features of this theory can be fully caught only if we relate planetary model with Leibniz's mechanism in physics and, in particular, with his individual ideas on gravity (Garber 1985). Actually, planetary model is also tied to Leibniz's metaphysical convictions, but we have no room to face here this problem (Bussotti 2015, chap. 6).

With regard to gravity, the situation is different from that connoting planetary theory: starting from his early work *Hypothesis Physica Nova* (Leibniz 1671 [1860] [1962], pp. 17-80) Leibniz developed a series of considerations on gravity based on a profound reworking of vortex theory. An extensive letter to Honoratus Fabri in 1677 (Leibniz 1677 [1860] [1962], pp. 81-98), a conspicuous section of *Tentamen*'s *Zweite Bearbeitung* and, above all, *De causa gravitatis et defensio sententiae auctoris de veris naturae legibus contra Cartesianos* (Leibniz 1690, [1860], [1962], pp. 193-204) represent interesting documents, in which Leibniz's conception of gravity is developed, changing many ideas expressed in the *Hypothesis*, but always remaining within the boundaries of vortex theory. Other documents are many letters as well as some sections of the *Illustratio Tentaminis* (1706). In our opinion, starting from *Hypothesis* Leibniz, while referring to gravity, had the intention – although *in nuce* – to create a *Physical–structural model* of the Universe. This intention was clarified and refined along his scientific and philosophical career: planetary model is a part of this general picture, the part in which the kinematical and dynamical bases of the system are provided. The considerations on gravity connected to planetary model is the part of Leibniz's physics, which gives planetary model its physical–structural profoundness as intended above. This does not mean, obviously, that this

---

[16] The fundamental works by Aiton on Leibniz's planetary theory, which have completely changed the way in which such a theory has been perceived and interpreted are: Aiton 1960, Aiton 1962, Aiton 1964, Aiton 1965, Aiton 1972. As to Bertoloni Meli, see, among other good contributions, the magnificent Bertoloni Meli 1993. With regard to Koyré, see Koyré 1965.

[17] On Leibniz's mathematics studies fundamental is Knobloch 1973; see also recently Goethe, Beeley and Rabouin 2015.

plan was clear from early Leibniz's speculation. It was not. But it is true that Leibniz tried to develop his project keeping faithful to some of his early convictions.

2 A Physical–Structural Modelling

2.1 The Main Kinematical Features of Leibniz's Planetary Studies

First of all, we will present a synthesis of the way in which Leibniz faced the kinematics of planetary motion. His basic idea is to divide an eccentric motion into two main components:

1) Each planet had a vortex connecting the planet to the sun. It is not completely clear what geometrical position the sun had inside the vortex, namely, if it was in the geometrical centre or in the centre of the movement. Whatever the answer is, the vortex is moved by what Leibniz called *Circulatio harmonica* and defined as in the following:

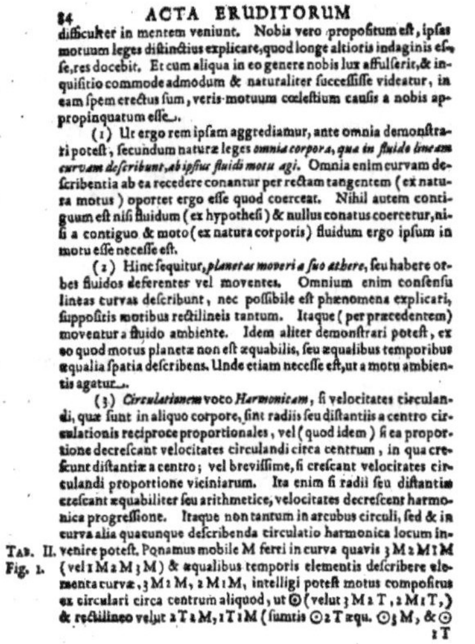
84 ACTA ERUDITORUM

difficulter in mentem veniunt. Nobis vero propoſitum eſt, ipſas motuum leges diſtinctius explicare, quod longe altioris indaginis eſſe, res docebit. Et cum aliqua in eo genere nobis lux affulſerit, & inquiſitio commode admodum & naturaliter ſucceſſiſſe videatur, in eam ſpem erectus ſum, veris motuum cœleſtium cauſis a nobis appropinquatum eſſe.

(1) Ut ergo rem ipſam aggrediamur, ante omnia demonſtrari poteſt, ſecundum naturæ leges *omnia corpora, quæ in fluido lineam curvam deſcribunt, ab ipſius fluidi motu agi*. Omnia enim curvam deſcribentia ab ea recedere conantur per rectam tangentem (ex natura motus) oportet ergo eſſe quod coerceat. Nihil autem contiguum eſt niſi fluidum (ex hypotheſi) & nullus conatus coercetur, niſi a contiguo & moto (ex natura corporis) fluidum ergo ipſum in motu eſſe neceſſe eſt.

(2) Hinc ſequitur, *planetas moveri a ſuo æthere*, ſeu habere orbes fluidos deferentes vel moventes. Omnium enim conſenſu lineas curvas deſcribunt, nec poſſibile eſt phænomena explicari, ſuppoſitis motibus rectilineis tantum. Itaque (per præcedentem) moventur a fluido ambiente. Idem aliter demonſtrari poteſt, ex eo quod motus planetæ non eſt æquabilis, ſeu æqualibus temporibus æqualia ſpatia deſcribens. Unde etiam neceſſe eſt, ut a motu ambientis agatur.

(3) *Circulationem* voco *Harmonicam*, ſi velocitates circulandi, quæ ſunt in aliquo corpore, ſint radiis ſeu diſtantiis a centro circulationis reciproce proportionales, vel (quod idem) ſi ea proportione decreſcant velocitates circulandi circa centrum, in qua creſcunt diſtantiæ a centro; vel breviſſime, ſi creſcant velocitates circulandi proportione viciniarum. Ita enim ſi radii ſeu diſtantiæ creſcant æquabiliter ſeu arithmetice, velocitates decreſcent harmonica progreſſione. Itaque non tantum in arcubus circuli, ſed & in curva alia quacunque deſcribenda circulatio harmonica locum invenire poteſt. Ponamus mobile M ferri in curva quavis 3M2M1M (vel 1M2M3M) & æqualibus temporis elementis deſcribere elementa curvæ, 3M2M, 2M1M, intelligi poteſt motus compoſitus ex circulari circa centrum aliquod, ut ⊙ (velut 3M2T, 2M1T,) & rectilineo velut 2T2M, 1T1M (ſumtis ⊙2T æqu. ⊙3M, & ⊙1T

TAB. II. Fig. 1.

(3) I call a *Circulation* a *Harmonic* one if the velocities of circulation in some body are inversely proportional to the radii or distances from the centre of circulation, or (what is the same) if the velocities of circulation round the centre decrease proportionally as the distances from the centre increase, or most briefly, if the velocities of circulation increase proportionally to the closeness[18].

Fig. 2. The definition of *Circulatio Harmonica* and its English translation. Image source: Google books – Public domain.

[18] "(3) Circulationem voco Harmonicam, si velocitates circulandi, quae sunt in aliquo corpore, sint radiis seu distantiis a centro circulationis reciproce proportionales, vel (quod idem) si ea proportione decrescant velocitates circulandi circa centrum, in qua crescunt distantiae a centro, vel brevissime, si crescant velocitates circulandi proportione viciniarum." (Leibniz 1689, pp. 84-85; *Id.*, 1689, [1860] [1962], VI, pp. 149-150. English Translation is adapted from Bertoloni Meli 1993, pp. 129-130).

Furthermore, Leibniz specified an important general law, which can be so synthetized:

> Harmonic circulation can connote any curved path, not only the circular one. (*Ibidem*).

In particular, it can characterize an elliptic path. Nonetheless, a question arises: *how is it possible that a non-circular path is generated?*
If a planet rotates in a vortex – in which each layer of infinitesimal thickness is animated by a harmonic circulation (Leibniz called it *harmonic vortex*), and if the sun is not the geometrical centre of the vortex, then the velocity of circulation will change. But the orbit will remain a circle. In effect, there is no sufficient reason – to use a fundamental principle of Leibniz's metaphysics – why it changes.

> 2) The orbit is an ellipsis because the planet has a second movement, independent from the *circulatio harmonica*: the *motus paracentricus*. Leibniz imagined that a planet, while moving in a layer of the vortex, moves, at the same time on a rigid ruler, along which it is animated by a rectilinear motion. If this rectilinear motion is directed towards the centre of movement, the planet approaches the centre, otherwise it moves away from the centre. This motion along the ruler is called, indeed, *motus paracentricus*.

Therefore, thanks to the *paracentric motion*, the planet passes through different layers of the harmonic vortex, so being animated by a harmonic movement in an ellipsis. This is the logic of Leibniz's reasoning because, as it is comprehensible, Leibniz started from the *datum* of the elliptic harmonic circulation of the planet to reach the idea that the planetary vortex is harmonic, too. As it is often the case, the physical–logical *iter* is traced taking into account a known *datum* (paths' ellipticity), which has to be explained. In the following observations and interpretations will be proposed.

Fig. 3. *Leibniz's planetary theory model.*

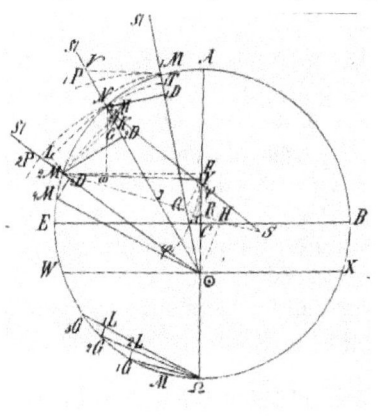 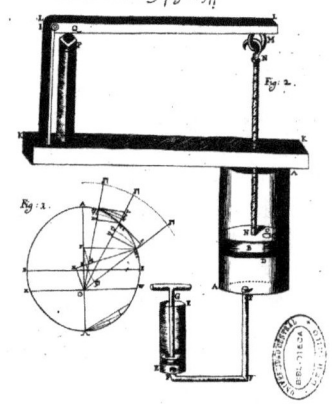

Fig. 3.1. This is a Leibnizian figure as proposed by Gerhardt (Leibniz 1689 [1860] [1962], [1971], VI, final diagrams). The diagram is unclear. There are many letters and this makes it difficult to clearly read the diagram. There is a typo because the $_2M$ written immediately over $_4M$ is a mistake. The right form is $_3M$. Furthermore there is the habit to write the index of a letter before the letter, while nowadays we write after the letter. Because of all these reasons – if we do not specify otherwise – we will refer to the figure 3.2 (see particulars), which is written in a more modern form but does not betray Leibniz's thought, at all. Image source–with permission of *Gottfried Wilhelm–Mathematische Schriften*, Georg Olms AG–Verlag).

Fig. 3.2. In *Tentamen* (1689) only two figures were presented by Leibniz Particularly see Fig. 1 (left-below: Leibniz 1689, p. 84). Image source: Google books – Public domain.

A discussion.

a) Leibniz does not clarify if the sun is in the geometrical centre of the harmonic vortex considered without the *motus paracentricus* (Leibniz 1689, § 3 p. 84; *Id.*, 1689 [1860] [1962], VI, § 3, pp. 149-150). The only figure (Figs. 3.1 and 3.2) that he added in the *Tentamen* is not of help because it is referred only to the actual planetary orbit, in which obviously, the sun is not in the centre. Thus, on our side, we propose to interpret the sun as is in the geometrical centre of the vortex. Let us consider the layers of the vortex included between aphelion and perihelion. They form a ring. Let us suppose that the planet is at the aphelion. Thanks to the inwards tendency of the *paracentric motion*, the planet approaches to the sun and enters into the ring. At the perihelion, the outwards component of the *paracentric motion* begins to prevail and the planet moves away from the sun, thus reaching once again the aphelion. From a geometrical point of view, this depends on the fact that an ellipsis can be constructed within a ring, as described. Therefore an elliptic orbit in which the sun is in the centre of movement (focus) can be designed within a Leibnizian perspective; and it can be designed also considering the sun in the geometrical centre of the harmonic vortex without *paracentric motion* which looks like geometrically and astronomically plausible.

b) Following Leibniz (Leibniz 1689, § 3, p. 84; *Id.*, 1689 [1860] [1962], VI, § 3, 149-150) and as explained in previous item *a)* we have used the idea of the vortex to introduce his planetary model. It is, however, necessary to point out that the same reasoning could be developed also without referring to the existence of a material vortex – though composed of aether – as Leibniz did.

c) Leibniz's concept of *velocitas circulandi* (Leibniz 1689, § 3, p. 84; *Id.*, 1689 [1860] [1962], VI, pp. 149-150) has been misunderstood for a long period. Newton and the Newtonians interpreted this notion by Leibniz as the module of the velocity in the planetary orbit. If this were the case, Leibniz would have made a serious mistake, because Newton had proved that the velocities' modules of the planets in their orbits, are as the square root of the distance sun-planet (Newton 1687, [1726] [1739-1742] [1822], I, prop. IV, cor. 6, p. 73). Koyré in the Appendix A of his excellent *Newtonian Studies* (Koyré 1965) adhered to such critics. However, Aiton's studies (Aiton 1960, 1962, 1964, 1965) proved that Leibniz's analysis is quite refined: the *velocitas circulandi* is not the module of velocity, but the transverse velocity, which, in the planetary motion, fulfils exactly Leibniz's law: it decreases as the distance from the sun. The *motus paracentricus* is the radial motion. Therefore Leibniz has reached an interesting physical perspective. He has analysed the motion from the point of view of the rotating planet using what nowadays we call polar coordinates. He has thence divided the velocity in the *transverse* and *radial component*[19], which is something important and significant from a physical point of view. No mistake. Not only. If a planet moves of *circulatio harmonica*, it is absolutely trivial to prove that in its orbit the area law is valid. From an epistemological point of view, one could object to Leibniz that his definition of harmonic circulation implies immediately area law. The hypothesis of the harmonica circulation is as strong as the thesis Leibniz had to prove.

---

[19] This discovery concerning Leibniz's planetary model is expounded in Aiton's contributions mentioned in the note 10.

2.2 The Main Dynamical Features of Leibniz's Planetary Model

The Leibnizian causes which determine the movement of the *Circulatio harmonica* depend on the material composition of the rotating vortex: the planet is afloat in the vortex and is carried by vortex's movement. This is an important aspect of the physical-structural modelling in Leibniz. As to the *paracentric motion*, according to Leibniz, it is due to two opposite tendencies (Leibniz 1689, § 9, p. 87; Leibniz 1689 [1860] [1962], § 9, p. 152):

1) *Impressio excussoria circulationis.*
2) *Attractio solaris.*

The *impressio excussoria circulationis*[20] is a physical cause, the centrifugal force (in the perspective of Newtonian physics a *fictitious force, pseudo*[21] *force*) due to the velocity[22] of a body in motion following the *Circulatio harmonica*. In Leibniz's view – along a curved path and independently from its origin – the effect produced by this apparent force is a tendency to recede along the tangent to the trajectory. This tendency exists in each point of the curved path considered as continuous curve. It is produced by this centrifugal force[23]. In his model, Leibniz is precisely interested to describe both a geometrical representation and an analytical expression for the instantaneous outward *impressio excussoria circulationis*. In the paragraphs 10 and 11 of the *Tentamen* (Leibniz 1689, § 10, pp. 87-88; Leibniz 1689, [1860], [1962], § 10, p. 152). Leibniz presents his geometrical representation of the *conatus centrifugus*. He named it *conatus centrifugus* or *conatus excussorius circulationis* (*Ibidem*).

---

[20] Bertoloni Meli translated it as "outward impression of the circulation" (Bertoloni Meli 1993, p. 132).

[21] In modern terms *impressio excussoria circulationis* is an external force so called nowadays *apparent* when viewed in the rotating reference frame itself.

[22] It could be considered like an angular velocity.

[23] It could be didactically presented such a physical model to describe the concept of acceleration, even if, the difference between *velocity–and–acceleration*, so *force–and–velocity* and *force–and–acceleration* such as *cause–and–effect* should be, *a priori*, historically and didactically well clarified.

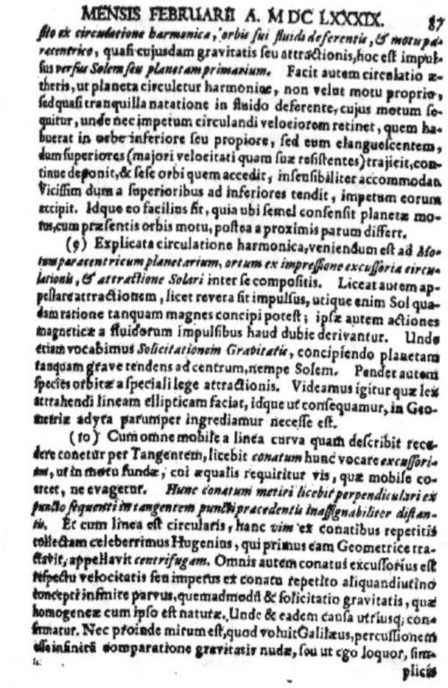

MENSIS FEBRUARII A. MDC LXXXIX. 87

ſito ex circulatione harmonica, orbis ſui fluidi deferentis, & motu paracentrico, quaſi cujuſdam gravitatis ſeu attractionis, hoc eſt impulſus *verſus Solem ſeu planetam primarium.* Facit autem circulatio ætheris, ut planeta circuletur harmonice, non velut motu proprio, ſed quaſi tranquilla natatione in fluido deferente, cujus motum ſequitur, unde nec impetum circulandi velociorem retinet, quem habuerat in orbe inferiore ſeu propiore, ſed eum elanguescentem, dum ſuperiores (majori velocitati quam ſuæ reſiſtentes) trajicit, continue deponit, & ſeſe orbi quem accedit, inſenſibiliter accommodat. Viciſſim dum a ſuperioribus ad inferiores tendit, impetum eorum accipit. Idque eo facilius fit, quia ubi ſemel conſenſit planetæ motus, cum præſentis orbis motu, poſtea a proximis parum differt.

(9) Explicata circulatione harmonica, veniendum eſt ad *Motum paracentricum planetarium, ortum ex impreſſione excuſſoria circulationis, & attractione Solari* inter ſe compoſitis. Liceat autem appellare attractionem, licet revera ſit impulſus, utique enim Sol quadam ratione tanquam magnes concipi poteſt; ipſæ autem actiones magneticæ a fluidorum impulſibus haud dubie derivantur. Unde etiam vocabimus *Solicitationem Gravitatis*, concipiendo planetam tanquam grave tendens ad centrum, nempe Solem. Pendet autem ſpecies orbitæ a ſpeciali lege attractionis. Videamus igitur quæ lex attrahendi lineam ellipticam faciat, idque ut conſequamur, in Geometriæ adyta parumper ingrediamur neceſſe eſt.

(10) Cum omne mobile a linea curva quam deſcribit recedere conetur per Tangentem, licebit *conatum* hunc vocare *excuſſorium*, ut in motu fundæ, cui æqualis requiritur vis, quæ mobile coërceat, ne evagetur. *Hunc conatum metiri licebit perpendiculari ex puncto ſequenti in tangentem puncti præcedentis inaſſignabiliter diſtantis.* Et cum linea eſt circularis, hanc *vim* ex conatibus repetitis collectam celeberrimus Hugenius, qui primus eam Geometrice tractavit, appellavit *centrifugam*. Omnis autem conatus excuſſorius eſt reſpectu velocitatis ſeu impetus ex conatu repetito aliquandiutino concepti infinite parvus, quemadmodum & ſolicitatio gravitatis, quæ homogeneæ cum ipſo eſt naturæ. Unde & eadem cauſa utriuſq; confirmatur. Nec proinde mirum eſt, quod voluit Galilæus, percuſſionem eſſe infinitæ comparatione gravitatis nudæ, ſeu ut ego loquor, ſimplicis

Fig. 4 On *Conatus centrifugus*. Image source: Google books – Public domain.

(10) [...] *This conatus is measured* [geometrically as a line] *by the perpendicular from the following point to the tangent at the inassignably distant preceding point.*[24]

Following Leibniz (Leibniz 1689, § 11, p. 88; p. Leibniz 1689, 1860, 1962, VI, § 11, p. 153) *conatus excussorius* can be represented by a segment *PN* (*Ibidem*; see also Fig. 7 below), namely the versed sine of the angle of circulation $M_1 \Theta N$. For, the versed sine – Leibniz continues:

> [...] is equal to the perpendicular drawn from one end-point of the arc of a circle to the tangent from the other end-point [...].[25]

The versed sine can be identified with D_1T_1 (see Fig. 7) the unassignable difference between two infinitely near radii-vector[26]. This means that, in general, the *conatus escussorius* can be represented by segments of the type D_iT_i (*Ibidem*) for every position of the radius vector. It is then evident to prove that the *conatus centrifugus* is equal to *PV* (Leibniz 1689, § 11, p. 88; Leibniz 1689 [1860] [1962], VI, § 11, p.

---

[24] Adapted by Bertoloni Meli's translation (Bertoloni Meli 1993, p. 132). "*Hunc conatum metiri licebit perpendiculari ex puncto seguenti in tangentem puncti praecedentis inassegnabiliter distantis.*" Leibniz's italic (Leibniz 1689, § 10, p. 87, Leibniz 1689, 1860, 1962, VI, § 10, p. 152).

[25] Adapted by Bertoloni Meli's translation (Bertoloni Meli 1993, p. 133). "[...] aequatur perpendiculari ex uno extremo arcus circuli puncto in tangentem alterius ductae [...]." (Leibniz 1689, § 11, p. 88; Leibniz 1689 [1860] [1962], § 11, VI, p. 153).

[26] With respect to Aiton's and Gerhardt's arrangements we used symbols with subscript on the right bottom–side of a letter (i.e., instead on the left bottom side such as Aiton used).

153). Here, Leibniz describes a trajectory as composed of an infinite number of infinitesimal circular arcs whose radii have infinitesimal differences and are all centred in the sun. Given this situation, the infinitesimal arcs of circumference can be considered as sides of a polygon[27].

Furthermore, Leibniz also presented an analytical expression of the *conatus centrifugus* (Leibniz 1689, paragraphs 11 and 12) in the following so synthetically summarized[28]. If the motion is circular and uniform, than the *conatus* is as V^2, where V is the transverse velocity, because the versed sine is as the square of the chord and the transverse velocity is as the chord. If two or more circles are considered in which the movement is uniform, then the *conatus* are as V^2/R, where R is the radius.

From this expression for the centrifugal force, Leibniz[29] deduced another expression, which is fundamental in his reasoning: if a body moves with a harmonic circulation, the *conatus centrifugus* is inversely proportional to the radius vector. This happens because of the inverse proportion between transverse velocity and radius vector in the *circulatio harmonica* and because of the following relation:

$$c = \frac{V^2}{R} \quad (c = centrifugal\ conate) \tag{1}$$

Leibniz considered a fixed elementary area, swept by the radius-vector in an infinitesimal time dt (the area law is valid), which he indicated by ϑa and assumed it is equal to the double of the elementary triangle

$$M_2M_3 \Theta \text{ namely equal to } D_2M_3 \cdot \Theta M_2$$

Particularly, the term ΘM_n can be indicated by r = radius, and because the difference between ΘM_i and ΘM_{i-1} is an infinitesimal value, then it can be neglected in this calculation. In addition, since $D_2M_3 = \vartheta a /r$ and the centrifugal conate[30] is obtained by $D_2T_2 = (D_3T_3)^2 / 2\Theta M_2$, in conclusion we have:

$$D_2T_2 = \frac{\vartheta^2 a^2}{2r^3} \tag{2}$$

---

[27] The idea to consider a curved trajectory as composed of right infinitesimal segments is typical of Leibniz and other scientists and mathematicians. We have here no room to go into this problem. For three partially different interpretations of this idea see: Bertoloni Meli 1993, pp. 75-84, Jauernig 2008, Bussotti 2015 § 3.2. As to other mathematicians, it is enough, only to give an example, namely to think of the way in which Newton proved the proposition I of the first *Principia*'s book. The same here, we have no further room to deal with this other interesting subject.

[28] Leibniz 1689, § 12, p. 88; Leibniz 1689 [1860] [1962], § 12, VI, p. 153.

[29] *Ibidem*.

[30] "(12) Conatus centrifugi mobilis harmonici sunt in ratione radiorum reciproca triplicata" (Leibniz 1689, § 12, p. 88; Leibniz 1689 [1860] [1962], § 12, VI, p. 153).

That is the centrifugal conate is as the inverse of the radius–cube such as Leibniz defined in his section 12 above cited.

This arguing is fully coherent with our previous observation:

> Leibniz is considering a non-inertial reference frame in polar coordinates, whose origin is posed in the rotating planet; that is rotating around the sun as its centre.

We add further reflections. From the point of view of the planet, the acceleration along the radius is given by two components:

- One outwards, which is the Leibnizian *conatus centrifugus* due to the harmonic circulation.
- The other one is due to gravity or levity.

Leibniz considered that this second component can be either inwards (gravity), which is the normal experienced case, or outwards (levity), which is a theoretical case. The acceleration along the radius is simply the algebraic sum of the two components, which is an arithmetical difference in case of gravity and an arithmetical sum in case of levity.

If one takes into account the case of gravity, and if the *conatus centrifugus* prevails[31], the radial acceleration is directed outwards. While, if the *solicitatio gravitatis* (Leibniz 1689, § 14, p. 89; Leibniz 1689, [1860] [1962], § 14, VI, p. 154) prevails, the radial acceleration is directed inwards. Finally, we have described how Leibniz represented the *conatus centrifugus*.

As to the *solicitatio gravitatis* (*Ibidem*, see Fig. 7), which, in this case, is the *attractio solaris*, Leibniz argued like this:

---

[31] Leibniz wrote "[…] differentia vel summa solicitationis paracentricae […] et dupli conatus centrifugi […]". (Leibniz 1689 [1860] [1962], VI, p. 154). For the problem of the "double centrifugal conate", Varignon revealed a mistake, which, however was easily corrigible. The correct version implies that "double centrifugal conate" is replaced by "centrifugal conate." (Bussotti 2015, § 2.2.2).

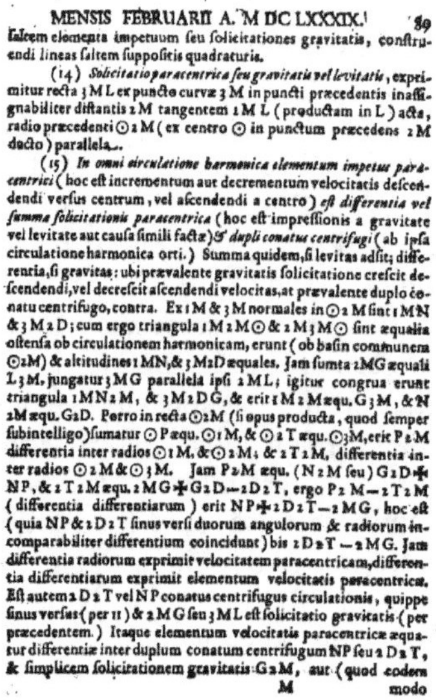
MENSIS FEBRUARII A. M DC LXXXIX. 89

ſaltem elementa impetuum ſeu ſolicitationes gravitatis, conſtruendi lineas ſaltem ſuppoſitis quadraturis.

(14) *Solicitatio paracentrica ſeu gravitatis vel levitatis*, exprimitur recta 3 M L ex puncto curvæ 3 M in puncti præcedentis inaſſignabiliter diſtantis 2 M tangentem 2 M L (productam in L) acta, radio præcedenti ⊙ 2 M (ex centro ⊙ in punctum præcedens 2 M ducto) parallela.

(15) *In omni circulatione harmonica elementum impetus paracentrici* (hoc eſt incrementum aut decrementum velocitatis deſcendendi verſus centrum, vel aſcendendi a centro) *eſt differentia vel ſumma ſolicitationis paracentricæ* (hoc eſt impreſſionis a gravitate vel levitate aut cauſa ſimili factæ) *& dupli conatus centrifugi* (ab ipſa circulatione harmonica orti.) Summa quidem, ſi levitas adſit; differentia, ſi gravitas: ubi prævalente gravitatis ſolicitatione creſcit deſcendendi, vel decreſcit aſcendendi velocitas, at prævalente duplo conatu centrifugo, contra. Ex 1 M & 3 M normales in ⊙ 2 M ſint 1 M N & 3 M 2 D; cum ergo triangula 1 M 2 M ⊙ & 2 M 3 M ⊙ ſint æqualia oſtenſa ob circulationem harmonicam, erunt (ob baſin communem ⊙ 2 M) & altitudines 1 M N, & 3 M 2 D æquales. Jam ſumta 2 M G æquali L 3 M, jungatur 3 M G parallela ipſi 2 M L; igitur congrua erunt triangula 1 M N 2 M, & 3 M 2 D G, & erit 1 M 2 M æqu. G 3 M, & N 2 M æqu. G 2 D. Porro in recta ⊙ 2 M (ſi opus producta, quod ſemper ſubintelligo) ſumatur ⊙ P æqu. ⊙ 1 M, & ⊙ 2 T æqu. ⊙ 3 M, erit P 2 M differentia inter radios ⊙ 1 M, & ⊙ 2 M; & 2 T 2 M, differentia inter radios ⊙ 2 M & ⊙ 3 M. Jam P 2 M æqu. (N 2 M ſeu) G 2 D + N P, & 2 T 2 M æqu. 2 M G + G 2 D − 2 D 2 T, ergo P 2 M − 2 T 2 M (differentia differentiarum) erit N P + 2 D 2 T − 2 M G, hoc eſt (quia N P & 2 D 2 T ſinus verſi duorum angulorum & radiorum incomparabiliter differentium coincidunt) bis 2 D 2 T − 2 M G. Jam differentia radiorum exprimit velocitatem paracentricam, differentia differentiarum exprimit elementum velocitatis paracentricæ. Eſt autem 2 D 2 T vel N P conatus centrifugus circulationis, quippe ſinus verſus (per 11) & 2 M G ſeu 3 M L eſt ſolicitatio gravitatis (per præcedentem.) Itaque elementum velocitatis paracentricæ æquatur differentiæ inter duplum conatum centrifugum N P ſeu 2 D 2 T, & ſimplicem ſolicitationem gravitatis G 2 M, aut (quod eodem

M modo

(14) *Paracentric solicitation, whether of gravity or levity* is expressed by the straight line M_3L drawn from the point M_3 of the curve to the tangent M_2L (produced to L), of the preceding unassignable distant point M_2 parallel to the preceding radius ΘM_2 (drawn from the centre to the preceding point M_2).[32] See also Fig. 7.

Fig. 5 On *Paracentric solicitation.* Image source: Google books – Public domain.

Leibniz suggested hence that, given an infinitesimal arc M_1M_2 – approximated by its chord – the inertial motion of a body moving in such an arc can be approximated by the prolongation of the chord (the *tangent* in the Leibnizian sense) rather than by the Euclidean *tangent,* without a detectable inaccuracy.

The section of the *Tentamen,* which concludes the part concerning the general properties of the paracentric motion is the 15th one (Leibniz 1689, § 15, p. 89; Leibniz 1689 [1860] [1962], VI, § 15, p. 154) where Leibniz determined geometrically the element of the *impetus paracentricus,* that is the instantaneous acceleration along the radius. Thus, he argued that in every harmonic circulation the element of *impetus paracentricus* is the difference or the sum of the *paracentric solicitation* and of the double centrifugal conatus.

[32] From Bertoloni Meli's translation (Bertoloni Meli 1993, p. 134). "*Solicitatio paracentrica, seu gravitatis vel levitates*, exprimitur recta $_3ML$ ex puncto curvae $_3M$ in puncti praecedentis inassegnabiliter distantis $_2M$ tangentem $_2ML$ (productam in L) acta, radio praecedenti Θ_2M (ex centro Θ in punctum precedens $_2M$ ducto) parallela". Leibniz's *italic.* As we above cited, we adopted modern quotation for subscripts (Leibniz 1689, § 14, p. 89; Leibniz 1689 [1860] [1962], VI, § 14, p. 154).

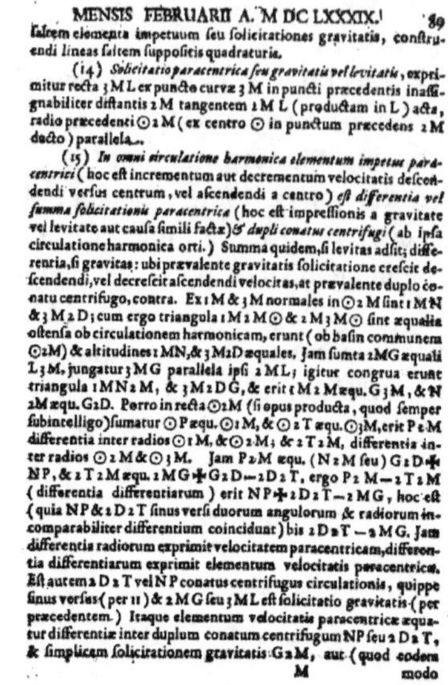
MENSIS FEBRUARII A. M DC LXXXIX. 89

ſaltem elementa impetuum ſeu ſolicitationes gravitatis, conſtruendi lineas ſaltem ſuppoſitis quadraturis.

(14) *Solicitatio paracentrica ſeu gravitatis vel levitatis*, exprimitur recta 3ML ex puncto curvæ 3M in puncti præcedentis inaſſignabiliter diſtantis 2M tangentem 1ML (productam in L) acta, radio præcedenti ⊙2M (ex centro ⊙ in punctum præcedens 2M ducto) parallela.

(15) *In omni circulatione harmonica elementum impetus paracentrici* (hoc eſt incrementum aut decrementum velocitatis deſcendendi verſus centrum, vel aſcendendi a centro) *eſt differentia vel ſumma ſolicitationis paracentricæ* (hoc eſt impreſſionis a gravitate vel levitate aut cauſa ſimili factæ) *& dupli conatus centrifugi* (ab ipſa circulatione harmonica orti.) Summa quidem, ſi levitas adſit; differentia, ſi gravitas: ubi prævalente gravitatis ſolicitatione creſcit deſcendendi, vel decreſcit aſcendendi velocitas, at prævalente duplo conatu centrifugo, contra. Ex 1M & 3M normales in ⊙2M ſint 1MN & 3M2D; cum ergo triangula 1M2M⊙ & 2M3M⊙ ſint æqualia oſtenſa ob circulationem harmonicam, erunt (ob baſin communem ⊙2M) & altitudines 1MN, & 3M2D æquales. Jam ſumta 2MG æquali L3M, jungatur 3MG parallela ipſi 2ML; igitur congrua erunt triangula 1MN2M, & 3M2DG, & erit 1M2M æqu. G3M, & N2M æqu. G2D. Porro in recta ⊙2M (ſi opus producta, quod ſemper ſubintelligo) ſumatur ⊙P æqu. ⊙1M, & ⊙2T æqu. ⊙3M, erit P2M differentia inter radios ⊙1M, & ⊙2M; & 2T2M, differentia inter radios ⊙2M & ⊙3M. Jam P2M æqu. (N2M ſeu) G2D + NP, & 2T2M æqu. 2MG + G2D − 2D2T, ergo P2M − 2T2M (differentia differentiarum) erit NP + 2D2T − 2MG, hoc eſt (quia NP & 2D2T ſinus verſi duorum angulorum & radiorum incomparabiliter differentium coincidunt) bis 2D2T − 2MG. Jam differentia radiorum exprimit velocitatem paracentricam, differentia differentiarum exprimit elementum velocitatis paracentricæ. Eſt autem 2D2T vel NP conatus centrifugus circulationis, quippe ſinus verſus (per 11) & 2MG ſeu 3ML eſt ſolicitatio gravitatis (per præcedentem.) Itaque elementum velocitatis paracentricæ æquatur differentiæ inter duplum conatum centrifugum NP ſeu 2D2T, & ſimplicem ſolicitationem gravitatis G2M, aut (quod eodem

M modo

Leibniz's reasoning is interesting due to two epistemological reasons:

A) It is an example of what one could call infinitesimal geometry applied to physics, that is both the finite and the infinitesimal quantities are represented by means of geometrical constructions and, at least in this section, there is not a transcription into analytical terms.

B) It is an example which clearly shows the use of differentials of different degree in a geometrical context (for more details see the next commentaries).

Fig. 6 On *Impetus paracentricus*[33] Image source: Google books – Public domain.

In the following we present an enlarged imagine of Leibniz's planetary model (see previous Fig. 3.1).

We present this imagine because in Aiton's the point G is not represented, while it is an important element in our context. We hope that this imagine can be of help to the reader to follow the mathematical reasoning developed in his running text. Let us remind the reader that the symbol ${}_2M_2$ near ${}_4M$ has to be replaced with ${}_3M$.

[33] Leibniz 1689 [1860] [1962], VI, § 14, p. 154; Leibniz 1689, § 14, p. 89; see also Bertoloni Meli 1993, p. 134.

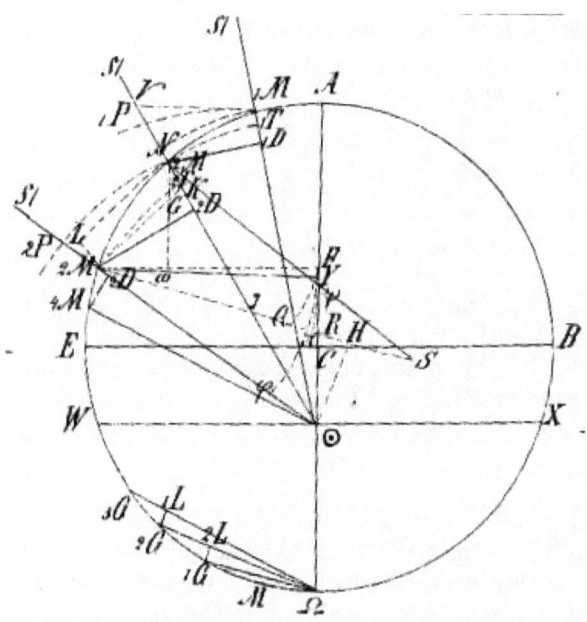

Fig. 7 Enlarged (see Fig. 3.1) image of Leibniz's Planetary model (Leibniz 1689 [1860] [1962], [1971], VI). Image source–with permission of *Gottfried Wilhelm–Mathematische Schriften*, Georg Olms AG–Verlag).

Leibniz reasons (*Ibidem*) like this:

1) Let M_1N and M_3D_2 be the perpendiculars from M_1 and M_3 to ΘM_2.

2) The circulation is harmonic, hence the triangles $M_1M_2\Theta$ and $M_2M_3\Theta$ are congruent. Therefore their altitudes M_1N and M_3D_2 are congruent.

3) Let M_2G be congruent to LM_3 and M_3G parallel to M_2L.

4) The triangles M_1NM_2 and M_3D_2G are congruent[34]. Therefore it is $M_1M_2 = GM_3$ and $NM_2 = GD_2$.

5) Let us assume $\Theta P = \Theta M_1$ and $\Theta T_2 = \Theta M_3$, so.

6) $PM_2 = (\Theta M_1 - \Theta M_2)$ and $T_2M_2 = (\Theta M_2 - \Theta M_3)$.

7) $PM_2 (= NM_2) = (GD_2 + NP)$ and $T_2M_3 = (GM_2 + GD_2 - T_2D_2)$. Hence.

8) $PM_2 - T_2M_2 = NP + T_2D_2 - M_2G$. But.

9) $NP = D_2T_2$ because they are the versed sinuses of two angles and radii whose differences are incomparable. Hence.

10) $PM_2 - T_2M_2 = 2T_2D_2 - M_2G$.

---

[34] We remind the reader that the two triangles are congruent because: a) $M_3D_2 = M_1N$; b) they are right triangles; c) For the angles the following identities are valid: $M_1NM_2 = D_2M_2L$ and $D_2M_2L = D_2GM_3$, because of the parallels GM_3 and M_2L. Thus, $M_1NM_2 = D_2GM_3$. Hence, the thesis follows.

11) The difference of the radii
$$PM_2 = (\Theta M_1 - \Theta M_2)$$
expresses the *paracentric velocity*; the difference of the differences expresses the element of the *paracentric velocity* (that is the *paracentric acceleration*). But T_2D_2 or NP is the centrifugal conatus of circulation and M_2G or M_3G is the *paracentric solicitation*.

Q. E. D.

In addition, we remark that the segments, one extremum of which is the centre of gravity Θ, are finite; all other elements used in the proof are infinitesimal. The quantities P_2M_2 and T_2M_2 are first differences and represent the instantaneous radial velocity; their difference

$$PM_2 - T_2M_2$$

is a second difference and represents the *radial instantaneous acceleration*.

By this proof, Leibniz completed the description and the explanation of the basic elements of his theory. He then applied these elements to the case of the elliptical orbits, the ones which are relevant for the planetary movements. In particular, Leibniz was able to determine both a geometrical and an algebraic-analytical form with regard to the *conatus centrifugus*, while, for the *solicitatio paracentrica*, he has only given the geometrical form.

His next step was to prove that such a solicitation is as the inverse of the square distance. Leibniz will succeed in proving the inverse square law inside his system. Nonetheless, we will focus neither on this problem nor on the critics Newton and the Newtonians addressed to Leibniz's concepts (i.e, see *Tentamen*, § 9) rather we underline that Leibniz had provided a dynamical explanation for the *motus paracentricus*, where the term *dynamics* has to be assumed in an etymological meaning: connected to the *dynameis*, the powers, the *forces* which determine the paracentric motion, that is *conatus centrifugus* and *solicitatio paracentrica gravitatis vel levitatis*.

Leibniz pointed out that such powers could be assumed as given, without looking for their physical origin. He assumed this position in the *Tentamen*. Notwithstanding, Leibniz dedicated a long series of speculations to the origin of such forces, and especially of gravity. The way in which he addressed the problem of gravity along his entire scientific career proves that his planetary model is a part of a wide and long-time reasoned program, although Leibniz had many uncertainties on the particulars and changed his mind more than once on such particulars.

2.3 Gravity: From *Hypothesis Physica Nova* (1671) to *Illustratio Tentaminis* (1706). Planetary Model and Gravity

The conceptualization of gravity was an important subject (Bussotti 2015, Chapter 5) in the construction of Leibniz's physical structural system. Taking into account this room here available, we will hence only focus on the connections

Leibniz posed between gravity and planetary model, or, if we refer to contributions preceding the *Tentamen*, between gravity and considerations on the sky movements; as it is not appropriate to speak of planetary model in Leibniz before 1689.

The general frame in which Leibniz's ideas on gravity are developed is the vortex theory designed by Descartes' *Principia Philosophiae*[35] (Bussotti and Pisano 2013), but Leibniz introduced fundamental modifications within such a theory. It is necessary to premise that, with regard to Leibniz, two different meaning of the term gravity has to be distinguished:

- a) The attraction exerted – directly or indirectly – by the sun on the planets. Starting from the *Tentamen* he called – as seen above – this attraction *solicitatio paracentrica seu gravitatis*;
- b) Gravity on the earth.

In spite of the fact that Leibniz argued (see references below) the similarities between these two actions, and although, along the years, he tended to assimilate them, it is still uncertain if he realized that, after all, they are the same attraction. Starting from the first section of *Hypothesis Physica nova – Theoria motus concreti* (Leibniz 1671 [1860] [1962], pp. 17-61); see also Duchesneau 1989, 1994) – Leibniz introduced a series of considerations, which induce us to believe he thought of a connection between gravity and planetary motion. Leibniz moved from two hypotheses:

1) The aether exists;
2) The celestial bodies move around an axis passing through their centres.[36]

As to our problem on physical-structural modelling, Leibniz's significant consideration is the following one: every motion, different from a motion which returns on itself (Leibniz is thinking of a closed trajectory), cannot be explained only by the two mentioned hypotheses. For, Leibniz wrote:

---

[35] Descartes 1897-1913, 1964-1974; very interesting are Schuster 2013a, Schuster 2013b.

[36] Leibniz 1671, [1860] [1962], § 3, p. 20.

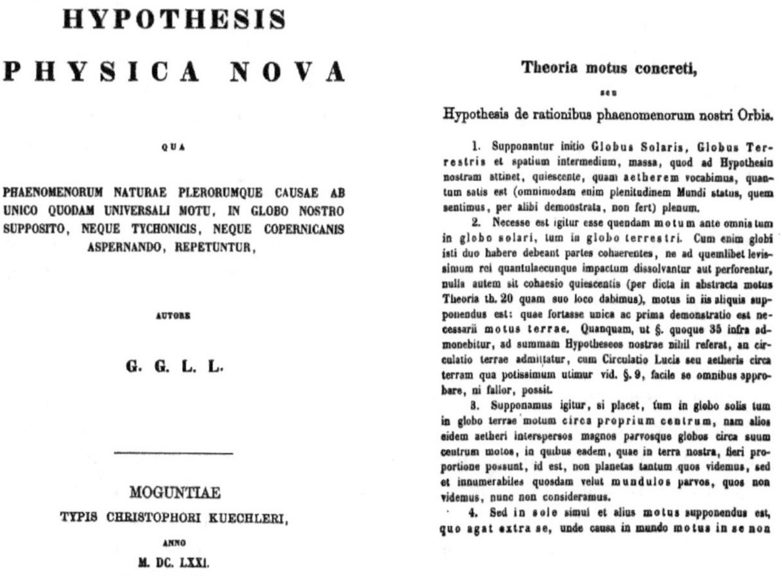

HYPOTHESIS
PHYSICA NOVA

QUA

PHAENOMENORUM NATURAE PLERORUMQUE CAUSAE AB UNICO QUODAM UNIVERSALI MOTU, IN GLOBO NOSTRO SUPPOSITO, NEQUE TYCHONICIS, NEQUE COPERNICANIS ASPERNANDO, REPETUNTUR,

AUTORE

G. G. L. L.

MOGUNTIAE
TYPIS CHRISTOPHORI KUECHLERI,
ANNO
M. DC. LXXI.

Theoria motus concreti,
seu
Hypothesis de rationibus phaenomenorum nostri Orbis.

1. Supponantur initio Globus Solaris, Globus Terrestris et spatium intermedium, massa, quod ad Hypothesin nostram attinet, quiescente, quam aetherem vocabimus, quantum satis est (omnimodam enim plenitudinem Mundi status, quem sentimus, per alibi demonstrata, non fert) plenum.

2. Necesse est igitur esse quendam motum ante omnia tum in globo solari, tum in globo terrestri. Cum enim globi isti duo habere debeant partes cohaerentes, ne ad quemlibet levissimum rei quantulaecunque impactum dissolvantur aut perforentur, nulla autem sit cohaesio quiescentis (per dicta in abstracta motus Theoria th. 20 quam suo loco dabimus), motus in iis aliquis supponendus est: quae fortasse unica ac prima demonstratio est necessarii motus terrae. Quanquam, ut §. quoque 35 infra admonebitur, ad summam Hypothesees nostrae nihil referat, an circulatio terrae admittatur, cum Circulatio Lucis seu aetheris circa terram qua potissimum utimur vid. §. 9, facile se omnibus approbare, ni fallor, possit.

3. Supponamus igitur, si placet, tum in globo solis tum in globo terrae motum circa proprium centrum, nam alios eidem aetheri interspersos magnos parvosque globos circa suum centrum motos, in quibus eadem, quae in terra nostra, fieri proportione possunt, id est, non planetas tantum quos videmus, sed et innumerabiles quosdam velut mundulos parvos, quos non videmus, nunc non consideramus.

4. Sed in sole simul et alius motus supponendus est, quo agat extra se, unde causa in mundo motus in se non

But in the sun, it is necessary to suppose another motion, too, which gets out from itself. This is the cause from which, in the world, the motion which does not return in itself derives: for, the motion around its centre does not get out from itself.[37]

Fig. 8 *Hypothesis Physica nova*[38] Image source–with permission of *Gottfried Wilhelm–Mathematische Schriften*, Georg Olms AG–Verlag).

Thence the movements whose trajectory is a closed line (we have the impression he is thinking specifically of the circular movements) can be explained relying upon the idea of a body which rotates around its axis and an aether which surrounds the body and moves circularly around the body itself. However, the movements in which a rectilinear component exists cannot. This way of reasoning is analogous to the one that induced Leibniz to distinguish between *circulatio harmonica* and *motus paracentricus*, coherently with what we have above described.

Now, if we transcribe the main concepts expressed in the *Hypothesis Physica nova* in the language of the *Tentamen*, we obtain that *motus paracentricus*, which is

---

[37] "Sed in sole simul et alius motus supponendus est, quo agat extra se, unde causa in mundo motus in se non redeuntis derivetur: motus enim circa proprium Centrum extra se non agit." (Leibniz 1671 [1860] [1962], § 3, pp. 20-21). The translation is ours.

[38] Leibniz 1671 [1860] [1962], [1971], Vi, pp. 17-61.

a rectilinear motion, cannot be caused only by circular movements. In his successive works (we will see the content of these works in details) Leibniz rejected many of the conceptions expressed in the *Hypothesis Physica nova*, in particular the bubbles theory, but the necessity to consider a motion different from aether's circulation remained a significant point in his theory.

The non-circular motion is originated by little particles of the sun (or of any radiating body) which leave sun's surface. Their action is propagated beyond sun's surface. For, Leibniz claimed that any aether's point is agitated by a ray of particles emitted from the sun. It is not necessary that the particles of the sun are present *in actu*, but that their action is present. He wrote:

> And, as no perceptible point around the sun can be given, which arrives until the earth and farther on, unless a sun-ray, namely the movement of the aether caused by a particle emitted by the sun in a straight line (if not the particle itself), reaches that point in any perceptible instant.[39]

The interactions between aether's circular movement and sun light's rays movement explain *the movement of the earth around the sun* and *of the moon around the earth* (*Ivi*, pp. 20-23). Leibniz explicitly asserts: as to the system of the world, the principal problem is to realize how the rotation of the sun around its axis and the rectilinear action of its particles towards the earth can produce the movement of the earth around the sun. A similar question is to explain the rotation of the moon around the earth. Indeed, we read:

> [...] this pertains to the doctrine of the system of the world; for, because of the same reason by which, from the rotation of the sun around its centre, which concurs with its rectilinear action towards the earth, the motion of the earth around the sun originates, and from the movements of the earth around its centre, which concurs with its rectilinear action of reflecting solar light towards the moon, moon's motion around the earth originates, it is probably allowed us to claim that, for the other planets, things proceed in the same way[40].

---

[39] "Et tot quidem, ut non possit dari punctum sensibile circa solem ad tellurem usque et ultra, ad quod non quolibet instanti sensibili radius aliquis solis, id est, aetheris agitatio per emissam a sole recta linea partem (etsi non pars ipsa) perveniat". (Leibniz 1671 [1860] [1962], § 3, p. 21). The translation is ours.

[40] "[...] pertinent talia ad doctrinam de systemate mundi; quemadmodum id quoque, qua ratione ex rotatione solis circa proprium centrum concurrente ejus actione rectilinea in terram oriatur motus terrae circa solem, et ex motu terrae circa proprium centrum, concurrente ejus lucem solarem reflectentis actione rectilinea in lunam, motus lunae circa terram; quae de caeteris planetis eadem probabilitate dicere licet [...]". (Leibniz 1671 [1860] [1962], § 3, pp. 22-23). The translation is ours.

Thus, though in the *Hypothesis Physica nova* Leibniz did not develop a planetary model, some of his conceptions were already present:

1) The rotational motion of the sun's vortex will become the *circulatio harmonica of the Tentamen*;
2) The actions of the particles responsible for the rectilinear movement originates what, afterwards, Leibniz called *motus paracentricus*.

The idea to consider the planetary motion as given by a rectilinear and a circular component is, thus, already expressed in the *Hypothesis Physica nova*, even though the general physical reference frame will change in the successive Leibniz's works.

With regard to the earth and to gravity on the earth, Leibniz hypothesized our planet is surrounded by a subtle aether, which, under the action of the sun light, rotates in the opposite direction of the earth movement around its axis. Hence, the aether moves from East to West. Due to the action of the sun rays, the aether enters in contact with earth's primitive matter and forms *bullae* (bubbles) of different density and size, which depend on the different movements impressed by the sun light to the aether. The air and all the other substances are heavy and composed of bubbes, while the aether is not heavy, but, because of its movement, it is the cause of gravity. Leibniz wrote:

> [...] I distinguish the air from the aether, as the air is heavy, while the aether produces gravity by means of its circulation[41].

This introduces the problem of gravity on the earth: according to Leibniz, the earth is composed of glass bubbles[42]. Gravity is considered by Leibniz a general affection of the earth (Leibniz 1671, [1860], [1962], § 15 p. 25, line 4). Given the importance of this affection, a duty of the physicists is to supply a mechanical explanation to gravity (*ivi*, p. 25).

In the section 16 (*Ivi*) Leibniz claimed that gravity derives from the circulation of the aether *around the earth, on the earth and across the earth*, according to the above described mechanism which is based on the interactions between rays of solar light, aether and original undifferentiated matter of the earth. The earth descends in the water as the earth contains less aether than water, analogously the water in the air. The aether is seen as a principle of *levitas*, since the lack of aether is the cause of bodies' relative gravity. Leibniz is clear, while writing:

> Gravity originates from aether's circulation around, inside and trough the earth. The cause of gravity was given in the previous § 9 and § 10. Furthermore, aether penetrates water and air because they are more porous. Therefore the earth descends in the water as it contains more

---

[41] "[...] aërem enim in eo ab aethere distinguo, quod aër est gravis, aether circulatione sua causa gravitatis" (Leibniz 1671 [1860] [1962], § 3, p. 24).
[42] *Ivi*, pp. 23-25.

aether, which is not in the appropriate place, than water. Due to the same reason, the water descends in the air.[43]

The aether is the element of *levitas*. It enters actively into the mechanism, which determines the fall of the bodies, as follows: usually water gravitates on the earth and air on the water. The circulation of the aether around the earth is not perturbed. If, for any reason, some particles of earth are – for example - in the air (which, in the same volume, contains more aether than water), aether's circulation is perturbed, but it tends to expel each object which could perturb it. Here Leibniz introduced an implicit principle of minimum: a direction exists in which the initial state without perturbations is restored as quickly as possible. Moreover, the body tends to be at rest where it encounters the least possible resistance by the aether. The direction of least resistance is downwards, thence the bodies tend downwards. This is why each body falls on the earth. The double role of the aether is to be:

I. Element of levity.
II. Cause of the expulsive motion of a body perturbing aether's circulation. Thus, aether determines gravity.

The relations between the *causes of planetary movements* and (origin) *causes of gravity* are evident:

1) The circulation of the aether around the sun determines the mean motion of the planets. This conception is outlined in the *Hypothesis Physica nova* and will be one of the fundamental statements of the theory explained in the *Tentamen*.
2) The rotation of the sun determines the manner in which the light rays leave the sun and reach the terrestrial aether.
3) The interaction between sun rays, terrestrial aether and earth's original matter determines the kind of bubbles of the terrestrial materials;
4) The specific weight of each material is given by the quantity of aether in its bubbles.
5) Gravity depends on the circulation of aether through the described mechanism. This means that the movements of the celestial bodies around the sun, the motion of the sun itself around its axis and the problem of gravity are related – not identified – from the beginning of Leibniz's philosophical and scientific speculation.

---

[43] "Gravitas oritur ex circulatione aetheris circa terram, in terra, per terram, de cujus causa supra § 9 et 10. Is porro maxime aquam at aërem penetrat, quippe porosiores. Unde terra in aqua, nisi cum plus aetheris superficiarii continet, quam ipsa aqua, aqua in aëre descendit" (Leibniz 1671 [1860] [1962], § 16, p. 25). Translation is ours.

The connections between planetary model and gravitational theory are so strong, that it is not believable to think that Leibniz developed the one without taking into account the other, too.

The way in which Leibniz began the *Conclusio* (Leibniz 1671, [1860] [1962], *Conclusio* pp. 58-59) to his work is the most significant document to testify what we have claimed.

He wrote that every *globus mundanus* (probably to translate as "celestial body") rotates around its axis; only the sun, by means of its light rays, exerts a rectilinear action. Leibniz wrote:

> I suppose a rotation of each globe of the world around its axis. Only the sun exerts a rectilinear action in our orbit out of itself [exactly a rectilinear action] [...][44].

The movement of the aether makes it possible to deduce the Copernican system. The physical basis of gravity and elastic force is related to the aether. Thence, the inclusion of the system of the world and of gravity in a theory was a conception to which Leibniz adhered from the initial phases of his speculation. This train of thought is prosecuted by Leibniz in a letter he addressed to Honoratus Fabri in 1677, (Leibniz 1677 [1860] [1962], pp. 81-106) though in a different physical context, as he replaced the *bullae* theory of the *Hypothesis* with the *guttae* theory[45].

The following fundamental six points remained unmodified in Leibniz's perspective:

1) The fluid surrounding us is responsible for the action of the planets and for the spread of the solar light;
2) It is moved by movements of various origins;
3) However: all these movements tend to the uniformity;
4) The most important of them is the movement by which the solar light surrounds the earth every day;
5) All these movements have to be explained by mechanical causes;
6) Gravity, elasticity and magnetism as attractions. Several of these properties are related to such movements.

We read:

> And so, every fluid surrounding us is agitated by movements, which are caused, first of all, by the action of the wandering celestial bodies and by sun light. They are different as to their origin, but, if considered together, tend to the equality. Among these movements, firstly, that rapid enough motion stands out, by which light turns around the earth every day. By means of laws drawn from mechanics, I wanted to look for the consequences of these causes, which are so powerful and largely

---

[44] "Suppono globorum mundanorum gyrationem circa proprium axem, et unius solis in nostro magno orbe actionem rectineam extra se [...]" (Leibniz 1671 [1860] [1962], p. 58). Translation of ours.

[45] Leibniz 1677 [1860] [1962], Propositions 7-9, p. 88.

spread. It seems to me that, among these consequences, I have also found Gravity, that force which is called elastic, the direction of the magnet, as well as many other natural phenomena[46].

Unitariness of Leibniz's project is evident from this quotation. It is developed as follows (Leibniz 1677 [1860] [1962], Props. 1-3, pp. 86-87, Prop. 6, pp. 87-88):

1) The world surrounding the planets is full (Prop. 1): light is visible everywhere, but light needs an intermediary means to be transmitted, thus the void does not exist.
2) Every motion in a full liquid is transmitted everywhere. If the motion is spread in closed trajectories, and the matter of the fluid is transported along the trajectories, each particle of the fluid pushes the following one. The movement is, thence, spread along the whole trajectory. When there is no transport of matter among the different layers of the fluid, but the motion is due to the rotation around an axis, the motion is diffused everywhere as a consequence of particles' tendency to escape along the tangent. The particles of a layer do not escape along the tangent because of the particles of the superior layers. (Prop. 2).
3) The motion is hence circular (Prop. 2).
4) The aether has an internal motion (Prop. 3).
5) Given a homogeneous fluid animated by a uniform circular motion, if a body perturbs this motion, a tendency to reach a new *status* of uniform motion exists (Prop. 6).

When Leibniz tried to answer Gregory's objections (for the references to Gregory's objections, see above), the fact that the *circulatio harmonica* restored its regular motion quite quickly played a pivotal role. Proposition 6 of the letter to Fabri expresses the same point of view in a slightly different manner. This means that Leibniz had reached this conception 29 years before the *Illustratio*, precisely in the same context: while referring to the motion of the sun aether.

As to gravity, in Proposition 10 (Leibniz 1677 [1860] [1962], p. 88) Leibniz offered a picture analogous to that of the *Hypothesis Physica nova*: if a body is in the atmosphere, it perturbs the circulation of the aether and, hence, due to the tendency to reach a new equilibrium, the body is pushed down by an action we call gravity. The different specific weights of the materials are also explained by the aether where the less aether one body contains, the heavier is the body (*ivi*, p. 88). However, Leibniz posed the following problem: if the earth rotates around its axis,

---

[46] "Itaque cum constet astrorum imprimis errantium actione atque luce solis fluidum omne circa nos motibus origine quidem variis, attamen in aequabilitatem compositis cieri, ex quibus ille imprimis motus eminet satis rapidus, quo lux quotidie tellurem ambit; volui harum causarum tam potentium tamque late fusarum consequantias scrutari adhibitis Mechanices legibus. Has inter consequentias visus sum mihi et Gravitatem et vim quam Elasticam vocant et Magnetis directionem, et multa alia naturae phenomena reperisse" (*Ivi*, p. 85). The translation is ours.

the bodies on its surface tend to recede along the tangent. Every theory, which admits the diurnal rotation of the earth has to face this problem. Therefore: *how can gravity oppose to this outwards tendency?* On that Leibniz wrote:

> Therefore, in this hypothesis, it is necessary that a force (*vim*), which maintains the objects on the earth, exists. This force must be stronger than earth's force of receding[47].

If the attractive effect is due to the rotation of the earth, that is if the attractive effect depends on the diurnal rotation of the earth, little solid and insensible corpuscles – composing the aether – necessarily exist around the earth. Though insensible, these corpuscles are dense (the adjective is *creber*; *Ivi*, p. 90). In the same volume, they are more numerous in the air than in a stone. They tend to be rejected by the earth movement around its axis more strongly than the more solid (the adjective is *crassus*) corpuscles, which compose the stones or other materials. These latter are, thus, pushed downwards and remain on earth's surface (*Ibidem*). The receding tendency of the subtle corpuscles is stronger than that of the less subtle ones. Leibniz imagined that this tendency outwards can produce an action directed inwards on the more solid material which is under the insensible corpuscles. Therefore the outwards tendency due to the rotation of the earth around its axis can produce an inwards tendency of the layers, which are nearer to earth's surface. Using other terms: the centrifugal tendency produces a centripetal action. However, Leibniz added, there are two problems:

1) Why do not the subtle corpuscles fly away along the tangent?
2) If the only acting force was that due to the diurnal movement of the earth, then gravity should be directed towards the axis and not towards the centre of the earth.

The conclusion of this long reasoning is quite interesting: gravity does not depend on the local effect of earth's diurnal motion, but on the general effect of the sun light on the aether. Leibniz had already sustained this hypothesis, but now he is sure that it is not a hypothesis, rather the mere truth. He wrote:

> Therefore, finally, it is necessary to resort to our cause of gravity by means of proposition 10, which is not based on a hypothesis, but on a sure demonstration. This cause not only formed the earth, but also keeps it unified, and constrains the whole material surrounding the earth inside narrow limits and joins it in a whole[48].

---

[47] "Necesse est ergo in ea Hypothesi esse vim retinentem vi terrae rejicientis fortiorem" (Leibniz 1671, [1860] [1962], proposition 15, p. 90). The translation is ours.

[48] "Itaque ad nostram tandem gravitatis causam confugiendum est per prop. 10, quae non hypothesi, sed certa demonstratione nititur, terramque non formavit tantum, sed et continet et quicquid ei circumfusum est arctis limitibus coercit atque

In propositions 16 and 17, Leibniz tried to explain why the action of sun light pushes the bodies towards the centre of the earth[49]. Thus, the letter to Fabri on May 1677 (see above) is a further evidence of the unitariness of Leibniz's project. It is, in fact, possible to speak of a very cosmology in Leibniz, in which the theory of planetary motion is connected to the theory of gravity in an entire cosmological vision, which was specified and modified in the course of the years, but whose bases remained unmodified:

1) The movements of the planets and gravity depend on the aethereal vortices surrounding the sun and the planets themselves respectively.
2) Every action is transmitted mechanically, that is by contact.
3) Every action needs a means to be transmitted, which is not the void.

Inside this general scheme, many changes occurred: in this initial, or almost initial, phase of his thought, Leibniz considered a sole aethereal vortex, which surrounds all the bodies of the solar system and which is responsible for the movements. We will see that, some years later, he differentiated two kind of vortices, at least. The intermediary of gravity is – in this phase – the solar light. No assertion is as clear as the following passage by Leibniz, where he:

a) Claimed that all movements depend on the motion of the stars and on light;
b) Remarked that the fixed stars also have an effect on the earth, even though their action is slow and difficult to be detected because of their distance;
c) Asserted that every property concerning the sun and the planets has to be transcribed into properties of light and movements by means of geometry and mechanics[50]. This is a very project to construct a system of the world, in which planetary model and gravity are among the most important subjects.

We read:

> I have no doubt that all the movements in the bodies, which are in front of us, derive from stars' movements and light. However, the great distance of the fixed stars is the cause why I believe that their movements, though having some effects, produce ones, which are slow

---

in unum compellit" (*Ivi*, pp. 90-91). The translation is ours. On Leibniz's principles see Duchesneau 1996.

[49] The problem that, if gravity were due to a centrifugal tendency of the aether, then gravity should be directed towards Earth's rotation axis and not towards Earth's centre, was an important element within the discussions on the origin of gravity developed during the period 1660-1690. It was a very challenge for the supporters of vortex theory (Bussotti 2015, Chapter 5).

[50] On mechanics and geometry around 18th and 19th see recently Pisano and Capecchi 2013.

and barely perceptible for us in the course of many centuries. The only possibility remains that every effect is transported by the light and by the movements of the sun and of the planets. These movements are not so numerous and complicated that we cannot hope the most skilled geometers and mechanicians reach to know them precisely enough[51].

The evolution of Leibniz's thought on gravity is hence connected to the evolution of his cosmology, of which, planetary model developed in the *Tentamen* and *Illustratio* is a section.

Around the 1690s Leibniz wrote probably his most significant contributions to gravity: the *Zweite Bearbeitung* of the *Tentamen* and *De causa gravitatis*. Here the two hypotheses on the origin of gravity are proposed[52]:

1) Gravity is caused by centrifugal force of aether (this is the old hypothesis);
2) Gravity is associated to *conatus explosivus*.

Let us first analyse the *Zeite Bearbeitung*.

In *Zweite Bearbeitung*, Leibniz also attempted to identify a relation between the fluid responsible for the *harmonic circulation* and the one on which *gravity on earth* depends. He wrote:

> It results that each body in the world, which is bigger than the others [of the same kind] has the force (*vim*) to attract (at least) the like bodies inside its sphere. As to the terrestrial bodies we call gravity this force, and we are transferring this name to the celestial bodies by means of an analogy[53].

That is:

---

[51] "Cum enim ego pro certe habeam, omnes motus in corporibus nobis obviis ab Astrorum motibus atque luce oriri, fixarum autem distantia causa sit cur credam, quae in ipsis fiunt, ea effectus quidem aliquos sed lentos tamen et multorum saeculorum decursu aegre sensibiles apud nos excitare; ideo superest, ut omnia solis et planetarum luci et motibus transscribantur. Hi motus neque tam multi neque tam implicati sunt ut a Geometriae et Mechanices intelligentibus accurate satis cognosci posse sit desperandum." (Leibniz 1677 [1860] [1862], p. 93). The translation is ours.

[52] Leibniz does not make distinction of gravity as *attraction* (by centrifugal force) and as a *property* (such as an intrinsic property from the beginning to the end of movement).

[53] "Constat [...] omne corpus mundanum majus [...] vim habere attrahendi cognata (minimum) corpora intra sphaeram suam, quam in terrestribus vocamus gravitatem, et analogia quadam ad sidera transferemus." (Leibniz 1690 [1860] [1862], p. 163). The translation is ours.

A) Gravity is an attraction exerted by a major body on a minor body, which works *at least* among *cognata*[54] (cognate, similar) bodies. Leibniz seems unsure whether gravity also acts between two not-*cognata* bodies. This qualitative reference to the composition of the bodies seems a negation of or, at least, a doubt on the universal character of gravity. This doubt is reinforced by the consideration that gravity seems an action of a bigger body on a smaller one.

B) We can extend gravity, *by analogy*, to the stars. The locution "by analogy" is meaningful: this means that, according to Leibniz, the attraction exerted by the stars is analogous to gravity on the earth, but, likely, it is not exactly the same attraction.

Leibniz's reasoning expounded in the *Zweite Bearbeitung* is connoted by the following properties:

1) *General consideration*: The attraction of gravity depends on a corporal radiation (*Ivi*, p. 163).
2) *Hypothesis of the conatus explosivus*: in an attractive body (supposed to be a sphere) there is a conate which tends to expel far from the sphere (*conatus explosivus*) the matter which perturbs the general motion. Leibniz used the expression "materiae inconvenientis sive perturbantis" (*Ivi*, p. 163). These words have to be referred to the fact that such matter perturbs the motion – supposedly of the aether -. This matter cannot hence move freely. Therefore a matter whose motion fits better with the global movement of the sphere and that, hence, perturbs it less, is attracted by an impulse coming from everywhere (we could say circular. Leibniz writes "*circumpulsio*"). The flame is an example of a physical phenomenon which behaves as described by Leibniz (*Ivi*, p. 164). He imagined a mechanism like this: the earth has a sort of gravific core which emits subtle particles composing the aether, which surrounds the earth. The bodies have pores, but are less tenuous than aether. If a body is surrounded by aether, the aether enters into the body by its pores, but it is expelled by other pores more quickly than it entered. This creates a sort of action-reaction mechanism, which pushes the body downwards[55]. This is gravity. The bodies have different reactions to such mechanism, according to the measures and density of their pores.
3) *Hypothesis that gravity is connected with aether centrifugal force*: each fluid has internal movements, which tend to become circular. The circles have the tendency to become as great as possible since, in this manner, they can better oppose to the conate to recede. According to

---

[54] The concept of *cognata corpora* was introduced, in connection with gravity, by Kepler in the *Astronomia Nova*. See, in particular, the subsection of the *Introductio*, entitled *Vera doctrina de gravitate* (Kepler 1937-2012, KGW, III, pp. 25-26).

[55] It is debatable whether Leibniz was thinking of a direct transport of matter or of the action due to a something like a wave (Bussotti 2015, § 5.4.1 and § 5.4.2).

Leibniz: the smaller the radius, the bigger the centrifugal force (*Ivi*, p. 164). If gravity has to be explained by means of the centrifugal force *a là* Kepler (in fact, *a là* Descartes or *à la* Huygens), then there could be the problem that gravity should act towards the terrestrial axis and not towards earth's centre. It is possible to avoid this obstacle (for the references see note 32). However, in *De causa gravitatis*, written in 1690, hence in the same year or the preceding year of Zwete Bearbeitung's composition, Leibniz rethought of this question and judged such an objection as a convincing critics to the theory gravity-deriving from a centrifugal force, thus adhering to the hypothesis of the *conatus explosivus*.

4) *Specific weight of the bodies*: Leibniz thought that specific weight depends on a second fluid. As seen, the pores of the bodies are different as to dimensions and as to their density inside the bodies. The second subtle fluid enters easier into those pores of the matter, which are smaller, but massed. The aether determining gravity cannot enter into these little pores, but the subtle aether can. Thence, it enters and, by a mechanism similar to that analysed for gravity, determines the specific weight (*Ivi*, p. 165).

In conclusion, there are three kinds of aether:

a) Aether responsible for the harmonic vortex. This aether exists in the whole solar system;
b) Aether responsible for gravity on the earth. This aether surrounds earth;
c) Aether responsible for the different specific weights of the bodies. This aether is more tenuous than that responsible for gravity on the earth which, in its turns is more tenuous than the aether composing the solar vortex.

The described picture is a further confirmation that Leibniz was going to frame his planetary model inside a general physical theory, of which cosmology is a part, where planetary movement is inscribed.

In *De causa* (see above) Leibniz specified his conception of gravity: it is due to the expulsion of a matter from the centre of the attracting body, which produces a radiation that, through the examined mechanism, produces gravity. Here Leibniz considered the origin of gravity on the earth and the origin of gravity on the other celestial bodies due to the same mechanism. Therefore, the *conatus explosivus* hypothesis can be seen as the origin, not only for terrestrial gravity, but for the attraction of the sun towards the planets, too. Leibniz wrote:

> Another cause can be given of this phenomenon [gravity], which is not involved with this difficulty [the difficulty behind the idea of gravity as due to the centrifugal force]. This can be obtained conceiving the explosion of a matter pushed everywhere from earth's globe or from *another celestial body*, which produces a certain radiation, analogous to light's radiation. For, in this way, we will have the receding of aethereal

matter from the centre, and, since the thicker (*crassiora*) bodies (as I will explain elsewhere) do not have the same force (*vim*) of receding, this mechanism pushes them towards the centre, or make them heavy[56].

In *De causa*, gravity theory and planetary model are even more thoroughly connected than in the previous works because Leibniz explicitly claimed that earth-gravity and sun–gravity are due to the same mechanism. Hence, this hypothesis allowed him to unify all actions responsible for the planetary movements and to offer a plausible hypothesis for gravity:

1) *Circulatio harmonica*: it is a movement, the circulation of the sun aether which transports the planets.
2) *Motus paracentricus*: *outwards tendency*, due to the fact that the motion is circular, hence it has a centrifugal force; *inwards tendency*, due to gravity of the sun whose mechanism is analogous to that producing gravity on the earth: the mechanism relying upon *conatus explosivus*.

The further important contribution as to gravity problem is the *Illustratio Tentaminis* (probably 1706). In this work, the identification of terrestrial gravity with the force responsible for the inwards tendency in the *motus paracentricus* is completed, at least as far as the action of this force is concerned (inverse square law). While, with regard to the fluids responsible for gravity in the solar system and gravity on the earth, Leibniz's thought is not completely explicit.

Leibniz referred to the two hypotheses on the origin of gravity (*vis centrifuga* and *conatus explosivus*). Although the objection against the latter hypothesis, according to which gravity would be directed toward celestial bodies' rotating axis rather toward their centre, in the § 16 of the *Illustratio*, Leibniz expounded an advantage of this hypothesis developing the following reasoning; if we suppose:

i) The origin of gravity is due to the centrifugal force caused by the rotation of the vortices.
ii) All planetary orbits have the same *vis viva*.

Then, the inverse square law and Kepler's third law can be easily deduced. The hypotheses i) and ii) allowed, thence, Leibniz to propose a unified and simple version of his theory because Leibniz wrote:

> In this way, the sole hypothesis of the concentric orbits having the same power of circulation, which, in itself, can be considered quite consistent

---

[56] "Alia ejusdem assignari posset causa non obnoxia huic difficultati, concipiendo displosionem materiae cujusdam ex globo telluris aut alterius sideris in omnes partes propulsae, quae radiationem quandam producat, radiationi lucis analogam; ita enim habebimus recessum a centro materiae aethereae, quae corpora crassiora eandem (ut alibi explicabo) vim recedendi non habentia versus centrum depellet, seu gravia reddet." (Leibniz 1690 [1860] [1962], p. 197.). The translation and italic are ours.

with the reasoning, would provide both gravity law and periodic times' law[57].

We will see in the conclusion which is the meaning of having accepted a problematic theory (gravity deriving from centrifugal force) and a strong and, at all appearances, strange hypothesis on planetary orbits' *vis viva*, for the concept of Leibniz's physical structural model. In this section we have pointed out the strong connections between gravity and planetary model in the attempt to create a very cosmology.

3 Concluding Remarks

In this paper, we have developed a research concerning some fundamental aspects of Leibniz's planetary theory: its idea to provide a *Physical–structural model* for the whole planetary movements. By *Physical–structural model*, we proposed a model which explains both the origins (in particular gravity both as attraction and force, to use modern terms) and the mechanical devices of such forces (relying on the vortices).

In our historical inquiring and epistemological–philosophical interpretation, we have distinguished five meta-levels in Leibniz's reasoning:

1) *Cosmological–Philosophical approach* (research of structures)
2) *Kinematical approach* (geometrical constructions)
3) *Dynamical approach* (research of causes)
4) *Physical approach* (physical–structural)
5) *Mechanical structural approach* (cosmological, kinematical and dynamical)

Particularly we discussed that the fourth–fifth, as *Physical–structural model* cosmologically also involves 1)–4) as well; and it generally describes both the causes of such forces and velocities (i.e., centrifugal force due to the *circulatio harmonica*), the causes of the *motus paracentricus* related to gravity and the origins of gravity itself. In particular we have focused on the fact that Leibniz's conception of *circulatio harmonica* and *gravity*, relies on a structure, or better a series of structures, of the world (typically the *vortices*).

We also remarked that, in Leibniz's perspective, the introduction of a plausible mechanical and *Physical–structural model* for the cause of gravity (attraction) was a fundamental step for the theory. On that, he presented the two analysed aspect of gravity:

---

[57] "Ita unica hypothesis orbium concentricorum potentia circulandi aequalium, quae per se rationi admodum consentanea judicari potest, simul legem gravitationis et legem temporum periodicorum daret." (Leibniz 1706 [1860] [1962], p. 268). The translation is ours.

i) Gravity related to centrifugal tendency of the vortex surrounding the earth.
ii) Gravity related to the *conatus explosivus*.

We also remark that Leibniz was well aware that a dynamical approach to gravity *à la Newton* was possible, namely an approach in which there is no reference to gravity's origin, but only to its action. As we have seen, he adopted such an approach in the *Tentamen*. Nonetheless, he continued to look for the cause of gravity and to insert this cause within a broad picture of which planetary model was a part. His project was grandiose. However,

> *What were the reasons, which induced Leibniz to venture in an enterprise having several weak or, at least, doubtful aspects?*

After all, Newton had already proved the inverse square law and the structure of Newton's *Principia* is incomparably better organised and framed than anything Leibniz had never written on planetary theory. Furthermore, Leibniz himself recognised that the hypotheses on the origin of gravity – though plausible – are just hypotheses.

A first answer is to interpret Leibniz's planetary theory and his speculations on gravity after 1689 as an attempt to provide a point of view alternative to Newton's. It is hard to think that, if Newton's results had not been published, Leibniz would had faced the details of planetary theory, though he claimed to have developed the bases of such a theory before his reading of the *Principia*. The stake was high: Newton's theory was based on the action at the distance, which was something unacceptable in Leibniz's perspective. Therefore, if he had been able to offer a theory, within which, not only Newton's dynamical results were achieved, but also the causes of gravity were offered, then Leibniz would have gone beyond Newton. He would have open the door to a physics without any action at a distance.

Planetary model and the connected problem of gravity are particularly significant in this perspective because among the most spectacular achievements of Newton's physics there were exactly his results in celestial mechanics. Thence the competition with Newton as well as Leibniz's aversion to the action at a distance played surely a role in the development of Leibniz's planetary theory and in his attempt to construct a *physical–structural model*. Notwithstanding, this is a part of the truth, though an important one. Indeed, Leibniz conceived the idea to unify the considerations on gravity and on the structure of the solar system before the *Tentamen*. These considerations were not systematic and well developed as in the last phase of his scientific career, but they existed.

Furthermore, the specific idea to separate the origin of the rectilinear movements from that of the curved movements was also conceived from the *Hypothesis Physica Nova*, albeit *in nuce*. This means that Leibniz aspired to construct a physical-structural system of the world from his youth. Obviously: at the beginning, the terms of his attempt were vague and not well defined. No Leibniz's planetary theory existed. Progressively such terms became clearer.

Planetary model assumed an important role in Leibniz's system because it represented a *tessera* in the mosaic of his physical structural system. The fact that

Leibniz defended his *circulatio harmonica* from the critics of scientists – as Huygens – who shared many of the other convictions and ideas by Leibniz is not a surprise: the *circulatio harmonica* is exactly that part of planetary theory in which the physical structure of the universe – the solar harmonic vortex – enters directly into the explanation of a component of the planetary motion: the mean motion of the planet around the sun, the deviation from the mean motion depending on the *paracentric motion*. This means that a structure of the universe enters directly in the explanation of the movement, which is – in Leibniz's opinion – a property differentiating his theory from Newton's with its void space and its action at a distance (at least in Leibniz's interpretation). It is hence clear why Leibniz defended the *circulatio harmonica*.

On the other hand, there is another proof that Leibniz's planetary theory is seen by his author as a part which is connected to the whole of his physical conception: Leibniz used a fundamental concept of his physics – that of *vis viva* – to explain Kepler's third law inside his planetary theory.

As we have seen, this happened in the *Illustratio tentaminis*. This attempt was based on the hypothesis that the orbits of the planets have the same *vis viva*, which induced Leibniz to prefer, at the end of his scientific career, the gravity theory of the *vis centrifuga* – to summarize – rather than that of the *conatus explosivus*, though this theory was preferable for other physical reasons. This shows the inner connections between planetary theory-ideas of gravity-concept of *vis viva*. The whole of these considerations makes the concept of physical-structural system a fundamental notion to grasp the main concepts as well as the nuances of Leibniz's physical works.

Finally, Leibniz's planetary production is extremely interesting because he tried to construct a system of the world different from Newton's. The fact that his system was methodologically coherent from a mathematical point of view, but did not provide the correct explanation of the celestial bodies' movements is a stimulating chapter in the complex history of the relationship between physics and mathematics in the history and philosophy of science.

As a final point on the subject – since the language and the conceptualization of Leibniz's planetary movements are far from those used by other scholars – in the following we list and summarize the meanings of the main concepts (and related our explanation) used by Leibniz in his planetary modelling above discussed:

Table 1. Main Leibniz's Concepts in his Planetary Model

Circulatio Harmonica = a movement [Harmonic Circulation]	This is the basic concept of Leibniz's planetary model. The Harmonic Circulation is a movement and has its origin in the fundamental *structure of the universe*, that is the aether's vortex which surrounds the sun and in which the planets are posed (Leibniz 1689, §§ 1, 2 and 3, pp. 84-85; see also Leibniz 1689, [1860], [1962], pp. 149-150). In our interpretation it is a physical structural-element. As far as the movements of the planets are concerned:

1) it is the cause of the mean motion of the planets;
2) it is the cause of the centrifugal tendencies which plays a role in the fact that the orbits of the planets are ellipses (*Ivi*, § 9, p. 87; see also p. 152);
3) it is the cause of the area law (today called Kepler's second law) (*Ivi* § 4 and 6, p. 85, 86, see also pp. 150-151)

Motus Paracentricus = a movement [Paracentric Motion]	It is a rectilinear movement. It indicates the approaching to (inwards tendency) or moving away (outwards tendency) of a planet from the Sun (*Ivi*, §§ 4 and 9, pp. 85 and 87, see also pp. 150 and 152). The outwards tendency is due to the centrifugal conate (*Ivi*, § 11, p. 88, see also p. 153) which, in its turn, depends on the solar vortex's *circulatio harmonica* (a physical–structural cause). The inwards tendency depends on the Solar Attraction (a cause) (*Ivi*, § 9, p. 87, see also p. 152). Appropriate values and relations of inwards tendency plus outwards tendency plus mean motion due to *circulatio harmonica* produce the elliptical motion (*Ivi*, § 9, p. 87, see also p. 152).
Impressio Excussoria Circulationis = a philosophical–physical cause [Outward Impression of the Circulation]	This is a possible cause of a movement: it is the centrifugal cause of the harmonic circulation. For Leibniz this tendency is a very force – not an apparent one. The *Impressio* is the physical cause of movement. The *Circulatio* (*Harmonica*) is a philosophical cause (namely the origin of a cause) depending on the harmonic vortex, that is a physical structure of the universe (*Ivi*, § 9, p. 87, see also p. 152).
Attractio Solaris = an attraction [Solar Attraction]	Leibniz is vague about solar attraction and its origin. He refers to something as a magnetic attraction, likely due to a vortex. But the origin of what we call Sun's gravity is not completely clear in Leibniz. Thence, for the centrifugal *conate* there is a clear physical-structural origin (the harmonic circulation), while for Sun's attraction there is not (*Ivi*, § 9, p. 87, see also p. 152).
Conatus Centrifugus or	It is a kinematic concept. In Leibniz's

Conatus Excussorius Circulationis = an acceleration [Centrifugal Conate or Excussory Conate of Circulation]	conception it is the instantaneous acceleration produced by the *Impressio Excussoria Circulationis*. (*Ivi* § 11, p. 88, see also p. 153).
Solicitatio Gravitatis = a force [Solicitation of Gravity]	It is a dynamical concept connected to Solar Attraction. Nowadays we call simply it as gravity force. It is likely that Leibniz distinguishes this from gravity in a proper sense because he was not sure that gravity on the earth was exactly the same attraction acting in the solar system. This is our interpretation (*Ivi*, §§ 9 and 14, pp. 87 and 89, see also pp. 152, 154).
Elementum Impetus = an acceleration [*Paracentrici*=element of Paracentric Acceleration]	It is a kinematical element. It is an instantaneous acceleration along the radius-vector Sun-planet. It is the algebraic sum of *Conatus Centrifugus* and *Solicitatio Gravitatis* (*Ivi*, § 15, p. 89, see also p. 154). In the *Tentamen* Leibniz wrote "twice centrifugal conate" and not "centrifugal conate" (*Ivi*). Due to a remark by Varignon (ref?), he recognized his mistake and in the following contributions corrected it (Cfr. Bussotti 2015 § 2.2.2).

Acknowledgements

We want to express our gratitude to anonymous referees for their blind peer–reviewed jobs and English proof–readers: their comments were of great help. We also want to express our warm thankfulness and appreciation to Christiane Busch (Rechte / Lizenzen, Georg Olms AG – Verlag) for giving us permission to use some *Gottfried Wilhelm–Mathematische Schriften*'s images.

References

Aiton, E. J. 1960. The celestial mechanics of Leibniz. *Annals of Science* 16(2):65-82.
Aiton, E. J. 1964. The celestial mechanics of Leibniz: A new interpretation. *Annals of Science* 20(2):111-123.
Aiton, E. J. 1965. An imaginary error in the celestial mechanics of Leibniz. *Annals of Science* 21(3):169-173.
Aiton, E. J. 1972. *The vortex theory of planetary movements*. New York: American Elsevier Publishing Company.

Aiton, E. J. 1962. The celestial mechanics of Leibniz in the light of Newtonian criticism. *Annals of Science* 18(1):31-41.
Aristotle 1955. *De Caelo*. Translation into English by Stocks J. L. Boston: The Tech Classics Archive of M.I.T.
Bertoloni Meli, D. 1993. *Equivalence and priority: Newton versus Leibniz*. Oxford: The Clarendon Press.
Bussotti, P. 2015. *The Complex Itinerary of Leibniz's Planetary Theory. Physical Convictions, Metaphysical Principles and Keplerian Inspiration*. Basel: Birkhäuser.
Bussotti, P., Pisano, R. 2013. On the Conceptual Frames in René Descartes' Physical Works. *Advances in Historical Studies* 2(3):106-125.
Bussotti, P., Pisano, R. 2014a. On the Jesuit Edition of Newton's Principia. Science and Advanced Researches in the Western Civilization. In: Pisano, R. (Ed), *Newton Special Issue: History and Historical Epistemology of Science. Advances in Historical Studies Special Issue* 3(1):33–55.
Bussotti, P., Pisano, R. 2014b. Newton's Philosophiae Naturalis Principia Mathematica "Jesuit" Edition: The Tenor of a Huge Work. *Accademia Nazionale Lincei-Rendiconti Matematica e Applicazioni* 25(4):13-444.
De Pace, A. 2009. *Niccolò Copernico e la fondazione del cosmo eliocentrico*. Milano: Mondadori.
Descartes, R. 1897–1913. *Œuvres de Descartes. Par Charles Adams et Paul Tannery*, 12 vols. Paris: Leopold Cerf. See also Descartes 1964–1974: *Id., Discours de la méthode et Essais, Specimina philosophiae*. Vol. VI; *Physico-mathematica,* Vol. X, *Le Monde ou Traité de la lumière*, Vol. XI.
Descartes, R. 1964–1974. *Œuvres. Adam J. et Tannery A. Nouvelle présentation par Rochet E, et Costabel P*, XI Vols. Paris: Vrin.
Dhombres, J. 1978. Nombre, mesure et continu : épistémologie et histoire. Nathan: Cedic.
Dhombres, J. 1998. Newton effaçant Descartes, Descartes parce que Newton. L'intérêt de bien choisir ses illustres, ou la déontologie de la profession intellectuelle au temps des Lumières. In: Pigeaud, J. (ed.). *Le culte des Grands Hommes au XVIIIè siècle, Entretiens de la Garenne*. Nantes: The University of Nantes Press, pp 119–138.
Donahue, W. H. 1981. *The dissolution of celestial spheres 1595–1650*. New York: Arno Press.
Duchesneau, F. 1996. Le principe de finalité et la science leibnizienne. *Revue philosophique de Louvain*, 94(3):387-414.
Duchesneau, F. 1998. Leibniz's Theoretical Shift in the Phoranomus and Dynamica de Potentia. *Perspectives on Science* 6(1):77-109.
Duchesneau, F. 1989. Leibniz's 'Hypothesis Physica Nova': A Conjunction of Models for Explaining Phenomena. In: Brown, J. R. and Mittelstrass, J. *Studies in the History and Philosophy of Science Presented to Robert E. Butts on his 60th Birthday*. Vol. 116 The Boston Studies in the Philosophy of Science. Dordrecht: Reidel–Springer, pp 153-170.
Duchesneau, F. 1994. *La dynamique de Leibniz*. Paris: Vrin.

Duhem, P. M. 1969. *To save the phenomena: An essay on the idea of physical theory from Plato to Galileo*. Translated by Donald, E., and Maschler, C. Chicago: The Chicago University Press.

Fichant, M. 1994. *Leibniz: La réforme de la dynamique*. Paris: Vrin.

Fichant, M. 2001. System of prestablished harmony" and criticism of occasionalism. *Shisō* 10(930):105-125.

Garber, D. 1985. Leibniz and the Foundations of Physics: The Middle Years. In: Okruhlik, K. and Brown, J. R. (Eds.). *The Natural Philosophy of Leibniz*. Dordrecht: Reidel, pp 27–130.

Goethe N. B., Beeley, P., Rabouin, D. 2015 (Eds.). *G. W. Leibniz, Interrelations between Mathematics and Philosophy*. Vol. 41. Dordrecht: Springer.

Goldstein, B. R., Barker, P. 1955. The Role of Rothmann in the Dissolution of the Celestial Spheres. *The British Journal for the History of Science* 28:385–403.

Goldstein, B. R. 1997. Saving the phenomena: the background to Ptolemy's planetary theory. *Journal for the History of Astronomy* 27(1):1-12.

Jauernig, A. 2008. Leibniz on Motion and the Equivalence of Hypotheses. *The Leibniz Review*. 18(2):1-40.

Kepler, J. 1609. *Astronomia Nova ΑΙΤΙΟΛΟΓΕΤΟΣ seu physica coelestis tradita de commentariis de motibus stellae Martis ex observationibus G. V. Tychonis Brahe*. In KGW, III.

Kepler, J. 1937-2012. Johannes Kepler, Gesammelte Werke–KGW. Van Dyck W, Caspar M. et al. (Eds.). Revised April 2013. 10 Vols. Deutsche Forschungsgemeinschaft und Bayerische Akademie der Wissenschaften. Beck'sche Verlagsbuchhandlung, München.

Knobloch, E. 1994. Harmonie und Kosmos: Mathematik im Dienste eines teleologischen Weltverständnisses. *Sudhoffs Archiv* 78(1):14-40.

Knobloch, E. 1973. The mathematical studies of GW Leibniz on combinatorics. Studia Leibnitiana Supplements 11.

Knobloch, E. 1995. Harmony and cosmos: mathematics serving a teleological understanding of the world. *Physis* 32(1):54-89.

Koyré, A. 1965. *Newtonian Studies*. Cambridge–MA: The Harvard University Press.

Leibniz G. W. 1690, [1860], [1962], [1971]. *Tentamen de motuum coelestium causis* [*Second Rework*] Hrsg. von C. I. Gerhardt (ed). *Zweite Bearbeitung*, VI:144-161.

Leibniz, G.W. 1690, [1860], [1962]. De Causa gravitatis et defensio sententiae Autoris de viris Naturae Legibus contra Cartesianos. In Leibniz 1849-1863, [1962], [1971], VI:193-201.

Leibniz, G. W. 1671, [1860], [1962]. *Hypothesis Physica nova* [...]. In Leibniz 1849-1863, [1962], [1971], VI:17-80.

Leibniz, G. W. 1689. *Tentamen de motuum coelestium causis*. In: ACTA ERUDITORUM Anno MDCLXXXIX publicata, ac *SERENISSIMO FRATRUM PARI* DN. JOHANNI GEORGIO IV Electoratus Saxonici Haeredi, & DN. FRIDERICO AUGUSTO, Ducibus Saxoniae &c.&c.&c. DICATA *Cum S. Casarea Majestatis & Potentissimi Eletoris Saxoniae Privilegiis*. *LIPSIAE*. Prostant apud J. GROSSIUM & J. F. GLEDITSCHIUM. Typis CHRISTOPHORI GÜNTHERI Anno MDCLXXXIX, pp 82-96.

Leibniz, G. W. 1677, [1860], [1962]. *Leibniz an Honoratus Fabri*. In Leibniz 1849-1863, [1962], [1971], VI:81-106.

Leibniz, G. W. 1689, [1860], [1962]. *Tentamen de motuum coelestium causis* In: *Acta Eruditorum Lipsiensium*. Hrsg. von Gerhardt, C. I. (Ed.). Zweite Bearbeitung, VI:161-187.

Leibniz, G. W. 1691, [1860], [1962]. *Essay de dynamique sur les loix du mouvement, ou il est monstré, qu'il ne de conserve pas la même quantité de mouvement, mais la même force absolue, ou bien la même quantité de l'action motrice*. In Leibniz 1849-1863, [1962], [1971], VI:215-230.

Leibniz, G. W. 1695, [1860], [1962]. *Specimen Dynamicum pro admirandis Naturae Legibus circa corporum vires et mutuas actiones detegentis et ad suas causas revocandis*. In Leibniz 1849-1863, [1962], [1971], VI:234-254.

Leibniz, G. W. 1706 [1860], [1962]. *Illustratio Tentaminis de Motuum Coelestium Causis*, Pars I et Pars II plus Beilage. In Leibniz 1849-1863, [1962], [1971], VI:254-280.

Leibniz, G. W. 1849-1863, [1962], [1971]. *Gottfried Wilhelm, Mathematische Schriften*, VII Vols. Gerhardt C. I. (Ed.). [Vol. VI: Halle 1860; Hildesheim 1971]. Hildesheim: Georg Olms Verlag.

Newton, I. (1712?, 1850) Epistola cujusdam ad amicum. *Correspondence of Sir Isaac Newton and Professor Cotes, including letters of other eminent men*. Edleston, J. (Ed.). London: Parker, pp. 308-314.

Newton, I. 1687 [1726], [1739-1742], [1822]. *Philosophiae naturalis principia matematica* [third edition], *auctore Isaaco Newtono, Eq. Aurato. Perpetuis commentariis illustrate, communi studio pp. Thomae le Seur et Francisci Jacquier ex Gallicana Minimorum Familia, matheseos professsorum. Editio nova, summa cura recemsita*. Glasgow: Duncan.

Newton, I. 1717 [1721]. *Opticks*. London: William and John Innys.

Peuerbach, G. (1473) Theoricæ novæ planetarum, id est septem errantium siderum nec non octavi seu firmamenti. Nürnberg.

Pisano, R, Capecchi, D. 2013. Conceptual and Mathematical Structures of Mechanical Science in the Western Civilization around the 18th century. *Almagest* 4/2:86-121.

Pisano, R., Bussotti P. 2013. Notes on the Concept of Force in Kepler. In: Pisano, R., Capecchi, D., and Lukešová A., (eds). *Physics, Astronomy and Engineering. Critical Problems in the History of Science. International 32nd Congress for The SISFA–Italian Society of Historians of Physics and Astronomy*. Šiauliai: The Scientia Socialis UAB & Scientific Methodical Centre Scientia Educologica Press, pp. 337–344.

Pisano, R., Bussotti, P. 2017. On the Conceptualization of Force in Johannes Kepler's *Corpus*: an Interplay between Physics, Mathematics and Metaphysics. In Pisano, R., Agassi, J., and Drozdova, D. (Eds). *Hypotheses and Perspectives within History and Philosophy of Science. Hommage to Alexandre Koyré 1964–2014*. Dordrecht: Springer.

Pisano, R., Casolaro, F. 2011. An Historical Inquiry on Geometry in Relativity. Reflections on Early Relationship Geometry–Physics (Part one). *History Research* 1(1):47–60.

Pisano, R. 2011. Physics–Mathematics Relationship. Historical and Epistemological Notes. In: Barbin, E., Kronfellner M., and Tzanakis, C. (Eds.). *Proceedings of the ESU 6 European Summer University History and Epistemology in Mathematics*. Vienna: Verlag Holzhausen GmbH–Holzhausen Publishing Ltd., pp 457–472.

Pisano, R. 2013. Historical Reflections on Physics Mathematics Relationship In Electromagnetic Theory In: Barbin, E., Pisano, R., (Eds.). *The Dialectic Relation between Physics and Mathematics in The XIXth Century*. Dordrecht: Springer, pp 31–57.

Pisano, R. 2016. A Development of the Principle of Virtual Laws and its Framework in Lazare Carnot's Mechanics as Manifest Relationship between Physics and Mathematics. *Transversal International Journal for Historiography of Science*, in press.

Pisano, R., Bussotti, P. 2012. Galileo and Kepler. On Theoremata Circa Centrum Gravitatis Solidorum and Mysterium Cosmographicum. *History Research* 2(2):110-145.

Radelet–de Grave, P. 1996. Entries: Stevin, Kepler, Leibniz, Huygens. In: *Dictionnaire du patrimoine littéraire européen, Patrimoine Littéraire Européen*. Vol. 8. Avènement de l'Équilibre européen 1616–1720. Bruxelles: De Boeck Supérieur, p 18, pp 745–755, pp 1020–1027.

Radelet–de Grave P. 2007. Kepler (1571–1630) et Cavalieri (1598–1647) astrologues, ou le logarithme au secours de l'astrologie. In : Daelemans F., Elkhadem H. (Eds.), *Mélanges offerts à Hossam Elkadem par ses amis et ses élèves*. Bruxelles: Archives et Bibliothèques de Belgique, pp. 297–315k.

Rosen, E. 1985. The Dissolution of the Solid Celestial Spheres. *Journal of the History of Ideas* 46(1):13–31.

Schuster, J. A. 2013a. *Descartes–Agonistes. Physico–mathematics, Method and Corpuscular–Mechanism 1618–1633*. Dordrecht: Springer.

Schuster, J. A. 2013b. Cartesian Physics. In: Buchwald, J. Z. and Fox, R. (Eds.) *The Oxford Handbook of the History of Physics*. Oxford: The Oxford University Press, pp 56–95.

Paolo Bussotti, Udine University, Italy
paolobussotti66@gmail.com
Raffaele Pisano, Lille University, France
raffaele.pisano@univ-lille3.fr

Leibniz and the Impossibility of Squaring the Circle

Davide Crippa

Abstract. In this paper, I shall study the last proposition of the *De Quadratura Arithmetica* (1676) of G.W. Leibniz, where we find a proof that it is impossible to solve algebraically the quadrature of an arbitrary sector of the circle, the hyperbola and the ellipse. I shall deal with the quadrature of the circle only, and I shall put this mathematical result into its proper context, that is to say the controversy that had opposed James Gregory and Christiaan Huygens several years earlier. Probably under Huygens' guidance, Leibniz studied the documents related to the controversy and wrote interesting observations about it. These observations show that Leibniz was led to investigate the issue of impossibility with the hope of correcting some flaws in Gregory's reasoning, and eventually come up with an original impossibility theorem.

Keywords: Infinitesimal Geometry, Gregory, Huygens, Leibniz, impossibility, quadrature of the circle, infinite series.

1 Introduction

The *De Quadratura Arithmetica Circuli, Ellipseos et Hyperbolae, cujus Corollarium est Trigonometria sine Tabulis* (hereinafter, *De Quadrature Arithmetica*)[1] is the only complete treatise written by Leibniz in mathematics, and certainly his first major

---

[1] I will use the letter 'A', followed by a Roman and an Arabic numerals, in order to refer to the edition of Leibniz's collected works published in the Academy Edition of Leibniz's miscellaneous works. Thus, 'AVll6' will refer to the sixth volume of the seventh tome of the Edition of the *Akademie der Wissenschaften*, and 'AVll6, 51' will refer to the text number 51 contained in that volume. In particular, AVll6, 51 contains a new critical edition of the *De quadratura arithmetica*, with an additional passage with respect to the first edition made by E. Knobloch in 1993 (by simplicity, I will use the shorthand 'LKQ' in order to refer to Knobloch's edition). Finally I shall use the abbreviation 'LSG' for Gerhardt's historical edition of Leibniz's mathematical works published in seven volumes (1849-1863).

Davide Crippa (2017) Leibniz and the Impossibility of Squaring the Circle. In: Pisano R, Fichant M, Bussotti P, Oliveira ARE (eds.), The Dialogue between Sciences, Philosophy and Engineering. New Historical and Epistemological insights. Homage to Gottfried W. Leibniz 1646-1716. College Publications, London, pp. 93–120
©2017 College Publications Ltd — ISBN: 978-1-84890-227-5 www.collegepublications.co.uk

achievement in infinitesimal geometry.[2] Composed between 1673 and 1676, during Leibniz's stay in Paris, the treatise was never published during Leibniz's lifetime.[3]

One of the major results expounded in this treatise consists in the arithmetical quadrature of the central conic sections, namely the expression of the area of an arbitrary sector of a central conic section (the circle, the ellipse and the hyperbola) by an infinite convergent series of rational numbers. This result can be summarized by the following theorem:[4]

Theorem. Let $CAFE$ be a sector of a circle, an ellipse, or an hyperbola with centre C and semi-conjugate diameter AC. Let AT be a segment cut, upon the tangent to the curve at the point A, by another tangent to the same curve at the point E (Fig. 1, Fig. 2). Let the conjugate semiaxis be $BC = 1$, and $AT = t$, $t \leq 1$ then the area of the sector $CAFE$ will be equal to a rectangle whose sides have length equal, respectively, to CB and to the convergent series: $t \pm \frac{t^3}{3} + \frac{t^5}{5} \pm \frac{t^7}{7} + ...$ The symbol $\pm$ should be interpreted as a sum, in the case of the hyperbola, or a difference, in the case of the circle and the ellipse.

From this general result, Leibniz derived the following corollary:

Corollary. The Circle is to the circumscribed Square, or the arc of a quadrant to its diameter, as $\frac{1}{1} - \frac{1}{3} + \frac{1}{5} - \frac{1}{7} + \frac{1}{9} - \frac{1}{11}...$ to the unity.[5]

This corollary establishes the "true proportion of a circle to the circumscribed square, expressed in rational numbers", as the title of a famous paper by Leibniz recites.[6]

In this article, I shall not reconstruct Leibniz's discovery of the arithmetical quadrature of the circle, for which I invite the reader to consult, for instance, the valuable Dennis and Addington (2010) and Horvath (Horvath 1983).

---

[2]For a general overview of the text, see Knobloch (Knobloch 1989); for a mathematical account of its major results see, among others by Knobloch (Knobloch 2002).

[3]Leibniz's *De quadratura arithmetica* has a complex editorial history (Knobloch 1989; Probst 2006, 2008). According to Leibniz's words, the final version of the treatise was completed in the years 1675-1976 (Cf. LSG 5, p. 128). Before then, from the Autumn of 1673 to the Autumn of the next year, Leibniz had also composed four drafts in Latin (they are now published as: AVll4 42; AVll6 1, 3, 8 = Alll1, 39.1) and a French draft (Alll1, 39, sent to Huygens). Two drafts in French can be dated back to 1675: one was intended for La Roque (Alll1, 72) and the other supposedly for Gallois (Alll1, 73). Other drafts were composed from Spring to September 1676. These are: AVll6, 14 (fragmentary), AVll6, 20 28, and the final version, namely AVll6, 51. Conclusively, we can say that Leibniz worked on the problem of the quadrature of the circle from 1673 (AVll4, 42) to September 1676, the last month of his stay in Paris. We have to wait more than 300 years to see the first published critical edition, prepared by E. Knobloch, in 1993 (Leibniz 1993).

[4]See in particular AVll6, 51, XLIII. p. 618, in which Leibniz proves, in one and the same proposition, a "rule ("*regula*"), i.e. a theorem for the general quadrature of the conic sections having a determined centre". The theorem was also published in *Acta eruditorum*, April 1691 (Cf. (Leibniz 2011 p. 69).

[5]*De quadratura arithmetica*, XXXII: "Circulus est ad Quadratum circumscriptum, sive arcus Quadrantis ad Diametrum ut $\frac{1}{1} - \frac{1}{3} + \frac{1}{5} - \frac{1}{7} + \frac{1}{9} - \frac{1}{11}$ etc. ad unitatem ", AVll6, 51 p. 600. Cf. also: Alll1, 39, p. 165; AVll6, 4, p. 74; 7, p. 89.

[6]This is the "De vera proportione circuli ad quadratum circumscriptum in numeris rationalibus expressa", which appeared in the *Acta Eruditorum* in 1682. See Leibniz 2011, p. 7.

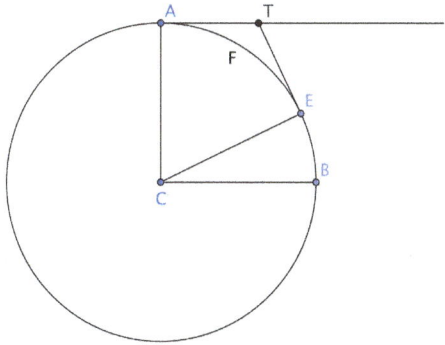

Fig. 1. A sector CAFE of a circle. Source: author's own drawing.

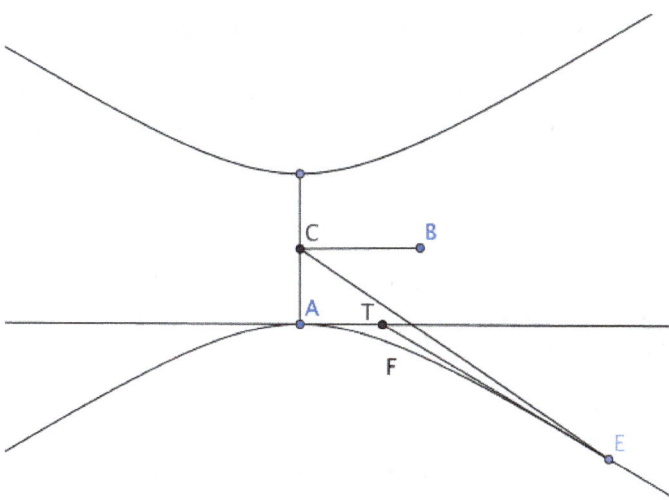

Fig. 2. A sector CAFE of an hyperbola. Source: author's own drawing.

Instead, I shall focus on a "negative" part of Leibniz's inquiry, as we may call his proof that it is impossible to solve the quadrature of any arbitrary sector of the circle algebraically. Leibniz did not formulate this impossibility theorem in these precise terms, but chose a more convoluted structure:

> It is impossible to find a better general quadrature of the circle, the ellipse or the hyperbola, or a relation between the arc and its chords [...] which is more geometrical than our own.[7]

As it will be made clear below, a "more geometrical quadrature" of a circular sector is obtained when the area of the sector, or the length of its bounding arc, can be expressed as an algebraic function of the radius and the tangent (or the sine) of the corresponding sector. In the *De quadratura arithmetica*, Leibniz proved that no algebraic function related the area or arc-length of a sector to its tangent or sine.

Leibniz was by no means the only one who, among his contemporaries, proved the impossibility of squaring a sector of the circle algebraically. In (Lützen 2014), for instance, Lützen has studied in detail four early-modern attempts to prove the same class of results, i.e. the impossibility of squaring a sector of the circle ("indefinite quadrature") or the whole circle ("definite quadrature"), by algebraic means. These attempts were made by John Wallis (1616-1703), James Gregory (1638-1675) Isaac Newton (1642-1727) and Leibniz himself.

Neither were these results developed independently one from the other. For example, it can be now ascertained, on the basis of a recent edition of Leibniz's mathematical works (AVll6, published in 2012), that James Gregory exerted, through the intermediary of Christiaan Huygens (1629-1695), a tangible influence on Leibniz's impossibility theorem.

In this paper, I shall reconstruct the connection between Gregory's impossibility results and Leibniz's proposition LI. I shall therefore argue, building upon a suggestion advanced by the historian of mathematics Joseph Hofmann (Alll1, p. LV, and (Hofmann 1975, pp. 63-64), that Leibniz, perhaps after Huygens' suggestion, was initially convinced that his own series for the arithmetical quadrature of the circle could be computed so as to yield a known rational or surd expression for the area of the circle (that is to say a rational or algebraic irrational value for π). However, from the beginning of 1675, and after having received non-enthousiastic responses from his Enligsh correspondent Henry Oldenburg, Leibniz examined more carefully Gregory's impossibility result. As a consequence of this critical examination he formulate, in 1676, the impossibility theorem that appears as the last proposition of the *De quadratura arithmetica*. After having studied the context of Leibniz's theorem, I shall examine its content and investigate its significance with respect to Leibniz's idea of geometrical exactness.

---

[7] AVll6, 51, p. 674: "Impossibile est meliorem invenire Quadraturam Circuli Ellipseos aut Hyperbolae generalem, sive relationem inter arcum et latera, numerumve et Logarithmum; quae magis geometrica sit, quam haec nostra est."

2 Leibniz's Acquaintance with Gregory's Works and his Criticism

2.1 General Background

In the treatise *Vera circuli et hyperbolae quadratura, in propria sua proportionis specie inventa et demonstrata* (Gregory 1667, hereinafter, *VCHQ*), Gregory sought to prove the impossibility of squaring any arbitrary circular (resp. elliptic or hyperbolic) sector analytically. The term "analytical" refers here to any magnitude obtained by any finite composition of the five arithmetical operations, according to the first paragraph of Descartes' *Géométrie* (1637): addition, subtraction, multiplication, division and root extraction (Descartes 1897-1913, 6, 369ff). Descartes' *Géométrie* represented therefore the background with respect to which Gregory judged the unsoulibity of the circle-squaring problem.

By generalising a classical Archimedean procedure, consisting in squeezing a curvilinear figure by a double sequence of in- and circumscribed polygons, Gregory managed to symbolize this iterative polygonal construction by means of a double converging sequence (that we shall indicate as: I_n, C_n), such that each couple, namely (I_n, C_n), could be expressed analytically from the previous couple (I_{n-1}, C_{n-1}). On this ground, Gregory reduced the geometric problem of squaring the sector to the algebraic problem of computing the limit of the double sequence (I_n, C_n).[8]

Gregory's impossibility argument boiled down to prove that the limit of the double sequence (I_n, C_n) could not be analytically computed, i.e. calculated by any finite sequence of additions, subtractions, multiplications divisions and root extractions from the terms of the double sequence. From the impossibility of squaring an arbitrary sector, Gregory also inferred the impossibility of the analytical quadrature of the whole circle as a simple corollary (Huygens 1888-1950, 6, p. 309; Dehn and Hellinger 1943, p. 475; Lüzen 2014, 224ff).

In the months following the publication of the *VCHQ*, Gregory's impossibility claim was severely criticized by Christiaan Huygens. Huygens' objections to Gregory, as well as Gregory's harsh response that followed, caused a controversy between the two mathematicians which lasted for few months, from Summer 1668 to the first months of 1669. Other outstanding mathematicians and scientific personalities became involved, like John Wallis, Henry Oldenburg, John Collins and Lord Brouncker.[9]

---

[8] Dehn 1943; Scriba 1983, pp. 13-27; Whiteside 1961, pp. 226-227.

[9] The main pieces of this controversy are reproduced in (Huygens 1888-1950, 6, nn. 1605, 1647, 1648, 1653, 1669, 1682, 1684, 1685). An older edition can be found in the volume *Christiani Hugenii Zulichemii, Dum viveret Zelemii Toparchae, Opera Varia. Volumen primum. Lugduni Batavorum*, 1724. In this work, the following pieces can be found under the title *De circuli et hyperbolae quadratura Controversia : Vera Circuli et hyperbolae Quadratura authore Jacobo Gregorio* (*Ivi*, 405-462); *Hugenii Observationes in librum Jacobi Gregorii, De Vera Circuli et hyperbolae quadratura* (*Ivi*, 463-466); *Domini Gregorii Responsum ad animadversiones Domini Hugenii, in ejus librum, De Vera Circuli et hyperbolae quadratura* (*Ivi*, 466-471); *Excerpta ex literis Domini Hugenii de responso* [...] (*Ivi*, 472-474); *Excerpta ex epistola D. Jacobi Gregorii, impressa in vindicationem* [...] (*Ivi*, 476–482). See also the documents and notes reproduced in (Turnbull 1939) and, for a reconstruction of the controversy, see (Dijksterhuis 1943; Lützen 2014; Crippa 2014).

Huygens disagreed with Gregory's arguments on the impossibility of the analytical quadrature of the central conic sections on at least two grounds. First, he claimed that Gregory's inference leading to the impossibility of the analytical quadrature was valid if one assumes that the limit of the sequence (I_n, C_n) of inscribed and circumscribed polygons could be computed *only* according to the special technique for the calculation of limits devised in the *VCHQ* (Huygens 1888-1950, 6, p. 220; Dijksterhuis 1943, p. 483; Lützen 2014, p. 228). Huygens refused to take this condition for granted, and objected that Gregory had not justified it well enough in his treatise. Without this justification, Gregory's impossibility proof would be incomplete. Second, Huygens objected to Gregory that he had not yet given a proof of the impossibility of squaring the *whole* circle. Huygens argued that one could not assume, as Gregory apparently did, that the impossibility of squaring any sector (namely, the "general", or "universal" quadrature) of the circle entailed the impossibility of squaring a definite sector of the circle, and hence the whole circle (the "particular" quadrature; Huygen, 1888-2950, 6, p. 273; Dehn and Hellinger 1943, p. 475; Lützen 2014, p. 230). Gregory replied to both these objections, but his answers were not considered satisfactory either by Huygens or by the other mathematicians who took part in the controversy as mediators, like Wallis. The controversy, which by the Winter of 1669 was running towards a dead-end, was then closed off almost by decree by Henry Oldenburg, the founding editor of the *Philosophical Transactions of the Royal Society* and first secretary of the Royal Society. It is worth nothing that, by that time, neither Gregory had managed to convince his adversary, nor vice-versa.

2.2 Leibniz's Acquaintance with Gregory's Work

Between 1672–1676, Leibniz learned about the controversy directly by Huygens, who was his mentor in Paris. Thanks to Huygens, he could get hold of a wealth of sources related to the circle-squaring problem a nd to the problem of its unsolvability.

Let us recall that Gregory had confided his reflections upon the impossibility of the algebraic quadrature of the conic sections not only to his *VCHQ*. Considerations on the same issue can be also found in the *Geometriae Pars Universalis* a treatise published by Gregory in Italy, in the first months of 1668 (hereinafter *GPU*), in the preface of the *Exercitationes Geometricae*, a booklet published in London, by the end of Summer 1668 (hereinafter *EG*) and in a number of letters written by Gregory between Summer 1668 and the subsequent Winter 1669. In particular, two of them had been published on the *Philosophical transactions of the Royal Society* with the titles: *Answer to the Animadversions of Mr. Hugenius upon his book De vera circuli et hyperbolae quadratura* (1668), and *An Extract of a Letter* (resp.3, Nr. 37 (1668); and 3, Nr. 44 (1669)) and were therefore accessible to Leibniz, as well as Huygens' answers published in the *Journal des Sçavans* (these are reproduced in Huygens 1888-1950, 6, 229ff, 273ff).

As a copy of a letter from Oldenburg to Curtius attests, [10] Leibniz was already

---
[10]The letter dates to 13th July 1668. See: All1, 9 pp. 17–18.

informed about Gregory's *Vera Circuli et Hyperbolae Quadratura* in 1668, before his arrival in Paris. Other cursory remarks, to be found in a letter sent to Jean Gallois in 1672 and in a manuscript from the end of the same year, reveal that Leibniz had already been aware of the general outlines of Gregory's method of quadratures since the beginning of his Parisian sojourn.[11]

In 1673-73, Huygens properly introduced Leibniz to the reading of Gregory's geometrical works (Cf. AVll1, 3). Huygens' notebooks record in fact that Leibniz borrowed a copy of the *De circuli magnitudine inventa*, Huygens' work on the approximate measurement of the circumference, a copy of Gregory's *Vera Circuli et Hyperbolae Quadratura* and possibly the relevant letters concerning the controversy between Huygens and Gregory (Huygens, vol. 20, p. 388) on 30th December 1673. On that occasion, Huygens probably instructed Leibniz about the issues at stake in that controversy and the reasons why he did not consider Gregory's arguments cogent enough. Afterwards, Leibniz began to study both Huygens' treatise and Gregory's *VCHQ*.[12]

On the other hand, a group of manuscripts now published in AVll4 (in particular numbers 31, 32) informs us about the way in which Leibniz came to know and read the *Exercitationes Geometricae*, the third booklet published by Gregory in the end of the Summer 1668, whose introduction contains important observations on the impossibility of the analytical quadrature of the circle and on Huygens' criticism. Leibniz certainly acquired Gregory's book during his first visit to London, in Spring 1673.[13] During the same visit, Leibniz bought and later perused two other mathematical works, Mercator's *Logarithmotechnia sive methodus construendi logarithmos nova, accurata & facilis* (1668) and Michelangelo Ricci's *Geometrica Exercitatio de maximis et minimis* (1668). Both texts were subsequently bound together with Gregory's *EG*, in a single volume which is still preserved in Leibniz's library in Hannover:[14]

The *Geometriae Pars Universalis*, completed in 1668, was read by Leibniz only starting from 1675 (Leibniz's first mention of this treatise occurs in a letter to Oldenburg, from March 1675, see AIII1 46, p. 202). Although Leibniz was very interested in this book, it seems that he did not work through it, but studied only certain of its propositions.[15] Since Leibniz's interest in this work is not directly related to impossi-

---

[11] See AIII, 1 p. 3, AVll3, 6, p. 65; AVll3, 20, p. 249.

[12] Early notes on Gregory's *VCHQ* can be found in AVll6, 2, 3, and in AVll5, 13. Many years later, Leibniz claimed in a letter to Wallis (28 May [or 7 June] 1697: AIII7, p. 428) that he had only skimmed through Gregory's book while in Paris. In the light of the aforementioned manuscripts this recollection appears to be entirely false, since it appears that Leibniz studied with a certain care at least the first part of Gregory's book, the one which deals with impossibility.

[13] As Gerhard recalls: "Leibniz paid two visits to London from Paris ... [the first] was from January 11 to the beginning of March 1673; the second was made on his way home to Germany, when he stopped in London for about a week, in October, 1676" (Child 1920, p. 159).

[14] See Hannover, Niedersächs. Landesbibl. (Ms IV 377; AVll4 3, p. 48; AVll5 47, p. 332.

[15] Hofmann argues that Leibniz did not possess a personal copy of this book while in Paris (Hoffman 1975, pp. 75-76). This conclusion is justified on the basis of the extant collection of books possessed by Leibniz, now in the Library of Hannover. There are in fact two copies of the *Geometriae Pars Universalis*; one came to Leibniz after Huygens' death (I point out that this copy contains Huygens' annotations too), which occurred in 1695, and the other originated in a similar way from Martin Knorre's library (Hofmann

bility results, but rather to the research on the quadrature of figures based on a method for transforming a given plane bounded region into an equivalent surface, I shall not consider the *GPU* any further in this inquiry.

The above survey thus confirms that from the time Leibniz began his studies on the circle-squaring problem, namely in 1673, he knew about Gregory's technique of quadratures, his attempts to prove the impossibility of solving this problem analytically and the controversy which ensued on this subject between Gregory and Huygens. This is a first, indirect confirmation that Leibniz's research in the circle-squaring problem and his parallel interest in the controversy on the impossibility of solving the problem analytically were intertwined.

However, the first fair copy of Leibniz's arithmetical quadrature of the circle, which Leibniz sent to Huygens in Autumn 1674 with the intention of having it published in the *Journal des Sçavans* afterwards, does not bear any mention of impossibility results. On the contrary, Huygens' enthusiastic response reveals that he considered Leibniz's arithmetical quadrature as a promising step which might lead to the discovery of the area of the circle in terms of rational or surd numbers.[16]

Leibniz had initially shared Huygens' optimism, and even tried to compute the sum of the arithmetical series basing himself on his studies on numerical progressions. However, he failed to achieve any significant advance.[17]

Gregory's impossibility theorem was explicitly mentioned during several exchanges between Leibniz and Henry Oldenburg, the secretary of the Royal Society and one of Leibniz's main recipients in Great Britain (Hofmann 1975, p. 95). In a letter to Oldenburg from October 1674, for instance, Leibniz boasted the originality and ingenuity of his discovery about the arithmetical quadrature of the circle, remarking how no one had given before him: "a progression of rational numbers, whose sum, continued to infinity, is exactly equal to the circumference of the circle".[18] Upon reading about Leibniz's solution to the quadrature of the circle, Oldenburg remained visibly unimpressed. Instead, he warned his correspondent about Gregory's impossibility result:

> And you truly say that no one has so far given a progression of rational numbers, whose sum, continued to infinity, is exactly equal to the circle [...] but I must add what I have recently received from a man expert

---

1975, p. 76; Mahnke 1925, p. 29). The copy from Knorre (catalogued as Marg. 98) contains also a reissue of the *VCHQ*, annotated by Leibniz. The annotations are obviously later than 1676, and may be dated to the end of XVIIth Century.

[16] I shall report Huygens' opinion in its full extent: "Je vous renvoie, Monsieur, Vostre escrit touchant la Quadrature Arithmetique, que je trouve fort belle et fort heureuse. Et ce n'est pas peu à mon avis d'avoir decouvert, dans un Probleme qui a exercé tant d'esprits, une voye nouvelle qui semble donner quelque esperance de parvenir a sa veritable solution. Car le Cercle, suivant vostre invention estant a son quarré circonscrit comme la suite infinie de fractions $\frac{1}{1} - \frac{1}{3} + \frac{1}{5} - \frac{1}{7} + \frac{1}{9} - \frac{1}{11}$ etc. à l'unitè, il ne paroistra pas impossible de donner la somme de cette progression ni par consequent la quadrature du cercle, apres que vous aurez fait voir que vous avez determinè les sommes de plusieurs autres progressions qui semblent de mesme nature". See also Hoffman 1975, p. 82.

[17] Leibniz's attempts to compute the infinite series are now published in the following tracts: AVII3 24, AVII6 7, p. 90, AVII6 11, p. 111.

[18] LSG, I, p. 53.

in these matters: in fact the aforementioned Gregory is already occupied with such a matter, so that he will show in one of his writings that the exactness of the quadrature cannot be obtained.[19]

Oldenburg's allegations seem to fall off the mark in the circumstances of his discussion. As we know, Gregory's theorem concerns the impossibility of squaring the circle analytically, i.e. by a finite algebraic expresison, whereas Leibniz was praising to Oldenburg only the virtues of his solution that amounted to an infinite series (Hofmann 1975, p. 100). We cannot exclude that at the source of Oldenburg's reservations towards the mathematical achievements of his colleague lay a misunderstanding about the meaning of exactness in mathematics. If an "exact" solution to a problem is identified with a solution that requires only algebraic curves or expressions, then Gregory's theorem implies, its correctness notwithstanding, that such an exact solution to the circle-squaring problem is wholly impossible. However, in the letter to Oldenburg, Leibniz certainly meant by an "exact solution" a solution expressed by an infinite series obeying a well-formed rule (see also section 3 below). Leibniz was thus justified in claiming that the circle-squaring problem could be exactly solved in the way he had discovered.

Probably alerted and interested by Oldenburg's reply, Leibniz started a systematic discussion on the degree of exactness that different solutions to the circle-squaring problem could attain (I will return to this topic in section 3), and in a parallel way a critical study of Gregory's arguments.

2.3 Leibniz's Criticism of Gregory

A series of drafts of a letter to Oldenburg written in March 1675 informs us more precisely about Leibniz's general dissatisfaction with the theorems of impossibility contained in the *VCHQ*. Even if Leibniz shared Huygens' dissatisfaction with Gregory's proof of his impossibility theorem, he admitted that Huygens' criticism was not persuasive enough and had not closed the question about the analytical or algebraic unsolubility of the circle-squaring problem. As a reaction, Leibniz promised new and original objections that could persuade mathematicians to investigate the circle-squaring problem and its impossibility further on:

> Besides the objections made by the celebrated Huygens, for which there is no general consensus, I have peculiar objections too, from which one can adequately conclude that the geometers must not give up this research.[20]

---

[19] The letter dates from 8th December 1674: "Quod vero ais, neminem hactenus dedisse progressionem numerorum rationalium cujus in infinitum continuata summa sit exacte aequalis circulo [...] supra dictum nempe Gregorium in eo jam esse, ut scripto probet, exactitudinem illam obtineri non posse" (LSG, p. 57). As Oldenburg's remarks confirm, a reissue of the *VCHQ*, which unfortunately did not survive till us, was being prepared around 1673-1674. This occasion in particular might have inspired the reference to Gregory's impossibility claim in the excerpt above.

[20] "[...] praeter objectiones ab illustri Hugenio factas, quibus nondum est satisfactum universis, habeo et ego peculiares, unde satis judicari potest, nondum geometras ab hac inquisitione desistere debere". A subsequent letter to Oldenburg, dating from 27 August 1676, shows how Leibniz had not abandon his

A detailed criticism of Gregory's impossibility arguments can be found in three manuscripts from 1676: *Quadraturae Circuli Arithmeticae Pars Secunda* (AVll6, n. 28, dated June or July 1676), *Series convergentes seu substitutrices* (AVll3, 60, from June 1676), and *Series convergentes duae* (AVll3, 64, June 1676). As the dates reveal, Leibniz wrote the aforementioned tracts while he was elaborating the ultimate version of the *De quadratura arithmetica*, the one including a theorem on the impossibility of squaring a sector of the circle and the other conic sections.

But there is more than a chronological coincidence between Leibniz's criticism of Gregory and the elaboration of the theorem of impossibility appended to the *De quadratura arithmetica*. The existence of a connection between Leibniz's own argument and his ongoing criticism of Gregory is confirmed especially by the manuscript AVll6, 28, a draft of the *De quadratura Arithmetica* from late Spring 1676. Leibniz concluded it with a *Scholium* containing a long critical discussion of the purported flaws in Gregory's impossibility argument.[21]

Leibniz's account of Gregory's errors begins by making the point about Gregory's strategy (his "*vis argumenti*") for proving the impossibility of squaring analytically a sector of a central conic. Leibniz correctly observed that the gist of Gregory's approach to the quadrature of the central conic sectors consisted in reducing the geometric problem of approximating the area of the sector by polygonal constructions to the problem of computing the limit of a certain convergent sequence (AVll6, 28, p. 352).

As I have surveyed in section 1, Gregory extrapolated, from a couple of polygonal sequences approaching from below and from above a sector of the circle (the ellipse and the hyperbola respectively) a pair of successions (I_n), (C_n), both monotonically increasing and convergent (cf. *VCHQ*, from propositions I to VI, pp. 11-19, AVll6, 28, p. 352. It should be pointed out that in Gregory's terminology, the term 'convergent' has a precise technical meaning, and indicates that the series of the differences $(C_n - I_n)$ becomes smaller than any given quantity as n grows. In other words, the double series (I_n, C_n) is a null series). Moreover, Gregory proved that the pair (I_n, C_n) is defined by the following recursive formula:

$$I_n = \sqrt{I_{n-1}C_{n-1}} \tag{1}$$

$$C_n = \frac{2(I_{n-1}C_{n-1})}{I_{n-1} + \sqrt{I_{n-1}C_{n-1}}} \tag{2}$$

---

conviction that Gregory's proofs of impossibility was imperfect and not fully rigorous: "[...] Ceterum ejus demonstrationi editae de impossibilitate quadraturae absolutae circuli et hyperbolae multa haud dubie desunt" (Alll1, 89, p. 580). Analogous remarks can be found in AVll6, 19, p. 175: "Hanc impossibilem esse asseruit ingeniosissimus Gregorius in libro de Vera Circuli Quadratura, sed demonstrationem tunc quidem, ni fallor, non absolvit."

[21] The *Scholium* does not figure in the final version of the *De quadratura arithmetica*. In removing the whole passage, Leibniz probably obeyed to the precise editorial policy, consisting in separating all historical digressions or philosophical notes from the mathematical content of the *De quadratura arithmetica*, and grouping them all together in an introduction, never finished, excerpts from which can be found in AVll6 39, 40 49.

Having obtained this analytical representation of the geometric polygonal construction, Gregory argued, in prop. XI of the *VCHQ* (*VCHQ*, p. 25) that the limit of the convergent series (I_n, C_n), which expresses the area of the sector, cannot be computed by a finite number of additions, subtractions, multiplications divisions and root extractions applied to the terms I_n and C_n.

In its general outlines, Gregory's argument proved the impossibility of finding an analytical composition f (namely a finite combination of additions, subtractions, multiplications divisions and root extractions) such that, applied to any pair (I_n, C_n) and to the sector S, will yield the same quantity K. In symbols: $K = f(I_0, C_0) = f(I_1, C_1) = ... = f(I_{n-1}, C_{n-1}) = f(I_n, C_n) = f(S, S)$....[22] If such a composition could be found, Gregory maintains, S could also be found as the root of the (algebraic) equation: $K = f(S, S)$.[23]

This result is supposed to prove the impossibility of the indefinite quadrature of the circle, since it holds for any sector S which can be approximated by an analytical sequence of polygons. As a corollary, Gregory stated the impossibility of the definite quadrature, that is to say, the quadrature of the whole circle too.[24]

After a sketchy presentation of Gregory's strategy and main result, the *pars destruens* of Leibniz's considerations properly begins. According to Leibniz, even if Gregory had presented in the *VCHQ* an ingenious procedure for computing the limit of convergent series and thus approximating the area of a conic sector, Gregory's impossibility result was vitiated by a logical flaw ("he somehow sinned in the form of reasoning": AVll28, p. 358).

In Leibniz's view, Gregory had grounded his impossibility proof on the assumption that a convergent sequence tended to an analytical limit only if this limit could be found according to the special method prescribed by Gregory, or that any method capable of computing the limit would be eventually reducible to Gregory's procedure. Since this assumption is by no means evident, Gregory's proof of impossibility as presented in the *Vera quadratura* was incomplete.

As we have mentioned above, this argument is by no means new, since Huygens, but also Wallis levelled the same criticism during their discussions of Gregory's impossibility theorems.[25]

Leibniz's criticism did not stop at this point, however. In fact, in the same tract AVll6, 28 and particularly in the contemporary manuscript *Series convergentes seu substitutrices* (AVll3, 60), he pushed on with and expanded his critical remarks by explaining Gregory's faulty arguments in the light of a mistaken distinction between "formula" and "quantity" (cf AVll3 60, p. 758-759). As we read in AVll6, 28:

> It seems to me that I see what has induced into error this very intelligent man, and I have serious reasons to doubt, which would not have

---

[22] Cf. Lützen 2014, pp. 225-226.
[23] *VCHQ*, XI, p. 25ff., and Lützen 2014, p. 226.
[24] *VCHQ*, p. 29.
[25] Huygens, 1888-1950, 3, p. 229. Wallis was especially outspoken in accusing Gregory of having committed a logical mistake, as pointed out by Beeley (Beeley 2012, 3, p. 47).

displeased Gregory himself if he were still alive. In fact he seems to have reasoned in this way [...] He will say that this is proven [i.e. that it is proven that the sector is not analytical with the sequence of inscribed and circumscribed polygons] since we have shown that an analytical formula formed by a and b, in the same way as from $\sqrt{ab}$ and $\frac{2ab}{a+\sqrt{ab}}$ cannot be given. I concede this. But if such a formula, analytically composed, is not given, then an analytical quantity expressed by this formula is not given either. It may be that the quantity is analytical and known, for instance a number; but the formula through which it is composed in the same way from the first pair and from the second pair of terms may be unknown and non-analytical.[26]

It seems to me that Leibniz was criticizing, in the passage above, Gregory's claim about the impossibility of the definite quadrature of the circle. Leibniz illustrated his objection with a simple numerical example: even if the number 3 is analytical both with respect to the numbers 4 and 6 and to the numbers 9 and 13, it could be obtained from the pairs $(4,6)$ and $(9,13)$ by means of a non-analytical, or "transcendental", composition. In other words, there are examples of transcendental compositions or functions, like the logarithms, which can take analytical, i.e. algebraic values for algebraic arguments.[27]

If we transpose this example to Gregory's result, then it appears that proving the non-existence of an analytical formula for computing the area of a sector of the circle (or of another conic) from the given polygonal series is not sufficient in order to prove that the area of a special sector, like the whole circle, is a non-analytical quantity with respect to the terms of the series (AVll3 60, p. 759; AVll6 28, p. 354).

We can compare Leibniz's objection with the content of Huygens' second reply to Gregory, from November 1668:

It is still uncertain whether the circle and the square on its diameter are not commensurable, that is, having the proportion of a number to a number; and similarly as regards to the hyperbola and its inscribed rectilinear

---

[26] AVll6, 28, p 354: "[...] nam et videre mihi videor, quod in errorem duxerit acutissimum Virum, et rationes dubitandi habeo graves, et ipsi ut arbitror Gregorio si in vivis esset, non displicituras. Itaque sic ille ratiocinatus esse videtur [...] Imo vero inquit, demonstratum est, quoniam ostendimus non posse dari formulam analyticam ex a. et b formatam, eodem modo quo ex $\sqrt{ab}$, $\frac{2ab}{a+\sqrt{ab}}$. Concedo. Si ergo non datur talis formula analytice composita; non datur quantitas analytica per hanc formulam significata. Potest enim fieri ut quantitas sit analytica et nota, verbi gratia numerus; formula autem secundum quam illa eodem modo componitur ex terminis duobus primis quo ex duobus secundis poterit esse ignota et non analytica".

[27] The same problem is discussed in other related tracts. A part the AVll6, 28, I refer also to: AVll6, 25, p. 297, and the later texts *De arte characteristica inventoriaque*, from 1678 (AVl4, 78, p. 331) and the *Symbolismus memorabilis calculi algebraici et infinitesimalis*, from 1710. We read in the latter, for instance: "uti impossibilitas extractionis in numeris rationalibus quasitae producit quantitates surdas; ita impossibilitas summationis in quantitatibus Algebraicis quaesitae, producit quantitates transcendentes [...] sane, ut saepe quantitates rationales per modum radicis seu irrationaliter exhibentur, etsi ad formulam rationalem reduci possint; ita saepe quantitates Algebraicae seu ordinariae per modum transcendentium exhibentur, etsi eas ad formulas ordinarias reducere liceat" (Leibniz 2011, p. 275).

figure. It is sufficient to remark that his [i.e. Gregory's] proposition XI and its supplement do not prove anything when we determine by rational or surd numbers the quantities a and b in his convergent series; because at that point the termination could be also some similar number, without us being able to prove, by this Proposition, that it not the case inasmuch as we won't able to tell how the termination is composed by the first and second terms. For instance, if a is 1; and b, 2, how shall we prove by his Proposition XI that the termination is not $\frac{3}{7}$? Hence, in order to conclude that the proportion of the circle to the square of its diameter is not analytical, one had to prove not only that the sector of the circle is not analytic indefinitely to its inscribed figure, although this proof still keeps a certain beauty, but that this is also true for any definite case.[28]

There are clear similarities between the two passages, to the point that one may consider Leibniz's observations as an attempt to make Huygens' original objection more precise and therefore more persuasive thanks to the conceptual distinction between quantities and formulas.

Thus, aware of the distinction between analytical and non-analytical formulas and quantities which in his opinion tainted Gregory's argument, Leibniz opted for a "wholly new approach" (AVll3, 60, p. 758) to the proof of the impossibility of squaring a circular, elliptical and hyperbolic sector. His new strategy is simple: whereas Gregory set out to solve the problem of determining the area of a sector from a given formula produced by one and the same operation, and ended up with its impossibility, Leibniz set out to solve the problem of determining the relation between the area of a sector and its tangent, and proved the non analytic, or transcendental nature of this relation.[29] As we would say today, Leibniz's result amounts to prove the non-algebraic nature of certain function (namely the trigonometric functions $\sin$ or $\arcsin$, and the logarithmic function, for what concerns the hyperbola).

In a note written between April and June 1676, titled: *Impossibilitas quadraturae circuli universalis*, Leibniz further clarified the meaning of the circle-squaring problem (and of its relative impossibility) in the following terms:

> The quadrature problem is twofold: there is a universal and a particular quadrature. The universal quadrature exhibits a rule with whose aid any

---

[28] "Il demeure encore incertain si le Cercle et le Quarré de son diametre ne sont pas commensurables, c'est à dire à raison de nombre à nombre; et de mesme en ce qui est d'une portion determinée de l'Hyperbole, et de sa figure rectiligne inscrite. Il suffit de remarquer que sa Proposition XI et son supplement ne prouvent rien lors qu'on determine les quantitez a et b dans sa progression convergente par des nombres rationels ou sourds; parce qu'alors la terminaison pourra aussi estre quelque nombre semblable, sans qu'on puisse demontrer le contraire par cette Proposition, d'autant qu'on ne pourra dire de quelle façon la terminaison est composée des premiers et des seconds termes. Par exemple, si a est 1; et b, 2; comment prouvera t on par sa Proposition XI que la terminaison n'est pas $\frac{3}{7}$? Pour conclure donc que la raison du Cercle au Quarré de son diametre nest pas analytique, il falloit demontrer non seulement que le Secteur de Cercle nest pas analytique indefinite à sa figure inscrite, quoyque cette demonstration ne laisse pas davoir sa beauté; mais que cela est vray aussi in omni casu definito [...]" (Huygens 1888-1950, 6, p. 273).

[29] AVll3, 60, p. 758.

portion of the circle can be measured, or with whose aid, from a given tangent (or sine) the arc or the angle can be found. And then there is the particular quadrature, which exhibits a certain part of the circumference (and those sectors, whose ratio with that part is known). Hence, if one exhibited the whole circle or the whole circumference, and nothing but these sectors whose ratio with the circumference is already known, one would not thereby achieve the desired universal quadrature.[30]

By distinguishing "universal" and "particular" quadratures,[31] Leibniz was perhaps among the first mathematicians who rendered explicit and precise, on a terminological level, the customary distinction, from the second half of XVIIth century onwards, between two sorts of problems related to the quadrature of a curve in general: on the one hand, the finding of the area included between the curve and two arbitrary coordinates (in the case of the circle, this problem was often understood as the problem of determining the area of an arbitrary circular sector); on the other hand, the determination of the area of the whole figure (the traditional circle-squaring problem is a problem of the latter sort).[32]

As the title of the piece AVII18 makes it clear, Leibniz claimed the impossibility of the universal quadrature of the central conic sections, by which he meant the impossibility of finding an algebraic relation between the arc and its corresponding tangent.

On the other hand, he maintained that the question of the possibility or impossibility of the particular quadrature – that is to say, the question of whether the circle might be analytical with respect to, or even commensurable with, the square constructed on its diameter – was not a question that had yet been settled.[33] This opinion persists in the *De quadratura arithmetica*, where the closing proposition only refers to the universal, or general, quadrature of the circle and of the other central conic sections.

---

[30] AVII6 18, p. 165: "Quadratura duplex est, universalis et particularis: Universalis, quae regulam exhibet cujus ope quaelibet Circuli portio possit mensurari, seu cujus ope ex data tangente (vel sinu) possit inveniri arcus sive angulus. Particularis , quae certam circumferentiae portionem, (: et eas, quarum ad hanc portionem nota est ratio:) exhibet. Unde et si quis totum circulum totamve circumferentiam exhiberet, non vero nisi eas partes, quarum ad circumferentiam nota jam tum est ratio, is quadraturam, qualis desideratur, Universalem non dedisset".

[31] The term "universal quadrature" was previously used by Mengoli to refer to Archimedes' quadrature of the parabola. Cf. *Novae quadrature arithmeticae* (1650): "Meditanti mihi persaepe Archimedis parabola Quadraturam, propterquam infinita triangula in continuè quadrupla proportione existentia certos limites quantitatis non excedunt; occurrit universalis illa Quadratura eiusdem argumenti occasione a Geometris demostrata, qua magnitudines infinita continuam quamlibet proportionem maioris inaequalitatis possidentes in praefinitas homogeneas quantitates colliguntur". On the intellectual relations between Mengoli and Leibniz, see Massa 2016.

[32] It should be pointed out that Leibniz did not strictly adhere to his own terminology, and sometimes employed the term "general" as a synonym for "universal". A notable case is AVII6, 51, prop. LI, as I plan to expound below.

[33] Regarding this concern, Leibniz affirmed in AVII6, 18: "Certas autem partes vel etiam totum Circulum (: sed non quamlibet ejus portionem:) analytice inveniri posse, nondum despero" ("I have not lost the hope yet that precise parts ("certas autem partes") or even the whole circle (but not any of its portions) can be found out analytically").

3 The Classification of Quadratures and the Impossibility of Squaring the Circle

3.1 A Classification of Quadratures

Aside from sparse notes from 1674 and 1675 (AVII3 39, p. 589; AVII5 26, p. 203), most of Leibniz's considerations on the impossibility of squaring a central conic section date back to 1676, where they appear in a number of manuscripts related to the quadrature of the circle (AVII6 18, p. 166; AVII6 19, p. 176; AVII6 28, 350ff; AVII3 60, 758ff.), and in a more complete form in proposition LI of the *De quadratura arithmetica*. We read there:

> It is impossible to find a better general quadrature of the circle, the ellipse or the hyperbola, or a relation between the arc and its chords, or between the number and its logarithm, which is more geometrical than our own. This proposition stands as the crowning of our theory.[34]

Leibniz's proposition is structured by an odd grammar which makes appeal not just to one but to two comparative forms: it is impossible - Leibniz wrote - to find a better quadrature (*"meliorem quadraturam"*), of the circle and the hyperbola, or a more geometrical relation (*"relationem quae magis geometrica sit"*) than the one presented in the treatise (and summarized above in section 1).

But in what sense might one solution to a quadrature problem be said to be better than another one? As regards the case of the circle (but the same reasoning can hold for the case of the ellipse and the hyperbola) an answer can be advanced considering Leibniz's attempt, consigned to a draft of the already mentioned letter to Oldenburg from 1675, to establish a hierarchy among several types of solution to the universal or particular circle-squaring problem according to their exactness.

Leibniz further refined this classification in a tract from 1676: *Praefatio opusculi de quadratura circuli arithmetica* (AVII6 19, p. 176-177), a purported preface, as its title indicates, to a contemporary draft of the *De quadratura arithmetica* now published in AVII6, 20.[35]

---

[34]"Impossibile est meliorem invenire Quadraturam Circuli Ellipseos aut Hyperbolae generalem, sive relationem inter arcum et latera, numerumve et Logarithmum; quae magis geometrica sit, quam haec nostra est. Haec propositio velut coronis erit contemplationis hujus nostrae" AVII6, 51, p. 674; LQK, p. 134. As it has been suggested by the editors of AVII6, Leibniz employs a similar construction in a letter to Oldenburg from August 1676, in which we read: "Non credimus, meliorem circuli quadraturam linearem quam haec est unquam datum iri" (AVII6 51 p. 520). For what concerns the relation between numbers and their logarithms, on the one hand, and the quadrature of conic sections, on the other hand, suffice it to say that Leibniz argued for the impossibility of finding an algebraic universal quadrature of the hyperbola on the grounds of the connection, discovered in 1647 by Grégoire of St. Vincent, between the hyperbolic areas of an equilateral hyperbola with equation: $xy = 1$ and the natural logarithm function. In short, the impossibility of finding a general, algebraic relation between any hyperbolic sector and its corresponding tangent can be derived from the impossibility of expressing the logarithmic function in algebraic terms.

[35]I point out that the *Praefatio* is not a preface intended for the *De quadratura arithmetica* (AVII6, 51). Interestingly, quadratures are classified along similar lines also in the later article *De vera proportione circuli ad quadratum in numeris rationalibus expressa*, published in 1682 (LSG5, p. 120).

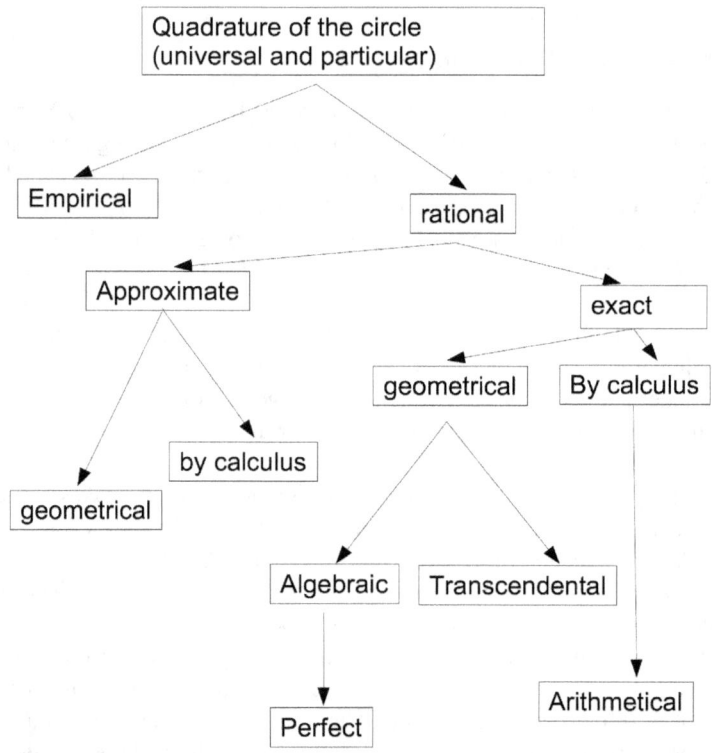

Fig. 3. Leibniz's classification of quadratures. Source: author's own drawing.

The classification presented in the *Praefatio*, the most elaborate of the two, follows both a conceptual and a historical rationale. Leibniz proceeds, in fact, by classifying, through successive dichotomies, actual solutions to the circle-squaring problem and merely conceivable ones. The first, most general subdivision demarcates quadratures obtained by tentative, empirical procedures, such as those obtained by unwinding a string around the circumference of a material circle, from "rational quadratures", which involve solutions obtained through proper mathematical methods ("*arte [...] et secundum regulam*"). Leaving "empirical" quadratures aside, Leibniz went on to clarify the variety of rational quadratures. Firstly, he distinguished two sub-species: "approximate" and "exact" quadratures. Both of them could be obtained either by geometric constructions ("*per ductum linearum*", AVll6, 19, p. 172) or by a numerical solution ("*per calculum*", AVll6 19, p. 172. See also Knobloch 1989, p. 130).

Among "approximate quadratures" Leibniz counted both geometric constructions gauging the circle in the style of the Archimedean polygonal method and computations which express π by irrational estimates (in Leibniz's terminology, these approximate quadratures are obtained "by a calculus").³⁶

By contrast, an "exact geometric" quadrature amounted, according to Leibniz's scheme, to a solution obtained by intersecting curves, either algebraic or transcendental. By virtue of the equivalence between geometrical and algebraic curves in Descartes' geometry (Bos 2001, p. 336), a geometric solution obtained by algebraic curves would stand on a par with an "analytical" quadrature, which expresses the area of the circle or of the sector as an algebraic function of the sine or the tangent and of the radius.

Leibniz acknowledged that solutions expressing the area of the circle (or of any circular sector) by a quantity or a sequence of quantities "whose nature and rule of continuation is known" should also be considered exact.³⁷ In particular, the solution presented in the *De quadratura arithmetica* belongs to this category (AVll6 19, p. 175).

Leibniz conceded that the arithmetical quadrature was not the most exact conceivable type of solution for the universal or particular circle-squaring problem, since one could certainly conceive (and some even tried to realize) the "analytical" or "geometric" quadrature of the circle and of all its sectors as the most exact or "perfect" quadrature, insofar as it does not make appeal to infinite expressions.³⁸

However the concept of a perfect quadrature is explicitly ruled out by Leibniz as contradictory: "it is impossible - Leibniz stated in the *Praefatio* - to express the general relation between a circular arc and its sine by an equation of a certain dimension".³⁹ From this assertion, Leibniz immediately derived the corollary:

> No full, analytical quadrature can be found which, while being expressed
> by an equation the values of the terms of which are rational numbers,
> would be more perfect than the one which we have given.⁴⁰

As we have seen, the same theorem is presented again with only minor variations, excepting a generalisation to the squaring of the hyperbola, in proposition LI of the *De quadratura arithmetica*, probably appended to the treatise in Summer 1676.⁴¹

3.2 An Impossibility Proof

---

³⁶Geometric and numerical quadratures were often intertwined. Among early modern representatives of approximate quadratures Leibniz names Metius, Snell and Van Ceulen, three renown Renaissance mathematicians who had worked on finding better approximate values for π, and a contemporary mathematician like James Gregory. Even if he does not mention him in the *Praefatio*, Leibniz probably had also Huygens in mind (AVll6 2).

³⁷"Valor exprimi potest exacte, vel per quantitatem, vel per progressionem quantitatum cujus natura et continunandi modus cognoscitur" (AVll6 19, p. 174).

³⁸"Perfecta autem Quadratura illa erit quae simul sit Analytica et linearis, sive quae lineis aequabilibus, ad certarum dimensionum aequationes revocabilibus, construatur" (AVll6 19, p. 175).

³⁹"Sed relationem arcus ad sinum in universum aequatione certae dimensionis explicari impossibile est" (AVll6 19, p. 175).

⁴⁰"Quadraturam plenam, analyticam, aequatione expressam, cujus terminorum dimensiones sint numeri rationales, perfectiorem, quam dedimus [...] reperiri non posse". AVll6, p. 176.

⁴¹See AVll6 28, p. 348 for an intermediate version from June or July 1676. This version does not contain any reference to the impossibility of squaring the hyperbola.

The earliest known proof that a perfect quadrature of the circle is impossible can be found already in the *Praefatio*. As regards its structure and content, the proof is very similar to the argument given in AVll51, which has recently been studied by Jesper Lützen (Lützen 2014, p. 233). In order to integrate the account given by Lützen into the present inquiry, I shall present here the version of the impossibility proof given in the *Praefatio*, which can be considered the earliest argument elaborated by Leibniz.

As in the *De quadratura arithmetica*, Leibniz reasoned by contradiction, and assumed that there exists an algebraic equation in the form: $P(\sin(v), v) = 0$, where P is a polynomial of finite degree m, expressing the relation between a circular arc v and its sine (AVll6, 19, p. 175). Therefore, the roots of the equation $P(\sin(v), v) = 0$ can be constructed, according to the Cartesian canon for the construction of equations, by intersecting algebraic curves.[42] The easiest way to perform this construction is by intersecting the curve associated with the (algebraic) equation $P(\sin(v), v) = 0$ with a straight line.

As Leibniz explained, a simple way to construct a curve associated with the equation $P(\sin(v), v) = 0$ requires a pointwise construction, obtained by applying ordinatewise each sine to successive arc-lengths.[43]

A construction of this curve, called *linea sinuum* or *curva sinuum*,[44]

Let the circular arc EFR be given (see below Fig. 4), wih radius ED and center D. Let an arc EF be taken on EFR, and let us take, or suppose given, a segment DB on DA, such that $DB = arc(EF)$ (notice that the construction of the curve of the sines requires a procedure for rectifying any arc of the circumference). From B, let us trace a segment BC, orthogonal to AB, and equal to the sine FH of the arc EF. If we repeat the same construction for any other arc, we will determine a collection of points: $C, C_1, C_2...$, each one corresponding to arcs $EF, EF_1, EF_2....$ The *linea sinuum* will be the locus of points C_n.[45]

Moreover, since the curve is supposed to be algebraic by virtue of the *reductio* assumption, the curve is receivable in Cartesian geometry.[46]

---

[42] For an overview and discussion on the history of the Cartesian technique for the construction of equations, see Bos 1984. Leibniz was certainly familiar with this technique, and he had made interesting contributions himself (as in the *De constructione*, AVI 3, 45).

[43] AVll6, 19, p. 175: "Hoc posito linea curva ejusdem gradus delineari poterit, ita ut abscissa exprimente sinus, ordinata exprimat arcus, vel contra. Hujus ergo lineae ope poterit arcus, vel angulus in data ratione secari, sive arcus, qui ad datum rationem habeat datam, inveniri sinus [...]".

[44] Leibniz probably came to know this curve from Honoré Fabri's treatise *Opusculum geometricum de linea sinuum et cycloide*, published in 1659 (Fabri 1659, p. 5, p. 10) is given in proposition XLVIII of the *De quadratura arithmetica*, according to the following procedure.

[45] AVll6 51, p. 642. The procedure explained by Leibniz corresponds, in a more modern guise, to the plotting in a Cartesian reference frame, of an arbitrary number of points whose abscissas correspond to the sines of given arcs, and whose ordinates express the corresponding arc-lengths: since in the setting of XVIIth Century geometry one coordinate was not preferred to another, the curve thus obtained might be identified with the graph of the arcsin function.

[46] Relying on Descartes' *Géométrie*, in fact, Leibniz accepted the alleged equipollence between the expressability of a curve through an algebraic equation and its constructibility by a system of "rulers and compasses intertwined, that push and guide each other" (AllI1 46, p. 204), namely articulated devices possessing one degree of freedom, so as to assure the unicity and continuity of the tracing motion. See also

Homage to Gottfried W. Leibniz 1646–1716

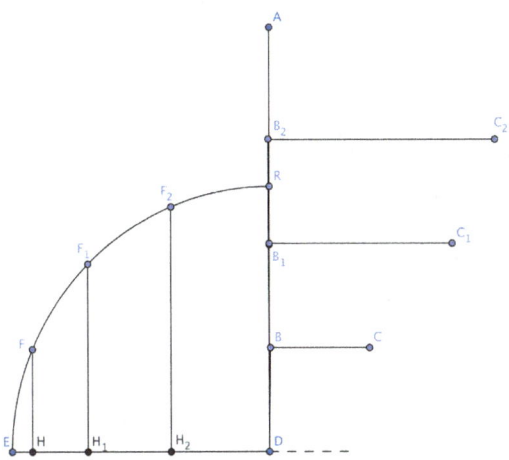

Fig. 4. A pointwise construction of the curve of the sines (Cf. *De quadratura arithmetica*, XLVIII). Source: author's own drawing.

The *linea sinuum*, as was promptly noted by Leibniz, can be successfully employed to divide any given arc not greater than a quadrant into n equal parts. For instance, if we want to divide the arc EF (Fig. 4) into n equal parts by means of this curve, it will be sufficient to trace its sine FH, and construct on the extension of the radius ED a segment $DK = FH$. Let then the perpendicular to DK be constructed on point K, and let C be the intersection point between the perpendicular and the curve of the sines (supposed to be traced): the normal to C on DA will cut this axis at a point B. We shall then have $DB = arc(EF)$, so that the arc EF is rectified. In order to solve the initial problem, it is sufficient to divide, by euclidean means, the segment DB into n parts, and find the corresponding sines.

Since this problem has been solved by the sole used of supposedly algebraic curves, the problem of dividing an angle into an arbitrary number of parts is algebraically solvable too. But this is absurd, Leibniz insisted, because:

> It is well-known indeed that so many are the various degrees of the problems, as many as are the (at least odd) numbers of the sections. Indeed, bisecting an angle is a plane problem, trisecting is a solid or conic problem, dividing the angle into five parts a supersolid problem, and so further on indefinitely. The higher is the problem, the greater is the number of equal parts in which the angle must be divided. This is admitted by the Analyticians, and it could be proved universally, if we had space. Thus,

(Descartes 1897-1913, 6, pp. 391-392).

it is impossible to express the relation between arc and sine universally, with a single equation of determinate degree.[47]

In the passage above, Leibniz referred to a result on the theory of angular sections to be found in François Viète's posthumous work, namely Viète (Viète 1646).[48] See (Viète 1983, pp. 418-450).

Viète's treatise deals with what we might call, in a modern mathematical terminology, the study of the trigonometric functions of the multiples of a given circular arc or angle. The presumed aim in his booklet *Ad angularium sectionum* was to give an algebraic treatment of the relations between trigonometric lines (sines and cosines) associated with arcs and angles. As a result, he consequently tabulated the coefficients of equations expressing the relations between the sine of an angle v and the sine of its submultiples $\frac{v}{n}$, for several n (*Ivi*, p. 295). The schema of coefficients, constructed according to a recursive rule, was to enable him to extrapolate the equations corresponding to the division of the angle into any number n of parts (where n is an integer). In this way, Viète claimed to have given the analytical translation of the more general problem of finding: "one angle to another as one number is to another", namely the problem of the general section of the angle (*Ivi*, p. 300).

One of Viète's main results obtained in the treatise on angular sections can be thus resumed, with the aid of a more expedient formalism than Viète's original symbolism:

$$sinv = \sum_{k=0}^{n} \binom{v}{n} cos^k(\frac{v}{n}) sin^{n-k}(\frac{v}{n}) sin(\frac{1}{2}(n-k)\pi)$$

As Leibniz remarked, the analytic treatment of the angular section problem involved the following corollary (not pointed out by Viète, by the way): following the procedure described in the *Sectiones angulares*, each instance of the problem of dividing an arbitrary angle into n parts can be associated with an equation of n-th degree at most. This fact can be gleaned through the examination of the equation above, relating $sin(v)$ to $sin(\frac{v}{n})$: at least if $n = 2k + 1$, the odd powers of cos^k in the previous expression will be equal to zero, and the even powers can be rewritten so as to eliminate the cosines thanks to trigonometric equalities (Lützen 2014, pp. 235-236).

On the basis of this result, Leibniz concluded that the problem of the general angular section could not be associated with a single polynomial equation in a finite degree. But this very problem is solved, for any n, by the *linea sinuum* (inasmuch as, for every n, this curve constructs, at most through the intersection with a straight line,

---

[47] AVI16, 19, p. 175-176: "Constat enim tot esse varios gradus problematum, quot sunt numeri (saltem impares) sectionum; nam bisectio anguli est problema planum, trisectio problema solidum sive Conicum, quinquesectio est problema surdesolidum, et ita porro in infinitum, altius problema prout major est numerus partium aequalium, in quas dividendus est angulus; quod apud Analyticos in confesso est, et facile probari posset universaliter, si locus pateretur. Impossibile est ergo relationem arcus ad sinum, in universum certa aequatione determinati gradus exprimi".

[48] The treatise was first published in 1615, with some additions by Alexander Anderson, and it was reprinted in 1646 with a slightly different title, in the edition of Viète's works edited by Frans van Schooten (Viète 1646).

the n-section of a given angle). Thus, it is not possible to associate this curve either with a polynomial equation in a finite, determinate degree.

There thereby arises a contradiction, from which it follows that: "it is impossible to express the relation between arc and sine universally, with a single equation of determinate degree".[49] As a consequence, not only could Leibniz conclude the impossibility of the universal quadrature of the circle, since the relation between an arc and its corresponding sine cannot be expressed by a final algebric equaton, but was able also to establish the transcendental nature of a curve, namely the *linea sinuum*.

The proof of the same theorem given in the *De Quadratura Arithmetica* follows an analogous structure, save for the dismissal of the curve of the sines, which is not strictly necessary for concluding the *reductio* argument,[50] and for a reference to the division of the angle into a prime number instead of an odd number of equal parts. This small but significant change is probably a consequence of the elementary fact that the problem of dividing an angle into an odd, non-prime number m of equal sectors can be further reduced to the more elementary problems of dividing the angle into each prime factor of m. For instance, solving the problem of dividing an angle into 15 equal parts amounts to solving the problem of dividing the original angle into five parts, and then to dividing each angle equal to one fifth of the original one into three equal parts (conversely, one could start by trisecting the angle and then proceed to divide into five parts each angle obtained from the trisection).

Leibniz took for granted that equations corresponding to divisions into a prime number of parts were irreducible to lower degree equations. From our viewpoint, this claim, crucial for the completion of the impossibility argument, is by no means obvious and needs also to be adequately proved.

However, it can be supposed that early modern and later mathematicians assumed Viète's insight into the algebraic structure of the angular section problem to be correct and definitive. Two historical clues can add plausibility to this supposition.

A first clue in support of this possibility can be found in a XVIIth Century treatise in analytic geometry, namely the *Analytic treatise on the conic sections* written by G. de l'Hôpital (1661-1704) and published posthumously in 1707 (De l'Hôpital 1707). In this text, the problem of the angular division is taken into account in connection with a discussion about the construction of regular polygons. Building directly on Viète's results, de l'Hôpital asserted without proof that the equations derived by Viète in order to deal with the problem of analytically dividing the angle into an arbitrary number of parts were: "the simplest ones, when the number of equal parts is a prime number".[51]

A second clue can be found in a later work, the *Histoire des recherches sur la Quadrature du cercle* written by J. E. Montucla (1725-1799) and originally published in 1758. Although this text appeared well into the XVIIIth Century, it is related to

---

[49] AVll6, 19, p. 176.
[50] Lützen reconstructed this proof (Lützen 2014, pp. 234-236).
[51] "Les plus simples qu'il est possible, lorsque le nombre des parties égales est un nombre premier" (de l'Hôpital 1707, p. 418).

our topic, since it contains one of the first published presentations of an argument for the unsolvability of the universal circle-squaring problem that sums up, in its essential outlines, Leibniz's argument from the *De quadratura arithmetica*, without however mentioning it. The argument related by Montucla takes the form of a *reductio*, and begins from the assumption that a circular sector or arc is a finite algebraic function of its corresponding cosine and conversely (Montucla 1831, pp. 108-109). Therefore, Montucla argued, the problem of finding the ratio between two arcs will be expressed by a finite algebraic equation too. In this way, the circumference can always be divided into two parts according to any given ratio between two integers resorting to an finite algebraic equation with a determinate degree n. But this is impossible, Montucla concluded, on the grounds of the theory of angular sections, since: "if the ratio between two arcs AB and AE, is expressed by two numbers, prime with respect to one another, and greater than n, the resulting equation will necessarily be of a degree higher than n" (*Ivi*, p. 109).

Not only did Montucla rephrase, in the language of 18th Century analysis, the same argument as Leibniz; he did not add anything to Leibniz's argument as regards the reducibilty of the equations associated with angular sections. On this subject, he merely remarked: "whatever may be the number n, it cannot be finite and determined, since it must respond to all the imaginable cases of angular sections, and there is an infinity of them leading to equations of an infinite degree".[52]

One may suppose that, had he or his contemporaries found any flaw in the use of angular sections to prove the impossibility of the universal quadrature of the circle, such a flaw would have been made explicit at some point in Montucla's account.

For an explicit criticism of Montucla's proof, we have to wait until the second edition of Montucla's *Histoire*, which appeared in 1831 with an *addendum* to the proof of the impossibility of solving the universal quadrature of the circle by algebraic means (Montucla 1831, p. 110). Since the second edition appeared posthumously, that addition was probably made by the editor S. F. Lacroix (1765-1843).

In particular, the author of the *addendum* observed that Montucla's reasoning could not yet be taken as a proof, although the numerous, failed attempts to solve the problem made the belief in the unsolvability of the circle-squaring problem by algebraic methods quite plausible. This remark was followed by a summary of Lambert's proof of the irrationality of π (a problem not tackled by Leibniz in the *De quadratura arithmetica*).

Although concise, the editor's note is telling. Since it was added for the second edition of 1831, it shows that by that time Montucla's argument (and Leibniz's argument *a fortiori*) were viewed as questionable. Unfortunately, the editor did not give details about why exactly that impossibility proof was no longer judged to be convincing.

Meanwhile, let us note that the impossibility of squaring the whole circle alge-

---

[52] "Quel que soit le nombre n, il ne peut donc être fini et déterminé, puisqu'il doit répondre à tous les cas imaginables des sections angulaires, et qu'il y en a une infinité qui conduisent à des équations d'un degré infini" (Montucla 1831, p. 110).

braically entails the impossibility of the universal quadrature of the circle. The reference to Lambert in the second edition of Montucla's book might suggest that mathematicians had by the beginning of XIXth century turned their attention to the problem of determining the nature of π, which is indeed sufficient to settle the question about the universal quadrature of the circle too. Eventually, the proof that π is a transcendental number, provided by Lindemann in 1882, put a final word to the questions about the possibility of the universal and particular quadratures of the circle.

It was probably as a consequence of these changes in the mathematical agenda that, in the course of XIXth Century, Montucla's and Leibniz's impossibility arguments would gradually glide towards the background and be forgotten, becoming not only unconvincing but also superfluous.

4 Concluding Remarks

In the historical setting of XVIIth Century geometry, the significance of the impossibility result proved by Leibniz is at first sight not obvious, since it seems to be at odds with respect to the main activity of mathematicians at the time. This consisted, in its general outlines, in the position of problems and in their solution through a geometric construction.

A first answer can be ventured in the light of the peculiar grammatical form in which the impossibility theorem presented in the *De quadratura arithmetica* is stated. In asserting that there is no more geometrical quadrature than his own, in fact, Leibniz set a clear-cut limit to the type of exactness with which a solution to the universal quadrature needed to be endowed, and at the same time advised the mathematician against searching any further for a "more geometrical" solution, which would exhibit the quadrature of a sector through an equation or through a construction by geometrical curves.

Yet was a solution of this kind, expressed by an infinite series, a solution at all? It certainly was not what one might have expected as a solution to a geometric problem, because it failed to provide a construction obtained by the intersection of curves, a traditional requirement that a solution to a geometric problem would normally have been expected to fulfill.[53]

Aware of this dilemma, Leibniz noted in the same *Scholium* to proposition XXXI of the *De quadratura arithmetica*:

> I don't even promise a quadrature by means of a geometrical construction, but via an arithmetical or analytical expression. Indeed the nature of a series, even infinite, can be understood even only a few terms are understood, provided the law of formation (*ratio*) of the series is evident. Once this is found, it is useless to continue the series, if the point is for clarifying our understanding instead of performing a mechanical operation.

---

[53]"At inquies magnitudo quaesita sic non potest exhiberi, quoniam in nostra potestate non est progredi in infinitum." (Schol. XXXI, AVII6 51, p. 600).

> If one asks for a true analytical and general relation which intervenes between the arc and the tangent, one can find in this proposition everything that can be done by Man, as I will prove below [namely, in the last proposition of the treatise]. One can find an equation of a very simple kind which expresses the dimension of the unknown quantity, whereas so far geometers have provided only approximations but not equations for the arc of the circle. I shall be silent on the fact that no one has given rational approximations to any arc or portion of the circle. Therefore, I am now the first by means of whose equation circular arcs and angles can be dealt with by an analytical calculus after the manner of straight lines.[54]

Thus, according to Leibniz, it is sufficient to know the law of formation of an infinite series for the whole series to be exactly known. One could certainly understand a series in geometrical terms, namely as a rule for performing approximate constructions. However Leibniz also made it clear that these constructions need not be executed in order to have a better understanding of the series itself, although they can also serve for pratical purposes. It was in this sense that I read Leibniz's remark above, concerning the uselessness of continuing (calculating) the terms of the series in order to elucidate our understanding of the series itself ("[...] *de mente illustranda*"). On the other hand, by calculating successive terms of the series (or performing the related constructions) one enters the realm of mechanical operations, useful for the practical goal of performing trigonometrical calculations without tables, and with an error as small as we please.[55]

With this consideration in mind, we might relate back to the impossibility of the universal quadrature of the central conic sections the following remarks, made by Leibniz to Conring while discussing the particular quadrature of the whole circle:

> Perhaps my quadrature shall be published one day in France, where I left my proofs. It is not the one desired by the vulgar mathematicians, but the one they should desire. Indeed it is impossible to express by one number the ratio between the circle and the square, but an infinite series

---

[54] "At inquies magnitudo quaesita sic non potest exhiberi, quoniam in nostra potestate non est progredi in infinitum. Fateor: neque enim eam constructione quadam geometrica exhibere promitto, sed expressione Arithmetica sive analytica. Seriei enim, licet infinitae, natura intelligi potest, paucis licet terminis tantum intellectis, donec progressionis ratio appareat. Qua semel inventa frustra progredimur, quoties de mente potius illustranda, quam de operatione quadam mechanica perficienda agitur. Itaque si quis veram relationem analyticam generalem quaerit quae inter arcum et tangentem intercedit, is quidem in hac propositione habet, quicquid ab homine fieri potest ut infra demonstrabo. Habet enim aequationem simplicissimi generis quae incognitae quantitatis magnitudinem exprimit cum hactenus apud geometras appropinquationes tantum, non vero aequationes pro arcu circuli demonstratae extent. Ut taceam ne appropinquationes rationales cuilibet arcui aut portioni circulari communes a quoquam fuisse datas. Quare nunc primum hujus aequationis ope arcus circulares, et anguli instar linearum rectarum analytico calculo tractari possunt." (AVII6, 51, p. 600).

[55] Cf. the same *Scholium*, AVII6, 51 p. 600: " et si quando contemplationem ad praxin referre licebit, operationes trigonometricae, ingenti geometriae miraculo sine tabulis perfici poterunt, errore quantumlibet parvo."

of numbers continued to the infinite is necessary, and I think a simpler series than mine cannot be given.[56]

The impossibility of finding a perfect quadrature, that is to say the one actually desired by the "vulgar" practitioner (although even more refined mathematicians, like Huygens, Leibniz's mentor and correspondent, believed in the possibility of the perfect quadrature of the circle) establishes that the arithmetical quadrature is the solution that mathematicians *should* desire, not the one they do desire.[57]

Let us recall that, according to the procedure established in Descartes' *Géométrie*, an equation was by no means the solution of a geometric problem, but was rather the last step in its analysis, which needed to be supplemented by a construction through the intersection of geometrical curves. By contrast, in the case of the quadrature of the sectors of a central conic, an equation, i.e. an infinite series, constitutes in itself the solution to the original geometric problem. It certainly has a geometric, constructive meaning, but this is downplayed as a mere aid to practice.

This is perhaps one of the major conclusions that Leibniz derived from the impossibility of providing a "perfect" quadrature of the circle and the other central conic sectors: even in the case of those problems where a solution complying with the traditional requirement of exhibiting the sought-for object through a finite stepwise construction was not attainable, it is still not impossible to achieve exact results- provided we rethink both our concepts of exactness and geometricity in order to legitimate infinite series both as instruments in problem-solving and, in some cases like that of the universal quadrature of the circle, as solutions to problems.

---

[56] The letter was written on 19 March 1678. Cf. AIII, p. 606: "Tetragonismus meus edetur fortasse aliquando in Gallia, ubi demonstrationes reliqui. Non est qualem desiderant Mathematici vulgo, sed qualem desiderare debent; nam rationem inter Circulum et Quadratum uno numero explicare impossibile est, opus est ergo serie numerorum in infinitum producta, nec puto simpliciorem dari posse quam mea est". It should be pointed out that no conclusions can be drawn, on the ground of the *De quadratura arithmetica*, concerning the quadrature of the whole circle.

[57] In this sense, Leibniz anticipates a viewpoint on the role of impossibility statements that would be emphasised in XVIII century, with Montucla and Condorcet (Lützen 2014, pp. 244-245).

Acknowledgments

The author is greatly thankful to Prof. Raffaele Pisano for his support and to the anonymous blind-referees for their suggestions.

References

Beeley, P., Scriba, C. 2012. *The Correspondence of John Wallis (1616-1703): volume III (October 1668-1671)*. Oxford: Oxford University Press.

Bos, H. 1984. Arguments on Motivation in the Rise and Decline of a Mathematical Theory; the "Construction of Equations", 1637 - ca 1750. *Archive for the History of Exact Sciences* 30:331-380.

Bos, H. 2001. *Redefining geometrical exactness*. New York, Berlin, Heidelberg: Springer-Verlag.

Child, J. M. 1920. *The Early Mathematical Manuscripts of Leibniz*. Chicago: Open Court.

Crippa, D. 2014. *Impossibility results: from geometry to analysis. A study in early modern conceptions of impossibility*. Phd. Dissertation. Université Paris Diderot, Paris [unpublished].

Descartes, R. 1897-1913. *Oeuvres complètes*. 12 Vols. Tannery P. and Adam J. (Eds.). Paris: Cerf.

Dehn, M., Hellinger, E. 1943. On James Gregory's "Vera quadratura". In: Turnbull, H. W. (Ed.). The James Gregory tercentenary memorial volume. Edinburgh: The Royal Society of Edinburgh Press, pp. 468-478.

Dennis, D., Addington, S. 2010. *Pascal and Leibniz: Sines, Circles, and Transmutations*, retrieved from: http://www.quadrivium.info/MathInt/Notes/Transmutation.pdf. Last access: 17/01/2017.

Descartes, R. 1659-1661. *Renati Descartes Geometria Editio Secunda*. Ed. by Frans van Schooten et alii. 2 Vols. Amsterdam: Apud Ludovicum et Danielem Elzevirios.

Dijksterhuis, E. J. 1943. James Gregory and Christiaan Huygens. In: Turnbull, H. W. (Ed.). The James Gregory tercentenary memorial volume. Edinburgh: The Royal Society of Edinburgh Press, pp. 478-480.

Fabri, H. 1659. *Opusculum geometricum de linea sinuum et cycloide*. Roma: Corbelletti.

Gauss, C. F. 1801. *Disquisitiones Arithmeticae*. Leipzig: Fleischer [Translated by Arthur A. Clarke. 1966. New Haven: The Yale University Press].

Gregory, J. 1667. *Vera circuli et hyperbolae quadrature in propria sua proportionis specie, inuenta, & demonstrata*. Patavii: ex typographia Iacobii de Cadorinis.

Hofmann, J. 1975. *Leibniz in Paris (1672-1676). His growth to mathematical maturity.* Translated by A. Prag and D. T. Whiteside. Cambridge: Cambridge University Press.

De l'Hôpital, G. 1707. *Traité analytique des sections coniques et de leur usage pour la résolution des équations dans les problèmes tant déterminés qu'indéterminés.* Paris: Montalant.

Horvath, Miklos. 1983. On the Leibnizian quadrature of the circle. *Annales universitatis scientiarum budapestiensis (Sectio computatorica)* 4:75-83.

Huygens, C. 1888-1950. *Oeuvres complètes publiées par la Société hollandaise des sciences.* 22 Vols. Bierens de Haan (Ed.). The Hague: M. Nijhoff.

Knobloch, E. 1989. Leibniz et son manuscrit inédit sur la quadrature des sections coniques. In *Proceedings of the Leibniz Renaissance International Workshop (Florence 2-5, 1986).* Firenze: Olschki, pp. 127-151.

Knobloch, E. 2002. Leibniz's rigorous foundation of infinitesimal geometry by means of Riemannian sums. *Synthese* 133(1):59-73.

Leibniz, G. W. 1993. *De quadratura arithmetica circuli ellipseos et hyperbolae cujus corollarius est trigonometria sine tabulis.* Knobloch E. (Ed.). Göttingen: Vandenhoeck und Ruprecht.

Leibniz, G. W., Hess, H.-J., Babin Malte-L. (Eds). 2011. *Gottfried Wilhelm Leibniz: Die mathematischen Zeitschriftenartikel.* Hildesheim, Zürich, New York: Georg Olms Verlag.

Lützen, J. 2014. 17th century arguments for the impossibility of the indefinite and the definite quadrature of the circle. *Revue d'histoire des mathématiques* 20: 211-251.

Mahnke, D. 1926. Neue Einblicke in die Entdeckungsgeschichte der höheren Analysis. *Abhandlungen der Preussischen Akademie der Wissenschaften, Physikalisch-Mathematische Klasse* 1.

Massa Esteve, M. R. 2016. Mengoli's mathematical ideas in Leibniz's excerpts. *BSHM Bulletin.* DOI: 10.1080/17498430.2016.1239807

Montucla, J.-E. 1831. *Histoire des recherches sur la quadrature du cercle, avec une addition concernant les problèmes de la duplication du cube et de la trisection de l'angle*, ed. by François S. Lacroix. Paris: Bachelier père et fils.

Probst, S. 2006. Zur Datierung von Leibniz' Entdeckung der Kreisreihe. In: Breger H., Herbst J. and Erdner, S. *Einheit in der Vielheit. VIII. Internationaler Leibniz-Kongress.* Hannover: Gottfried-Wilhelm-Leibniz-Gesellschaft, pp. 813-817.

Probst, S. 2008. Neues über Leibniz' Abhandlung zur Kreisquadratur. In: Hecht, H., Mikosch, R., Schwarz I., , Siebert, H., and Werther, R. *Kosmos und Zahl.* Stuttgart: Franz Steiner Verlag, pp. 171-176.

Scriba, C. 1983. Gregory's converging double Sequence. A New look at the controversy between Huygens and Gregory over the "analytical" quadrature of the circle. *Historia Mathematica* 10:274-285.

Turnbull, H. W. 1939. *The James Gregory tercentenary memorial volume.* Edinburgh: Royal Society of Edinburgh Press.

Viète, F. 1646. Ad angulares sectiones theoremata katholikwtera. In: Schooten, van F. *Francisci Vietae opera mathematica.* Lugd. Batav.: ex officina Johannis Elsevirii, pp. 286-304.

Viète F. 1983. *The Analytic Art: Nine Studies in Algebra, Geometry and Trigonometry from the Opus restitutae Mathematicae Analyseos, seu Algebrâ Novâ,* ed. by Richard Witmer. New York: Dover.

Whiteside, D. T. 1961. Patterns of mathematical thought in later 17th century mathematics. *Archive for the History of Exact Sciences* 1:179-388.

Davide Crippa, Research Fellowship Academy of Sciences, Praha, Czech Republic
davide.crippa@gmail.com

A Scientific Re–Assessment of Leibniz's Principle of Sufficient Reason

Antonino Drago

Abstract. A comparison of all the scientific theories that have been presented non-deductively shows that they make use of intuitionist logic in their reasoning to obtain a concluding predicate. By implicitly appealing to the Principle of Sufficient Reason (PSR) the said predicate is changed into an affirmative proposition, in order to obtain one more hypothesis to be exploited for further deductive development; in such a way the PSR changes the logic of the theory to the classical one. This change is formalized in mathematical logic. Among the above-mentioned theories, Markov's theory of computable numbers suggested two constraints on the application of PSR.

It is shown that under these constraints the PSR cannot be applied to metaphysical questions; moreover, this PSR is a better improvement of Leibniz's original version than those suggested by Heidegger, Sleigh and Pruss.

Keywords: Leibniz, Calculus, Infinite analysis, Principle of Sufficient Reason, relationship, Doubly Negated Statements, Logic.

1 Introduction

The study of the Principle of Sufficient Reason (PSR) is cumbersome because the supporters of it do not share a common appraisal of its content or range of application. This situation is illustrated in philosophical terms in section 2.

I take advantage of my discovery that several scientific theories were presented by their authors non-deductively. A comparison of these presentations suggests a common model of organization of a theory, which Leibniz also attempted to suggest. In each of these theories the PSR plays a crucial role. The author's reasoning makes use of propositions which at present we recognize as belonging to intuitionist logic. He eventually obtains a concluding universal predicate. In order to obtain one more hypothesis to be exploited for further deductive development, the author, by implicitly appealing to PSR, changes this intuitionist predicate into the corresponding affirmative proposition, hence into a proposition belonging to classical logic.

Sect. 3 illustrates the scientific evidence for the new model of organization of a scientific theory and also its development.

Antonino Drago (2017) A Scientific Re-Assessment of Leibniz's Principle of Sufficient Reason. In: Pisano R, Fichant M, Bussotti P, Oliveira ARE (eds.), *The Dialogue between Sciences, Philosophy and Engineering. New Historical and Epistemological insights. Homage to Gottfried W. Leibniz 1646-1716.* College Publications, London, pp. 121-139
© 2017 College Publications Ltd | ISBN: 978-1-84890-227-5 www.collegepublications.co.uk

Sect. 4 formalizes the logical properties of the PSR. Actually, the application of the PSR – changing an intuitionist universal predicate in the corresponding affirmative predicate of classical logic – proves to have the same logical formula as the change of this concluding predicate into the corresponding classical predicate. Yet, the conclusive predicate resulting from the application of the PSR changes the entire logic of the previous development theory into the classical logic of the subsequent theory.

Section 5 examines one of the above-mentioned theories, Markov's theory of computable numbers. It suggested two constraints to the application of the PSR.

In sect. 6 the traditional applications of PSR will be evaluated in the light of the above two constraints. It is shown that under these constraints the PSR cannot be applied to metaphysical questions.

Sect. 7 compares the new version of this principle with the versions suggested by some recent authors. It proves to be a better improvement of the original version than those suggested by Heidegger, Sleigh and Pruss.

2   Towards a New Interpretation of Leibniz's PSR

Various versions of the PSR occur in the writings of several philosophers, including some ancient ones. Leibniz suggested several versions of it; he has the merit of having - at least in the latter period of his philosophical activity - recognized in the PSR a basic principle of human reasoning[1].

Leibniz grounded it on physical experience. Yet, at the same time he considered it to be a metaphysical principle. The applications of the PSR to both levels produced so many and such extraordinary results as to seem unbelievable. Moreover his main follower, Wolff, claimed that he had reduced the said principle to a consequence of the non-contradiction principle (Cassirer 1952-1958, 595ff.) It is not surprising that there was widespread disbelief in such a principle among the subsequent scholars.

All scholars recognized that in Leibniz's thinking the PSR constitutes a very cumbersome conceptual problem. He attributed several meanings to it, all of which represent an "interpretative labyrinth" (Sleigh 1983, 193-196, 213-214). Couturat, who introduced the logicist interpretation of Leibniz's PSR, remarked that Leibniz's writings present two kinds of application.

> Here we meet a difficulty [suggested by Boutroux]: it seems that Leibniz changed his mind about the way the two rational principles correspond to the different kinds of truths. Sometimes the PSR is applied to all truths, both the necessary and the contingent ones.
> Sometimes only the principle of contradiction is enough to deal with logical and mathematical truths, while (the) physical, metaphysical and moral truths are dealt with by the PSR only. These two theses appear in contemporary writings [...] (Couturat 1901, 216).

---

[1] According to Kant's pre-critical texts, there exist at least four kinds of PSR. In the following I will mainly consider the epistemological one (*ratio cognoscendi*).

Cassirer (Cassirer 1986, 390) added that Leibniz's entire philosophy cannot be consistently interpreted without selecting those of his almost innumerable writings that are most suitable. In sum, an interpretation of Leibniz's writings inevitably pays the price of leaving aside a part of them. In the following, I will do so with the aim of achieving a well-defined rational version of the PSR which is at the same time a modern re-interpretation of Leibniz's main suggestions.

I quote a specific enunciation of PSR by Leibniz, which in my opinion is the most accurate one into original language. I will divide it into three parts A, B and C:

> [...] deux vérités primitives, savoir, en premier lieu, le principe de contradiction: car autrement si deux contradictoires peuvent être vraies en même temps, tout raisonnement devient inutile; et, en deuxième lieu, que [A:] rien est sans raison, [B:] ou que toute vérité a sa prévue *a priori*, tirée de la notion des termes, [C:] quoiqu'il ne soit pas toujours en notre pouvoir de parvenir à cette analyse (Leibniz 1686; see also Leibniz 1714, §§ 31-32).

Let us begin an interpretation of it by first removing one difficulty. Contrary to the opinion of the best known of Leibniz's followers, Wolff, the PSR is not reducible, in Leibniz's view, to the principle of non-contradiction. Cassirer stated:

> Everywhere Leibniz does opposes the [logical] connection suggested by the PSR to the connection established by the principle of non contraddiction. (Cassirer 1986, p. 404).

Hence by assuming that the PSR plays a role that is independent of the other principle, let us remark that several scholars developed the PSR in order to obtain a version which they could analyse more easily. E.g., Couturat, by taking as fundamental Leibniz's idea that the predicate is included in the subject, stated apodictically the PSR as follows:

> Every truth is analytic (Couturat 1961, x-xi; see also Rescher 1979, 23).

This proposition corresponds to part (B), i.e. the affirmative part of Leibniz's quotation above; the PSR is meant by him as an affirmative proposition belonging to an apodictically organised theory (AO) theory. In fact, Couturat's interpretation of Leibniz's philosophy is usually seen as a logicist interpretation, of course, within classical logic which he conceived as a deductive theory. Yet, after Quine's criticism of the analytic/synthetic distinction (Quine 1951), Couturat's proposition cannot be accepted as a basic tenet.

Some scholars preferred to enunciate the PSR by substituting "cause" for "reason". This preference reveals the prejudice for a deterministic philosophy; or even, in theoretical terms, for classical logic. Actually, Leibniz himself stated – although in his early writings – the radical difference between the two above notions on the grounds that a cause is less frequent than a reason:

> Surely, nothing is without a reason, but this does not mean that nothing is without a cause (Leibniz 1923, VI 4 1360).

Some scholars substituted "explanation" for "reason". Unfortunately, the new word denotes an ambiguous and context-sensitive notion[2]. Anyway, it implies that one is capable of exhibiting a theory. Yet, linking the PSR to produce a specific theory makes the scope of PSR excessive and hence makes more complex our evaluation of its validity.

One of Ruin's remarks is enlightening. Leibniz's principle goes

> [...] in two different directions: first, the principle is said to mean that every existing fact must have a sufficient reason for its existence. Secondly, it speaks of statements and their truth. The first aspect could be considered as simply another version of the principle of sufficient or efficient cause, i.e. the maxim that for every existing being or state of affairs there must have been a prior different state of affairs from which it necessarily arose. The meaning of the second formulation, however, does not lend itself as easily to an interpretation. (Ruin 1998, 50; see also Piro 2002, 3-18).

I will investigate the second formulation, concerning the truths of statements only, despite its complexity.

One more clarification comes from Wiggins: "[...] two different ways of applying the principle calling for different kinds of justification [...] either *theological-cum-cosmological* [...] or else it is *methodological*" (Wiggins 1996, 120).

He discusses four reasons for preferring a "plain methodological interpretation" of the PSR, which he defines as follows:

> [...] Sufficient Reason simply furnishes a maxim that prompts and conditions the search for explanations. When a generalization arises from an explanation that is generated in this way, it becomes a candidate for testing, confirmation, disconfirmation, qualification and all the rest. [...] simply [PSR is] giving voice to a certain methodology (Wiggins 1996, 123)[3].
> We [hence] can dispense with any use of the idea of the world as a totality. In this way we can dispense with the idea on which the Kantians will place the blame for antinomies concerning the age and the size of the world (Wiggins 1996, 125).

In sum, following Wiggins, I dismiss the "theological-cosmological issues and investigate the methodological aspect of the PSR.

Dascal, on the other hand, warned that Leibniz's thinking includes several essential oppositions.

---

[2] For an attempt to clarify it, see Hughes 1993.

[3] A century ago Enriques (Enriques 1909, 19) had already suggested a similar appraisal: the PSR "[...] expresses the *conditions of legitimacy of abstraction, constructing concepts*, leading to representing reality in that way which constitutes the theoretical knowledge ." From my previous researches (Drago 1998).

First, two separate realms correspond to the necessary vs. contingent divide; the set of all possible worlds vs. the one existing actual world. Second, two kinds of truth: the truths of reason and the truths of fact (*Monadology*, §§ 33, 34). Third, two great principles upon which our reasonings are based: the Principle of Contradiction and the PSR (*Ivi*, 31, 32). Fourth, two logics: "Just as the mathematicians have excelled above the other mortals, in the logic, i.e. in the art of reason, of the necessary, so too the jurists did it in the logic of the contingent (Leibniz 1686; Leibniz 1948, p. 303; Quotation from Daseal 2008, 60).

Of the above divides, I follow Cassirer's suggestion that fundamental to Leibniz's thinking is a logical divide between a traditional logic and a new kind of logic, a divide as is suggested by the scientific development of modern times: He suggests:

> [...] an idea of a reform for logic, which has to fill up with the real contents of sciences. From science of "the forms of the thinking", logic has to change to the science of the objective knowledge (Cassirer 1986, 91).

Indeed, in one of Leibniz's fragments his desire to define a new logic was expressed:

> To clarify the understanding, it is necessary to perfect the art of reasoning, that is, the method of judging [true or false] and the method of founding out ["d'inventer"], the latter one being the true logic and the source of all knowledge (Leibniz 1975, ff.).

All of this suggests that we explore Leibniz's search for a new logic. Here we are confronted by what has constituted a difficult problem for most philosophers, whether it is admissible that the human mind may work with two completely different logics. For instance, the celebrated philosopher of science, Karl Popper, who introduced an absolutely new philosophy of science, was wedded to classical logic which he believed was the only valid logic for reasoning about the real world:

> [...] note that [in my philosophy of science] there is only one type of argument which proceeds in an inductive direction: the deductive *modus tollens* (Popper 1959, p. 314).

which - he often emphasized - belongs to "classical logic" (Popper 1959, p. 41 and p. 76). He claimed that classical logic is more capable of answering to the needs of science:

> [...] I suggest that there are good reasons why, for most purposes of science, the classical concept [of negation] should be preferable to others - simply because it is stronger, more explicit (*Ibidem*).

Here Popper does not give a scientific criterion but rather two vague words, "explicit" and "stronger", and a common sense argument ("good reasons"). Ironically, a scholar showed that his basic arguments belongs to minimal logic (Tennant 1985); a similar conclusion was reached concerning his entire philosophy of science (Drago and Venezia 2007).

Indeed, at present time we can – as in the following - accurately define a logical divide in a well-formalized field, i.e., mathematical logic. This formalization will clarify in what way logic may change into "the science of the objective knowledge", as Cassirer suggested in the above.

3   A New Kind of Organization of a Scientific Theory

In order to analyse Leibniz's logic of science, in particular his PSR, it is important to recognize that there are two fundamentally different kinds of the organization of a theory.

Already Rescher stressed that Leibniz wanted to abandon the deductive model of the organization of a theory.

> For him, the ideal model of a cognitive system was provided not by the geometry of the ancient Greeks [...] In his view, the calculus [...] provided a new model of rational systematization [of a theory] [...] [which was aimed at solving a problem, as he did by systematizing] the calculus of variations [...] We once again see at work in the thought of Leibniz a Renaissance-inspired busting of the classical bonds (Rescher 1981, 122; see also Fichant 1998, 265-266).

Regarding this I discovered that the original presentations of several theories – among which Lazare Carnot's mechanics and Sadi Carnot's thermodynamics (Gillispie and Pisano 2014) Einstein's special relativity, Kolmogorov's foundation of non-classical logic, Church-Turing's thesis on computable functions, Markov's theory of computable numbers, etc. – did not conform to the deductive model of a theory. A comparative analysis of all these theories shows the characteristic features of their common model of organization, which proves to be alternative to the deductive one (Drago 2012)[4].

According to such a model of organization, a theory is aimed at finding out a new method for solving an apparently irresolvable problem (PO). An examination of the original texts of such theories shows that they include doubly negated

---

[4] Two centuries before Leibniz, this duality had already been suggested by Cusanus (1401-1464). He opposed, to the *theologia adfirmativa*, his *theologia copulativa* (overcoming also the *theologia negativa*). He considered this logic to be as valid as classical logic, which was advocated by the group of theologians, who Cusanus used to call *Aristotelis secta*. He distinguished two faculties of the mind, the discursive *ratio*, relying on this kind of logic, and the *intellectus*, advancing through surmise Cassirer (Cassirer 1927, 31) hinted that the search for a new kind of logic was a basic issue of Cusanus' theology. I showed that Cusanus made a correct use of intuitionist logic (Drago 2010).

propositions which are not equivalent to the corresponding affirmative ones due to the lack of evidence to support the latter (hereafter DNPs)[5]; in logical terms, here the double negation law fails and thus these propositions belong to intuitionist logic (Prawitz and Melmnaas 1968; Grize 1970, 206-210; Dummett 1977, 17-26; Troelstra and van Dalen 1988, 56ff.)[6].

They may construct *ad absurdum* arguments whose conclusions work as premises to the following arguments[7]. The final *ad absurdum* argument concludes (again a DNP which is) a universal predicate concerning the stated problem and all related problems. Let us consider some instances – illustrated by Drago (Drago 2012; see also Gillispie and Pisano 2014) of this final predicate.

S. Carnot's thermodynamics:

> [...] to make caloric yield a <u>greater</u> amount of motive power than was produced in our first sequence of operations [...] is <u>inadmissible</u> (Carnot S. 1824, 20-21).

Lobachevsky on the new hypothesis of two parallel lines:

---

[5] In the following each of the two negative words in a DNP will be underlined to facilitate the reader's recognition of the nature of the proposition.

[6] I remark that the current usage of the English language exorcises DNPs as representing a characteristic feature of a primitive language. A reason is related to unclear aspects of the DNPs. This long tradition is called by L. Horn a "dogma" (Horn 2002, 79ff; 2010, pp. 111-112). The following three well-known DNPs belonging to mathematics, physics and classical chemistry show that DPNs have pertained to scientific research since its origin. In Mathematics it is usual to assure that a theory is "<u>without</u> <u>contradictions</u>"; to state the corresponding affirmative statement, i.e. the consistency of the theory, is impossible, owing to Goedel's theorems. In theoretical physics it is usual to study the <u>in</u>-<u>variant</u> magnitudes, which does not mean that the magnitudes stay fixed. In order to solve the problem of what the elements of matter are, Lavoisier suggested defining these unknown entities by means of DNPs. "If [...] we link to the name of elements or principles of corps the idea of the last term to which [through chemical reactions] the analysis arrives, all the substances which we were <u>not</u> capable of <u>decomposing</u> through any tool are for us, elements", where the word 'decomposable' naturally carries a negative meaning and stands for '<u>non</u>-ultimate' or '<u>non</u>-simple'. (Lavoisier 1862-1892, p. 7). As a matter of fact, Grzegorczyk (Grzegorczyk 1964) independently proved that scientific research may be formalized through propositions belonging to intuitionist logic, i.e. DNPs.

[7] The first propositions of Leibniz's *Monadology* (Leibniz 1714) conforms to the first part of a PO theory (Drago 2005) Notice that it is commonly maintained that the *ad absurdum* proofs can all be translated into direct proofs (Gardiès 1991). Unfortunately, it is rarely remarked that this translation is possible only by applying classical logic to its conclusion, i.e. the absurdity of the negated thesis Ts that one wants to prove, $\neg\neg Ts$. Yet, as remarked in the above, the law of double negation is precisely what best distinguishes classical from most of non-classical logics, above all intuitionist logic; hence, precisely this last step, the translation of $\neg\neg Ts$ in Ts, is prohibited by intuitionist logic.

[...] [it] can likewise be *admitted* <u>without</u> leading to any <u>contradiction</u> in the results (Lobachevsky 1840, 19).

Kolmogorov on the alternative logic:

> The use of the principle of excluded middle <u>never</u> leads to a <u>contradiction</u> (Kolmogorov 1924-1925, 431).

Kleene's statement on Turing-Church's thesis:

> Every general recursive function <u>cannot</u> <u>conflict</u> with the intuitive notion which is supposed to complete [...] (Kleene 1952, 318-319).

Each author of these PO theories then deduces in classical logic all the relevant consequences from a predicate which is easily recognised to be the affirmative corresponding one to previous DNP. Respectively: "The maximum caloric yield of motive power is that produced in our first sequence of operations". "The new hypothesis can be consistently admitted." "The use of the principle of excluded middle is consistent." "Every general recursive function equates the corresponding intuitive function".

This author's move appears to be justified by his having collected by means of all possible arguments the maximum evidence for the final predicate. The move can be considered also as an application of the PSR: there is <u>no</u> reason <u>preventing</u> us from assuming as one more hypothesis the affirmative predicate corresponding to the final DNP.

Notice that the resulting affirmative predicate of a PO theory is in no way considered by the author to be an asserted cause, but rather a new hypothesis to be either tested against experimental data, or (as e.g. Lobachevsky did after proposition no. 22) added to the previous ones to derive from all them as many possible consequences in order to verify that the new hypothesis does not lead to (any) contradiction.

4   Logical Properties of the Principle of Sufficient Reason

By accepting previous Cassirer's invitation - to discover a new kind of logical reasoning by "filling up logic with the real contents of sciences" - I formalize the logical arguing of the scientific theories listed in previous section, and hence also the PSR, which governs this arguing.

Let us start by formalizing in logic Leibniz's basic distinction between contingent and necessary propositions. About their distinction Leibniz suggested a parallelism with the incommensurability relationship between rational and irrational numbers:

> [...] Just as [even] a large [rational] number contains another [irrational number] which is incommensurable with it, although even if one continues to infinity with a resolution one will never arrive at a common measure, so in the case of a contingent truth you will

never arrive at a demonstration [of it, as in the case of a necessary proposition] [...]. The sole difference is that in the case of such relations we can, none the less, establish demonstrations, by showing that the error involved is less that any assignable error; but in the case of the contingent truths not even this is conceded to a created mind (Leibniz 1975, 18; 1973, 97).

Cassirer added:

The abyss between the necessary and the contingent cannot be filled up. (Cassirer 1986, 404)

On this subject Leibniz (Leibniz 1686b, § 13) wrote a basic proposition:

a necessary proposition is one whose <u>contradictory</u> implies <u>contradiction</u> [...]. But in the case of contingent truths [...] the proposition [cannot] ever be reduced to an [...] identity (Leibniz 1961, 181-182; quotation from Leibniz 1973, 108-109]).

Given that Leibniz's proposition is a DNP, I suggest that this necessary/contingent divide corresponds to the divide between classical propositions and intuitionist propositions[8].

In order to translate this divide into formal terms, let us consider that a necessary proposition N is contradictory with $\neg N$ hence, a necessary proposition belongs to classical logic. On the other hand, a contingent proposition C is not in contradiction with $\neg$ C. Notice that the same holds true for its negative proposition $\neg$ C.; i.e. it is not contradictory with $\neg\neg$ C; at last, $\neg\neg C \neq C$ (otherwise, C would belong to classical logic); that is possible only if C itself is a DNP. Hence, a contingent proposition belongs to intuitionist logic. In fact, Leibniz represented through single proposition a mutual incommensurability between two entire theories, i.e. the old logical theory and the new one.

Since in previous sect. 3 we saw that the two kinds of logic govern respectively the two organizations of a scientific theory, this logical divide is the same as the AO/PO divide.

In sect. 3 we also saw that a characteristic feature of a PO theory is to include, beyond DNPs, ad absurdum proofs; each of them concludes a universal predicate, again a DNP. Let us translate into logical formulas the above-listed four instances of this predicate. By calling $T(x)$ a quantifier-free proposition concerning the variable x, we can represent each of the first three predicates by means of the logical formula $\neg \exists x \neg T(x)$; and the last predicate through the formula $\forall x \; \neg\neg T(x)$. Dummett (Dummett 1977, p. 29) offered a table summarizing all relationships of equivalence and implications between two predicates belonging to intuitionist logic. (His

---

[8] To the latter ones belong the experimental propositions, owing to the inaccuracy of the results of all measurements; indeed their results are inaccurate rational numbers (which subsequently a theoretical physicist, in order to make use of the easier mathematical tool – continuous calculus –, changes into accurate real numbers).

notation is slightly different since he dismisses the commas of the argument of the proposition). In the left top case of the table, the above two formulas (the fifth one and the third one) are mutually equivalent. In sum, all predicates concluding OP theories have the same logical formula, which I call U^I (where the apex I stands for intuitionist).

As a subsequent step, the author of a PO theory, in order to then argue in the easier classical logic, changes U^I into the corresponding affirmative predicate U, whose logical formula is $\forall x\, f(x)$. What does this change amount to?

Let us remark that Dummett's above-mentioned table is composed of cases, corresponding to the four theses of Aristotle's square of opposition. There one sees that the predicate U^I implies the existential one, located under the former one; hence, the change of the former implies the change in the corresponding classical predicate of the latter one, whose negative predicate, being negated once only, is a classical predicate; in other words, the predicate located in the right case is also changed into the corresponding classical one E; and since it implies the predicate on its lower case, also the latter one is changed in the classical one. In sum, the change of U^I changes at the same time all intuitionist predicates on the same subject to the corresponding classical ones. Hence, the author's change of the final predicate of a PO theory allows him subsequently to reason about any classical predicate whose scientific content T(x) is the same as that of U.

Now we are in the position to analyse PSR in logical terms.

Notice that part (A) of the above quotation from Leibniz's PSR, "<u>Nothing</u> is <u>without</u> a reason", includes two negative words. Incidentally, almost all PSR's versions, which have been suggested by both Leibniz and other scholars, include two negative words. This fact qualifies part (A) of PSR as pertaining to contingent truths. Strangely enough, the literature on the PSR does not notice this logical fact[9].

In Leibniz's mind part (A) was a contingent proposition, whereas part (B) was clearly a necessary proposition. Part (C) states that a contingent proposition cannot imply a necessary proposition; i.e. the words ("although we are not […]") stress that it is not always possible to obtain part (B) ("[…] every truth has an a priori proof […]") from part (A). In other words, owing precisely to the limitation expressed by part (C), it is part (A) that holds true, not part (B). Hence, the relationship between (A) and (B) is not a logical implication. Anticipating the birth of intuitionist logic, Leibniz's addition of part (C) explains that here the double negation law fails; thus, part (A) is essentially a DNP.

Unfortunately, this fact is ignored, so that most scholars dismiss the last part (C) of Leibniz's version of PSR referred to above.

As a further characterization of part (A) of PSR let us also remark that the application of PSR within an AO theory is useless. Given that an axiom is a primitive idea, one cannot demand its logical reason. Moreover, the propositions which are classically deduced from axioms are necessary propositions whose reasons are known a priori, i.e. their reasons are the respective classical derivations

---

[9] Only Heidegger (Heidegger 1996, end page of Lesson 5) remarks that in the classical version of PSR one can stress the two words "<u>Nothing</u>" and "<u>without</u>" and then the words "All" and "has" of the corresponding affirmative version. But he went no further in analysing what the logical relationship between these two propositions is.

from the given axioms. Hence in an AO theory the application of PSR adds nothing. This logical situation is well illustrated in philosophical terms by Sleigh:

> Leibniz regarded [PSR] as providing a deep semantic analysis of contingent truths [i.e. DNPs], just as he regarded the principle of contradiction as providing a deep semantic analysis of necessary truths (Sleigh 1983, 206).

In order to understand this "deep semantic analysis" provided by PSR let us consider its role inside within the alternative kind of organization of scientific theory, i.e. a PO theory.

Being a DNP, part (A) of PSR belongs – equally to all basic propositions of a PO theory – to intuitionist logic. The logical formula of part (A) is: $\neg\exists x\neg R(x)$. As remarked in the above, in intuitionist logic this predicate is equivalent to the predicate $\forall x \neg\neg R(x)$. Remarkably, these two equivalent formulas are the same as those for the final predicates of PO theories. This fact suggests an accurate context for the application of the PSR. The aim attributed to PSR, i.e. to discover reasons or explanations, appropriately represents the final step of the arguments of a PO theory aimed at discovering a solution to a given problem.

From part (A) of PSR Leibniz wants to achieve part (B), whose logical formula is $\forall x\, Rf(x)$, i.e. the formula obtained by dropping the two negations of the formula $\forall x \neg\neg R(x)$ expressing part (A). We saw in sect. 3 that each author of a PO theory wanted to overcome the same logical gap dividing (B) from (A). The above quotation of Leibniz's PSR, when translated into logical terms, mirrors the previously illustrated change of the last predicate of a PO theory.

Since there is no law of either classical logic or intuitionist logic allowing this change, such a change occurs only by means of *a methodological principle*, that kind of principle that Wiggins rightly maintained was a characteristic feature of PSR.

In agreement with Wiggins' methodological interpretation, this is the very role played by the PSR, instead of – as it is commonly considered – either stating proposition (A) only, or affirming proposition (B) only, to be then investigated analytically.

Hence, the discussions about what reason part (B) of the PSR suggests, are pointless. The PSR does not suggest a new, surprising idea, although it adds a novelty; it is rather the assumption of a relationship the evidence for which is not assured, but only subjectively evaluated as sufficient for exploring the consistency of a new deductive development. Moreover, the PSR does not imply the search for an antecedent (as is the search for a cause in an AO theory); rather it is a forward step for overcoming a gap, the gap separating a contingent proposition from the corresponding necessary one. In other words, the PSR does not represent a new certainty, but a rule regulating a process of reasoning.

Let us remark that, owing to the same logical reasons concerning the above translation of U^I, the application of PSR translates all other intuitionist predicates into classical logic, whatever their arguments are.

Hence, PSR does not pertain to a specific logic, but to the relationship between classical and intuitionist logic; it is a logical translation between two kinds of logic[10]. It performs a logical translation in the inverse direction of the negative translation from classical logic to intuitionist logic (Troelstra and van Dalen, 57-59)[11].

In particular, an author of a PO theory - who translates the final predicate in the corresponding affirmative one and then proceeds by applying classical logic - is justified in performing this translation by an application of the PSR. In this sense the author's strategy is inspired by the PSR.

But the PSR does not depend on any specific scientific argument of a PO theory; i.e., it causes the logical translation of any final predicate of a PO theory; or equally, it changes the logic of all the predicates in all scientific arguments, since the PSR generalizes the change in logic of any U^1. Hence, rightly he PRS is named a "principle" in the sense that its level of generality is greater than a single theory; it constitutes an architectonic principle for all PO theories.

In sum, the application of the PSR pertains to a kind of reasoning which is logically rational, but it does not pertain to any kind of logic; precisely as Leibniz stated, it belongs to the universal principles of thinking[12]. No surprise if in the past the interpreters of this principle met with great difficulties in conceiving this super-logical role played by the PSR.

A similar principle is Markov's principle stating: $\neg\neg \exists x\ M(x) \Rightarrow \exists x\ M(x)$. Since the existential quantifier $\exists x$ is implied by the corresponding universal quantifier $\forall x$ Markov's principle does not translate the entire logic[13].

Nor is "Peirce's principle of abduction": $\exists x\ \neg\neg P(x) \Rightarrow \exists x\ P(x)$, which I recognised in Peirce's writings, sufficient (Drago 2013). Since its first part implies the first part of Peirce's principle (see the second and the third line in the left bottom case of Dummett's table).

In conclusion, among the "principles", translating intuitionist predicates into the corresponding classical predicates, the PSR is the most relevant, since it alone translates all the other predicates into the corresponding classical ones.

5 Markov's Two Constraints on the Application of Leibniz' PSR

The analysis of PO scientific theories provides one more piece of information about the change of the final predicate. In his PO theory of computable numbers

---

[10] This point is discussed by several scholars in terms of the attainability or not of the last term approached by an infinite series of reasons of a contingent proposition (e.g. see Sleigh 1983, 206ff); in logical terms, this point was decided by a paper by Gödel (Gödel 1933), which proved that there exists an infinite series of kinds of logic between the intuitionist one and the classical one.

[11] Notice that this way of translating intuitionist into classical logic by dropping the two negations is the second way out of the four ones (Tennant 1990, 57; Hand 1999, 188). However, these authors do not mention the PSR.

[12] Already Enriques (Enriques 1909, 19) stated that the PSR "is located not in [a single] logic (considered in the formal sense), but in the theory of knowledge."

[13] However, if one obtains the completeness of intuitionist predicate logic, this translation becomes a valid intuitionist implication (Dummett 1977, 247, 256).

Markov introduced what at present is called his "principle". He admitted that this principle cannot be accepted by the intuitionists; however, he justified it as being an application of the PSR ("I see no reasonable basis for rejecting it [= the resulting affirmative predicate]") by subjecting this move to two constraints on the predicate at issue: *i*) to be the result of an *ad absurdum* proof; *ii*) to be a decidable predicate[14]. (Markov1971, 5)

Actually, Markov's two constraints are fulfilled by the final predicate of any scientific PO theory, since *i*) the nature of scientific theory requires that its arguments have to rely on operative[15] - hence decidable – propositions (otherwise, to allow metaphysical ideas - e.g. the magic wand – into a PO theory makes it easy to solve any basic problem); *ii*) this final predicate of a PO theory results from (the last) *ad absurdum* proof.

Hence, when he appealed in philosophical terms to the PSR in order to translate his predicate, Markov merely described the two characteristic features of his scientific theory.

Incidentally, in such a way all the characteristic features of the model of a PO theory have been recognized. This result achieves the goal of a millennial search for discovering a formally detailed model of an organization of a scientific theory which is alternative to the Aristotelian, apodictic model (Beth 1959, I. 2).

6 About Traditional Applications of the PSR

As Boutroux and Couturat remarked (see the first quotation in Sect. 2) in Leibniz's writings sometimes the PSR "is applied to all truths", sometimes "to physical, metaphysical and moral truths" only. We have to conclude that, lacking the suitable formal tools, Leibniz's intuitive reasoning was insufficient to improve the distinction between contingent and necessary propositions and obtain a complete characterization of the PSR. In fact, it is notorious that his applications of the PSR are questionable.

Some of Leibniz's applications of the PSR concerned physical phenomena, likewise Archimedes applied it in order to base his statics on levers. Given that the propositions concerning phenomena are decidable, the applications of the PSR after proving *ad absurdum* theorems fulfil the two constraints. This conclusion also applies to physical symmetries, which are correctly considered as governed by the principle of sufficient reason (Morrison 1958)[16].

---

[14] Let us recall that a predicate P is called decidable when one can decide through an algorithm whether an element x belongs or not to the predicate P. Of course, this property holds true in the case the predicate P concerning an operational content since measurements can decide the question.

[15] Kant was close to imposing the requirement of operationalism, because he severely restricted the PSR to the case in which it is applied to the experience, otherwise in his opinion it leads to contradictory result (Kant 1787, "Second analogy of experience").

[16] In the literature some specific objections to the PSR have been suggested. E.g., Hume-Edwards' objection (Edwards 1959), which relies on the thesis that a bunch

A celebrated application by Leibniz of the PSR was to state the non-existence of in-discernibles. But here the scholars do not notice that Leibniz had previously referred to "absolute" beings:

> There are not in nature two real absolute beings which are in-discernibles. (Leibniz 1956, Fifth Paper, 21-26).

Notice that our operational means allow relative results only; hence, the word "absolute" stands for a negation: "not relative". Hence, the quotation is a DNP, where the word in-discernibles has to be understood as operatively decidable – as Leibniz stresses. By applying (apart from an *ad absurdum* proof) the PSR to the first two negations, we obtain: "Relative in-discernibles exist", which is a mere hypothesis to be tested against reality. If data, e.g. bosons in quantum mechanics, deny it, the hypothesis fails.

The same holds true for Leibniz's proposition: "No absolute existence of in-divisibles", i.e. a-toms". Yet, relative atoms exist, as chemistry and physics stated, although subsequently they proved to be further sub divisible. Hence, Leibniz's above quotations that apply PSR are valid methodological principles, but they are not certainly true.

When exported into a metaphysical realm the application of the PSR without adding the operationalism constraint represented an encouragement to speculate about new metaphysical characteristics of God. The most famous application by Leibniz of the PSR concerns the existence of God. In terms of the two above constraints, the first one, an *ad absurdum* proof, was suggested by Anselm, whereas the constraint of decidability does not apply. Or rather, it is obvious that a believer bases his religious tenets on (both intimate and social) experiences; which he can decide by means of his psychological feelings; in such a psychological sense the constraint of decidability holds true for him and hence an *ad absurdum* proof of the existence of God is valid. Yet, for an atheist disregarding religious experiences, the decidability constraint is not fulfilled; hence, he rejects the application of the PSR whatever *ad absurdum* proof he meets. In sum, this proof is valid only for confirming the persuasion of those who are already believers in God[17].

---

of contingent beings can explain their own existence; hence, they obtain this necessary proposition. But its thesis is not acceptable, since the DNP nature of all contingent propositions prevents us from obtaining a necessary proposition, i.e. a proposition in classical logic without the application of the PSR. One more objection comes from the indeterminism of quantum mechanics, which denies that there are reasons for some microscopic phenomena. But part (C) warns us that the determination of the reason is not always possible (at a given time and in a specific theoretical context).

[17] Rather it is an interesting fact that Cusanus accomplished his long search for a better name of God by suggesting the name *Possest*, i.e. *Posse* = *est* (Cusanus 1460). Being a modal word, *posse* is equivalent to a DNP (Chellas 1980, 76ff), as part A of the PSR is; the second part of God's name is *est*, i.e. the corresponding affirmative predicate of *posse*, as part (B) of the PSR corresponds to part (A). Hence, the name *Possest* represents the most general metaphysical idea suggesting the metaphysical application (without constraints) of the PSR (Drago 2015).

7   Comparison with Other Scholars' Versions

Does the above formalization of the PSR and its constraints represent a novelty in current literature on the PSR? In the following, I will review the most notable suggestions for achieving a clear-cut meaning of the PSR.

Let us start with *Heidegger's* 'strict' version of PSR:

> Nothing exists whose sufficient reason for existing cannot be rendered (Heidegger 1996, "Lecture Five", pp. 32-40).

This version is composed of two propositions; which however are equivalent to "There is nothing for which it is impossible to render a reason". This DNP is similar to part (A) of Leibniz's version. Yet, the addition of the verb "render" implies an actual result in all cases, that contradicts part (C) of the PSR.

Sleigh suggested a "deep form" of PSR:

> P is contingently true if and only if there is a proof sequence for p but there is no a priori proof of p (Sleigh 1983, 206).

Since the word "contingently" is a modal word, it is equivalent *via* the S4 model of modal logic, to a DNP (Chellas 1980, 76ff.), just as part (A) of the PSR is. This version refers to Leibniz's procedure for proving directly any proposition in either a finite ("a priori proof") or infinite sequence ("proof sequence") of steps. Nothing is said about whether the latter may be substituted by an *ad absurdum* proof, likewise in antiquity it was proved that the area of a circle is the result of the infinite sequence of areas of the approximating polygons. However, the result of such a version is the opposite of Leibniz's PSR, i.e. a contingent proposition instead of a necessary one, as part (B) is (which is excluded by the last part of Sleigh's version).

Wiggins suggested the following version of PSR:

> Nothing is true or is actual, but that there is a proper reason for this truth [a full explanation could be found for this] (Wiggins 1996, 118; Wiggins' insertion).

Here the word "proper" adds to the application of the PSR the two requirements, to be able to give a reason for p, and also a reason why p rather than q; and this latter reason has to be different from the reason for p. The first requirement is too strong with respect to decidability and the second one is too general with respect to the constraint of an *ad absurdum* proof.

Pruss suggested the following version:

> [...] necessarily, every contingently true proposition has an explanation (Pruss 2004, 168).

The modal word "contingently" being equivalent to a DNP, it, together with the remaining part, reiterates part (A) only of Leibniz's PSR. Yet, the premise "necessarily" contradicts part (C). Moreover, Pruss also prefers to "reason" the word "explanation"; which however does not improve our knowledge, but rather it makes more difficult to conceive the reason (a new model? A new theory?).

He suggested also a "restricted" version.

> RPSR If *p* is a true proposition and possibly *p* has an explanation, then *p* actually has an explanation (Pruss 2004, 167).

Here the premise "If *p* is true" attributes *p* to an AO theory; this trivializes the investigation into its reason, which is easy to discover, since it is derived from the axioms. The premise is acceptable only if for "true" one means a DNP: "<u>non-false</u>". Yet, these two negative words make superfluous the subsequent modal word "possibly", which alone is equivalent to a DNP. However, Pruss means by "possibly" the following: *i*) provided that an explanation exists ("If one knew that *p* could not have an explanation; the request for an explanation would be flawed"); *ii*) there exists "an explanation *q* in a possible world *w* at which *p* is true [and] [...] *q* explains *p*."; *iii*) "nothing is said about whether [*q* is] the same" as the explanation in the actual world. Of course, request *i*) a priori includes most of the answer. Moreover, request *iii*) seems to render useless *ii*) requirement. Last but not least, the words "then actually" seem to state a logical implication, which however is not proved at all.

In conclusion, all previous scholars have suggested new versions of the PSR by investigating it through intuitive reasoning. Although all they wanted to add some requirements to the application of the PSR, they ignored part (C) of the above-quoted version and moreover they did not suspect Markov's two constraints.

8   Conclusions

The above interpretation appears to be a continuation of what Cassirer stated a century ago, i.e. that Leibniz' intellectual effort began to establish the basic categories of our knowledge:

> The *Characteristica universalis* appeared to us as essentially an attempt for achieving *a complete system of categories*, in which one scientifically isolates and represents the basic relationships among the contents of the knowledge (Cassirer 1986, 401; *my emphasis*).

More specifically about logic he stated:

> I agree with Couturat in admitting that logic constituted the formal framework on which Leibniz wanted to build his system, yet I have to remark, on the other hand, that the matter for such a construction has to be taken from the inquiry on positive science (Cassirer 1952-1958, vol. II, V, II, fn. 1).

In the above we have seen that an accurate inspection of "positive science" suggested "a formal framework on which Leibniz" wanted "to build his system", whose basic logical principle was the PSR, by which he meant a well-defined principle of the same relevance as the non-contradiction principle ("Our reasonings are based on two great principles [...]"; Leibniz 1714, § 31). In the above we have seen that the PSR can be accurately formalized in mathematical logic, formalized as a change of the kind of logic, from the logic of possibility to the logic of certainty; a fact surprisingly ignored for three centuries.

In fact, it was scientific research that, by applying the PSR to the theories concerning the real world, suggested, through Markov, the suitable constraints which qualify the PSR as a valid principle of reason, thus leaving aside as mere guesses all applications of the PSR to the most celebrated, i.e. the metaphysical, subjects.

Acknowledgement

I am grateful to Prof. David Braithwaite for having revised my poor English and to an anonymous referee for some important suggestions. I Acknowledge also an anonymous referee who suggested a great improvement of the paper.

References

Beth, E. W. 1959. *Foundations of Mathematics*. Amsterdam: North-Holland.
Cassirer, E. 1927. *Individuum und Kosmos in der Philosophie der Renaissance*. Darmstadt: Wissenschaftliche Buchgesellschaft, 1963.
Cassirer, E. 1986. *Cartesio e Leibniz*. Bari: Laterza.
Cassirer, E. 1952-1958. *Storia della filosofia moderna*. Torino: G. Einaudi.
Chellas, B. F. 1980. *Modal logic*. Cambridge: Cambridge University Press.
Couturat, L. 1961. *La logique de Leibniz après des documents inédits* (orig. 1901), Hildesheim: Georg Olms.
Dascal, M. 2008. A two-Pronged Dialectic. In *Leibniz: What Kind of Rationalist?* Dascal, M. (Ed.), Berlin: Springer, pp. 37-72.
Drago, A. 1998. Il ruolo del principio di ragione sufficiente nella scienza secondo Federico Enriques. *Federico Enriques. Filosofia e storia del pensiero scientifico*. Livorno: Belforte, pp. 223-265.
Drago, A. 2005. La monade di Leibniz alla luce dello sviluppo della scienza moderna. In *Monadi e monadologie. Il mondo degli individui tra Bruno, Leibniz e Husserl*. Mannelli: Rubbettino, pp. 291-313.
Drago, A. 2010. Dialectics in Cusanus (1401-1464), Lanza del Vasto (1901-1981) and beyond. *Epistemologia* 33:305-328.
Drago, A. 2012. Pluralism in Logic: The Square of Opposition, Leibniz' Principle of Sufficient Reason and Markov's principle. In Béziau, J. Y. and Jacquette, D. (Eds.). *Around and Beyond the Square of Opposition*, Basel: Birkhaueser, pp. 175-189.
Drago, A. 2013a. A Logical Model of Peirce's Abduction as Suggested by Various Theories Concerning Unknown Entities. In Magnani, L. (Ed.). *Model-Based Reasoning in Science and Technology. Theoretical and Cognitive Issues.* Berlin: Springer, pp. 315-338.

Drago, A. 2013b. The emergence of two options from Einstein's first paper on quanta. In Pisano, R., Capecchi, D., and Lukesova, A. *Physics, Astronomy and Engineering. Critical Problems in the History of Science and Society* (eds). Siauliai: Scientia Socialis Press, pp. 227-234.

Drago, A. 2015. From Aristotle's square of opposition to the "Tri-Unity's concordance": Cusanus' non-classical arguing. In Béziau J.-Y. and Basti G. (Eds.) *The Square of Opposition: A Cornerstone of Thought*, Berlin: Springer, pp. 53-78.

Drago, A., Venezia A. 2007. Popper's falsificationism interpreted by non-classical Logic. *Epistemologia* 30:235-264.

Dummett, M. 1977. *Elements of Intuitionism*. Oxford: The Claredon University Press.

Enriques, F. 1909. Il principio di ragion sufficiente nella costruzione scientifica. *Scientia* 3(5):1-20.

Fichant, M. 1988. *Science et Metaphysique dans Descartes et Leibniz*. Paris: PUF.

Gillispie, C. C., Pisano, R. 2014 *Lazare and Sadi Carnot. A Scientific and Filial Relationship*. 2nd edition. Dordrecht: Springer.

Gardiès, J.-L. 1991. *Le raisonnement par l'absurde*. Paris: PUF.

Goedel, K. 1933. *Collected Works*. Oxford: The Oxford University Press.

Grize, J.-B. 1970. Logique. In: Piaget, J. (Ed.). *Logique et la connaissance scientifique*, dans *Encyclopédie de la Pléyade*. Paris: Gallimard, pp. 135-288.

Grzegorczyk, A. 1964. Philosophical plausible formal interpretation of intuitionist logic. *Indagationes Mathematicae* 26:596-601.

Hand, M. 1999. Antirealism and Falsity. In: Gabbay, D. M. and Wansing, H. (Eds.). *What is Negation?* Dordrecht: Kluwer, pp. 185-198.

Heidegger, M. 1996. *The Principle of Reason*, Indiana: The Indiana State University.

Horn, L. 2002. The Logic of Logical Double Negation. In: Kato, Y. (ed.), *Proceedings of the Sophia Symposium on Negation*. Tokyo: The Sophia University Press, pp. 79-112.

Horn, L. 2010. Multiple negations in English and other languages, *The Expression of Negation*, Mouton: de Gruyter, pp. 111-148.

Hughes, R. I. G. 1993. Theoretical explanation. In eds. French, P. A, et al., *Midwest Studies in Philosophy* 18:133-153.

Kant, I. 1781. *Kritik der Reine Vernuft*. Riga: The University of Riga Press.

Kleene, S. C. 1952. *Introduction to Metamathematics*. Princeton: Van Nostrand.

Lavoisier, A. L. 1862-92. *Oeuvres de Lavoisier*. Paris: Imprimerie Impériale, Tome 1.

Leibniz, G. W. 1686. Letters to Arnauld, July 4th-14th. A version can be found at: http://www.earlymoderntexts.com/assets/pdfs/leibniz1686a_2.pdf

Leibniz, G. W. 1714. *Monadologie*. Paris: 1714.

Leibniz, G. W. 1923. *Saemtliche Schriften und Briefe*. Herausgegeben von der Preußischen (jetzt Deutschen) Akademie der Wissenschaften, Darmstad und Berlin.

Leibniz, G. W. 1956. *The Leibniz-Clarke Correspondence*. Manchester: Manchester University Press.

Leibniz, G. W. 1975. Memoire: pour les personnes éclairés et de bonne intention, *Lettres et opuscule inédits de Leibniz* Paris: Libraire Philosophique de Ladrange, 1854, reprint by Hildesheim: G. Olms.

Markov, A. A. 1962. On Constructive Mathematics, *Trudy Math. Inst. Steklov*, 67:8-14; [English translation in: *American Mathematical Society Translations* [1971] 98(2):1-9.

Morrison, P. 1958. Approximate Nature of Physical Symmetries. *American Journal of Physics* 26:358.

Piro, F. 2002. *Spontaneità e ragion sufficiente. Determinismo e filosofia dell'azione in Leibniz.* Roma: Edizioni di storia e letteratura.

Popper, K. 1959. *The Logic of Scientific Discovery.* New York: Harper.

Prawitz, D., Melmnaas. P.-E. 1968. A survey of some connections between classical intuitionistic and minimal logic. In: Schmidt, H.A., Schütte K., and Thiele H.-J. (Eds.). *Contributions to Mathematical Logic,* Amsterdam: North-Holland, pp. 215-229.

Pruss, A. B. 2004. A restricted Principle of Sufficient Reason and the cosmological argument. *Religious Studies* 40:165-179.

Pruss, A. B. 2006. *The Principle of Sufficient Reason. A Reassessment.* Cambridge: The Cambridge University Press.

Quine, W. v. O. 1951. Main Trends in Recent Philosophy: Two Dogmas of Empiricism. *The Philosophical Review* 60:20–43.

Rescher, N. 1979. *Leibniz. An introduction to His Philosophy.* Oxford: Basil Blackwell.

Rescher, N. 1981. Leibniz and the Concept of a System. *Studia Leibnitiana* 13:114-122.

Ruin, H. 1998. Leibniz and Heidegger on Sufficient Reason. *Studia Leibnitiana* 30:49-67.

Sleigh, R. Jr. 1983. Leibniz on the two great principles of all our reasonings. In French, P. A. et al. eds. *Midwest Studies in Philosophy* 8:193–216.

Tennant, N. 1985. Minimal Logic is adequate to Popperian Science. *British Journal of Philosophy of science* 36:325-329.

Tennant, N. 1990. *Natural Logic*, Edinburgh, Edinburgh University Press.

Troelstra, A., van Dalen, D. 1988. *Constructivism in Mathematics*, vol. I. Amsterdam: North-Holland.

Wiggins, D. 1996. Sufficient Reason: a principle in diverse guises, both ancient and modern. *Acta Philosophica Fennica* 61:117-132.

Antonino Drago, Napoli Federico II University, Italy
drago@unina.it

Leibniz's Theory of Series

Giovanni Ferraro

Abstract. In this paper I examine the use of infinite series in Leibniz's work and highlight the conceptual aspects of the Leibnizian approach to the theory of series. Initially, I investigate how Leibniz used the limit process in summing series. Although he employed the sequences of partial sums to calculate the sum, he thought that the limit process did not define the sum but was only a method for operating upon series. In his opinion, the sum of a series was coincidental with the actual aggregation of all the infinite terms of series. I also point out that Leibniz obtained some theorems by using divergent series. I then clarify the relationship between calculus and the theory of series in Leibniz's work. Series theory originated in a geometrical context; the expansion of a geometric quantity had certain peculiarities with respect to the modern expansion of a function. In modern mathematics, functions are expanded under *a priori* conditions which are intrinsic to the nature of the function: for instance, for instance, $f(x) = \frac{a}{b+x}$ is equal to $\frac{a}{b} - \frac{ax}{b^2} + \frac{ax^2}{b^3} + \cdots$ under the condition $|x| < b$. However, Leibniz thought that the quantity $a:(b+c)$ could be expanded as both $\frac{a}{b} - \frac{ac}{b^2} + \frac{ac^2}{b^3} + \cdots$ and $\frac{a}{c} - \frac{ab}{c^2} + \frac{ab^2}{c^3} + \cdots$. Only, a posteriori, at the moment of the numerical determination of the given quantity, could one choose the development that was appropriate (in other words, convergent) in a certain interval. The conditions of convergence were viewed as extrinsic to the procedure of expansion: they were inherent to the specific geometric problem. In the last section of the present paper, I discuss Leibniz's attempt to ascribe a meaning to equation $1 - 1 + 1 - 1 + \ldots = \frac{1}{2}$.

Keywords: Leibniz, *Acta Eruditorum*, Series, Calculus, Convergence, Grandi, Bernoulli.

1 Introduction

Leibniz began to study higher mathematics in 1672 and, in a short time, became interested in infinite series. Between 1675 and 1676, he wrote a treatise, entitled *De quadratura arithmetica circuli ellipseos et hyperbolae cujus corollarium est trigonometria sine tabulis* (Leibniz 1993), which aimed to provide the quadrature of certain curves by means of series. Leibniz wrote at least six versions of this treatise, which nevertheless remained unpublished through his life[1]. In the years that followed, Leibniz published many of the results of the *De quadratura arithmetica* (but not the proofs and the solution methods) in *De vera proportione circuli ad quadratum circumscriptum in numeris rationalibus expressa* (1682) and in *Quadratura arithmetica communis sectionum conicarum* (1691). Then Leibniz wrote other important papers concerning series, in particular *Supplementum geometriae practicae sese ad problemata transcendentia extendens, ope novae methodi generalissimae per series infinitas* (1693) and *Epistola ad V. Cl. Christianum Wolfium um Grandi* (1710).

In this paper, I aim to investigate some aspects of the Leibnizian theory of series and to highlight the underlying principles and internal motivations[2]. In particular, I will clarify the relationship between convergence and a formal manipulation in Leibniz's work. I will show that, although Leibniz employed the sequences of partial sums to calculate the sum, he thought that the limit process did not define the sum but was only a method for operating upon series. In his opinion, the sum of a series was coincidental with the actual aggregation of all the infinite terms of series. Then, I will deal with the use of power series in Leibnizian calculus: I will show that the conditions of convergence of a certain series were viewed as extrinsic to the procedure of expansion of that series; rather such conditions were inherent to the specific geometric problem that originated that series. Finally, I will discuss Leibniz's attempt to ascribe a meaning to equation $1 - 1 + 1 - 1 + \ldots = \frac{1}{2}$.

2 Convergence and Formal Manipulation

I begin by illustrating Leibniz's notion of the sum, which relied on both an intuitive idea of convergence and a formal manipulation of series. As concerns convergence, my thesis can be summarised as follows:

(a) Leibniz had a *quantitative* conception of the sum of a series, namely, in his opinion, the equality

$$S = \sum_{k=0}^{\infty} a_k \tag{1}$$

was to be intended in the sense that $\sum_{k=0}^{\infty} a_k$ and S denoted the same quantity.

(b) The statement "$\sum_{k=0}^{\infty} a_k$ and S denoted the same quantity" meant that the series $\sum_{k=0}^{\infty} a_k$ was *convergent* to the quantity denoted by S.

---

[1] Only recently has Eberhard Knobloch published its last and most extensive version, which consists of 51 propositions and many scholia. For an analysis of the manuscript, see Knobloch (Knobloch 1989, 1991).

[2] For a general discussion of the theory of series in 17th and 18th century, see Ferraro (Ferraro 2000, 2002, 2007a, 2008a, b) and Ferraro and Panza (Ferraro and Panza 2003, 2012).

(c) However, the limit process did not define the sum but was only a method for operating upon series. Leibniz thought that that the sum of a series was coincidental with the actual aggregation of all the infinite terms of series. This means that, according to Leibniz, a series $\sum_{k=0}^{\infty} a_k$ converged to a quantity S if and only if the sequence of the nth sums

$$S_n = \sum_{k=0}^{n} a_k$$

approached S indefinitely when n increased so that S_n was ultimately equal to S, when n was infinite.

As concerns formal manipulation, which consisted in using procedures or rules applied to (finite or infinite) analytical expressions $A(x, y, ...), B(x, y, ...), ...$, regardless of the actual meaning of such expressions, I observe that it stemmed from the lack of any distinction between infinite and finite series and was based upon the following assumption :

If a rule R was valid for finite expressions or if a procedure P depended on a finite number n of steps $S_1, S_2, S_3, ..., S_n$, then it was legitimate to apply the rule R and the procedure P to infinite expressions and in an unending number of steps $S_1, S_2, S_3, ...$ [3]

For instance, since the following rules hold

$$\sum_{n=1}^{k} (a_n + b_n) = S_1 + S_2$$

$$\sum_{n=1}^{k} \tau a_n = \tau S_1$$

$$\left(\sum_{n=1}^{k} a_n\right)\left(\sum_{n=1}^{k} b_n\right) = S_1 S_2$$

$$\sum_{n=2}^{k} a_n = S_1 - a_1$$

$$\sum_{n=1}^{k} a_n = \sum_{n=1}^{k} a_{j(n)} = S_1,$$

---

[3] This principle was of great importance in the early series and largely shared by the main mathematicians (Ferraro 2007b, 2008a, b).

when S_1 and S_2 were the sum of the finite series $\sum_{n=1}^{k} a_n$ and $\sum_{n=1}^{k} a_n$, τ was a number and $a_{j(n)}$ was any rearrangement of the finite sequence a_n, it seemed natural to extend these rules to infinite series as well. Consequently, it was assumed that

$$\sum_{n=1}^{\infty} (a_n + b_n) = S_1 + S_2$$

$$\sum_{n=1}^{\infty} \tau a_n = \tau S_1$$

$$\left(\sum_{n=1}^{\infty} a_n\right)\left(\sum_{n=1}^{\infty} b_n\right) = S_1 S_2 \qquad (2)$$

$$\sum_{n=2}^{\infty} a_n = S_1 - a_1$$

$$\sum_{n=1}^{\infty} a_n = \sum_{n=1}^{\infty} a_{j(n)} = S_1,$$

when S_1 and S_2 were the sum of the infinite series $\sum_{n=1}^{\infty} a_n$ and $\sum_{n=1}^{\infty} a_n$, τ was a number and $a_{j(n)}$ was any rearrangement of the infinite sequence a_n.

To make these points clear, I examine some theorems which are found in Leibniz's writings. In *De quadratura arithmetica*, Leibniz expounded the so-called Mercator's method of the long division as follows. The fraction

$$\frac{a}{b+c}$$

where *a, b,* and *c are* positive quantities, can be write in the form

$$\frac{a}{b} - \frac{ac}{b^2 + bc}$$

If one replaces

$$ac, \quad b^2, \quad bc$$

by

$$A, \quad B, \quad C$$

in the equality

$$\frac{a}{b+c} = \frac{a}{b} - \frac{ac}{b^2 + bc}$$

one obtains
$$\frac{A}{B+C} = \frac{A}{B} - \frac{AC}{B^2+BC}$$

Hence
$$\frac{a}{b+c} = \frac{a}{b} - \frac{ac}{b^2} + \frac{AC}{B^2+BC}.$$

In a similar way, one derives
$$\frac{a}{b+c} = \frac{a}{b} - \frac{ac}{b^2} + \frac{ac^2}{b^3} + \frac{\alpha\gamma}{\beta+\beta\gamma}$$

Since this reasoning can be repeated indefinitely, one finds that the fraction $\frac{a}{b+c}$ is equal to an expression of the type
$$\frac{a}{b} - \frac{ac}{b^2} + \frac{ac^2}{b^3} + \cdots + R,$$

where R is what today is named the remainder of the series. Leibniz stated that the remainder decreases as desired (*quantumlibet decrescentes*) when the procedure is continued to infinity (Leibniz assumes $c<b$) and, therefore, at the infinity, $\frac{a}{b+c}$ becomes equal to the expansion
$$\frac{a}{b} - \frac{ac}{b^2} + \frac{ac^2}{b^3} + \cdots + R \qquad (3)$$

(see Leibniz 1993, pp. 76–77).

In other terms, the sum of the series $\frac{a}{b} - \frac{ac}{b^2} + \frac{ac^2}{b^3} + \cdots$ is $\frac{a}{b+c}$, since the remainder of the series decreases as desired and thus the series approached the sum increasingly: it is evident that Leibniz had a clear but intuitive idea of intuitive idea of convergence.

Other evidences of this intuitive notion of convergence can be found in his writings; however, these evidences also show that Leibniz's notion of the sum presented some aspects that made it different from the modern one.

The first aspect concerns the relationship between a series and its sum. According to Leibniz, a series could express the sum in an exact form (and not merely in an approximated form) if it is considered a whole and not a limit process. For instance, in *De vera proportione*[4], Leibniz justified that
$$1 - \frac{1}{3} + \frac{1}{5} - \frac{1}{7} + \cdots$$

was equal to $\frac{\pi}{4}$ by observing that:

---

[4] See Leibniz 1682, p. 44.

- ✓ if one took the first term of this series (namely, 1), then $\frac{\pi}{4}$ was approximated with an error less than 1/3,
- ✓ if one took the first two terms of this series (namely, $1 - \frac{1}{3}$), the error was less than 1/5,
- ✓ if one took the first three terms of this series ($1 - \frac{1}{3} + \frac{1}{5}$), the error was less than 1/7,
- ✓ etc.

According to Leibniz, if the series was continued to infinity, the error became less than any given quantity, however, only the whole series, i.e., the actual infinity of the terms of series, contains all approximations and expresses the exact value. Moreover, although one could express the sum of the series by means of no (rational) number and this series was produced in infinitum, one conceived the whole series in his mind, since the law of progression was known:

> [...] Therefore, the whole series contains all the approximations likewise either with the values just greater or just less: for it is understood that as it is made to continue for a long time, the error will be less than the fraction given, and thence smaller than any given quantity. Therefore, the whole series contains all the approximations likewise either with the values just greater or just less: for it is understood that as it is made to continue for a long time, the error will be less than the fraction given, and thence smaller than any given quantity. Whereby the whole series expresses the exact value. Although it shall not be possible to express the sum of this series by a single number, even if the series may be produced indefinitely, because yet it agrees with a single law of progressing, the whole may be understood well enough. For if indeed the circle is not commensurable with the square, it cannot be expressed by a single number, but it must be shown by necessity by a series in terms of rational numbers, just as with the diagonal of a square, cut both in the extreme and mean section, and which some call the divine ratio[5], and many other quantities, which are irrational[6].

---

[5] I.e., the golden section.

[6] Tota ergo series continet omnes appropinquationes simul sive valores justo majores et justo minores : prout enim longe continuata , intelligitur, erit error minor fractione data, ac proinde et minor data quavis quantitate. Quare tota series exactum exprimit valorem. Et licet uno numero summa ejus seriei exprimi non possit, et series in infinitum producatur, quoniam tamen una lege progressionis constat, tota satis mente percipitur. Nam siquidem circulus non est quadrato commensurabilis, non potest uno numero exprimi, sed in rationalibus necessario per seriem exhiberi debet, quemadmodum et diagonalis quadrati, et sectio extrema et media ratione facta, quam aliqui divinam vocant, aliaeque multae quantitates, quae sunt irrationales. (Leibniz (GMS, 5:120; see translation in B14b, pp. 3-4).

In other terms, *the sum of a series was achieved when the aggregate of the series* (namely, the infinite terms of the series) *was all together taken into account.* The partial sums could even be used to find the sum of a series or to prove a result; however, in order to sum a series, one had to reach *the ultimate value* of series.

According to Leibniz, a series $\sum_{k=0}^{n} a_k$ had the following property:

If the sequence of the nth sums $S_n = \sum_{k=0}^{n} a_k$ approaches S indefinitely when n increases, then the difference between S_n and S (in absolute value) becomes less than any given quantity.

However, this property was not thought to be the definition of the sum of the series; it was only a trivial consequence of the fact that the sequence S_n approaches S indefinitely[7].

Leibniz went on stating that a circle is equal to an infinite series in the same way as the segment of length 1 is equal to the sum of the segment $\frac{1}{2} + \frac{1}{4} + \frac{1}{8} + \frac{1}{16} + \frac{1}{32} + \frac{1}{64} + \cdots$ since (see Fig. 1) the segment $AB = 1$ could be obtained by the juxtaposition of the segments $AC = \frac{1}{2}$, $CD = \frac{1}{4}$, $DE = \frac{1}{8}$, ... Here is his words:

> Moreover, lest anyone little experienced in these matters should think, that a series constructed from an infinite number of terms should not to be equal to the circle, which is a finite quantity, it is required to know that for many series with an infinite number of terms, the sum to be in a finite quantity. For to put in place the easiest example, there

---

[7] While an infinite series was equal to a quantity exactly only if it was conceived globally, the limit process of the partial sums provided approximations of the quantity represented by the series. Leibniz stressed the importance of series as the instrument that made it possible for numerical results to be computed with as small a margin of error as desired (Leibniz–KQA, 79). According to Leibniz, if one moved from theoretical observations to practice, the results of *De quadratura arithmetica circuli ellipseos et hyperbolae cujus corollarium est trigonometria sine tabulis* (the title means: "On the arithmetical quadrature of the circle, the ellipse, and the hyperbola. A corollary is a trigonometry without tables") gives a significant indication of the main scope of the treatise) enabled trigonometric operations to be performed without tables with as small a margin of error as desired; he argued that this was an incredible sign of the power of geometry (Leibniz–KQA, 79). In a letter to Gallois written in December 1678, Leibniz wrote "I left my manuscript on the arithmetical quadrature in Paris. The theorems which are contained in it are considerable in theory and very useful in practice. Since if one memorises just two very simple progressions which I give and which are almost impossible to forget, once one has learnt them, all problems of trigonometry could easily be solved without tables, without instruments, and without books, with the exactness one wishes [...] Having some tables is a convenience, but not being able to solve problems without tables is a defect of science, for which I claim to have found a remedy" (Leibniz–GMS 1:186).

may be the series of the geometrical progression starting from one decreasing geometrically by two

$$\frac{1}{8}+\frac{1}{16}+\frac{1}{32}+\frac{1}{64} etc.$$

indefinitely, which still does not make more than 1. For added to the right line AB which shall be 1, there will be AC 1/2 , and on bisecting the remainder (CB) at D, you will have CD 1/4; and on bisecting the remainder (DB) at E, you will have DE 1/8; and on bisecting at F the remainder (EB), you will have EF 1/16; and thus by continuing without end, you will never advance to the end B[8].

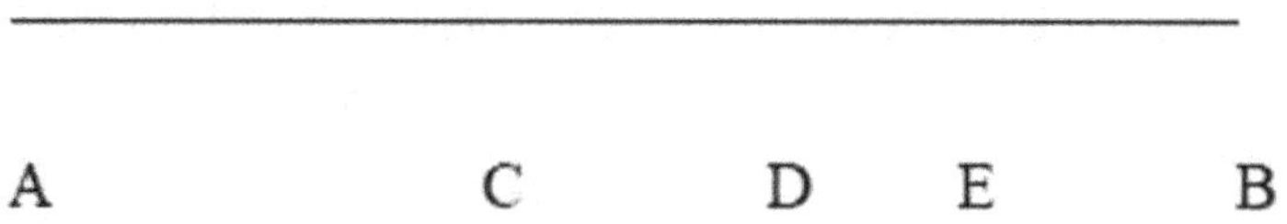

A   C   D   E   B

Fig. 1. Source: author's own drawing.

Another aspect that makes Leibniz's notion of the sum different from the modern one is the way Leibniz handled series. For instance, in *De quadratura arithmetica* (Leibniz 1993, pp. 88–90), Leibniz considered the hyperbola *GCH* (see Fig. 2) and observed that the area of the square *ABCD* is

---

[8] Ne quis autem in his parum versatus putet, seriem ex in…nitis terminis constantem non posse aequari circulo, qui est quantitas …nita, sciendum est, multas series numero terminorum infinitas esse in summa quantitates finitas. Exempli facillimi loco sit series ab unitate decrescens progressionis geometricae duplae
$$\frac{1}{8}+\frac{1}{16}+\frac{1}{32}+\frac{1}{64} etc.$$
in infinitum, quae tamen non facit plus quam 1. Nam in adjecta linea recta AB quae sit 1, erit AC 1/2, et residuum (CB) bisecando in D, habebis CD 1/4; et residuum (DB) bisecando in E, habebis DE 1/8; et residuum (EB) bisecando in F, habebis EF 1/16 ; et ita continuando sine fine, nunquam egredieris terminum B (Leibniz–GMS, 5:120–121; translation in: B14b, 4).

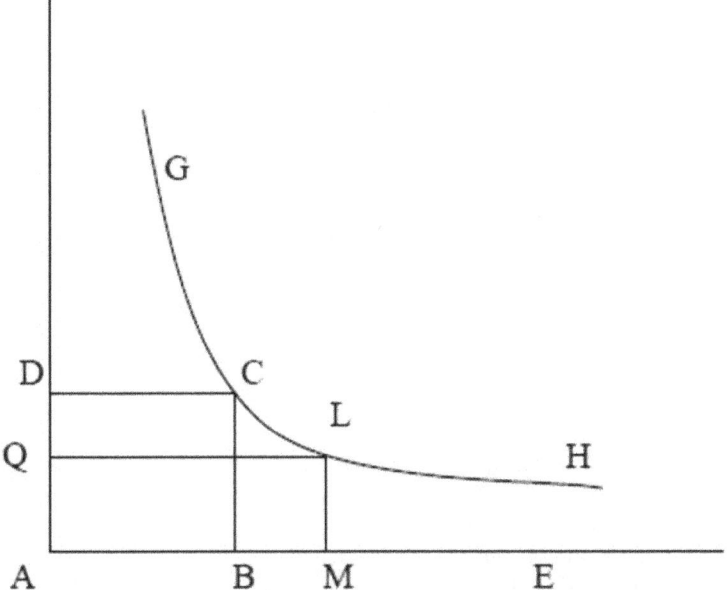

Fig. 2. Source: author's own drawing.

equal to the rectangle $AMLQ$ by the nature of the hyperbola. Therefore,

$$AM \cdot ML = AB^2$$

and

$$ML = \frac{AB^2}{AM} = \frac{AB^2}{AB + BM}$$

By using the expansion $\frac{a}{b} - \frac{ac}{b^2} + \frac{ac^2}{b^3} + \cdots$ of $\frac{a}{b+c}$, with $a = AB^2$, $b = AB$, and $c = BM$, one had

$$ML = AB - BM + \frac{BM^2}{AB} - \frac{BM^3}{AB^2} + \frac{BM^4}{AB^3} - \frac{BM^5}{AB^4} + \cdots$$

under the condition $BM < AB$.

At this point, Leibniz applied some theorems that he had proved earlier in *De quadratura arithmetica* and showed that the area of the figure $BEHC$ (which is equal – he said – to the sum all the segments ML from BC to EH) was given from the sum of the areas of the curves having ordinate equal to $\frac{BM^n}{AB^m}$ (with M varying from B to E). In this way, he obtained

$$\text{Area } BEHC = AB \cdot BE - \frac{BE^2}{2} + \frac{BE^3}{3AB} - \frac{BE^4}{4AB^2} + \frac{BM^5}{5AB^3} - \frac{BE^6}{6AB^4} + \cdots \quad (4)$$

In a modern form, if B is the origin of the axes, $AB = b, BE = c$, and $BM = x$, Leibniz's reasoning is equivalent to considering the hyperbola

$$y = \frac{b^2}{b+x},$$

expanding it into the power series

$$b - x + \frac{x^2}{b} - \frac{x^3}{b^2} + \frac{x^4}{b^3} - \frac{x^5}{b^4} + \cdots$$

and integrating it term by term in order to obtain

$$\text{Area } BEHC = \int_0^c \frac{b^2}{b+x} dx = bc - \frac{c^2}{2b} + \frac{c^3}{3b^2} - \frac{c^4}{4b^3} + \frac{c^5}{5b^4} - \frac{c^6}{6b^4} + \cdots$$

If one chooses BE appropriately so that $BE<AB$, then (4) makes it possible for any finite part of the given hyperbola to be squared: this means that, although the interval of convergence is finite, one can compute any finite part of the area by means of a change in the system of coordinates.

Even if Leibniz knew that (4) converged only if $BE<AB$, he posed $AB=BE$ and obtained

$$\text{Area } BEHC = AB^2 - \frac{AB^2}{2} + \frac{AB^2}{3} - \frac{AB^2}{4} + \frac{AB^2}{5} - \frac{AB^2}{6} + \cdots$$

and

$$\text{Area } BEHC = \left(AB^2 - \frac{AB^2}{2}\right) + \left(\frac{AB^2}{3} - \frac{AB^2}{4}\right)$$
$$+ \left(\frac{AB^2}{5} - \frac{AB^2}{6}\right) + \cdots$$
$$= \frac{AB^2}{2} + \frac{AB^2}{12} + \frac{AB^2}{30} + \cdots$$

Hence, he found that the area of $BEHC$ for $BE = \frac{1}{2}$ is equal to

$$\frac{1}{4}\log 2 = \frac{1}{8} + \frac{1}{48} + \frac{1}{120} + \cdots$$

At a first glance, one could think that, in deriving the last formula, Leibniz did not apply the logical rule according to which all the consequences of a theorem are subject to the same conditions of that theorem. However, it is more probable that Leibniz did not break this law; rather, he thought that one could operate on a series independently of the conditions of convergence. For Leibniz,

- one could use the symbol $a_1 - a_2 + a_3 - a_4 + \ldots$ independently of the possibility of associating it with a finite number or a finite quantity or

before proving that it was possible to associate a number with certain series;
- one could operate upon these series by assuming that if an algebraic operation (and also differentials or integrals, if one considers power series) could be performed upon finite series then it could be performed upon infinite series;
- what was important was that the results had a numerical or geometrical meaning, whereas the steps of the formal procedure that yielded a certain result might lack a numerical or geometrical meaning.

To illustrate these statements, I examine the summation of the series $\sum_{n=1}^{\infty} \frac{1}{(2n)^2-1}$. In his *De quadratura arithmetica* (Leibniz 1993, p. 82), Leibniz considered the series

$$\sum_{n=1}^{\infty} \frac{1}{2n+1}, \quad \sum_{n=1}^{\infty} \frac{1}{(2n)^2-1}, \quad \sum_{n=1}^{\infty} \frac{2}{(2n)^2-1}$$

and set

$$\frac{1}{1} + \frac{1}{3} + \frac{1}{5} + \frac{1}{7} + etc. = A$$

$$\frac{1}{3} + \frac{1}{15} + \frac{1}{35} + \frac{1}{63} + etc. = B$$

$$\frac{2}{3} + \frac{2}{15} + \frac{2}{35} + \frac{2}{63} + etc. = 2B$$

Subtracting term by term, he derived

$$C = \frac{1}{3} + \frac{1}{5} + \frac{1}{7} + \frac{1}{9} + etc. = A - 2B = A - 1$$

Leibniz handled $A - 2B$ and $A - 1$ as if A and B were numbers and

$$A - 2B = A - 1$$

an ordinary equation in the unknown B; thus, he derived $B = \frac{1}{2}$ and

$$\frac{1}{3} + \frac{1}{15} + \frac{1}{35} + \frac{1}{63} + etc. = \frac{1}{2} \tag{5}$$

Leibniz often used this method. For instance, he set[9]

$$\frac{1}{1} + \frac{1}{2} + \frac{1}{3} + \frac{1}{4} + etc. = A$$

$$\frac{1}{1} + \frac{1}{3} + \frac{1}{6} + \frac{1}{10} + etc. = 2B$$

$$\frac{1}{2} + \frac{1}{6} + \frac{1}{12} + \frac{1}{20} + etc. = B$$

and derived

$$\frac{1}{2} + \frac{1}{3} + \frac{1}{4} + \frac{1}{5} + etc. = A - B = A - 1$$

Hence, $B = 1$ and

$$\sum_{n=1}^{\infty} \frac{2}{2n+1} = 2$$

(Leibniz 1993, p. 83).

Nowadays we realize that this method worked well because Leibniz employed series $\sum_{n=1}^{\infty} a_n$ such that $\lim_{n \to \infty} a_n = 0$ and transformed them into series with the nth term of the type

$$c_n = a_n - a_{n+s},$$

where $s \geq 1$ is a fixed integer. Under this constraint, one has

$$\sum_{n=1}^{\infty} c_n = \sum_{n=0}^{\infty} (a_n - a_{n+s}) = \sum_{n=1}^{s} a_n - \lim_{k \to \infty} \sum_{n=1}^{s} a_{k+n} = \sum_{n=1}^{s} a_n.$$

Leibniz was aware, at least intuitively, that the condition

$$a_\infty = 0$$

was necessary to sum a series $\sum_{n=1}^{\infty} a_n$. This condition was explicitly formulated by Jacob Bernoulli, who showed what would happen if this was not satisfied. Bernoulli set

$$S = \frac{2a}{c} + \frac{3a}{2c} + \frac{4a}{3c} + \frac{5a}{4c} + \cdots$$

and

---

[9] Leibniz knew that the harmonic series was divergent (Leibniz–KQA, pp. 103-104).

$$T = \frac{3a}{2c} + \frac{4a}{3c} + \frac{5a}{4c} + \cdots$$

and observed that the application of the above method led to

$$Q = S - T = \frac{2a}{c},$$

whereas the sum is

$$Q = S - T = \left(\frac{2a}{c} + \frac{3a}{2c} + \frac{4a}{3c} + \frac{5a}{4c} + \cdots\right) - \left(\frac{3a}{2c} + \frac{4a}{3c} + \frac{5a}{4c} + \cdots\right)$$
$$= \left(\frac{2a}{c} - \frac{3a}{2c}\right) + \left(\frac{3a}{2c} - \frac{4a}{3c}\right) + \left(\frac{4a}{3c} - \frac{5a}{4c}\right) + \cdots$$
$$= \frac{a}{2c} + \frac{a}{6c} + \frac{a}{12c} + \cdots = \frac{a}{c}.$$

Bernoulli explained that the method failed because Q was equal to the first term of S (namely $\frac{2a}{c}$) minus the last term of T, which is not 0, but equals $\frac{a}{c}$ (Bernoulli 1689–1704, pp. 252–262).

One could think that the use of divergent series in the above proofs was due to difficulties in symbolism or an inadequate formalization of the notion of convergence. In reality, Leibniz thought that series could be handled independently of any preliminary meaning that might be attributed to them. For instance, in *De vera proportione* (Leibniz 1682, p. 121), Leibniz disregarded the fact that

$$\sum_{k=0}^{\infty} \frac{1}{1+4k}$$

and

$$\sum_{k=0}^{\infty} \frac{1}{3+4k}$$

were divergent and stated

$$\frac{\pi}{4} = \frac{1}{1} + \frac{1}{5} + \frac{1}{9} + \frac{1}{13} + \cdots - \left(\frac{1}{3} + \frac{1}{7} + \frac{1}{11} + \frac{1}{15} + \cdots\right) \qquad (6)$$

Leibniz proved this formula by rearranging the series $1 - \frac{1}{3} + \frac{1}{5} - \frac{1}{7} + \cdots$ in order to obtain (6) (Leibniz 1993, p. 81).[10]

---

[10] Equation (6) can be interpreted in the sense that the difference between

$$Q_1(i) = \sum_{k=0}^{i} \frac{1}{1+4k}$$

The famous convergence criterion for alternating series is another example of how Leibniz's conception differed from the modern one. In 1713, Leibniz formulated it thus:

> If the terms of a series are continuously decreasing and, alternatively, positive and negative, then the series is "advergent" (namely, convergent)[11].

One can translate this statement into modern terms, by stating that if a_n is a decreasing sequence, $a_n > 0$, and a_n goes to 0 when $n \to \infty$, then the series $\sum_{n=1}^{\infty}(-1)^n a_n$ is convergent.

Leibniz's proof runs as follows. First, even if he had not demonstrated the series $\sum_{n=1}^{\infty}(-1)^n a_n$ was convergent, he set

$$S = \sum_{n=1}^{\infty}(-1)^n a_n$$

Then, he stated that

$$S_{2n-1} = a_1 - a_2 + a_3 - a_4 + \cdots + a_{2n-1}$$
$$> a_1 - a_2 + a_3 - a_4 + \cdots + a_{2n-1} + (-a_{2n} + a_{2n+1})$$
$$+ (-a_{2n+2} + a_{2n+3}) + \cdots = S$$

(S_n is the nth sum) since $a_n \geq a_{n+1}$. In the same way, Leibniz asserted that

$$S_{2n} = a_1 - a_2 + a_3 - a_4 + \cdots - a_{2n}$$
$$> a_1 - a_2 + a_3 - a_4 + \cdots - a_{2n} + (a_{2n+1} - a_{2n+2})$$
$$+ (a_{2n+3} - a_{2n+4}) + \cdots = S$$

At this point, he stated that S was finite, since
$$S_{2n} < S < S_{2n+1}.$$

Finally, Leibniz observed that

and

$$Q_2(i) = \sum_{k=0}^{i} \frac{1}{3+4k}$$

approaches $\frac{\pi}{4}$ closer and closer and it is ultimately

$$Q_1(i) - Q_2(i) = \frac{\pi}{4}$$

Today we should say that the sums $Q_1(i)$ and $Q_2(i)$ have the same asymptotic behavior (with respect to $\log i$).

[11] See Leibniz GMS, 3:926. A first version of Leibniz's criterion is found in *De quadraturae arithmetica* (Leibniz–KQA, p. 115). On Leibniz's criterion, see Knobloch (Knobloch 1993).

$$S - S_{2n} < a_{2n+1} \text{ and } S_{2n-1} - S < a_{2n}$$

and, therefore, the difference between the finite series s_n and the infinite series S could be made smaller than a_{n+1}. In other terms, it became as small as desired, provided one considered a sufficient number of terms.

Today, the statement "S is the sum of the series $\sum_{n=1}^{\infty}(-1)^n a_n$" means that the difference between s_n and S is less than any quantity when n goes to infinity. Therefore, if we examine the Leibniz's demonstration from a modern point of view, it is a vicious circle. In reality, to understand Leibniz's reasoning, we must admit that, in his opinion, the sum S of the series $\sum_{n=1}^{\infty}(-1)^n a_n$ was not defined by

$$S = \lim_{m \to \infty} \sum_{n=1}^{m}(-1)^n a_n$$

For Leibniz, the sum S of a series was coincidental with the actual aggregation of all the infinite terms of series. The fact that this sum S is the value to which the series $a_1 - a_2 + a_3 - a_4 + \cdots$ approaches closer and closer had to be proved.

3   Power Series

In the second part of this paper, my aim is to clarify the relationship between the theory of power series and the calculus in Leibniz's work. Power series theory originated in a geometrical context: power series were always derived as the expansion of the analytical expression of given curves and were used to obtain geometric results. An example is the above-illustrated quadrature of hyperbola. However, it is to be noted that the way Leibniz expanded a quantity representing a curve differs from the modern expansion of a function. In modern mathematics, functions are expanded under a priori conditions, which are intrinsic to the nature of the function: for instance,

$$f(x) = \frac{a}{b+x}$$

is equal to

$$\frac{a}{b} - \frac{ax}{b^2} + \frac{ax^2}{b^3} + \cdots$$

under the condition $|x| < b$. Given a function $y = f(x)$, we know from the start the constants, the dependent variable and the independent variable. Leibniz did not consider a function $y = f(x)$; he reasoned upon a certain figure F and adapted an appropriate system of coordinates (origins, unit, axes) to the figure F. According to the specific circumstances, he chose what were to be considered as variables and what were to be considered as constants. To compute an area or length connected to the figure F, the series that expressed the area or length had to be convergent, but convergence could be obtained simply by changing constants into variables, modifying the value of the constant in an appropriate way, or choosing between different expansions regarding the given quantities. Thus, Leibniz derived two different expansions of $\frac{a}{b+c}$: in fact, the quantity $\frac{a}{b+c}$ could be expanded as both

$$\frac{a}{b} - \frac{ac}{b^2} + \frac{ac^2}{b^3} + \cdots$$

and

$$\frac{a}{c} - \frac{ab}{c^2} + \frac{ab^2}{c^3} + \cdots \qquad (7)$$

(Leibniz 1993, pp. 76–77). Expansion (7) was obtained by reversing b by c in the above reasoning. Leibniz thought that there was no reason to refuse one of two expansions a priori. Rather, a posteriori, one could choose between two different developments according to a given geometric situation. Leibniz (*Ibidem*) applied these two expansions to the curve represented by the equation

$$z = \frac{t^2}{1+t^2}$$

In modern symbols, his results can be expressed as follows. Expansion (3) yielded

$$\frac{t^2}{1+t^2} = t^2 - t^4 + t^6 - t^8 + \cdots$$

Hence, by integrating term-by-term, one obtained

$$t - \arctan t = \frac{t^3}{3} - \frac{t^5}{5} + \frac{t^7}{7} - \cdots$$

$$\arctan t = t - \frac{t^3}{3} + \frac{t^5}{5} - \frac{t^7}{7} - \cdots \qquad (8)$$

If one applied (7), one had

$$\frac{t^2}{1+t^2} = 1 - \frac{1}{t^2} + \frac{1}{t^4} - \frac{1}{t^6} + \cdots$$

By integrating

$$t - \arctan t = t - \frac{1}{t} + \frac{1}{3t^3} - \frac{1}{5t^5} + \frac{1}{7t^7} - \cdots$$

and

$$\arctan t = \frac{1}{t} - \frac{1}{3t^3} + \frac{1}{5t^5} - \frac{1}{7t^7} + \cdots \qquad (9)$$

According to Leibniz, formula (8) served when t was less than 1; instead, formula (9) served when t was greater than 1 (*prior servit cum t minor quam 1, posterior servit cum major est 1*). Leibniz (*Ibidem*) explicitly mentioned Newton, who gave a similar example in his writings. Indeed, in order to find the area enclosed by the curve

$$y = \frac{1}{1+x^2}$$

and the axis of abscissa, Newton (OO, 1:264) observed that

$$1 : (1+x^2)$$

was equal to $1 - x^2 + x^4 - x^6 + \cdots$ by continued division. By integrating term by term, it was easy to express the area by the series

$$x - \frac{1}{3}x^3 + \frac{1}{5}x^5 - \frac{1}{7}x^7 + \cdots$$

According to Newton, however, one could also divide

$$1 : (x^2 + 1).$$

In this case, the result was $x^{-2} - x^{-4} + x^{-6} + \cdots$ and the area was given From

$$-x^{-1} + \frac{1}{3}x^{-3} - \frac{1}{5}x^{-5} + \cdots$$

The former expansion, Newton said, was to be used when x was small (*satis parva*) enough whereas the latter was to be used when x was large enough (*satis magna*). The difference of the developments $1:(1+x^2)$ and $1:(x^2+1)$ was not viewed as a defect but as a useful tool for facilitating the computation of areas.

A preliminary analysis of convergence seemed not only superfluous but even counterproductive. It was preferable to expand a series formally: after determining the coefficients and form of development, one verified if the interval of convergence was suitable for the specific problem and chose the series that fitted it best. Convergence concerned the moment of the application, namely when one computed the values of the given series in order to find the values of a certain geometrical quantity. It was the geometrical context of *De quadratura* (and more generally of the early calculus) that made this approach possible. Another example of the importance of geometric references (and, in particular, of diagrammatic representations) in Leibniz's theory of series concerns the expansion of logarithms, which modernly can be written as

$$\log(1+x) = x - \frac{x^2}{2} + \frac{x^3}{3} - \cdots \qquad (10)$$

Leibniz showed (Leibniz 1993, 94–101) that

$$\log_c \frac{AX}{AB} = \log_c \frac{AB + BX}{AB} = \log_c \frac{b+n}{b}$$

and

$$\log_c \frac{AZ}{AB} = \log_c \frac{AB-BZ}{AB} = \log_c \frac{b-m}{b}$$

(where $AB=b$; $BX=n$; $BZ=m$) were proportional to the decreasing series

$$n - \frac{n^2}{2b} + \frac{n^3}{3b^2} - \frac{n^4}{4b^3} + \frac{n^5}{5b^4} - \frac{n^6}{6b^5} + \cdots \qquad (11)$$

and

$$m + \frac{m^2}{2b} + \frac{m^3}{3b^2} + \frac{m^4}{4b^3} + \frac{m^5}{5b^4} + \frac{m^6}{6b^5} + \cdots \qquad (12)$$

respectively (Fig. 3).

On p. 121 of *De quadratura*, Leibniz observed that if $n > b$, then the series (11) was not decreasing. However, in order to compute $\log_c \frac{b+n}{b}$, namely to calculate the area *BXLK*, one could take the figure *BKHZ* such that the area *BXLK* was equal to *BKHZ* and calculate the area *BKMZ* (which is equal to $\log_c \frac{b-m}{b}$). Indeed, if $n>b$, then $m<b$, and (12) was decreasing.

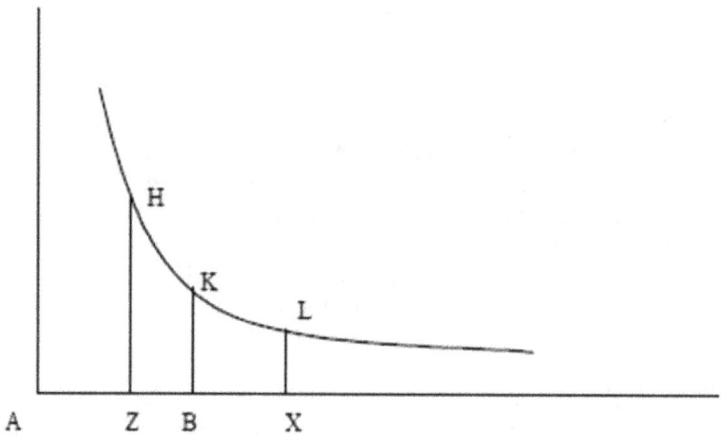

Fig. 3. Source: author's own drawing.

In *Quadratura arithmetica communis sectionum conicarum*, a short paper written in 1691, Leibniz published his main results on the series expansions of geometrical quantities. In this article, apart from the expansions of arctangent and logarithm (see Eqs. (8) and (10)), he provided the expansions of sine, cosine, and exponential quantities:

$$\sin x = x - \frac{1}{3!}x^3 + \frac{1}{5!}x^5 - \cdots,$$

$$1 - \cos x = \frac{1}{2!}x^2 - \frac{1}{4!}x^4 - \cdots,$$

$$e^x - 1 = x + \frac{1}{2!}x^2 + \frac{1}{3!}x^3 + \cdots.$$

Later, in *Supplementum geometriae praticae* (1693), Leibniz presented a method that made it possible to solve differential equations using series. This method, which is known today as the method of indeterminate coefficients, is based upon the following principle:

A series

$$\sum_{k=0}^{\infty} b_k x^{\alpha_k}$$

is equal to 0 for every k on an interval I if and only if all the coefficients b_k (b_k, $k = 0,1,2, ...$) are separately equal to zero.

Leibniz illustrated the method by means of several examples. I will give two of them here. The first again concerns the expansion of logarithm. In *Supplementum geometriae practicae* (Leibniz 1693, p. 286), Leibniz stated that if

$$y = a \log \frac{a+x}{a},$$

then

$$dy = \frac{a\,dx}{a+x}.$$

Hence,

$$a\frac{dy}{dx} + x\frac{dy}{dx} - a = 0. \tag{13}$$

He set

$$y = Bx + Cx^2 + Dx^3 + \cdots$$

(he assumed the arbitrary constant to be equal 0). He differentiated term by term and obtained

$$\frac{dy}{dx} = B + 2Cx + 3Dx^2 + \cdots \tag{14}$$

By replacing (14) into (13), he derived

$$(aB - a) + (2aC + B)x + (3aD + 2C)x^2 + (4aE + 3D)x^3 + \cdots = 0.$$

Equating the coefficients to zero, he obtained

$$aB - a = 0$$
$$2aC + B = 0$$
$$3aD + 2C = 0$$
$$4aE + 3D = 0$$
$$\cdots$$

Hence,

$$B = 1$$
$$C = -\frac{1}{2a}$$
$$D = \frac{1}{3a^2}$$
$$E = -\frac{1}{4a^3}$$
$$\cdots$$

and

$$y = x - \frac{x^2}{2a} + \frac{x^3}{3a^2} - \frac{x^4}{4a^3} + \cdots \tag{15}$$

The second example concerns the equation

$$a^2 dx^2 = a^2 dy^2 + y^2 dx^2. \tag{16}$$

This equation has an easily understood geometric meaning. Indeed, given any circle, if the sine, the arc and the radius are denoted by y, x, and a, then (16) represents the sine in terms of the arc, and radius. In *Supplementum geometriae practicae* (Leibniz, 1693, p. 287), Leibniz differentiated (16) and derived

$$0 = 2a^2 dy d^2 y + 2y dy d^2 x$$

(he supposed that dx was a constant, namely, in more modern terms, that x was the independent variable). Hence,

$$a^2 \frac{d^2y}{d^2x} + y = 0. \tag{17}$$

Leibniz set
$$y = Bx + Cx^3 + Dx^5 + Ex^7 + \cdots$$
and obtained
$$\frac{d^2y}{d^2x} = 2 \cdot 3Cx + 4 \cdot 5Dx^3 + 6 \cdot 7Ex^5 + \cdots. \tag{18}$$

Substituting (18) into (17), he had

$$(2 \cdot 3a^2C + B)x + (4 \cdot 5a^2D + C)x^3 + (6 \cdot 7a^2E + D)x^5 + \cdots = 0.$$

By applying the principle of indeterminate coefficients and assuming $B=1$, he obtained

$$B = 1, \quad C = -\frac{1}{2 \cdot 3a^2}, \quad D = -\frac{C}{4 \cdot 5a^2}, \quad E = -\frac{D}{6 \cdot 7a^2}, \quad \ldots,$$

and

$$y = x - \frac{x^3}{2 \cdot 3a^2} + \frac{x^3}{2 \cdot 3 \cdot 4 \cdot 5a^4} - \frac{x^7}{2 \cdot 3 \cdot 4 \cdot 5 \cdot 6 \cdot 7a^6} + \cdots. \tag{19}$$

In *Supplementum geometriae praticae*, Leibniz again stated that a series provided the exact solution to a differential equation only if it was considered as a whole; however, if one considered the partial sums of the series, then one obtained a solution that could be approximated as desired (Leibniz 1693, 286). Leibniz made no further mention of convergence in this article, unlike in *De quadratura* and also *Quadratura arithmetica communis sectionum conicarum* (Leibniz 1691).

4   The Sum of a Divergent Series and Concluding Remarks

In the final part of the present paper, I briefly consider the problem of Grandi's series $1 - 1 + 1 - 1 + \cdots$. We saw that Leibniz's conception admitted the possibility of using divergent series as a means of deriving information about certain quantities; however, he did not consider the problem of summing a divergent series, namely the problem of effectively associating a quantity with a divergent series. It seemed obvious that a series such as $1 - 1 + 1 - 1 + \cdots$ had no sum, as Jacob Bernoulli had explicitly noted in 1696, since the condition $a_\infty = 0$ was not verified.

The situation changed when Guido Grandi proved a geometric reasoning that can be translated and summarized in analytical terms as follows. Let us consider the equation

$$y = \frac{a^3}{a^2 + x^2}$$

of the witch of Agnesi (where a is the diameter of the circle generating the witch). If this equation is expanded into series, one obtains

$$\sum_{n=1}^{\infty}(-1)^k \frac{x^{2k}}{a^{2k-1}};$$

By setting $a = x = 1$, one has

$$1 - 1 + 1 - 1 + \cdots = \frac{1}{2} \qquad (20)$$

(see Grandi 1703, 29).

Series (20) posed a serious problem. Equation $1 - 1 + 1 - 1 + \cdots = \frac{1}{2}$. was quantitatively false, in the sense that $1 - 1 + 1 - 1 + \cdots$. did not express the same quantity as $\frac{1}{2}$; however, the derivation of (20) involved the law of continuity, which was a basic notion of Leibniz's mathematics. In the Preface to the *Nouveaux essais sur l'entendement humain*, Leibniz expressed this law by stating that nothing takes place suddenly and that nature never makes leaps
(A VI vi 56). In *Cum Prodiisset*, more precisely, Leibniz stated:

> In any supposed continuous transition, ending in any terminus, it is permissible to institute a general reasoning, in which the final terminus may also be included.[12]

Equation (20) was obtained by applying this principle and, therefore, Leibniz could not merely reject (20) by asserting that it was expressed a false relation between quantities. One can note that the idea that (20) was a false equation was the viewpoint of various mathematicians of the time. For example, on June 1, 1712, Hermann wrote to Leibniz, stating that Grandi's idea that infinite zeros could be gathered to form a finite quantity was clearly ridiculous.[13] In his opinion, the mistake was due to the fact that one dealt with series $1 - 1 + 1 - 1 + \cdots$.as though it were a convergent series (see Leibniz 1849-1863, 4:369–370). Some years later, Varignon (Varignon 1715, pp. 203–225) considered the equality

$$\frac{a}{m \pm n} = \sum_{i=0}^{\infty}(\mp 1)^i \frac{an^i}{m^{i+1}}$$

for $m>0$ and $n>00$ and stated that

---

[12] Proposito quocunque transitu continuo in aliquem terminum desinente, liceat raciocinationem communem instituere, qua ultimus terminus comprehendatur (Leibniz 1846, p. 40).

[13] Hermann referred to the fact that the sum could be read as $(1-1) + (1-1) + \cdots = 0 + 0 + 0 + \cdots$, and it seemed trivial that $0 + 0 + 0 + \cdots$ was equal to 0. Therefore equation $1 - 1 + 1 - 1 + \cdots = \frac{1}{2}$ gave rise the absurd equality $0 = \frac{1}{2}$.

- if $m > n$, then equality (21) was true;
- if $m < n$, then equality (21) was false;
- if $m = n$ and the denominator was the difference $m - n$, then equality (21) is simply $\infty = \infty$: it was true but does not provide information;
- if $m = n$ and the denominator was the sum $m + n$, then equality (21) was false.

Varignon's proof consisted in verifying that (21) was quantitatively valid in case 1 (and in case 3) and that the first and the second sides of (21) were not equal in the other cases. In particular, in case 4, Varignon stated that the sum of $1 - 1 + 1 - 1 + \cdots$ could never be $\frac{1}{2}$.

Leibniz (Leibniz 1713) attempted to justify equation (20). He stated that (20) could be derived by using the same procedure commonly employed to derive valid results about convergent series. He considered the following series:

$$\frac{1}{1+x} = 1 - x + x^2 - x^3 + \cdots$$

$$\frac{1}{1+x^2} = 1 - x^2 - x^4 + \cdots$$

$$\arctan x = \int_0^x \frac{1}{1+x^2} = x - \frac{x^3}{3} - \frac{x^5}{5} + \cdots$$

The third series can be derived from the first two. For $x=1$, the first (and second) series furnished

$$\frac{1}{2} = 1 - 1 + 1 - 1 + \cdots,$$

whereas the third gave

$$\frac{\pi}{4} = 1 - \frac{1}{3} + \frac{1}{5} - \frac{1}{7} + \cdots$$

(Leibniz 1849-1863, 5:382–383). For Leibniz, this was a good justification for $1 - 1 + 1 - 1 + \cdots = \frac{1}{2}$ (Ivi, 5:385).

Leibniz sought to provide other arguments for equation (20). Thus, he stated that the partial sums of the series $1 - 1 + 1 - 1 + \cdots$ were 0 or 1, attributing an even or odd number of terms to each sum, but if the series $1 - 1 + 1 - 1 + \cdots$ was continued up to the infinite, where the nature of the number vanished, then the possibility of distinguishing between odd and even vanished as well (Ivi, 1849-1863, 5:386). According to Leibniz, the sum of the "whole" series could not be either 0 or 1; since 0 and 1 had the same probability when the number of terms was finite, their arithmetic mean[14] provided the sum of the series when the number of term was infinite (*Ibidem*).

Finally, I observe that Leibniz refused to justify (20) in an exclusively formal manner and rejected certain methods that could be used to sum other divergent series such as (Leibniz 1860, pp. 143–149):

$$1 - 2 + 4 - 8 + 16 - 32 + \cdots$$

He proposed some criteria to examine the correctness of the sum of a series, such as the existence of a geometrical proof. It is precisely because there existed a geometric demonstration of $1 + 1 - 1 + \cdots = \frac{1}{2}$ that Leibniz believed that (20) was well-founded (Leibniz 1860, p. 147). This argument shows the strongly geometrical character of Leibnizian analysis. For Leibniz, indeed, an analytical expression, such as $\frac{1}{1+x}$ was the symbolical representation of a geometrical quantity and, consequently, an analytical argumentation had to be reduced to a geometrical proof. At the same time, this argument shows that, even if Leibniz's approach to series theory had a certain degree of formalism, it differed considerably from Euler's approach, which was prevalent in the 18th century.[15]

References

Abbreviations

[A] Leibniz, G. W. *Sämtliche Schriften und Briefe.* Darmstadt, Leipzig, Berlin, 1923.
[B14a] Leibniz, G. W. *The Arithmetic Quadrature of common Conic Sections... From Actis Erudit. Lips. April. 1691; Transl. with notes by Ian Bruce*, 2014. Retrieved via: http://www.17centurymaths.com/contents/Leibniz/ae5.pdf
[B14b] Leibniz, G. W. *Concerning the true proportions of a circle. From Actis Erudit. Lips. Feb. 1682; Transl. with notes by Ian Bruce*, 2014. Retrieved via: http://www.17centurymaths.com/20 contents/ Leibniz/ ae6.pdf
[LB] Leibniz, G.W.–Bernoulli, Johann *Virorum celeberr. Got. Gul. Leibnitii et Johan. Bernoulli Commercium philosophicarum et mathematicarum*, Lousannae et Genevae: M.M. Bousquet, 1745.
[CE] Newton, I. *Commercium epistolicum d. Johannis Collins et aliorium de analysi promota*, London: Iussu Soc. Regoae, typis Paeersoniani, 1712.
[OO] Newton, I. *Opera quae exstant omnia commentariis illustrabat Samuel Horsley*, Londini: J. Nichols, 1779–1885, 5 vols.

---

[14] On the modern use of the arithmetic mean in the summation of series, see Ferraro 1999.

[15] On Euler's series theory, see Ferraro 1998, 2007a, 2008a, 2008b, 2010, 2012.

Bernoulli, D. 1728. *Observationes de seriebus quae formantur ex additione vel subtractione quacunque terminorum se mutuo consequentium, Commentarii academiae scientiarum imperialis Petropolitanae* 3, pp. 85–100.
Bernoulli, D. 1771. *De summationibus serierumquarunduam incongrue veris earumque interpretatione atque usu, Novi Commentarii academiae scientiarum imperialis Petropolitanae,* 16, pp. 71–90 [Summarium pp. 12–14].
Bernoulli, J. 1689–1704. Positiones arithmeticae de seriebus infinitis, earumque summa finita, J. Conradi a Mechel, Basileæ. In: *Jacobi Bernoulli basileensis Opera*, Genevae: Cramer et fratum Philibert [1744], 2 vols., vol. I: pp. 375–402 and 517–542, vol. 2: pp. 745–767, 849–867 and 955–975.
Bernoulli, Jacob [P] 1713. *Ars conjectandi. Opus posthuma. Accedit Tractatus de seriebus infinitis et Epistola Gallice scripta de ludo pilae reticularis*, Basilae, Thurnisiorum Fratrum.
Ferraro, G. 1998. Some aspects of Euler's series theory. Inexplicable functions and the Euler–Maclaurin summation formula. *Historia Mathematica* 25:290–317.
Ferraro, G. 1999. The first modern definition of the sum of a divergent series. An aspect of the rise of the 20th century mathematics. *Archive for History of Exact Sciences* 54:101-135.
Ferraro, G. 2000a. Functions, functional relations and the laws of continuity in Euler. *Historia Mathematica* 27:107–132.
Ferraro, G. 2000c True and fictitious quantities in Leibniz's theory of series. *Studia Leibnitiana* 32:43–67.
Ferraro, G. 2001. Analytical symbols and geometrical figures in Eighteenth Century Calculus. *Studies in History and Philosophy of Science Part A*, 32:535-555.
Ferraro, G. 2002. Convergence and formal manipulation of series from the origins of calculus to about 1730. *Annals of Science* 59:179–199.
Ferraro, G. 2004. Differentials and differential coefficients in the Eulerian foundations of the calculus. *Historia Mathematica* 31:34-61.
Ferraro, G. 2007a. Convergence and formal manipulation in the theory of series from 1730 to 1815. *Historia Mathematica* 34:62–88.
Ferraro, G. 2007b. The foundational aspects of Gauss's work on the hypergeometric, factorial and digamma functions. *Archive for History of Exact Sciences*, 61:457-518.
Ferraro, G. 2008a. D'Alembert visto da Eulero. *Bollettino di Storia delle Scienze Matematiche* 28:257-275.
Ferraro, G. 2008b. *The rise and development of the theory of series up to the early 1820s.* New York: Springer.
Ferraro, G. 2010. Pure and Mixed Mathematics in the Work of Leonhard Euler in *Computational Mathematics: Theory, Methods and Applications*, edited by Peter G. Chareton. New York: Nova Science Publishers, pp. 35-61.
Ferraro, G. 2012. Manuali di aritmetica, algebra, trigonometria e geometria analitica nella Napoli preunitaria. History of Education & Children's Literature 7:415-435.
Ferraro, G., Panza, M. 2003. Developing into Series and Returning from Series. A Note on the Foundation 18th Century Analysis. *Historia mathematica* 30:17-46.

Ferraro, G., Panza, M. 2012. Lagrange's theory of analytical functions and his ideal of purity of method. *Archive for History of Exact Sciences* 66:95-197.

Knobloch, E. 1991. Leibniz and Euler: Problems and solutions concerning infinitesimal geometry and calculus. In: Galuzzi M. (ed.). *Giornate di Storia della Matematica*, Commenda di Rende: Editel, pp. 271–293.

Knobloch, E. 1989. Leibniz et son manuscrit inédit sur la quadrature des sections coniques. In: Mugnai M. (ed.). *The Leibniz Renaissance*. Firenze: Olschki, pp. 127–151.

Leibniz, G. W. 1682. De vera proportione circulis ad quadratum circumscriptum in numeris rationalibus expressa. In: *Acta Eruditorum* 41–46. In Leibniz 1849-1863, 5:118–122.

Leibniz, G. W. 1684. *Meditationes de cognitione, veritate et ideis*, Acta Eruditorum, pp. 537–542.

Leibniz, G. W. 1691 Quadratura arithmetica communis sectionum conicarum, quae centrum habent, indeque ducta trigonometria canonica ad quantamcumque in numeris exactitudinem a tabularum necessitate liberata, cum usu speciali ad lineam rhomborum nauticam, aptatumque illi planisphaerium. In: *Acta Eruditorum* 1691, pp. 178–182. See also Leibniz 1849-1863, 5:128–132.

Leibniz, G. W. 1693. Supplementum geometriae practicae sese ad problemata transcendentia extendens, ope novae methodi generalissimae per series infinitas. In. *Acta Eruditorum* 1693, pp. 178–180. See also Leibniz 1849-1863, 5:285–288.

Leibniz, G. W. 1710. Symbolismus memorabilis calculi algebraici et infinite simalis in comparatione potentiarum et diɔerentiarum, et de lege homogeneorum transcendentali, *Miscellanea Berolinensia*, pp. 160–165. See also Leibniz 1849-1863, 5:377–382.

Leibniz, G. W. 1713. Epistola ad V. Cl. Christianum Wolfium, Professorem matheseos Halensem, circa scientiam infiniti, *Acta Eruditorum Supplem* 5, pp. 264–270. See also Leibniz 1849-1863, 5:382–487.

Leibniz, G. W. 1768. *Gothofridi Guillemi Leibnitii Opera Omnia, nunc primum collecta, in classes distribuita, praefationibus et indicibus exhornata, studio Ludovici Dutens*, VI Vols. Genevae: Fratres de Tournes.

Leibniz, G. W. 1875–1890. *Die philosophischen Schriften von G.W. Leibniz*, Gerhardt C.I. (ed.). Berlin: Weidmann.

Leibniz, G. W. 1846. Cum prodiisset atque increbuisset Analysis mea infinitesimalis in *Historia et Origo calculi differentialis a G. G. Leibnitio conscripta*, Gerhardt, C. I. (Ed.) Hannover: Hahn'sche Buchhandlung, pp. 39–50.

Leibniz, G. W. 1849-1863. Leibnizes mathematische Schriften, VII Vols. Gerhardt C. I. Berlin: Asher & Comp. 1849–1850: vols. I–II; Halle: H. W. Schmidt, 1855–1863: vols. III–VII.

Leibniz, G. W. 1860. *Briefwechsel zwischen Leibniz und Christian Wolff*. Gerhardt, C.I. (ed.). Halle: Schmidt [reprinted Hildesheim: 1963].

Leibniz, G. W. 1993. *De quadratura arithmetica circuli ellipseos et hyperbolae cujus corollarium est trigonometria sine tabulis*. Knobloch E. (ed.). Göttingen: Vandenhoeck & Ruprecht.

Varignon, P. 1715 Précautions à prendre dans l'usage des suites ou series infinies résultantes, tant da la division in [...] nie des fractions, que du développement à l'infini des puissance d'exposants négatifs entiers. *Mémoires de l'Academie Royal des Sciences. Mémoires Mathématiques et Physique,* pp. 203–225.

Giovanni Ferraro, Molise University, Italy
gm.ferraro@alice.it

Prospects for an Idealist Interpretation of Leibniz's Dynamics

Glenn A. Hartz

Abstract. I argue that one mainstream interpretation – "Idealism" (according to which simple substances and their states are all that exist in the world) – makes Leibniz's dynamics unintelligible. And when an interpretation entails turning hundreds of pages of some of Leibniz's most important scientific work into gibberish, it is much too expensive. In the first part of the paper I examine some of the principal claims of Leibniz's dynamics. I then argue that, to be understood properly (that is, as Leibniz intended them), they require mind-independent bodies.
 I examine two Idealist analyses of the dynamics advanced by the prominent commentators Donald Rutherford (Rutherford 1995) and Robert M. Adams (Adams 1994). They consider bodies mere phenomena in minds and then try to explain how for Leibniz bodies could also be congeries of monads.
  In the second part I explore some assumptions behind "monolithic" interpretations like the Idealist one. I ask why they hold that sciences like the dynamics need to be taken over by the philosophy and re-written to reflect a single, favoured metaphysic, and why they think one needs to alter every area of Leibniz's philosophy so they all tell the same theoretical story.
  Keywords: Leibniz, Dynamics, Idealism, Adams, Rutherford, Body, Primitive Active Force, Derivative Active Force, Extended, Berkeley, Kant, Mechanics.

1   Introduction

From a philosophical standpoint, Leibniz's dynamics is his most important scientific endeavour. For it is the dynamics which dovetails most directly with the theory of substance he propounded in his philosophical works over the last thirty years of his life. Moreover, the dynamics was an answer to Descartes' natural philosophy, where bodies are pure extension, as well as Newton's, according to which bodies are inherently passive. At the same time, it represents Leibniz's alternative to Occasionalism, where all activity in bodies is attributed to God. Those systems held that is there no active force in bodies considered in themselves. Thus they stranded the physical world without a naturalistic explanation of the obvious activity of objects in it, while Leibniz maintained that there was always a naturalistic explanation and the "derivative active force" of bodies provided it. Thus it is a pressing matter for Leibniz scholarship to ask what kind of metaphysic his dynamics needs to be rendered intelligible. (I do not say 'rendered plausible' because that is probably not attainable, given that its central claim – that derivative forces in bodies arise from active substances contained in them – involves metaphysical matters.) I argue that one mainstream interpretation – "Idealism" (according to which simple substances and their states are all that exist in the world) – makes the dynamics unintelligible. And when an interpretation entails turning hundreds of pages of some of Leibniz's most important scientific work into gibberish, it is much too expensive.

In the first part of the paper I examine some of the principal claims of Leibniz's dynamics. I then argue that, to be understood properly (that is, as Leibniz intended them), they require mind-independent bodies. I examine two Idealist analyses of the dynamics advanced by the prominent commentators Donald Rutherford (Rutherford 1995) and Robert M. Adams (Adams 1994). They consider bodies mere phenomena in minds and then try to explain how for Leibniz bodies could also be congeries of monads.

In the second part I explore some assumptions behind "monolithic" interpretations like the Idealist one. I ask why they hold that sciences like the dynamics need to be taken over by the philosophy and re-written to reflect a single, favored metaphysic, and why they think one needs to alter every area of Leibniz's philosophy so they all tell the same theoretical story.

2   The Dynamics in the Idealist Interpretation: Rutherford and Adams

In order to incorporate Leibniz's dynamics into its system, an interpretation must be able to explain how all of the following claims can be true. (This is not an exhaustive list. "Dead forces" and other specific features of the dynamics are left aside.)

1. Bodies exist.
2. Bodies are extended.
3. Bodies move.
4. Bodies have derivative active force, one manifestation of which is the quantity conserved in the universe in mechanical interactions – mv^2.
5. Bodies have derivative active force because they are composed of, and contain, an infinity of substances (variously referred to as "individual substances," "corporeal substances," or "monads"), each of which has primitive active force.

On the sort of Idealism I'm considering, there is no problem with 1. – 3. Assume that there are perceiving minds. Then bodies *exist* as appearances in those minds. The appearances are stretched out in the visual or tactile fields of those minds; that's how bodies are *extended*. And appearances can change their relative positions in those fields, so bodies can *move*.

But when Idealism confronts 4. and 5., there seems no way forward. How will mind-dependent appearances possess mind-independent forces (often manifest as conserved quantities in nature)? How could a mind-dependent appearance be composed of or contain mind-independent substances with their own primitive forces? Surely the mind-dependent appearance-bodies of 1. – 3. cannot possess the properties Leibniz attributes to bodies in 4. and 5.

Prima facie one might conjecture that Leibniz is offering two different analyses. Sometimes the Idealist elements predominate, and other times Realist features show up alongside the Idealist scheme or on their own. But, while this "theory-pluralism" is often evident in the larger metaphysical scheme (Cfr. Hartz 2007), it won't work here. To preserve the integrity of the theory, one needs 1. – 5. to be about the *same sort of bodies*. Despite this bleak outlook for the coherence of 1. – 5., Rutherford and Adams do not abandon the dynamics as incompatible with Idealism. I will now evaluate their attempts to accommodate the dynamics within their interpretations.

Rutherford recognizes trouble created by the doctrine of corporeal substance for the view that bodies are mere phenomena:

> If bodies exist only as phenomena – things perceived by monads – then we might wonder whether the organism or soul-body composite is even a coherent ontological type or whether it is, instead, a sort of category mistake: an entity formed by illicitly combining what is mere appearance (the body) with what is truly real (the soul). To answer this doubt, we must [note] [...] that bodies are themselves pluralities of monads. For Leibniz, the term "body" does not refer merely to an appearance of phenomenal object. Instead, a body *is* a plurality of monads, which happens to give the appearance of being an extended object when apprehended by other finite monads (Rutherford 1995, p. 218).

Again Rutherford asks how bodies can be appearances *and* aggregates of monads:

> Given the radically disparate properties of bodies and monads, how could bodies possibly be aggregates of monads? (*Ivi*, p. 219).

Despite the apparent problems, Rutherford believes there is a solution. It involves transforming the identity claim between phenomena and aggregates of monads into a claim about "essences" or "types of being." The identity of appearance-bodies and aggregate-bodies is an "ontological thesis" which holds that the "*qua* species of being, matter is essentially an infinite plurality of monads" (*Ivi*, p. 220). Again "any material thing is by nature a plurality of monads" (*Ivi*, p. 222) and "a distinct understanding of the essential properties of matter (its multiplicity, force, and resistance) reveals bodies to be multitudes of unextended, active substances" (*Ivi*, p. 226).

Given this move, Leibniz's dynamics is concerned with the *concept* of body rather than bodies themselves. We understand bodies as having forces, and that understanding is grounded in what we grasp of monadic reality. As Rutherford says, "The force that belongs to the essence of body is ascribed to the fact that the matter of bodies is constituted from substances that are by nature principles of action" (*Ivi* , p. 242). After quoting a passage from *Specimen Dynamicum* (GM 6, 235/L 435) in which Leibniz unequivocally attributes force to bodies, Rutherford offers this gloss:

> [Leibniz's] principal claim is that there is something else "in" body besides extension: a force introduced or implanted (*idita*) by God. In the 1698 essay *De ipsa natura*, he speaks of the innate or inborn (*insita*) force of bodies. This force, a *conatus* or "striving," is assigned

to the nature of substance, but is also "the inmost nature of body." [...]. [T]he significant point is that the force associated with body – what is required in order render intelligible its action – is referred back to the force or power that is an essential property of substance (*Ivi*, pp. 242-243).

To elude the Realism about bodies in this passage, Leibniz's *in* becomes "merely metaphorically in." And while Rutherford quotes Leibniz as saying "there is something besides extension in corporeal things" (*Specimen Dynamicum* 1695, GM 6, 235/L 435), Rutherford reads this as "there is something besides extension in the *nature of* corporeal things." This reading does great violence to Leibniz's intentions. Leibniz is talking in that text about *specific bodies*, while Rutherford has him talking about *body in general*. Indeed, Rutherford explicitly addresses the point of taking "in" metaphorically:

> The spatial terminology that infects Leibniz's own account of body – most prominently in his claim that unextended monads are "in" or "diffused throughout" bodies – must be interpreted figuratively. The monads constitutive of a body are not its ultimate spatial parts, but its "immediate requisites." As we have seen, this expression connotes a priority of essence or being. [...] Monads supply a ground for the existence and properties of bodies, insofar as they are involved in our conception of what matter is; they are not to be regarded as spatial components of bodies (*Ivi*, p. 251).

Now there are several problems with Rutherford's Idealist rescue of the dynamics.

First, there are *Leibniz's Law* problems. The claim: appearance-body x = aggregate-body x seems possible because for Rutherford the "=" is not numerical identity, but is identity in a qualified sense. It seems to mean "identity at the level of essence" or something like that. But, since we are never told what it is for two things to be identical at the level of "essence," "being" or "nature," we can't judge whether the identity claim holds.

Leibniz himself always has numerical identity in mind. Given the logically incompatible properties of appearance-bodies and aggregate bodies, the numerical identity claim is necessarily false. Leibniz explicitly warns us against confusing continuous phenomena with discrete multitudes – or what he calls "actual substantial things" or "substantial realities" (c. 1695, G 4, 491-92/Leibniz 1898, pp. 329-330; for a full treatment, see Hartz and Cover 1988).

The doctrine of the continuum clearly proscribes this identity claim. A mind-dependent, continuous, arbitrarily divisible extended appearance-body cannot be numerically identical with a mind-independent aggregate-body which is an actually divided, discrete multitude of unextended substances. So there are two different objects here.

Still, it must be admitted that it sometimes looks as if Leibniz identifies them because he carelessly uses 'phenomena' sometimes to refer to the appearance-body, sometimes the aggregate-body. But instead of following this tendency to use 'phenomena' to blur the border between two items, we should keep them clearly distinguished. As we might say, appearance-bodies have more to do with the epistemic status of bodies as encountered in experience, while aggregate-bodies represent the metaphysical account of what's there in nature.

Second, Rutherford's "rescue" is not a rescue of the *science of dynamics* at all. For dynamics studies individual bodies, not their essence. If every claim of the dynamics must be translated into generic claims about the nature of body, there is nothing left. The famous "notable error of Descartes" concerning the quantity of motion assumes that bodies are treated individually and different quantities of weight, mass, forces and effects can be attributed to bodies. (Thus it is mv^2 which is conserved *in the universe* (and not at the level of "nature" or "essence") as bodies follow the laws God established for them.) If the bodies involved were conceived only at the level of "nature" or "being," everybody (considered only as an instance of our concept of body and its features) would possess the same generic qualities (e.g., is conceived as having has force and motion). And the very idea of measuring *those* features is laughable.

For the science of dynamics to remain a feather in Leibniz's cap, the metaphysic in the background has to let him talk literally and without embarrassment about bodies rather than first filtering such claims through an inimical, monolithic theory. Idealisms of this sort become a hindrance to our understanding of Leibniz's scientific work, because they force Leibniz to hew to a favoured metaphysic and "really" mean what he could not have meant.

Third, neither is it a rescue of the *metaphysics of the dynamics*. If you take 1. – 3. to be about appearance-bodies, but 4. and 5. to be metaphorical (that is, about the essence of bodies), you equivocate on 'body' and the analysis is ruined. If you hold that 1. – 5. are all to be taken metaphorically as about the essence of bodies, the entire metaphysic beneath the dynamics is metaphorical. And of course a purely metaphorical account has no business masquerading as a grounding for sober science. So either way, the analysis collapses.

Adams offers an account of his own, to which I now turn. He begins his treatment of body thus:

> The most fundamental principle of Leibniz's metaphysics is that "there is nothing in things except simple substances, and in them perception and appetite" (G II, 270/L 537). It implies that bodies, which are not simple substances, can only be constructed out of simple substances and their properties of perception and appetition.
> A construction of the whole of reality out of perceiving substances [...] exemplifies a broadly idealist approach to metaphysics. Leibniz was the first of the great modern philosophers to develop an idealist metaphysics.
> How can bodies be both mere phenomena and aggregates of substances? [...] Leibniz believed (rightly or wrongly) that the two theses [...] are consistent, and he held both of them throughout the mature period [...] (Adams 1994, pp. 217-218)

Adams's claim of consistency faces worries similar to those Rutherford faced with respect to what I called "appearance-bodies" and "aggregate-bodies":

> The apparent conflict between the thesis that bodies are phenomena and the thesis that they are aggregates of substances springs from the

assumption that an aggregate of *F*s must have the same ontological status as the *F*s. This is at best a controversial assumption. There is nothing at all odd about Leibniz's rejecting it. [...]. We should not expect it to be assumed without argument that an army [...] has the same ontological status as the soldiers that are its elements. In fact, Leibniz makes clear repeatedly that he believes that all aggregates, as such, are at most phenomena, and hence that an aggregate of substances does not have the same ontological status as the substances (*Ivi*, pp. 244-245).

But Leibniz does *not* (most of the time) reject that assumption. *Aggregate-bodies*, unlike appearance-bodies, are "real" because they contain "real" substances. When Leibniz says that aggregates are "at most phenomena," he has to be talking about (or interpreted as talking about) appearance-bodies. Otherwise the doctrine is at odds with the entire mature continuum doctrine.

Adams stops short of declaring numerical identity between the two kinds of bodies: "I think Leibniz believed that the two accounts [of harmonious phenomena and aggregates of substances] are at least materially equivalent" (*Ivi*, p. 260). But the claim appearance-body *x* exists iff aggregate-body *x* exists is, like the identity claim, necessarily false. For two objects so related cannot have logically incompatible properties.

The doctrine of corporeal substance which Rutherford found problematic is also an obstacle for Adams, who tries to square it with strict Idealist principles:

> A mass of secondary matter, as such, is thus merely a phenomenon *because* it is an aggregate of substances [...]. According to Leibniz, *every* created monad has an organic body of this sort, with which it forms a corporeal substance. The monad *always* has its body, and hence the organic body is an enduring, though constructed and merely phenomenal, object permanently attached to its dominant monad. The substances that are included in an organic body can be replaced with other substances as long as the body retains the necessary organs and the same dominant monad (*Ivi*, pp. 266-267).

The view as described seems impossible. If aggregates which serve as organic bodies are "constructed and merely phenomenal," how can they be attached in any intelligible sense to an unconstructed and not-merely-phenomenal dominant monad? How can they have the "permanence" of the monad when they depend on a finite "constructing" mind? How can a mere phenomenon have substances "included in it?"

To sum up this section: pure Idealist schemes will never successfully assimilate Leibniz's dynamics into their systems.

3 Motivations for Idealism

What fuels Idealisms of this sort?

It is partly motivated by the belief that Leibniz offered only one "real" metaphysic and that, when not talking about his beloved monads, he had to condescend to the level of mechanism and couch his claims in terms of ontologically negligible items like bodies, motion and physical forces.

That is a tension we do not observe in Leibniz himself. He is happy to talk about the deepest matters from a God's–eye perspective, but he emerges from that and begins, without qualification or circumlocution, treating in realist terms mass, motion, and dynamical features of bodies.

The practice of imposing Idealism on sciences appears saliently in theorists in the broadly Idealist tradition quite independently of Leibniz – for example, in Berkeley and Kant. Berkeley thought he had to replace the scientific study of vision and perspective (i.e., optics) with the study of these subjects which takes account of the Idealist claim that we never see a physical world in three dimensions (*An Essay Towards a New Theory of Vision* [1709] 1965). And Berkeley criticizes Newton and Leibniz for postulating "occult" forces in objects rather than simply noting regularities in motions of observed appearances (*De Motu* [1721] 1965). (We should, I suppose, abandon mechanics and dynamics in favour of a phoronomy of appearances.) Indeed, this last point underscores the vast difference between Berkeley and Leibniz. Berkeley has no patience for qualities of bodies (e.g., forces accounting for activity and inertia) which aren't manifest as qualities of sense-data; Leibniz does and centres the entire dynamics on them.

For his part, Kant's architectonic is so broad it conquers, subdues, and subsumes all the sciences in its path. Kant thinks the dynamics of bodies needs to find a "metaphysical foundation" in considerations about the *concept* of body as seen through the lens of the categories of pure reason (Kant [1786] 1985). Only then, Kant seems to think, will it be rationally respectable and defensible. But Leibniz would never make a claim like Kant's proposition that "there are only two moving forces that can be thought of […], to which as such all the forces of motion in material nature must be reduced" (*Ivi*, p. 43). The "a priori philosophy," dictating conceptual constraints on the physical world, is not a Leibnizian legacy. Kant's approach would reach its zenith in the largely regrettable era of *Naturphilosophie.*

Berkeley and Kant maintain a "colonialist" attitude toward science. Leibniz does not. Leibniz had an Idealist metaphysic, but he did not regard it as threatening the autonomy of sciences like optics, dynamics and mechanics. He did not subject them to a hostile takeover, assuming that philosophy was obviously deeper and better and had privileged access to the truth. For Leibniz, the scientists can talk about bodies, motion, mass, and force without fear that the Idealist metaphysician is anxiously looking over their shoulder, eager to get back to the Ultimate Reality of world–apart, windowless substances.

What makes Leibniz's work so interesting is that he moves effortlessly between philosophy and science, as he does in his undisputed masterpiece,

Discourse on Metaphysics. After spending lots of time at the dizzyingly abstract level of "complete concepts" and world-apart substances, he moves to weighted bodies, motion, and conserved forces. The metaphysical preamble is borne out by the manifest properties of bodies. Everything fits.

As R. S. Westfall observes, "One of the measures of Leibniz's clarifying genius was his ability to distinguish this ultimate metaphysical level [of living force] from the phenomenal realm of actualised mechanics" (Westfall 1971, p. 288), and even *that* level would not be ultimate until it was grounded in substance (*Ivi*, pp. 311-315). Indeed, one can distinguish three levels: mechanics as the science of measurement, the metaphysics of bodies and forces beneath it, and the metaphysics of substance beneath that.

On the philosophical front, Idealism wants its favoured metaphysic to colour and qualify every other area of Leibniz's thought. The assumption is that if it does not, Leibniz will be inconsistent. Leibniz's silence about discord between the theories is taken to mean that he thought they are consistent, and the commentator's job is to show how they are.

That is a crucial mistake. I believe Leibniz ignores the inconsistencies between schemes because his goal is to have as many options as possible. Thus he has a very rich array of views to consult to solve problems. His Renaissance-inspired ideal was to combine the apparently discordant elements into one even if he did not see how the details were going to be settled. The most important belief running the show for Leibniz was the belief that God had created the inventors of these alternative theories and welcomed them into the best world. Thus Leibniz was open-minded to a fault: that is, the cost of being open to nearly everyone is inconsistency. I think Leibniz loved this universal perspective more than "consistency across systems" and so we find a huge array of theories lined up in the corpus.

Leibniz accords the other metaphysical options their own proper sphere and initial claim to legitimacy. Rarely is one given an initial advantage as the highest, though there are texts where he seems to say "now I'm laying my cards on the table" and, caught up in a momentary rapture, he speaks in exclusive terms about his adherence to a given view. He does that just often enough to give the Idealist a few texts (like Adams' "most fundamental principle") to stand by and quote in favour of their monolithic view.

In the case of Rutherford and Adams, the attempt to export Idealism into every part of Leibniz's thought is ruinous to those parts. Thus the doctrine of corporeal substance "must" somehow be made to jibe with monads-only Idealism even if that doctrine thereby becomes metaphorical or unintelligible. All claims about bodies must be reconciled with (as Adams says) "the most fundamental principle of Leibniz's metaphysics" (G 2, 270/L 537). But how can any interpreter make that call? Out of the hundreds of thousands of sentences Leibniz wrote, *one* is exalted as the most fundamental. All the rest have to hew to it. Sentences like this remark to De Volder are relatively rare in the corpus. They are swamped by the number of realist "aggregates of substances" texts which are completely inconsistent with it.

When all the rewriting and redacting and adjusting is done, what is the prize the Idealist holds up? They have saved Leibniz from inconsistency by destroying half his philosophy!

The grip of Idealism is waning, by all accounts. And that is a good thing. One of the first contemporary commentators to make the bold move of declaring Leibniz's various schemes inconsistent was, in Anglo-commentary, Catherine Wilson (Wilson 1989). In relation to the matter of consistency, she writes

The unspoken assumption is something like this: the probability of a *famous logician and mathematician* having been guilty of serious, fundamental inconsistencies is *low*, while the probability of *one of a teeming mass of commentators* lacking [...] the logical and linguistic skills to decode it, or the insight to realize its significance is *high*. [I]t presupposes that Leibniz approached philosophy [including his logical theory, his theory of substance, space and time, his theory of pre-established harmony, his physics, his eschatology, etc.] with the same expectations and critical checks as the modern, professional, specialized, full-time metaphysician does when trying to articulate and defend against criticism a single isolated thesis or two! (Wilson 1994, pp. 7-8)

Thus Wilson notes that there was not a single, all-encompassing intention in Leibniz's system – any more than there is (as supposed by some interpreters) one "monological" theme in *The Changeling* or *Dr. Faustus*.

Idealism is, in a word, a *reading*. It asks whether we can make sense of the corpus viewed through the lens of Idealism. Any doctrines that survives will be exalted. Those that do not will be gracefully and quietly discounted.

But the cost is much too high. To accomplish this, Idealist commentators either allow Leibniz to break logical rules (in which case why is inconsistency a problem?) or half the corpus is rendered unintelligible (and inconsistency is not a problem because the system isn't worth saving anyway).

4   Conclusion

In his philosophy, Leibniz is magnanimous, as everyone knows who has studied him. Among competing philosophical accounts, Leibniz rules out only those parts of systems which detract from the proper concept of God (as in the case of Hobbes, Spinoza, Descartes, Newton and Pufendorf).

With respect to science, he does not make invidious assumptions about philosophy's deserving special treatment as the "highest" discipline. Mechanism is allowed to remain a separate, autonomous, worthy realm of investigation. That there is also a story about those same events to be told in terms of monads and final causes does not detract from the legitimacy of studying nature mechanically.

Supposing otherwise diminishes Leibniz's reputation as a philosopher and a scientist.

References

Adams, R. M. 1994. *Leibniz: determinist, theist, idealist*. New York: The Oxford University Press.

Berkeley, G. 1965. *Berkeley's philosophical writings*. Armstrong, D. M (Ed.). New York: Collier Books.

Hartz, G. A. 2007. *Leibniz's final system: monads, matter and animals*. London: Routledge.

Hartz, G. A., Cover, J. A. 1988. Space and time in the Leibnizian metaphysic. *Nous* 22:493-519.

Kant, I. [1786] 1985. *Metaphysical foundations of natural science*. In Ellington, J. W. (ed.). *Philosophy of Material Nature*. Indianapolis: Hackett.

Leibniz, G. W. 1849-1863. *Mathematische Schriften von Gottfried Wilhelm Leibniz*. Gerhardt, C. I. (Ed.). 7 vols. Berlin: A. Asher [abbreviated as 'GM'].

Leibniz, G. W. 1875-1890. *Die philosophischen Schriften von Gottfried Wilhelm Leibniz*. Gerhardt, C. I. (ed.) 7 vols. Berlin: Weidmann (abbreviated as 'G').

Leibniz, G. W. 1898. *The monadology and other philosophical writings*. Latta, R. (Ed.). Humphrey Milford: Oxford University Press.

Leibniz, G. W. 1969. *Philosophical papers and letters*. Loemker, L. E. (Ed.). Dordrecht: Kluwer [abbreviated as 'L'].

Rutherford, D. 1995. *Leibniz and the rational order of nature*. Cambridge: The Cambridge University Press.

Westfall, R. S. 1971. *Force in Newton's physics*. London: Macdonald.

Wilson, C. 1989. *Leibniz's metaphysics: a comparative and historical study*. Princeton: The Princeton University Press.

Wilson, C. 1994. Reply to Cover's 1993 review of *Leibniz's metaphysics*. *Leibniz Society Review* 4:5-8.

Glenn A. Hartz, Ohio State University, USA
hartz.1@osu.edu

Asymmetric, Pointless and Relational Space: Leibniz's Legacy Today

Joseph Kouneiher

It is worth noting that notation facilitates discovery. This, in a most wonderful way, reduces the mind's labors. We need an analysis which is of geometric nature and describes physical situations as directly as algebra express quantities
Gottfried Wilhelm Leibniz (1646–1716)

Abstract. In his Monadology Leibniz advanced the principle of individuation. On this basis he tried to find a geometric tool to describe the phenomena. So we find the principle Leibniz took to be the one that brings the experienced world into existence, is much more closely related to the principle of individuation. That is the foundation of his philosophy. According to this principle, individuals are distinguished by "Differentiation". The seeking approach depict a relational Geometric calculus tool.
In this paper, we will stress Leibniz vision of a universal tool of calculus, and see how such program finds a resurgence in the new mathematical physics methods today.

Keywords: Monadology, foundations of Physics, Geometric Calculus, Relational space, Variety or variance, Ausdehnungslehre, Grassmann, Cartan, Differential forms, Cohomology, Gauge theory, noncommutative geometry.

1 Introduction

Geometry is a mathematical model for describing both invariant geometric properties and their representation by local coordinates. In ancient times, one only considered invariant geometric properties. The description of geometric properties by coordinates dates back to René Descartes (1596–1650). In 1667 Descartes published his *Discours de la méthode* which contains, among a detailed philosophical investigation and its application to the sciences, the foundation of analytic geometry (e.g., the use of Cartesian coordinates).

Until the middle of the nineteenth century, one of the essential and fundamental problem of geometry was to relate geometry and numbers. As we know, the modern concept of real number, achieved essentially by Simon Stevin around 1600 wasnecessary to resolve this problem. The integration of real numbers into geometry already began with Descartes and Fermat in 1630, and achieved with the introduction of the analytic geometry, at the end of eighteenth century. From the analysis point of view focusing on functions, this was satisfactory; but from the geometry point of view, it was not. Indeed, the method used to attach numbers and geometric entities was awkward. Leibniz, in 1679, has mused upon the possibility of a universal algebra.

In 1679, in a letter to Huygens, Leibniz expresses his dissatisfaction with algebra to describe the phenomena and seek to find a geometrical analysis to express the situations. In his letter[1] on September 8, 1679, Leibniz wrote:

> I am still not satisfied with algebra [...] I believe that, so far as geometry is concerned, we need still another analysis which is distinctly geometrical or linear and which will express situation directly as algebra expresses magnitude directly (Leibniz to Huygens on September 8, 1679).

and he added

> I believe that by this method one could treat mechanics almost like geometry, and one could even test the qualities of materials I have no hope that we can get very far in physics until we have found such a method (*Ibidem*).

Leibniz never discovered the "new method" he was envisioning.

His letter[2] to Huygens was published in 1833, and hence had little influence on

---

[1] The letter can be found at Leiden, coll. Huygens. It was published by P.J. Uylenbroek et C.I. Gerhardt. Chr. Huygens's answer conserved under the No. 2203

[2] The original letter was written in french :

A Hanover ce 8 de Sept. 1679.
MONSIEUR

Un de mes amis, nommé M. Hansen, qui a eu l'honneur de vous parler, me mande, que vous continués d'avoir de bons sentiments pour moy, de quoy je vous suis fort obligé, et j'en ay voulu prendre l'occasion de vous témoigner combien j'honnore vostre merite extraordinaire, que tout le monde reconnoist avec moy, et qui vous met au premier rang.

[...] Mais apres tous les progres que j'ay faits en ces matieres, je ne suis pas encor content de l'Algebre, en ce qu'elle ne donne ny les plus courtes voyes, ny les plus belles constructions de Geometrie . C'est pourquoy lors qu'il s'agit de cela, je croy qu'il nous faut encor une autre Analyse proprement geometrique ou lineaire, qui nous exprime directement, situm, comme l'Algebre exprime magnitudinem.

Et je croy d'en voir le moyen, et qu'on pourroit representer des figures et mesme des machines et mouvemens en caracteres, comme l'Algebre represente les nombres ou grandeurs; et je vous envoye un essay qui me paroist considerable; il n'y a personne qui en puisse the historical development. By the 1830s, others were already working on "Linear Algebra".

One of the first systems in which geometrical entities (points) were operated on directly, was Möbius' Barycentric Calculus (1827) (Möbius 1885–1887). His approach was similar to Giusto Bellavitis' *"Calculus of Equipollences"*.

Möbius showed how to add collinear line-segments, but gave no addition rules for non-collinear segments and no multiplication (Möbius 1885–1887, 1976). Such a system was first constructed by Hermann Grassman[3].

In the Foreword of *Die Lineale Ausdehnungslehre, ein neuer Zweig der Mathematik (Linear Extension Theory, a new branch of mathematics)*, Grassmann wrote:

> [...] geometry can in no way be viewed, like arithmetic or the theory of combinations, as a branch of mathematics; instead, geometry relates to something already given in nature, namely, space. I also had realised that there must be a branch of mathematics which yields in a purely abstract way laws similar to those of geometry, which is limited to space. By means of the new analysis it is possible to form such a purely abstract branch of mathematics; indeed this new analysis, developed without assuming any principles established outside its own domain and proceeding purely by abstraction, was itself this science (Grassmann 1839, Foreword)

Leibniz's approach to find his universal calculus falls rather within the modal metaphysics approach[4]. His calculus is rather geometric and relational. Indeed, if we read the Monadology[5] carefully, we find that the principle Leibniz took to be the

---

mieux juger que vous Monsieur et vostre sentiment me tiendra lieu de celuy de beaucoup d'autres.

je croy qu'on pourroit manier par ce moyen la mecanique presque comme la geometrie et qu'on pourroit mesme venir jusqu'à examiner les qualites des materiaux, par ce que cela depend ordinairement de certaines figures de leur parties sensibles. Enfin je n'espere pas qu'on puisse aller asses loin en Physique avant que d'avoir trouver un tel abrege pour soulager l'imagination. (annexe de la lettre) No 2192. G.W. Leibniz à Christiaan Huygens. 8 septembre 1679.

[3]Hermann Gunther Grassmann born April 15, 1809, in Stettin (Szczecin). His father Justus Grassmann, a gymnasium teacher, was the first to invent a purely geometrical product ("Raumlehre"; Grassman 1824).Hermann studied Theology and Philosophy in Berlin (1827–1830). He returned to Stettin to become a teacher, in the footsteps of his father. He never had any formal education in Mathematics and he was outside the mathematical establishment all his life.

He was only 23 when he discovered the method of adding and multiplying points and vectors which was to become the foundation of his *Ausdehnungslehre*. In 1839 he composed a work on the study of tides entitled *Theorie der Ebbe und Flut* (Grassmann 1894–1911), which was the first work ever to use vectorial methods. In 1844 Grassmann published his first *Ausdehnungslehre* (*Die lineale Ausdehnungslehre ein neuer Zweig der Mathematik*; Grassmann 1839, Foreword).

[4]Indeed, it was space itself which became an appearance. True substances were explained as metaphysical points which, Leibniz asserted, are both real and exact — mathematical points being exact but not real and physical ones being real but not exact. Clearly, besides metaphysics, the developing of calculus had also provided some grounds for seeking universal elementary constituents.

[5]The monad, the word and the idea, belongs to the western philosophical tradition and has been used

one that brings the experienced world into existence (rather than some other possible world) is much more closely related to the principle of individuation (see section). That is the foundation of his philosophy. He argued that any contingently existing thing must be described by its attributes.

According to this principle, individuals are distinguished by "Differentiation". The very essence of being is "Differentiation". His bold conclusion was that, in reality, actual things are simply descriptions of the universe from different perspectives, like all the different views of a city.

As we shall see, the Geometric calculus introduced by Grassmann[6] to generalize the ideas of complex structure, vectors, quaternions, consists of an algebraic calculus on geometric entities (points, oriented segments, oriented triangles and so on), differently from what happens in analytical geometry, that is based on an algebraic manipulation of coordinates.

Grassmann's principal contribution to the physical sciences was his discovery of a natural language of geometry from which he derived a geometric calculus of significant power. For a mathematical representation of a physical phenomenon to be "correct" it must be of a tensorial nature and since many "physical" tensors have direct geometric counterparts, a calculus applicable to geometry may be expected to find application in the physical sciences.

The word "Ausdehnungslehre" is most commonly translated as "theory of extension", the fundamental product operation of the theory then becoming known as the exterior product. The notion of extension has its root in the interpretation of the algebra in geometric terms : an element of the algebra may be "extended" to form a higher order element by its (exterior) product with another, in the way that a point may be extended to a line, or a line to a plane, by a point exterior to it. The notion of exteriorness is equivalent algebraically to that of linear independence. If the exterior product of element of grade 1 (for example, points or vectors) is non-zero, then they are independent.

A line may be defined by the exterior product of any two distinct points on it. Similarly, a plane may be defined by the exterior product of any three distinct points on it, and so on for higher dimensions. This independence with respect to the specific points chosen is an important and fundamental property of the exterior product. Each time a higher dimensional object is required it is simply created out of a lower dimensional one by multiplying by a new element in a new dimension. Intersections of elements are also obtainable as products.

Simple elements of the Grassmann algebra may be interpreted as defining subspaces of a linear space. The exterior product then becomes the operation for building higher dimensional subspaces (higher order elements) from a set of lower dimensional independent subspaces. A second product operation called the regressive product may

---

by various authors. Leibniz used it himself round the mid-1696 when he was sending for print his New System.

[6]To learn more about Grassmann's work and life see Flament and Bekemeier 1997; Flament, Garma and Navarro 1994; Flament 2008.

then be defined for determining the common lower lower dimensional subspaces of a set of higher dimensional non-independent subspaces.

Most importantly however, Grassmann's contribution has enabled the operations and entities of all these algebras to be interpretable geometrically, thus enabling us to bring to bear the power of geometric visualization and intuition into our algebraic manipulations. Finally, it should be noted that the Grassmann algebra subsumes all the real algebra, the exterior product reducing in this case to the usual product operation among real numbers.

Grassmann's work gave birth to the *differential forms calculus* and later to the *exterior forms calculus* by Elie Cartan. Differential forms are central and ubiquitous to the modern formulation of Mathematical Physics, where manifolds and Lie groups are employed to describe the configuration and time evolution of mechanical systems. The mathematical description of observables as infinitesimal measurements in modern mechanics makes differential forms are thereby a natural choice. This modern language makes the foundations of physical models clear and precise. Surprisingly, it also makes the computations clearer. Further, the coordinate's free language turns out to be very easy to illustrate.

Cartan's calculus is the proper language of generalizing the classical calculus due to Newton and Leibniz to real and complex functions with n variables. The key idea is to combine the notion of the Leibniz differential with the alternating product $a \wedge b$ due to Grassmann. Cartan's calculus has its roots in physics. It emerged in the study of point mechanics, elasticity, fluid mechanics, heat conduction, and electromagnetism. It turns out that Cartan's differential calculus is the most important analytic tool in modern differential geometry and differential topology, and hence Cartan's calculus plays a crucial role in modern physics (Gauge theory, theory of General Relativity, the standard Model in particle physics). In particular, the language of differential forms shows that Maxwell's theory of electromagnetism fits Einstein theory of special relativity, whereas the language of classical vector calculus conceals the relativistic invariance of the Maxwell equations.

Our aim in this paper is to produce a personal account of how Leibniz's ideas might be used as a basis for a contemporary account of the principles of philosophy with the minimum of modification and to indicate where Leibniz's insights seem compatible with specific aspects of modern physics. In many cases this confirms the validity of Leibniz's approach.

In this paper we will explore the genesis and the conceptual foundations of those ideas. One aspect of our work is to show how many of the modern mathematical approach of physics realize a Leibniz vision of physics. Starting from this idea we want to show how differential forms give us a new insights to the foundations of physics.

2 Leibniz's Legacy: A Pointless and Relationsal Space

In his Monadology[7] Leibniz argued that any contingently existing thing must be described by its attributes. According to this principle, individuals are distinguished by its *variance* their *difference*. The very essence of being is *difference* or *variance*. What one means by "good" is notoriously difficult to define. How can one maximize something one cannot define? In contrast, something that can be defined and maximized is variance. His bold conclusion was that, in reality, actual things are simply descriptions of the universe from different perspectives, like all the different views of a city.

Indeed, it is clear that this is the deeper meaning of Leibniz's scheme, for in paragraphs 57 and 58 of the Monadology we read:

> [...] and just as the same town, when looked at from different sides, appears quite different and is, as it were, multiplied in perspective, so also it happens that because of the infinite number of simple substances [monads], it is as if there were as many different universes, which are however but different perspectives of a single universe in accordance with the different points of view of the monads. And this is the means of obtaining as much variety as possible, but with the greatest order possible; that is to say, it is the means of obtaining as much perfection as possible.

In his letter to Clarke about his Third Paper Leibniz wrote:

> As for my own opinion, I have said more than once, that I hold space to be something merely relative, as time is, that I hold it to be an order of coexistences, as time is an order of successions (Third Paper, paragraph 4; G VII.363/Alexander 2526).

Leibniz held that the entire world consists of nothing but distinct individuals, and that the sole essence of these individuals is to have perceptions (not all of which they are distinctly aware of).

The most radical element in the Monadology, postulated rather than explained or made directly plausible, is the claim that the perceptions of any one monad — its defining attributes — are nothing more and nothing less than the relations it bears to all the other monads.

In his fifth Paper Leibniz introduces a helpful example that illustrated more intuitively his main idea, there he suggests the family tree as model for that space and time. Unlike the relationship between, say, a mighty oak and its leaves, a genealogical tree is not something which exists as a thing independently of, and prior to, its members, but is itself rather something like an abstract system of relations holding between

---

[7]Leibniz's aim when he introduced the idea of the Monadology was to put an end from a monist point of view to the main question of what is reality, and particularly to the problem of communication of substances, or what we called Descartes mind-body dualism. Thus, Leibniz offered a new solution to mind and matter interaction by means of a pre-established harmony expressed as the best of all possible worlds form of optimism; in other words, he drew the relationship between *"the kingdom of final causes"*, or teleological ones, and *"the kingdom of efficient causes"*, or mechanical ones.

brothers, sisters, parents, children, aunts, uncles, etc. Analogously for Leibniz, space and time are not to be thought of as containers in which bodies are literally located and through which they move, but rather as an abstract structure of relations in which actual (and even possible) bodies might be embedded.

By contrast, Newton's cosmos, space and time provide a fixed, immutable and eternal background, with respect to which particles move. Space and time are a stage. Like actors on that stage, particles move, exert forces onto each other, while the stage itself does not change.

Newton's view of the universe is manifestly background dependent — whether or not a particle is moving or at rest is determined in relation to Newton's absolute space and time.

From a relationist point of view, there is no background of absolute space and time, and the fundamental properties of the elementary entities consist entirely in relationships between those elementary entities. Time is nothing but changes in these relationships, and consists of nothing but the ordering of these changes.

Leibniz's central idea was that our world is constituted by dynamic units of what we might now call *energy* and not by extended matter', nor by ideas' (Edward 1714). This simple message, which is very close to what modern physics would say, has been obscured by commentators convinced that Leibniz was proposing some strange supernatural realm, outside space and time.

As Richard Arthur addresses in his Leibniz (Arthur 2014),

> [...] the accusation that Leibniz is fanciful or metaphysical' in a pejorative sense is simply a reflection of the academic philosophical community's inability to understand Leibniz's agenda. If anything Leibniz is the most cold-blooded of all natural philosophers, for whom there is only one sort of goings on' in the world (*Ibidem*).

Through the idea of monads Leibniz wants to resolve two very different problems: the nature of the most basic entities of our world and second the relation of subjective experience, or having a point of view, to the world that is experienced. Leibniz sees that these two problems should have a single solution, which relies on the correct definition of the fundamental units of dynamic relation.

For Leibniz metaphysics is the framework of getting your basic concepts right before you try to build a 'physics'. Neither the monads, nor indeed God, are 'outside physics', or outside space and time, for Leibniz. He wanted to base his idea of the real explanation for everything on the evidence manifest in the regularity of events. In comparaison of Spinoza who says the mind is the '*idea*' of the body, with no explanation and Newton who proposes an absolute space, like some invisible graph paper for matter to move on, Leibniz is ruthlessly minimalist. And yet that does not stop him dealing with the subjective and even with morality.

The key motivation for attempting an updating of Monadologyis to explore just how much more there may be to learn from Leibniz's principles, which even now may not have been fully appreciated. The central suggestion is that Leibniz's monad is a very reasonable stab at identifying what we now know to be the indivisible dynamic units of our world the modes of excitation described by modern field theory. Leibniz's application needs modification, but my suggestion is that if one returns the basic principles that he uses to infer the nature of the monad the principles themselves still look very good and have the potential to form the basis of a fully contemporary account of how the dynamics of the world relate to our perceptions of it which is the essential subject matter of science.

Leibniz assumes that our changing thoughts reflect perception of a real world, so this is a point of view on a world. So from this point of view, Leibniz is not an idealist in the sense of not believing in any real world. His analysis is more practical but also more subtle. Leibniz accepts also that there are many points of view on the world and that these points of view must be based in real entities of some sort. He was convinced that points of view had to be real entities, to explain each being distinct. Even if these entities are in some way just instances of relation back to the world they must be real instances rather than arbitrarily defined aspects. Much of Leibniz's philosophy arises from arguments about what real entities, or identities, must entail.

Leibniz might seem to have already saddled himself with some sort of 'substance dualism' or 'theory of two types of entity' with the world on the one hand and points of view on the other. A point of view does not immediately strike one as an element of physics challenging the idea that physics is about 'things' made of 'matter'. His solution is to say that everything has to be seen in terms of dynamic relation. The nature of an entity is the way it relates dynamically to everything else. That seems to do well for a point of view and also works very well for the subject matter of physics, which is about the way entities relate dynamically to the world. Terms like mass and charge are shorthand for dispositions to interact with the world — to attract or repel or to resist change. This point of view became possible because shortly after Descartes's death it became clear to physicists that the extension' of matter arose from internal forces. Understanding nature in terms of constituent forces presents quite different possibilities from having to understand it in terms of abutting parts.

Leibniz was convinced that the basic nature of an entity is what he calls force' but which is closer to what we now call 'energy'. That seems entirely in agreement with modern physics. Intuitively, we tend to think that energy must be underpinned by some 'stuff' aspect, often thought of in terms of 'mass'. However, for a physicist, mass has always been just another disposition to dynamic relation, and with the Higgs mechanism probably confirmed any sense that mass is the basic stuff' has evaporated. Note that when Leibniz insists on the fact that entities are entirely determined in terms of their relational power, he does not getting rid of relata. He is simply indicating that their only knowable nature is in their relation to the universe in fact just as the equations of physics have it.

The monad can be equated to an indivisible dynamic mode, as described by modern physics. Indeed, dynamic modes are now defined by *Quantum Field Theory* (QFT), and are often called 'particles' but are better thought of as dynamic patterns or 'units of action', examples being electron orbitals, phononic modes, or photonic modes (excitations of the EM field), with a certain content of energy and domains that are contingent on their environment. Thus it may be said that a monad can only come into being or come to an end all at once; that is, it can come into being only by creation and end by annihilation, while that which is compound comes into being or comes to an end by parts. Nevertheless, he does recognize heterogeneity of monads. This makes his view both monistic and pluralist. Since there is constant change in the universe we also take as agreed that every created being, and consequently the created monad, is subject to change, and further that this change is continuous in each. The form of a monad is always that of a progression in spacetime.

The change in the monad is a manifestation of an internal principle. Change is intrinsic to all monads. This fits with the dynamic description of modes in modern physics[8]. The equations that describe modes indicate constant change, or progression with time. What may confuse is that what is observed is often independent of time because stages of progression cannot be distinguished by observation. The presence of relations within a monad in the absence of parts to relate may, again, seem contradictory. Heil (Heil 2012) has indicated how one can resolve the problem by distinguishing substantial and spatial parts.

With 'Necessary Being' substituted for 'God', Leibniz seems to be arguing, reasonably, that although the events going on around us form a causal network of combinatorial complexity they must yet be anchored in some common source of regularity or reason. Modes do not just go off on their own doing what they please. In a sense the vast array of individual events must merely be manifestations of some 'grand schema' that we can call the Universe. This is consistent with the view in modern physics that all modes are, in a sense, asymmetries of the universe progressing according to regularities of the whole, rather than whims of individual modes. And just as for the individual mode, the ultimate fate of the universe appears to be pre-determined by some unknown source of 'sufficient reason'.

On the other hand, Leibniz saw the monads as true individual substances. Whether this is relevant to modern physics is unclear but it is of note that while the dynamic laws of physics appear to be continuous, their instantiation does appear to be by discrete (and 'actual' in Whitehead's terms) individual dynamic units ('quanta'). However, Leibniz would still agree with modern physics that no further material substance is required, it is arguable that physics has to postulate some sort of prior overarching 'necessity' to make sense of regular dynamics at all. This is in keeping with motivation in modern physics to minimize the need for arbitrary or restricted parameters in the universe. There is a constant search for a framework, such as supersymmetry, that minimises apparently arbitrary parameters (e.g. mass and charge of particles). Where arbitrary parameters, such as the cosmological constant, do seem to remain some suggest that we should consider our environment as one of a set of universes, a Multiverse, in which all possible values of the parameter are expressed somewhere at

---

[8]The fact that Leibniz presages the basic principles of quantum field theory seem not to be commonly recognized. The immediate reaction you have when you read the Monadology is that Leibniz was neatly describing modes of excitation of fields.

some time. An alternative, perhaps closer to Leibniz is that 'reality' in modern physics may in a sense entail knowledge and only certain values for parameters could give rise to our level of knowledge.

Among other things, Leibniz gives a more detailed account of perfection, being the greatest possible combination of variety and order or symmetry. This contrasts with the individual monad, which being 'just one point of view necessarily lacks variety, and also concomitent symmetry. It is consistent with the identification of individual modes with asymmetries in modern field theory. In mathematical terms Leibniz's argument seems durable.It follows also that created beings derive their perfections, certainties or symmetries, from the Necessary Being, but that their imperfections, uncertainties or asymmetries, reflect their own nature, which is incapable of being unconstrained (being constrained by the laws of possible relation to the Universe). An instance of this original imperfection of created beings may be seen in the natural inertia of bodies. The distinction that Leibniz makes here between the 'imperfect' modes and the unconstrained perfection of the universe echoes well with modern physics in terms of the concept of asymmetry. Moreover, whereas the regularities that underlie modes are constant and involve continuous variables, instances of modes themselves, coming as they do in discontinuous units, cannot be described entirely in terms of the constant regularities. They are significantly unpredictable. It is intriguing that Leibniz considers inertia an 'imperfection', but it is intriguing that inertia is now looking to be related to the way modes interact with what would otherwise be a symmetrical uniform Higgs field, generating asymmetries not only in the Higgs field but in spacetime itself. Note that inertia was about the only physical property other than shape and motion that was well recognized in Leibniz's time.

So Leibniz's physical universe is asymmetric one: to accomplish this universe the symmetries between the individualities must be broken.

3 From Leibniz to Grassmann and Cartan

We need a mathematics which describes the *relational aspect* of the *spaces, symmetry* and the *broken of the symmetry* naturally. *As the complex structure unifies the plan and rotational symmetry in this plan, the quaternions the 3-spaces and rotations in three dimensions*, but here we should consider all the symmetries.

3.1 Grassmann's Breakthrough

In 1844 Grassmann published his first Ausdehnungslehre (Die lineale Ausdehnungslehre ein neuer Zweig der Mathematik) and in the same year won a prize[9]

---

[9]We have no certainty that Grassmann at that time had knowledge of this part of the work of Leibniz; However, it is known that Grassmann, informed by Möbius, participated and won the 4 July 1846 a price that was created on the occasion of the second centenary of the birth of Leibniz. The *Fürstlich Jablonowski'schen Gesellschaft Leipzig* proposed as a subject: "Reconstruct and develop the geometric calculation invented by Leibniz, or institute a similar calculation". Grassmann found in Leibniz a predeces-

for an essay which expounded a system satisfying an earlier search by Leibniz for an 'algebra of geometry' (Grassman 1847). Despite these achievements, Grassmann received virtually no recognition. This was no doubt in some part because his ideas and notation were too new and complicated.

In 1862 Grassmann re-expounded his ideas from a different viewpoint in a second Ausdehnungslehre (*Die Ausdehnungslehre. Vollständig und in strenger Form*; Grassmann 1862). Again the work was met with resounding silence from the mathematical community, and it was not until the latter part of his life that he received any significant recognition from his contemporaries. Of these, most significant were J. Willard Gibbs[10] (Gibbs 1886) who discovered his works in 1877 (the year of Grassmann's death), and William Kingdon Clifford (Clifford 1878) who discovered them in depth about the same time. Both became quite enthusiastic about this new mathematics.

According to Grassmann, a significant elements of the first draft of such a theory go back to 1832:

> As I read the excerpt from your memory concerning the geometric differences and addition which was published in the Proceedings [Volume 21, 1845], I was struck by the wonderful resemblance that exists between the results, which are communicated and mu own discoveries since 1832; [...] I have conceived the idea of the first geometric difference and addition of two or more lines as well as the geometric product of two or three lines in that year, it's an identical idea in all respects to that shown in the excerpt from your memory.
>
> [...] It is also in the same year 1832 that I got the idea of extending the use of these algebraic signs to the geometric operations that we use in mechanics when treating lines or areas, but I have nothing published before 1845.

Grassmann wrote this letter to Saint-Venant in 1947 to show that he anticipated some of Saint-Venant's ideas on vector addition and multiplication. But, not knowing Saint-Venant's address, Grassmann wrote to Cauchy in April 1847, asking him to forward his letter to Saint-Venant, along with one of the two copies of the Ausdehnungslehre of 1844. Cauchy never did so. And six years later Cauchy's paper (Cauchy 1953) appeared in *Comptes Rendus*. Grassmann's moment was that, on reading this, "*I recalled at a glance that the principles which are there established and the results which are proved were exactly the same as those which I published in 1844, and of which I gave at the same time numerous applications*". An investigating committee of three members of the French Academy, including Cauchy himself, never came

---

sor for his own creation: his *Ausdehnungslehre*, new mathematical discipline whose first application is the "*theory of space*" (Raumlehre), had already been reduced to one more modest "geometric calculus".

[10]Gibbs publishes one of his most important and creative papers in mathematics. Entitled "*On Multiple Algebra*," it makes a case for increased attention to multiple algebra, praises Grassmannian methods, and concludes with the famous line "*We begin by studying multiple algebras; we end, I think, by studying MULTIPLE ALGEBRA.*"

to a decision on the question. On June 17, 1847, Saint-Venant replied to Grassmann who sent him another letter, along with several papers, in January 1848 (Gesammelte mathematische und physikalische Werke, 3, Leipzig, 1911, pp. 120–122; see also (Grassmann 1894–1911).

We can find traces of his work in several papers of his, especially that of 1839 (Grassmann 1839), and in his studies on the theory of tides, where in its *Theorie der Ebbe und Flut* (Grassmann 1840) he uses methods and concepts that are explained and justified later in the 1844's *Ausdehnungslehre*. This works sign, as in Möbius and Hamilton, the emergence of the intrinsic, which is specific to the geometry which is indifferent to the coordinate system used. We see also emerge the idea of *invariance*.

So the pure mathematics from Grassmann's point of view are a *theories of Forms*:

> As I began then to rework the results obtained in their vein, I try from the beginning to not call to any theorem proved in any branch of mathematics, it appeared that the analysis that I discovered had stood not only, as it seemed at first, in the field of geometry; but I soon discovered that I had reached here the ground of a new science, the geometry itself is only a particular application.
>
> There must be [...] a branch of mathematics which generates intrinsically, in a purely abstract way, some similar laws like those, in geometry, appear to be related to space. The possibility of developing such a purely abstract branch of mathematical was given by the new analysis; better, this analysis, when it was developed independently of any theorem demonstrated elsewhere using the pur in abstraction, was that science itself. The main advantage obtained by this interpretation was, for the form, that all the principles that express the visions of space completely disappeared in this way and thus the beginning became as obvious as that of the arithmetic; however, for the content, the advantage was that the limitation to three dimensions became obsolete.

In 1850 Hamilton knew about Grassmann's work and in a set of letters from 1852 to 1853 to A. De Morgan[11] wrote (so we can see the evolution of Hamilton appreciation of Grassmann's work)

> [...] I have recently been reading [...] more than a hundred pages of Grassmann's Ausdehnungslehre, with great admiration and interest. Previously I had only the most slight and general knowledge of the book, and thought that it would require me to learn to smoke in order to read it. If I could hope to be put in rivalship with Descartes on the one hand, and with Grassmann on the other, my scientific ambition would be fulfilled! But it is curious to see how narrowly, yet how completely, Grassmann failed to hit off the Quaternions. He published in 1844, a little later than

---

[11] Lettre to A. De Morgan (Cf. Graves 1891, III, 424, 421, 442, 444, 425, 70, 87).

myself, but with the most abvious and perfect independence (28 Janvier 1853; Hamilton 1853) [...].

I am not quite so enthusiastic to-day about Grassmann as I was when I last wrote. But I have read through nearly all of what I could procure of his writings, including a subsequent commentary (in German) by Möbius. Grassmann is a great and most german genius; his view of space is at least as new and comprehensive as mine of time, [] I must say that I should not fear the comparison (2 Février 1853; *Ivi*).

Finally in his letters to Georges Salmon (23 juin 1857) related to Grassmann, Hamilton wrote (see Graves 1891; Hamilton 1853):

It is fair to say that (when too late) I found that Grassmann had independently (but perhaps not quite so soon — still is not a matter worth contesting) arrived at the same conception and notation, respecting the difference of two points (B–A), regarded as their directed distance — what he calls *strecke* and I *vector*. But this is merely a preparation for quaternions, and not as yet in any degree the Doctrine of the Quaternions themselves. I admire Möbius very much indeed, but he has (I think in his Barycentric Calculus, etc.) approached less nearly to the quaternions than Grassmann in his Ausdehnungslehre, p. 102 (*Ibidem*).

3.2 Cartan's Breakthrough

Later on Elie Cartan[12], starting from the Grassmann's Geometric calculus, develop the theory of exterior differential forms.

Cartan's mathematical work can be roughly classified under three main headings: group theory, systems of differential equations, and geometry. These themes are, how-

---

[12] About Elie Cartan, Shiing-Shen Chern and Claude Chevalley (1952) write:

Cartan was an excellent teacher; his lectures were gratifying intellectual experiences, which left the student with a generally mistaken idea that he had grasped all there was on the subject. It is therefore the more surprising that for a long time his ideas did not exert the influence they so richly deserved to have on young mathematicians. This was perhaps partly due to Cartan's extreme modesty. Unlike Poincaré, he did not try to avoid having students work under his direction.

And about Cartan's recognition Dieudonné writes

Cartan's recognition as a first rate mathematician came to him only in his old age; before 1930 Poincaré and Weyl were probably the only prominent mathematicians who correctly assessed his uncommon powers and depth. This was due partly to his extreme modesty and partly to the fact that in France the main trend of mathematical research after 1900 was in the field of function theory, but chiefly to his extraordinary originality. It was only after 1930 that a younger generation started to explore the rich treasure of ideas and results that lay buried in his papers. Since then his influence has been steadily increasing, and with the exception of Poincar and Hilbert, probably no one else has done so much to give the mathematics of our day its present shape and viewpoints.

ever, constantly interwoven with each other in his work. Almost everything Cartan did is more or less connected with the theory of Lie groups.

The idea of studying the abstract structure of mathematical objects which hides itself beneath the analytical clothing under which they appear at first was also the mainspring of Cartan's theory of differential systems. He insisted on having a theory of differential equations which is invariant under arbitrary changes of variables. Only in this way can the theory uncover the specific properties of the objects one studies by means of the differential equations they satisfy, in contradistinction to what depends only on the particular representation of these objects by numbers or sets of numbers. In order to achieve such an invariant theory, Cartan made a systematic use of the notion of the exterior differential of a differential form, a notion which he helped to create and which has just the required property of being invariant with respect to any change of variables.

Cartan's theory of exterior differential systems[13] was deeply rooted in the historical context of the late nineteenth century theory of partial differential equations. Indeed, Cartan's works were placed at the cross-road of two closely related fields of research: the theory of not completely integrable systems of Pfaffian equations as developed by Engel and von Weber[14] after Frobenius, Clebsch, Forsyth and Darboux, and the theory of general systems of partial differential equations that was the main focus of attention of Méray, Riquier and Delassus among others.

Cartan's great genius was to reinterpret them systematically in a new and powerful geometrical language whose central core was represented by his exterior differential calculus. At the same time, the very emphasis placed by him on the language of exterior differential forms made his achievements inaccessible. For instance, Vessiot in 1924, while praising the beauty of Cartan's integration theory, felt the necessity to translate exterior Pfaffian forms with the notion of faisceau of infinitesimal transfor-

---

[13] Generalized later by Kähler (1934) to differential systems of any degree.
[14] Some of von Weber's idea was to assume a role of outstanding importance in Cartan's theory. Motivated by the possibility of developing fruitful applications in the realm of the theory of general partial differential equations, he introduced further hypotheses which guarantee the existence of integral varieties of increasing dimensions. Indeed, he explained, if one supposes that the system S_ν

$$\frac{\partial x_{m+h}}{\partial u_r} = \Sigma_{i=1}^m a_{ih} \frac{\partial x_i}{\partial u_r} \quad (r = 1, \ldots, \nu; h = 1, \ldots, n-m) \tag{1}$$

$$= \Sigma_{i=1}^m \Sigma_{k=1}^m a_{ikh} \frac{\partial x_i}{\partial u_r} \frac{\partial x_k}{\partial u_s} = 0 \quad (r, s = 1, \ldots, \nu; h = 1, \ldots, n-m) \tag{2}$$

can be put in a canonical passive form simply by solving it with respect to certain derivatives $\frac{\partial x_i}{\partial u_\nu}$ and, furthermore, if one supposes that from S_ν and from equations obtained from it by differentiation and elimination no relation can be deduced among the variables $x_i, \frac{\partial x_i}{\partial u_1}, \ldots, \frac{\partial x_i}{\partial u_\nu}, (i = 1, \ldots, n)$, already contained in $S_{\nu-1}$, and so on for the systems $S_{\nu-1}, , \ldots, S_1$, then every 1-dimensional integral variety M_1 of a system of nm Pfaffian equations :

$$\nabla_s = dx_{x+s} - \Sigma_{i=1}^m a_{si} dx_i = 0, \quad (s = 1, \ldots, n-m)$$

belongs at least to one 2-dimensional integral variety M_2, etc. and, finally, every $\nu - 1$-dimensional integral variety $M_{\nu-1}$ belongs at least to one ν-dimensional integral variety M_ν.

mations.

As Kähler remarked (Kähler 1934), such a double historical origin was reflected in the twofold virtue of the theory: on one hand, The emphasis on exterior forms provided Cartan with the necessary tools for subsequent applications to geometry (namely, the method of moving frames) as well as to the theory of infinite continuous Lie groups. On the other hand, it offered a deeper insight into the machinery of partial differential equations.

Starting from 1894, Cartan was concerned by working the applications of the his theoretical results. The theory of partial differential equations became one of the main fields of his interest. Cartan's work (Cartan 1899) on Pfaffian forms was part of this interest. Indeed, the integration of partial differential equations and the integration of Pfaffian forms were considered as equivalent formulations of the same problem.

To appreciate the efficiency of the exterior forms formulation. Let us describe E. Cartan deductive approach to resolve Pfaff problem. He started usually by giving a symbolic definition of what a differential expression in n variables is; this was defined as a homogenous expression built up by means of a finite number of additions and multiplications of the n differentials $dx_1, \ldots, dx_n$ as well as of certain coefficients which are functions of $x_1, \ldots, x_n$. In this way, a Pfaffian expression was defined as a differential expression of degree one such as $A_1 dx_1 + \ldots, +A_n dx_n$; a differential form of degree two was given, for example, by $A_1 dx_2 \wedge dx_1 + A_2 dx_3 \wedge dx_2$.

At this point, Cartan defined two differential expressions of degree h to be equivalent if their value is the same independently of the choice of parameters $\alpha_1, \ldots, \alpha_h$. In this way, he was able to establish Grassmann's well-known multiplication rules, which were to be interpreted, in Cartan's view, as equalities between equivalence classes of exterior differential forms. For example, one has $dx_1 \wedge dx_2 = -dx_2 \wedge dx_1$ or Cartan did not employ the wedge product symbol $\wedge$. One can find the germs of this crucial notion already in some of Poincaré's work on integral invariants.

The theory of exterior differential systems plays a crucial role in Cartan's Whole mathematical achievements. It provided him with powerful technical tools useful in many different fields such as the theory of partial differential equations, the theory of infinite dimensional Lie groups (Lie pseudogroups) and differential geometry. Cartan's work on Pfaffian systems (what we would nowadays call exterior differential systems) is not only a landmark of Cartan own mathematical achievements but of the development of twentieth century mathematics, such as the general theory of partial differential equations, Lie groups and the theory of equivalence and differential geometry, to mention only a few, also.

4 Differential Forms and the Realization of Leibnizian Project

Elie Cartan, using the Geometric calculus of differential forms, introduced the structure of a *soldering 1-form*, which is a generalization of the idea of moving frame[15]

---

[15] The idea of moving frames is a generalization of Darboux kinematical theory.

(Cartan 1935). The soldering form associate to each point of basis manifold an attributes, for instance, a metric in the tangent bundle. As we will see this idea of associating a metric to the tangent bundle, is essential in general relativity. So, general relativity realizes the Leibnizian idea which is a point is determined by its attributes, in our case the attribute is the metric tensor. In general case we can associate a p-forms to a point, generalizing the concept of momentum-energy tensor.

4.1 The Idea of Soldering Form

In Riemannian[16] geometry people follow Levi-Civita's approach where a connection on a differentiable manifold is essentially has the meaning of parallel transport along curves. This notion of parallel transport then gives rise to the concept of covariant derivative, etc. The principal bundle connected with this approach is the bundle of orthonormal frames. In the classical (19th century) theory of surfaces, there was another notion intimately related to parallel transport, and that is the development of a surface on a plane along a curve. This idea was introduced by Felix Klein in 1872 in his work on homogeneous space (Klein 1892–1893), who considered the quotient space G/G_y, where G_y is stabilizer of some point $y \in Y$. This given us the isomorphism $Y \cong G/G_y$ as G-spaces, allow us then to study algebraic structure associate with the geometry. Note that since $G_{gy} = gG_y g^{-1}$, all the algebraic descriptions are G-equivariant. Intuitively speaking, one rolled the surface on the plane along the curve, maintaining first-order contact. This gives an identification of the tangent space to the surface at each point of the curve with the plane. Then parallel translation in the Euclidean geometry of the plane gives the parallel transport in the sense of Levi-Civita along the curve. From this point of view the notion of development is central, and the crucial property of the plane is that it is a homogeneous space in the sense that it admits a transitive group of isometries that includes the full rotation group as the isotropy group of a point.

Later on the Klein idea is extended by Elie Cartan to the case of *non-homogeneous* manifold M (Sharpe 1996; Cartan 1904). But this time the broken symmetry in a G connection induces an isomorphism $e : T_x M \Rightarrow \mathfrak{g}/\mathfrak{g}_y$ for each tangent space, which is the *coframe-soldering form* identifying $T_x M$ with $\mathfrak{g}/\mathfrak{g}_y \cong T_y Y$, thus solder a copy of Y to M, at each point x. All theses copies are related by Cartan connection.

Unfortunately, the success of the Levi-Civita connection to study the general relativity has overshadowed the importance of Cartan geometry. It reappeared in the framework of Yang-Mills theory via Ehresmann connection formalism, lacking the *broken symmetry mechanism analogue* in Cartan original formulation (Ehresmann 1936). Here the geometry involved is that of internal spaces, a *bundle* over the space-time rather than the space-time itself. Recently many people start using Elie Cartan's original formulation to study gravity (Wise 2010).

5 The Geometrization of Physics

---
[16]and semi-Riemannian.

Homage to Gottfried W. Leibniz 1646–1716

The essence of the method of physics is inseparably connected with the problem of interplay between local and global properties of the universe. Physics started its triumphant progress when people like Galileo and Newton succeeded in isolating free fall of a stone from the network of interactions shaping the structure of the world. On the other hand, the question imposes itself: Is the "whole of the universe" a sum of its parts (or aspects) or perhaps "something more", something that cannot be reconstructed by investigating only "local details"? It seems that the essence of the method of physics is inseparably connected with the problem of interplay between local and global aspects of the world's structure.

One cannot grasp modern physics without understanding Gauge theory, which tells us that the fundamental interactions in nature are based on parallel transport, and in which forces are described by curvature, which measures the path-dependence of the parallel transport the three-dimensional Euclidean space of our intuition. Indeed, in 1917, it was discovered by Levi-Civita that the study of curved manifolds in differential geometry can be based on the notion of parallel transport of tangent vectors (velocity vectors).

Gauge theory is the result of a fascinating long-term development in both mathematics and physics. Gauge transformations correspond to a change of potentials, and physical quantities measured in experiments are invariants under gauge transformations. Let us briefly discuss this. Gauss discovered that the curvature of a two-dimensional surface is an intrinsic property of the surface. This means that the Gaussian curvature of the surface can be determined by using measurements on the surface (e.g., on the earth) without using the surrounding three-dimensional space. The precise formulation is provided by *Gauss' theorema egregium* (the egregious theorem). Bernhard Riemann (1826–1866) and Elie Cartan (1859–1951) formulated far-reaching generalizations of the *theorema egregium* which lie at the heart of modern differential geometry (the curvature of general fiber bundles), and modern physics (gauge theories).

Interestingly enough, in this way, Einstein's theory of general relativity (the curvature of the four-dimensional space-time manifold), and the Standard Model in elementary particle physics (the curvature of a specific fiber bundle with the symmetry group $U(1) \times SU(2) \times SU(3)$) can be traced back to *Gauss' theorema egregium*.

In classical mechanics, a large class of forces can be described by the differentiation of potentials. This simplifies the solution of Newton's equation of motion and leads to the concept of potential energy together with energy conservation (for the sum of kinetic and potential energy). In the 1860s, Maxwell determined that the computation of electromagnetic fields can be substantially simplified by introducing potentials for both the electric and the magnetic field (the electromagnetic four-potential).

Gauge theory generalizes this by describing forces (interactions) by the differentiation of generalized potentials (also called connections). The point is that gauge transformations change the generalized potentials, but not the essential physical effects. Physical quantities, which can be measured in experiments, have to be invariant under gauge transformations. Parallel to this physical situation, in mathematics the Riemann curvature tensor can be described by the differentiation of the Christoffel symbols (also called connection coefficients or geometric potentials). The notion of the Riemann curvature tensor was introduced by Riemann in order to generalize *Gauss' theorema egregium* to higher dimensions. In 1915, Einstein discovered that the Riemann curvature tensor of a four-dimensional space-time manifold can be used to describe gravitation in the framework of the theory of general relativity. The basic idea of gauge theory is the transport of physical information along curves (also called parallel transport). This generalizes the parallel transport of vectors in Einstein geometrized gravitation in his 1915 theory of general relativity. Quantum mechanics was geometrized by Dirac, as a unitary geometry of Hilbert spaces. In the introduction to his book *The Principles of Quantum Mechanics* (Clarendon Press, 1930), the young Dirac (1902–1984) wrote:

> The important things in the world appear as invariants [...] The things we are immediately aware of are the relations of these invariants to a certain frame of reference [...].. The growth of the use of transformation theory, as applied first to relativity and later to the quantum theory, is the essence of the new method in theoretical physics (*Ibidem*).

Finally, note that the Standard Model in particle physics starts from a classical field theory which is closely related to the geometry of specific fiber bundles.

5.1 Cohomology and Invariants

The other aspect of the modern Mathematics and physics is the studies of the invariants[17]. Indeed, Classify the invariants became a central issue in physics and mathematics, a fundamental tool which will play an essential role to find and calculate those invariants is the Cohomology and homology.

Indeed, Cohomology[18] plays a fundamental role in modern mathematics and physics.

---

[17] Differential forms provide a modern view of calculus. They also give you a start with algebraic topology in the sense that one can extract topological information about a manifold from its space of differential forms, it's called cohomology. For instance, de Rham cohomology is a good way to detect the "shape" of a domain. Indeed, we say that a vector field F in $\mathbb{R}^3$ is conservative if $F = \nabla f$ for some scalar-valued function f. This has natural applications in physics (e.g. electric fields). It's easy to see this happens if and only if line integrals of F are path independent, if and only if line integrals around closed loops vanish, etc. On a simply connected domain, F is conservative if and only if $\nabla \times F = 0$ (use the freshman version of Stokes' theorem). On a non-simply connected domain, this may fail (e.g. $\mathbb{R}^3$ minus a line). The extent to which it fails is of course the de Rham cohomology of the domain. Notice also that the differential forms are necessary for the development of cohomology theory in the context of manifolds without getting into the aspects which depend on metric notions.

[18] Often in math you wish something were true, but in general it is not. But if you can quantify how badly it fails, that's a good step towards finding out a more precise statement that holds generally. In algebra, geometry, and topology, a good method for doing this is usually to express "failure" as the non vanishing of a cohomology class. The size (or dimension) of the corresponding cohomology group is a measurement of how many ways things can go wrong. If it is nice or if you can understand it completely, then you may be able to analyze all the possible failure modes exhaustively, and use that to prove something interesting. This idea can be applied in an amazingly broad set of contexts.

Homage to Gottfried W. Leibniz 1646–1716

Cohomology is an example of a local — global structural connection that permeates mathematics.

Roughly, Poincare duality connects the local statements of cohomology to the global statements of homology. Homology detects some types of holes on manifold. It's a limited form of homotopy, where a person will push and pull on circles and spheres like lassos until they hit an obstruction — a hole. So the mathematical idea is that cohomology is equivalent to a limited form of homotopy, namely homology, that detects holes in manifolds. It turns out that cohomology and homology have their roots in the rules for electrical circuits formulated by Kirchhoff in 1847.

Cohomology is used in physics to compute topological structure of gauge fields, like the electromagnetic field in the Ahranov-Bohm effect. Here, the electron encircles a magnetic flux, which you can measure in the self interference pattern of the electron. That is amazing because the electron never actually passes through a magnetic field. Maxwell's equations tell us that the only interaction between the magnetic field and the electron is local — when the electron passes through the field. So what is going on?

The magnetic field is the curvature of a vector potential, which is a gauge connection and "lives" in the cohomology of the underlying manifold. Because of its cohomology, it is a real thing with measurable consequences in physics experiments. The magnetic flux plays the role of the hole or obstruction. A single electron may encircle a single flux in quantum physics because the electron can be in two places at once — it can take the path to the left of the flux and to the right, at the same time, and meet itself at the other side. The result is a full circle. Quantum mechanics also says that the electron picks up a phase from the vector potential, which physically manifests itself in self interference patterns of the electron. But why would an interference pattern emerge when the electron only travels through space in regions where the magnetic field is zero?

To understand this we will use Stoke's theorem. The field strength B is the curvature of the vector potential A, $B = dA$. The magnetic vector potential A is closed $[B = dA = 0]$ but not exact (i.e. A cannot be written as the curvature of another form everywhere i.e. A not equal dQ, though $A = dQ$ in patches and each patch contains a different Q, which must be patched together like a quilt). The vector potential adds up along the path(s) in the phase of the electron interacting with the electromagnetic gauge field. That is, around a path encircling the magnetic flux, you add up A, which by Stoke's theorem equals adding B in the enclosed disk, which is non zero because of the enclosed magnetic flux.

Differential cohomology is the group structure built up from the vector space of forms that are closed modulo the vector space of forms that are exact. The fact that idea manifests itself in physics was a huge surprise in the 50s, 60s and 70s. We still seem to be continually surprised by cohomology popping up in physics everywhere we look. That is a deeply intuitive way to see cohomology. The electron and the field interact locally in a trivial way yet sense non trivial global structure of the system, namely the encircled magnetic flux, which acts as an obstruction, when the path of the electron in spacetime forms a closed circle around it. Adding up local variations yields global data, even far away from where the global data is most manifest, namely at the obstructions and holes.

This helps to explain why the Maxwell equations in electrodynamics are closely related to cohomology, namely, de Rham cohomology based on Cartan's calculus for differential forms and the corresponding Hodge duality on the Minkowski space. Since the Standard Model in particle physics is obtained from the Maxwell equations by replacing the commutative gauge group $U(1)$ with the noncommutative gauge group $U(1) \times SU(2) \times SU(3)$, it should come as no great surprise that de Rham cohomology also plays a key role in the Standard Model in particle physics via the theory of characteristic classes (e.g., Chern classes which were invented by Shing-Shen Chern in 1945 in order to generalize the Gauss-Bonnet theorem for two-dimensional manifolds to higher dimensions).

It is very clear now that the gauge-theoretical formulation of modern physics is closely related to important long-term developments in mathematics pioneered by Gauss, Riemann, Poincaré and Hilbert, as well as Grassmann, Lie, Klein, Cayley, Elie Cartan and Weyl. The prototype of a gauge theory in physics is Maxwell's theory of electromagnetism. The Standard Model in particle physics is based on the principle of local symmetry. In contrast to Maxwell's theory of electromagnetism, the gauge group of the Standard Model in particle physics is a noncommutative Lie group. This generates additional interaction forces which are mathematically described by Lie brackets.

We also emphasize the methods of invariant theory. In terms of physics, different observers measure different values in their experiments. However, physics does not depend on the choice of observers. Therefore, one needs both an invariant approach and the passage to coordinate systems which correspond to the observers, as emphasized by Einstein in the theory of general relativity and by Dirac in quantum mechanics. The appropriate mathematical tool is provided by invariant theory.

One final remark concerns the cohomolgical nature of the masse: "it parametrizes the extensions of the Galileo group". Indeed, classical mechanics, the Galilei group acts on the symplectic manifold of states of a free particle. But in quantum mechanics, we only have a projective representation of this group on the Hilbert space of states of the free particle. The cocycle is the particle's mass. In other words, you can't see the mass of a free classical particle by just watching its trajectory, since it goes along a straight line at constant velocity no matter what its mass is. But you can see the mass of a free quantum particle, because its wave function smears out faster if it's lighter! So there's some difference between classical and quantum mechanics. Ultimately this arises from the fact that the latter involves an extra constant, Planck's constant. In slight disguise, one can see this cocycle also control already the classical free non-relativistic particle, in the sense that its action functional is of the form of a 1d WZW model with that cocycle being the "WZW term" that however comes down to be the ordinary free action.

Homage to Gottfried W. Leibniz 1646–1716

6 The Road to a Generalized Mathematical Framework for Modern Physics

The idea of a soldering form on a vector bundle is a subtle variant of the notion of a moving frame (Cartan 1935) $(e_1, \cdots, e_m)$ on a manifold M of dimension m and of its dual coframe $(\alpha^1, \cdots, \alpha^m)$ (Hélein and Kouneiher 2004), in which we are considering an auxiliary vector bundle $W\mathcal{M}$ which is isomorphic to $T\mathcal{M}$ and hence in particular has the same dimension m as $\mathcal{M}$ (for more details see (Kouneiher and Barbachoux,2015)). Here the identity automorphism $Id_{T\mathcal{M}} : T_M\mathcal{M} \longrightarrow T_M\mathcal{M}$ is replaced by an isomorphism α from $T\mathcal{M}$ to $W\mathcal{M}$, i.e. a section of $W\mathcal{M} \otimes_\mathcal{M} T^*\mathcal{M}$, the vector bundle of linear maps from $T\mathcal{M}$ to $W\mathcal{M}$, such that, $\forall M \in \mathcal{M}$, α_M is an isomorphism. Though a fixed metric g on $W\mathcal{M}$, induces automatically a Riemannian metric α^*g on $\mathcal{M}$ defined by

$$\forall M \in \mathcal{M}, \forall v, w \in T_M\mathcal{M}, \quad (\alpha^*g)_M(v,w) := g_M(\alpha_M(v), \alpha_M(w)).$$

Note that α may alternatively be viewed as a 1-form on $\mathcal{M}$ with values in $W\mathcal{M}$. This 1-form is called the *soldering form*. This construction is related to the moving frame description since, a moving frame $(E_1, \cdots, E_m)$ on $W\mathcal{M}$, give automatically a moving frame $(e_1, \cdots, e_m)$ on $T\mathcal{M}$ defined by :

$$\forall M \in \mathcal{M}, \forall i = 1, \ldots, m, \quad \alpha_M(e_i) = E_i. \tag{3}$$

Our idea is to generalize this structure. and use the collection $(\mathcal{M}, W\mathcal{M}, g, \nabla, \sigma)$, which we will called soldered space. Where $\mathcal{M}$ is a manifold, $W\mathcal{M}$ is a vector bundle over $\mathcal{M}$ isomorphic to $T\mathcal{M}$, g is a metric on $W\mathcal{M}$, ∇ is a connection structure associated with Dirac operator or spin structure for instance, on $W\mathcal{M}$ compatible with g and σ is a 1-form on $\mathcal{M}$ with values in $W\mathcal{M}$ forms.

Our aim now is to use the generalized p-soldering forms *together with the connection to solder together the fibers of $W\mathcal{M}$*, i.e. to sew together $W_M\mathcal{M}$'s. Hence the main object in our geometry is not $\mathcal{M}$ and its points, but the collection of all fibers $(W_M\mathcal{M})_{M\in\mathcal{M}}$ i.e. the origins of fibers $W_M\mathcal{M}$, soldered together by the connection ∇ and the soldering form α. From this point of view the resulting Riemannian manifold inherits his rigidity from the fibers $W_M\mathcal{M}$ which are a rigid objects.

7 Physical Applications: General Relativity as Leibnizian Project

In the relationist version, only the objects are real. The lattice has no independent existence. Even without the lattice, it does make sense to talk about the distance between two given objects. But it does not make sense to talk about the distance between two space points where there are no objects at all. Today, closely related to background independence is a basic ingredient of general relativity, known by the imposing name diffeomorphism invariance. It concerns the coordinates physicists use to describe space and time.

The principle of diffeomorphism invariance implies that, unlike in theories prior to general relativity, there are no additional structures in physics that allow us to distinguish preferred coordinate systems. As far as the laws of physics are concerned, no coordinate system is better than another, and one is free to choose. So, a point in space is defined only by the presence of an object, not by its location according to some special set of coordinates.

So, in general relativity: *A point in spacetime (an event) is defined only by what physically happens at it, not by its location according to some special set of coordinates. The physics does not depend on what lattice you choose.*

With its dynamical geometry and the inability of pinning geometry down in terms of a rigid lattice (diffeomorphism invariance), general relativity is at least a partly relational theory. In fact, the relational/absolute debate is the basic historical and philosophical context for Einstein's work.

In General Relativity, the manifold $\mathcal{M}$ and the metric should be built simultaneously[19] when solving equation (4)

$$R_{ij} - \frac{1}{2}Rg_{ij} = T_{ij}, \tag{4}$$

From this point of view, the only 'kinematic' condition imposed is that at each point of space-time, the tangent space is endowed with a metric (which is a Minkowski metric in the physical case of pseudo-Riemannian manifolds and an Euclidean one in the Riemannian analogous problem).

Then the field (g_{ij}) describes the way these metrics depend on the point in a smooth way and the Einstein equation (4) is the 'dynamical' constraint on g_{ij}. So we have to imagine an infinite continuous family of copies of the same Minkowski or Euclidean space and to find a way to sew together these infinitesimal pieces into a manifold, by respecting (4).

The Euler–Lagrange system (4) is replaced by the system

$$\begin{cases} d^\nabla \sigma & = 0 \\ \lambda_\ell := \epsilon_{ijk\ell}\Omega^{ij} \wedge \sigma^k & = 0, \end{cases} \tag{5}$$

where $\epsilon_{ijk\ell}$ is the completely skewsymmetric tensor such that $\epsilon_{1234} = 1$, $\Omega^{ij} := \Omega^i_k g^{kj}$ and $\Omega^i_j := d\omega^i_j + \omega^i_k \wedge \omega^k_j$ is the curvature 2-form of the connection ∇.

Note that (5) is the Euler–Lagrange equation of the Palatini action

$$\mathcal{P}[\nabla, \sigma] := \int_\mathcal{M} \epsilon_{ijk\ell}\Omega^{ij} \wedge \sigma^k \wedge \sigma^\ell \tag{6}$$

A variant of this formulation is the Ashtekar action. As we have seen the first equation in (5) is a compatibility condition between ∇ and σ, the torsion free condition, whereas, once we know that $\sigma^*\nabla$ is torsion free, the second equation of (5) reads $R_{ij} - \frac{1}{2}R(\sigma^*g)_{ij} = 0$, i.e. the Einstein equation (4) in a vacuum.

To understand the intuition behind this, it is useful to go back to the "points" of $(\mathcal{M}, V\mathcal{M}, \nabla, \varphi)$. Then the 1-form φ and the connection ∇ provide informations about how to connect a *pair of points* in the geometry: if $M_0 \in \mathcal{M}$ and $M_1 \in \mathcal{M}$, then any $\mathcal{C}^1$ path $\gamma : [0,1] \longrightarrow \mathcal{M}$ which connects M_0 to M_1 can be lifted using φ and ∇.

---

[19] Each monad mirrors the entire universe. *"Every body responds to all that happens in the universe"* (61)

The two structures 1-form φ and the connection ∇ realize Leibniz's idea: although bodies may be held to stand in spatial and temporal relations to one another, Leibniz claims, *space and time themselves must be considered abstractions or idealizations with respect to those relations. For while relations between bodies and events are necessarily variable and changing, the relations constituting space and time must be viewed as determinate, fixed, and ideal.*

Indeed, the attributes of the monad (or points) have two substructures, one changing and the other is fixed substructure, and which represent space and time. This substructure is realized by the soldering form which associate to each point a metric in the tangent bundle. The rest of the structure will describe the matter attributes associate to each point of our soldered space or Monadology.

8 Conclusion

Leibniz was the first to consider categorizing and divisions between reality and knowledge were artificial because they are established by men, reality and knowledge are one, everything is united and bound in a reality that has the same source: the universal reason of God. Leibniz was concerned by the idea of unifying sciences and reconciling knowledge. Leibniz mathematical competence combined with the philosophical erudition gives his metaphysical system a particular shade: his metaphysics is logic, consistency and rigorous.

The modern Mathematical-Physics thought questions the essential of interrelationship between matter, space and time. This is also the case of Leibniz during his last years; Leibniz, in letters to many scholars, took some important points of his system; with Father B. des Bosses, it deals with the Monad, the material of the body and the bodily substance; with Bourguet, perception and the increasing perfection of the creatures; with Clarke, God, space and time. Questioning the nature of space and time is not, strictly speaking, the cause of Leibniz's philosophy. The founding stone of this edifice is rather the principle of sufficient reason and the system of pre-established harmony.

One essential pillar of Leibniz's positive account of space and time is rooted in his view that — even understood as systems of relations — space and time as "beings of reason" are in a sense at least two steps removed from the monads of his mature metaphysics.

Differential forms are central to the modern formulation of classical mechanics where manifolds and Lie groups are employed to describe the configuration and time evolution of mechanical systems (Abraham and Marsden 1978; Arnold 1989; Cartan and Kouneiher 2006; Holm, Schmah and Stoica 2009). One of the principal applications of differential forms in modern mechanics is the mathematical description of observables: infinitesimal measurements, which, when integrated, yield a value that can be verified through real world experiments, at least in principle. Differential forms are thereby a natural choice when one requires that measurements satisfy:

1. covariance, that is invariance under coordinate transformations;

2. covariance under differentiation, which is crucial since the time evolution of most systems is described by differential equations;

3. measurements are obtained by integration from the infinitesimal quantities employed to describe time evolution.

The first requirement implies that differential forms have to be tensors, objects whose physical manifestation does not change under coordinate transformations, and the second requirement implies that these have to be anti-symmetric, leading to differential forms, anti-symmetric tensors that are "[...] ready (or designed, if one prefers) to be integrated" (Desbrun, Kanso and Tong 2006). Differential forms can hence be seen as a modern formulation of classical infinitesimals and, as we will see in the following, a formulation that adds much insight and efficacy to the concept.

The calculus of differential forms is not only vital to the mathematical description of mechanical systems in the continuum limit but, as began to be understood only recently, it is also crucial for numerical computations (*Ibidem*; Arnold, Falk and Winther 2010). First applications were in electromagnetism (Deschamps 1981; Bossavit 1997), but their relevance for many other systems has been demonstrated.

Cartan's theory of exterior differential forms is NOT just another fancy notational system. Indeed, the algebra of differential forms along with the exterior derivative defined on it is preserved by the pullback under smooth functions between two manifolds. This feature allows geometrically invariant information to be moved from one space to another via the pullback, provided the information is expressed in terms of differential forms[20]. As a particular example, the change of variables formula for integration becomes a simple statement that an integral is preserved under pullback.

So they can be used to describe situations that involve topological evolution, a irreversible or non conservative physics while the usual contra-variant tensor fields of particle mechanics are restricted (usually) to a reversible evolution that preserves topological features. In a certain sens, physics seems to be a conditions of contingencies of a new geometry.

---

[20] One of the main reasons the cotangent bundle rather than the tangent bundle is used in the construction of the exterior complex is that differential forms are capable of being pulled back by smooth maps, while vector fields cannot be pushed forward by smooth maps unless the map is, say, a diffeomorphism. The existence of pullback homomorphisms in de Rham cohomology depends on the pullback of differential forms.

References

Abraham, R., Marsden, J. E. 1978. *Foundations of Mechanics*. Massachusetts: Addison-Wesley Publishing Company, Inc.

Arnold, V. I. 1989. *Mathematical Methods of Classical Mechanics*. 2nd. Graduate Texts in Mathematics. New York: Springer-Verlag.

Arnold, D. N., Falk, R. S., Winther, R. 2010. Finite element exterior calculus: from Hodge theory to numerical stability. *Bulletin (New Series) of the American Mathematical Society* 47:281–354.

Arthur, R. 2014. *Leibniz*. Cambridge: Polity Press.

Barbour, J., Smolin, L. 1992. Extremal Variety as the Foundation of a Cosmological Quantum Theory, available via: arXiv:hep-th/9203041v1

Barbour, J. 1994. On the origin of structure in the universe. In: Rudolph, E., Stamatescu, I.-O., (Eds.), *Philosophy, Mathematics and Modern Physics*, Berlin: Springer-Verlag, pp. 120–131.

Bossavit, A. 1997. *Computational Electromagnetism: Variational Formulations, Complementarity, Edge Elements*. San Diego: Academic Press.

Cartan, E. 1904. Sur la structure des groupes infinis de transformations. *Annales Scientifiques de l'cole Normale Suprieure* 21:153–206.

Cartan, E. 1935. *La Méthode du Repère Mobile, la Théorie des Groupes Continus et les Espaces Généralisés*, Exposés de Géométrie. Vol. 5, Paris: Hermann.

Cartan, H., Kouneiher, J. 2006. *Differential Forms*. Paris: Hermann.

Cauchy, A. L. 1953. Sur les clefs algèbriques. *Comptes Rendus de l'Académie des Sciences*, 36:70–75; 129–136; 161–169.

Chern S. S., Chevalley, C. 1952. Elie Cartan and his mathematical work. *The Bulletin of the American Mathematical Society* 58:217–250.

Clifford, W. K. 1878. Application of Grassmann's Extensive Algebra. *American Journal of Mathematics* 1:350–358.

Desbrun, M., Kanso, E., Tong, Y. 2006. Discrete Differential Forms for Computational Modeling. In: *SIGGRAPH '06: ACM SIGGRAPH 2006 Courses*. New York: ACM Press, pp. 39–54.

Deschamps, G. A. 1981. Electromagnetics and Differential Forms. *Proceedings of the IEEE 69*, pp. 676–696.

Edward, J. C.W. 2014. *A 21st Century Monadology or Principles of Philosophy. A 300th anniversary recasting of, and tribute to, the text of Gottfried Leibniz 1714*. Available via https://www.ucl.ac.uk/jonathan-edwards/monadology.pdf

Ehresmann, C. 1936. Sur les espaces localement homogenes, *Annales scientifiques de l'École normale supérieure — Mathématiques* 35:317–333.

Flament, D. 2008. *Théorie des formes et avènement d'une nouvelle discipline des mathématiques pures, selon Hermann Günter Grassmann (1809-1877)*. Revista Brasileira de História da Ciñcia 1(2):178-210.

Flament, D., Bekemeier B. 1997. *Le nombre une hydre à n visages; entre nombres complexes et vecteurs*. Paris: Ed. Maison des Sciences de l'Homme.

Flament, D., Garma, S., Navarro, V. (Eds.). 1994. Hermann Günther Grassmann. *La science de la grandeur extensive. La lineale Ausdehnungslehre*. Paris: Blanchard.

Gibbs, J. Willard. 1886. *On Multiple Algebra*, Mass: Salem.

Grassman, J. 1824. *Raumlehre für die untern Klassen der Gymnasien, und für Volksschulen*. Berlin: Ebene raeumliche Groessenlehre.

Grassman, H. 1839. *Ableitung der Krystallgestalten aus dem allgemeinen Gesetze der Krystallbildung. Programm der Ottoschule*. In: Grassmann 1894–1911, II(2), pp. 115–146.

Grassman, H., 1840. *Theorie der Ebbe und Flut*. In: Grassmann 1894–1911, III(1), pp. 1–238.

Grassman, H. 1847. *Die Geometrische Analyse geknüpft an die von Leibniz erfundene geometrische Charakteristik*. In: Grassmann 18941911, I(1), pp. 322–399.

Grassman, H. 1862. Ueber die Verbindung der Konsonanten mit folgenden j und die davon abhängigen Erscheinungen. *Zeitschrift fr vergleichende Sprachforschung* 11:1–52, 81–103.

Grassman, H., 1894–1911. *Gesammelte mathematische und physikalische Werke*. Vol. 3. Leipzig: Teubner [New York: Chelsea Publishing Company 1969; New York: Johnson Reprint Corporation 1972].

Graves, R. P. 1891. *Life of Sir William Rowan Hamilton*. Vol 3. Dublin: Hodges Figgis, and Co.

Hamilton, W.R. 1853. *Lectures on Quaternions*, Dublin: Hodges and Smith.

Hélein, F., Kouneiher, J. 2004. Covariant Hamiltonian formalism for the calculus of variations with several variables: Lepage–Dedecker versus De Donder–Weyl, *Advances in Theoretical and Mathematical Physics* 8(3):565-601.

Heil, J. 2012. *The Universe As We Find It*. Oxford: Oxford Uiversersity Press.

Holm, D. D., Schmah, T., Stoica, C. 2009. *Geometric Mechanics and Symmetry: From Finite to Infinite Dimensions*. Oxford texts in applied and engineering mathematics. Oxford: Oxford University Press.

Käher, E. 1934. Einführung in die Theorie der Systeme von Differentialgleichungen [Introduction to the theory of systems of differential equations]. *Hamburger Mathematische Einzelschriften* 16.

Klein, F. 1892–1893. A comparative review of recent researches in geometry. *The Bulletin of the American Mathematical Society* 2:215-249.

Kouneiher, J., Barbachoux, C. 2015. Cartan's soldered spaces and conservation laws in physics. *The International Journal of Geometric Methods in Modern Physics* 12.

Möbius, A. F. 1847. *Die Grassmann'sche Lehre von Punktgrössen und den davon abhangigen Grössenformen*. In: Grassmann 1841–1911, I(2), pp. 613-633.

Möbius, A. F. 1885–1887. *Gesammelte Werke*. Vol. 4. Leipzig: Hirzel.

Möbius, A. F. 1827. *Der Barycentrische Calcul*. Leipzig: Verlag von Johann Ambrosius Barth [Hildesheim-New York, Georg Olms 1976].

Sharpe, R. W. 1996. *Differential Geometry: Cartan's Generalization of Klein's Erlangen Program*. Berlin: Springer-Verlag.

Wise, D. K. 2010. MacDowell–Mansouri gravity and Cartan geometry. *Classical and Quantum Gravity* 27. Available via: arXiv:gr-qc/0611154

Joseph Kouneiher, Nice and Sophia Antipolice University, France.
joseph.kouneiher@unice.fr

We Live in the Best of Possible Worlds: Leibniz's Insight Helps to Derive Equations of Modern Physics

Vladik Kreinovich and Guoqing Liu

Abstract. To reconcile the notion of a benevolent and powerful God with the actual human suffering, Leibniz proposed the idea that while our world is not perfect, it is the best of *possible* worlds. This idea inspired important developments in physics: namely, it turned out that equations of motions and equations which describe the dynamics of physical fields can be deduced from the condition that the (appropriately defined) action functional is optimal. This idea is very helpful in physics applications, but to fully utilize this idea, we need to know the action, and there are many possible action functionals. Our idea is to apply Leibniz's insight once again and to assume that (similarly) on the set of all action functionals, there is an optimality criterion, and the actual action functional is optimal with respect to this criterion. This new idea enables us to derive the standard equations of General Relativity, Quantum Mechanics, Electrodynamics, etc. only from the fact that the corresponding expressions for action are optimal. Thus, the physical equations describing our world are indeed the best possible.

Keywords: Leibniz, Best of Possible Worlds, Fundamental Physics, Gravity, Einstein's Equations, Electrodynamics, Maxwell's Equations, Quantum Physics, Schroedinger's Equations, Cosmological Λ-Term, Scalar-Tensor Theories of Gravity, Brans-Dicke equations

1 Introduction

1.1 Leibnitz's Idea

Many religious philosophers have been trying to reconcile the notion of a benevolent and powerful God with the actual human suffering. Leibniz's idea of solving this problem is to conjecture that while our world is not perfect, it is the best of possible worlds (see, e.g., Leibniz 1996).

1.2 Lebniz's Idea and Physics

Leibniz's idea of optimality of our world inspired not only interesting philosophical discussions, it also inspired important developments in physics: namely, it turned out that equations of motions and equations which describe the dynamics of physical fields can be deduced from the condition that the action functional

$$S = \int L(x)\,dx,$$

as determined by the corresponding *Lagrange function* $L(x)$, is optimal (see, e.g., Feynman, Leighton and Sands 1989; Landau and Lifschitz 1987).

In other words, there is an optimality criterion on the set of all trajectories, and the actual trajectory is optimal with respect to this criterion. There is an optimality criterion on the set of all possible fields, and the actual field is optimal with respect to this criterion.

This idea turned out to be very successful in physics applications, since it is often easier to solve optimization problems than to solve complex systems of partial differential equations (Feynman, Leighton and Sands 1989).

Comment. It is important to mention that while the physical optimality idea was inspired by Leibniz's thoughts, this idea did not come from Leibniz itself. This idea is a (later) *interpretation* (and simplification) of Leibniz's concept of the best possible world. The main motivation for this physical interpretation is *not* that it claims to adequately describe Leibniz's reasoning — it is rather removed from his reasoning — but the pragmatic fact that this Leibniz-inspired idea is successful in describing the physical world.

1.3 Applying Leibniz-Inspired Idea to Physics: the Main Challenge

The above application is interesting but not always very very helpful: to find the equations, we need to know the Lagrange function, and there are many possible Lagrange functions. How should we select the most appropriate one?

1.4 Our Idea

Our idea is to apply Leibniz's insight once again and to assume that (similarly) on the set of all Lagrange functions, there is an optimality criterion, and the actual Lagrangian is optimal with respect to this criterion.

1.5 What We Get by Using This Idea: a Brief Description

This idea enables us to derive the equations of physics only from the fact that they are optimal. Specifically, under reasonable conditions on the optimality criterion, this

Homage to Gottfried W. Leibniz 1646–1716

approach leads to the standard Lagrange functions for General Relativity, Quantum Mechanics, Electrodynamics, etc.

Thus, the Lagrange functions (and hence equations) of our world are indeed the best possible.

2. An Optimality Criterion on the Set of All Lagrange Functions General Requirements

2.1 What Is an Optimality Criterion

When we say an *optimality criterion* is defined on the set of all possible Lagrange functions, we mean that on the set of all such functions, there must be a relation $\geq$ describing which Lagrange function is better or equal in quality.

This relation must be transitive (if L is better than L', and L' is better than L'', then L is better than L''). This relation is not necessarily asymmetric, because we can have two Lagrange functions of the same quality.

Definition 1. *Let $\mathcal{A}$ be a set; elements of this set will be called* alternatives. *By an* optimality criterion, *we mean a transitive relation $\geq$ on the set $\mathcal{A}$.*

2.2 Optimality Criterion Must Be Final

We would like to require that this relation be *final* in the sense that it should define a unique *best* Lagrange function L_{opt} (i.e., the unique Lagrange function for which $\forall L\, (L_{\text{opt}} \geq L)$. Indeed:

- If none of the Lagrange functions is the best, then this optimality criterion is of no use, so there should be *at least one* optimal family.

- If *several* different Lagrange functions are equally best, that means that this optimality criterion is not sufficient to determine the actual Lagrange function: we must still select between the several "best" ones. As a result, the original optimality criterion was not final: we get a new criterion ($L \geq_{\text{new}} L'$ if either $L \geq_{\text{old}} L'$ in the sense of the old criterion, or if $L \sim_{\text{old}} L'$ and L is better according to some additional criterion), for which the class of optimal Lagrange functions is narrower. We can repeat this procedure until we get a final criterion for which there is only one optimal Lagrange function.

Definition 2. *We say that an optimality criterion $\geq$ on a set $\mathcal{A}$ is* final *if there exists one and only one* optimal *alternative, i.e., an alternative a_{opt} for which $\forall a\, (a_{\text{opt}} \geq a)$.*

2.3 Optimality Criterion Must Be Scale-Invariant

It is reasonable to require that the relation $L \geq L'$ should not change if we simply change the units in which we measure length, i.e., if we change the length *scale*.

In other words, we want the optimality criterion to be *scale-invariant*.

Definition 3. *Let G be a group of transformations from $\mathcal{A}$ to $\mathcal{A}$. We say that a criterion $\geq$ is G-invariant if for every two alternatives a and a', and for every transformation $g \in G$, $a \geq a'$ implies $g(a) \geq g(a')$.*

Comments.

- Symmetry ideas are known to be very useful in physics and in foundations of physics (see, e.g., Finkelstein 1997; Finkelstein, Kosheleva and Kreinovich 1996, 1997, 1997a,1986; Finkelstein and Kreinovich 1985; Moshinsky, Wolf and Frank 1992; Krienovich 1976; Nguyen and Kreinovich 1973; Olver 1995; Wolf, Seligman, Frank and Moshinsky 1992)), so it is reasonable to apply these ideas to our problem as well.

- In particular, when the optimality criterion $\geq$ is G-invariant with respect to a transformation group G describing *scalings*, we will say that $\geq$ is *scale-invariant*

Let us describe how these requirements apply to different fundamental physical fields.

3 First Case: Optimal Lagrange Function for Gravitation

3.1 Lagrange Function for Gravitation: General Definition

A physical field which describes gravitation is the metric field $g_{ij}(x)$. So, a general Lagrange function for gravitation can depend on the values of this field and of its derivatives of all orders. In the absence of the gravitation, $g_{ij} = \eta_{ij} = \text{diag}(1, -1, -1, -1)$, and all partial derivatives of metric are equal to 0: $g_{ij,k} \stackrel{\text{def}}{=} \frac{\partial g_{ij}}{\partial x_k} = 0$, $g_{ij,kl} = 0$, etc. It is therefore reasonable to require that the Lagrange function L be *analytical* in terms of the differences between the actual values $g_{ij}(x)$, $g_{ij,k}, \ldots$ of the field and of its derivatives, and the values $\eta_{ij}, 0, \ldots$ corresponding the null-field (absence of gravitation). In other words, it is reasonable to require that L is an analytical function of $g_{ij} - \eta_{ij}$, $g_{ij,k}$, $g_{ij,kl}$, $\ldots$ Thus, we arrive at the following definition:

Definition 4. *By a gravitational Lagrange function L, we mean a generally covariant analytical function of the differences $g_{ij}(x) - \eta_{ij}$ and of the derivatives $g_{ij,k}(x)$, $\ldots, g_{ij,k\ldots l}(x)$ in the same point x:*

$$L(x) = L(g_{ij}(x) - \eta_{ij}, g_{ij,k}(x), g_{ij,kl}(x), \ldots). \tag{1}$$

3.2 What Does Scale Invariance Mean for Gravitation?

When we say that a Lagrange function must be generally covariant, we mean that its value should not depend on the choice of coordinate system (i.e., it should not change if we change a coordinate system). This guarantees that the resulting action $S = \int L \cdot \sqrt{-g}\, d^4x$ will also be generally covariant, and so the resulting field equations will be generally covariant.

In addition to changing coordinates, we can also change the unit of length. From the physical viewpoint, if we change a unit of length, the physical space-time will not change. However, from the mathematical viewpoint, the space changes: if we change the unit of length to a unit which is λ times smaller, then the numerical value of the length

$$ds = \sqrt{\sum g_{ij} \cdot dx^i \cdot dx^j} \tag{2}$$

will change to $ds' = \lambda \cdot ds$, i.e., we will get $ds' = \sqrt{\sum g'_{ij} \cdot dx^i \cdot dx^j}$ with a new metric field

$$g'_{ij} = \lambda^2 \cdot g_{ij}. \tag{3}$$

How can we best describe this transformation in physical terms? From the purely mathematical viewpoint, we can simply keep the same coordinate system x^i; then the corresponding scaling transformation can be simply described as a transformation (3) for the metric tensor and, correspondingly, a similar transformation

$$g'_{ij,k...l} = \lambda^2 \cdot g_{ij,k...l} \tag{4}$$

for its derivatives.

However, from the physical viewpoint, this description (3), (4) would be rather unnatural, because coordinates are usually assigned based on distances, and therefore, if we change the unit for length, the coordinates should also change accordingly: from x^i to

$$x'^i = \lambda \cdot x^i. \tag{5}$$

In this case, if we change both the metric ds to $ds' = \lambda \cdot ds$ and coordinates from x^i to $x'^i = \lambda \cdot x^i$, then, from (2), we can conclude that $ds' = \sqrt{\sum g_{ij} \cdot dx'^i \cdot dx'^j}$, i.e., that the metric does not change:

$$g'_{ij} = g_{ij}. \tag{6}$$

Correspondingly, due to (5) and (6), the derivatives of the metric get transformed as

$$g'_{ij,k} = \lambda^{-1} \cdot g_{ij,k}; \tag{6a}$$

$$g'_{ij,kl} = \lambda^{-2} \cdot g_{ij,kl}; \tag{6b}$$

etc.

How does the Lagrange function change under this transformation? From the physical viewpoint, action $S = \int L \cdot \sqrt{-g}\,\mathrm{d}^4 x$ is energy×time. We are considering a relativistic theory, and moreover, we are following the tradition of gravitation theory in using the units in which distance and time are measured by the same unit, i.e., in which the speed of light c is equal to 1 (and so, $h_{ij} = \mathrm{diag}(1, -1, -1, -1)$). In such units, energy $E = m \cdot c^2$ is described in the same units as mass, and time in the same units as distance, so action changes as mass×distance.

If we change a unit of length, how will the corresponding unit of mass change? To describe this change, it is sufficient to look at the known approximate gravitational theory: Newtonian gravitation. In Newtonian gravitation, the force $F = m \cdot a$ with which a body of mass M attracts a body of mass m is proportional to

$$m \cdot a = G \cdot \frac{m \cdot M}{r^2},$$

hence

$$a = \frac{G \cdot M}{r^2}. \tag{7}$$

When we change a unit of length (and the corresponding unit of time), we get $r' = \lambda \cdot r$, $t' = \lambda \cdot t$, $a' = r'/(t')^2 = \lambda^{-1} \cdot a$ and therefore, to preserve the above relation (7), we must have $M' = \lambda \cdot M$.

So, the mass (hence, the energy) transforms as $M \to \lambda \cdot M$; we already know that time t transforms as $t \to t' = \lambda \cdot t$. Hence, the action (energy×time) transforms as $S \to S' = \lambda^2 \cdot S$, and therefore, the Lagrange function L, which is defined as the density of the action, i.e., as $L \sim S/r^4$, is transformed as

$$L \to L' = (\lambda^2/\lambda^4) \cdot L = \lambda^{-2} \cdot L.$$

Hence, after scaling, the old Lagrange function (1) transforms, in the new units, into the expression

$$L'(x) = \lambda^{-2} \cdot L(g_{ij}(x) - \eta_{ij}, g_{ij,k}(x), g_{ij,kl}(x), \ldots).$$

This expression describes L' as a function of values $g_{ij}, g_{ij,k}, g_{ij,kl}, \ldots$, expressed in the old units. We want to get an expression of L' in terms of $g'_{ij}, g'_{ij,k}, g'_{ij,kl}, \ldots$, i.e., in terms of the field values and derivatives expressed in new units. From (6), (6a), (6b), etc., we can conclude that $g'_{ij} = g_{ij}$, $g'_{ij,k} = \lambda \cdot g_{ij,k}$, $g'_{ij,kl} = \lambda^2 \cdot g_{ij,kl}$, etc. Therefore, in the new unit, the Lagrange function is expressed as:

$$L' = g_\lambda(L) = \lambda^{-2} \cdot L(g_{ij} - \eta_{ij}, \lambda \cdot g_{ij,k}, \lambda^2 \cdot g_{ij,kl}, \ldots). \tag{8}$$

So, for gravitational Lagrange functions, scale transformation means going from L to $L' = g_\lambda(L)$, and scale-invariance means invariance with respect to such transformations.

3.3 Main Result for the Gravity Field

Now, we are ready to describe our main result for gravitation.

Theorem 1. *For every scale-invariant final optimal criterion on the set of all gravitational Lagrange functions, the optimal Lagrange function has the form $L = b \cdot R$, where b is a constant, and R is the scalar curvature.*

In other words, for any reasonable optimality criterion, General Relativity is the best of all possible Lagrange functions. To be more precise, of all Lagrange functions in which gravitation is described by a single field: metric tensor field g_{ij}; alternative gravitation theories are described in Sections 6 and 7.

Proof. This proof is similar to the proofs from (Nguyen and Kreinovich 1973); the second part is similar to the proofs from (Finkelstein and Kreinovich 1985; Finkelstein, Kreinovich and Zapatrin 1986).

$1°$. Let us first show that the optimal Lagrange function L_{opt} is itself scale-invariant, i.e., that for every $\lambda > 0$, $g_\lambda(L_{\text{opt}}) = L_{\text{opt}}$.

Indeed, let $\lambda > 0$ be an arbitrary positive number. Since L_{opt} is optimal, for every other Lagrange function L, we have $L_{\text{opt}} \geq g_{1/\lambda}(L)$. Since the optimality criterion $\geq$ is invariant, we conclude that $g_\lambda(L_{\text{opt}}) \geq g_\lambda(g_{1/\lambda}(L)) = L$. Since this is true for every Lagrange function L, the Lagrange function $g_\lambda(L_{\text{opt}})$ is also optimal. But since our criterion is final, there is only one optimal Lagrange function and therefore, $g_\lambda(L_{\text{opt}}) = L_{\text{opt}}$. In other words, the optimal Lagrange function is indeed invariant.

$2°$. Let us now show that $L = b \cdot R$.

From Part 1 of this proof, we conclude that $g_\lambda(L) = L$, i.e., that

$$\lambda^{-2} \cdot L(g_{ij} - \eta_{ij}, \lambda \cdot g_{ij,k}, \lambda^2 \cdot g_{ij,kl}, \ldots) = L(g_{ij} - \eta_{ij}, g_{ij,k}, g_{ij,kl}, \ldots). \quad (9)$$

Let us consider an arbitrary point A and normal coordinates in it (see, e.g., (Misner, Thorne and Wheeler 1973)). It is known that in some neighborhood of A, $g_{ij}(B) = h_{ij}+$ some analytical function of $B^i - A^i$ with coefficients which polynomially depend on curvature tensor $R_{ijkl}(A)$ and its covariant derivatives of arbitrary orders. Therefore $g_{ij}(A)$ and every derivative $g_{ij,i_1\ldots i_p}(A)$ are also such polynomial functions. If we substitute these expressions into L, then L will become an analytical function of the curvature tensor R_{ijkl} and of its covariant derivatives $R_{ijkl;m}, R_{ijkl;mn}, \ldots$, i.e. a sum of infinitely many monomials of the variables $R_{ijkl;m}, R_{ijkl;mn}, \ldots$:

$$L = L(g_{ij} - \eta_{ij}, R_{ijkl}, R_{ijkl;m}, R_{ijkl;mn}, \ldots).$$

Let us express (9) in terms of these new variables. With respect to scale transformations,

$$R_{ijkl} \to R'_{ijkl} = \lambda^{-2} \cdot R_{ijkl}, \text{ and}$$

$$R_{ijkl;i_1...i_p} \to R'_{ijkl;i_1...i_p} = \lambda^{-(2+p)} \cdot R_{ijkl;i_1...i_p}.$$

Therefore,
$$R_{ijkl} = \lambda^2 \cdot R'_{ijkl},$$

$$R_{ijkl;i_1...i_p} = \lambda^{2+p} \cdot R'_{ijkl;i_1...i_p},$$

and (9) turns into

$$\lambda^{-1} \cdot L(g_{ij} - \eta_{ij}, \lambda^2 \cdot R_{ijkl}, \lambda^3 \cdot R_{ijkl;m}, \lambda^4 \cdot R_{ijkl;mn}, \ldots) =$$

$$L(g_{ij} - \eta_{ij}, R_{ijkl}, R_{ijkl;m}, R_{ijkl;mn}, \ldots). \tag{10}$$

Expressions on both sides of (10) are sums of similar monomials. Since the two analytical functions coincide, this means that all the coefficients at the corresponding monomials must coincide.

Each monomial in the right-hand side does not depend on λ; the corresponding monomial in the left-hand side of (10) is multiplied by $\lambda^{2n_R+n_D-2}$, where n_R is a total number of all curvature tensors and their covariant derivatives in this monomial, and n_D is a total number of all differentiation indices in it. Since the coefficients must coincide, we conclude that the function L can only have monomials with

$$2n_R + n_D = 2.$$

Both numbers n_R and n_D are non-negative integers, so there are only two possibilities for $2n_R + n_D = 2$: when $n_R = 1$ and $n_D = 0$, and when $n_R = 0$ and $n_D = 2$.

In the first case, L is a linear function of R_{ijkl}, so, since L is generally covariant, we have $L = b \cdot R$.

In the second case, there is no curvature tensor in L, and covariant differentiation is applied only to g_{ij}, therefore the result is zero ($g_{ij;kl} = 0$).

So, $L = b \cdot R$. The theorem is proven.

Comments.

1. We have shown that *if* a Lagrange function is optimal with respect to some scale-invariant final optimality criterion, then it is $L = b \cdot R$, but we have not yet proven the existence of such criteria. The following simple example proves this existence: we can define an optimality criterion according to which $R > L$ for any $L \neq R$, and $L \sim L'$ for every two $L, L' \neq R$. This criterion is clearly scale-invariant and final.

2. The requirements that the Lagrange function L is analytical and that the optimality criterion is scale-invariant are both essential:

- If we do not require that L is analytical, then we can have the Lagrange function $L_0 = \sqrt{R_{ij} \cdot R^{ij}}$, and an optimality criterion according to which $L_0 > L$ for any $L \neq L_0$, and $L \sim L'$ for every two $L, L' \neq L_0$. This criterion is scale-invariant and final, and the corresponding optimal Lagrange function is $L_0 \neq b \cdot R$.

- If we do not require that the optimality criterion is scale-invariant, then we can take a Lagrange function $L_1 = R + R^2$, and an optimality criterion according to which $L_1 > L$ for any $L \neq L_1$, and $L \sim L'$ for every two $L, L' \neq L_1$. This criterion is final, and the corresponding optimal Lagrange function is $L_0 \neq b \cdot R$.

3.4 Fundamentality Principle

In the previous text, we used transformational properties of L with respect to scaling, which were deduced from physical arguments. In the present section we show that we can eliminate these arguments, if we use the following *fundamentality principle*:

A phenomenon is called *fundamental* if it can be explained without using other phenomena. In our case, it means that transformation law for L must be chosen in such a way that field equations are uniquely determined by optimality requirement, i.e. L must be determined uniquely modulo multiplicative constant.

To be more precise, we define scale transformations as

$$L' = g_\lambda(L) = \lambda^{-d} \cdot L(g_{ij} - \eta_{ij}, \lambda \cdot g_{ij,k}, \lambda^2 \cdot g_{ij,kl}, \ldots). \tag{8a}$$

for some unspecified value d.

Proposition. *The only value d for which all Lagrange functions which are optimal with respect to scale-invariant final optimal criteria lead to the same dynamical equations is $d = 2$.*

Proof. Similarly to Theorem 1, we conclude that for every d, the optimal Lagrange function is a sum of terms for which $2n_R + n_D = d$. The value d is a sum of two non-negative integers, so $d \geq 0$.

The Lagrange function is a scalar, so the total number of indices in every term P is even, hence n_D is even. So, d must also be even.

If $d = 0$, then $n_R = n_D = 0$, hence $L = $ const, and there are no variational equations at all.

If $d \geq 4$, then we can take terms $L = (R_{ij}R^{ij} + bR_{ijkl}R^{ijkl} + cR^2) \cdot R^{(d-4)/2}$. For different b and c, these Lagrange functions lead to different variational equations, and each of these function L_0 is optimal with respect to some scale-invariant final optimality criterion: namely, a criterion in which $L_0 > L$ for all $L \neq L_0$, and $L \sim L'$ for all $L, L' \neq L_0$.

Thus, only for $d = 2$, we get the desired uniqueness. The proposition is proven.

4 Second Case: Electromagnetic Field (in Curved Space)

4.1 Lagrange Function for Electromagnetic Field: General Definition

Electromagnetic field is described by a vector potential $A_i(x)$; its source is the 4-current j^i which satisfies the charge conservation law $j^i_{;i} = 0$. In classical electrodynamics, the vector potential does not have a direct physical meaning, only $F_{ij} = A_{i,j} - A_{j,i}$; therefore, it is normally assumed that the Lagrange function should be invariant under *gauge transformations* $A_i \to A_i - f_{,i}$ which preserve F_{ij} for an arbitrary function $f(x)$.

Definition 5. *By a* Lagrange function for electromagnetic field L, *we mean a generally covariant analytical function of the differences* $g_{ij}(x) - \eta_{ij}$, *of* $A_i(x)$, $j^i(x)$, *and of the derivatives* $g_{ij,k}(x), g_{ij,kl}(x), \ldots, A_{i,k}(x), A_{i,kl}(x), \ldots, j^i_{,k}, j^i_{,kl}(x), \ldots$, *in the same point* x:

$$L(x) = L(g_{ij}(x) - \eta_{ij}, g_{ij,k}(x), g_{ij,kl}(x), \ldots,$$

$$A_i(x), A_{i,k}(x), A_{i,kl}(x), \ldots, j^i(x), j^i_{,k}(x), j^i_{,kl}(x), \ldots),$$

for which the variational equations are gauge-invariant.

4.2 What Does Scale Invariance Mean for Electromagnetic Field?

In the Newtonian approximation, the force $F = q \cdot Q/r^2$ between the two charges is described by the same formula as the (gravitational) force between the two masses; therefore, if we want to preserve this approximation, then when we change the unit of length, we must transform charges in exactly the same way as masses, i.e., as $q \to q' = \lambda \cdot q$. Thus, the 4-current j^i (charge/length^3) should transform as $j^i \to j'^i = \lambda^{-2} \cdot j^i$. In Newtonian approximation, the electromagnetic potential is Q/r, so the potential A_i should not change under scale transformations. So, the expression

$$L'(x) = \lambda^{-2} \cdot L(g_{ij}(x) - \eta_{ij}, g_{ij,k}(x), g_{ij,kl}(x), \ldots,$$

$$A_i(x), A_{i,k}(x), A_{i,kl}(x), \ldots, j^i(x), j^i_{,k}(x), j^i_{,kl}(x), \ldots).$$

leads to the following scale transformation:

$$L' = g_\lambda(L) = \lambda^{-2} \cdot L(g_{ij} - \eta_{ij}, \lambda \cdot g_{ij,k}, \lambda^2 \cdot g_{ij,kl}, \ldots,$$

$$A_i, A_{i,k}, A_{i,kl}, \ldots, \lambda^2 \cdot j^i, \lambda^3 \cdot j^i_{,k}, \lambda^4 \cdot j^i_{,kl}, \ldots).$$

4.3 Main Result for Electromagnetic Field

Theorem 2. *For every scale-invariant final optimal criterion on the set of all Lagrange functions for electromagnetic field, the optimal Lagrange function has the form $L = b \cdot R + c \cdot F_{ij} \cdot F^{ij} + d \cdot A_i \cdot j^i$ for some constants b, c, and d.*

Thus, the Lagrange function corresponding to standard Maxwell's equations is indeed optimal.

Proof. Similarly to the proof of Theorem 1, we conclude that the optimal Lagrange function is scale-invariant, that it is an analytical function of the metric field, of curvature, of vector potential, of 4-current, and of their covariant derivatives, and, therefore, that it can only contain monomials which do not depend on λ. On the other hand, each monomial is proportional to $\lambda^{2n_R + n_D + 2n_J - 2}$, where n_R and n_D are defined as in the proof of Theorem 1, and n_J is the total number of currents and its derivatives in this monomial. Thus, we must have $2 = 2n_R + n_D + 2n_J$. Since all three numbers n_R, n_D, and n_J are non-negative integers, we have three possibilities:

- $n_R = 1, n_D = n_J = 0$;
- $n_R = n_D = 0, n_J = 1$;
- $n_R = n_J = 0, n_D = 2$.

In the first case, L contains either R, or the product of R_{ijkl} and terms A_i; this product leads to the terms in variational equations which are not gauge invariant, so it cannot be in L.

In the second case, due to the fact that L is a scalar, the total number of indices of all tensors (whose product constitutes the monomial) must be even; therefore the total number n_A of potentials and its derivatives in this monomial must be odd. If $n_A = 1$, the only possibility is $P = d \cdot j_i \cdot A^i$. If $n_A \geq 3$, the result of varying is not gauge invariant.

In the third case, n_A must also be even. If $n_A = 0$, then $P = g_{ij;kl} = 0$. If $n_A = 2$, then gauge invariance leads to $P = c \cdot F_{ij} \cdot F^{ij}$, and if $n_A \geq 4$, the result of varying is not gauge invariant. The theorem is proven.

5 Third Case: Non-Relativistic Quantum Mechanics

5.1 Lagrange Function for Non-Relativistic Quantum Mechanics General Definition

We want to obtain a Lagrange function describing the dynamics of a particle of mass m, described by a (complex-valued) wave function $\psi(x,t)$, in a field with a potential energy function $V(x,t)$. Since the Lagrange function must be real-valued, it can also depend on the complex conjugate values $\psi^*(x,t)$.

This Lagrange function should be rotation-invariant. There is one more invariance specific for non-relativistic quantum mechanics. Namely, it is known that in quantum mechanics, we can add a constant phase to all the values of $\psi(x,t)$ without changing the physical meaning. Thus, the Lagrange function should be *phase-invariant*, i.e., invariant with respect to the transformation

$$\psi(x,t) \to \exp(i \cdot \alpha) \cdot \psi(x,t)$$

for any real constant α.

Definition 6. *By a* Lagrange function for non-relativistic quantum mechanics L, *we mean a phase-invariant rotation-invariant real-valued analytical function of the mass* m, *its inverse* m^{-1}, *fields* $\psi(x,t)$, $\psi^*(x,t)$, *and* $V(x,t)$, *and their derivatives of arbitrary orders with respect to time and spatial coordinates:*

$$L(m, m^{-1}, \psi(x,t), \psi_{,k}(x,t), \dot{\psi}(x,t), \ldots, \psi^*(x,t), \psi^*_{,k}(x,t), \dot{\psi}^*(x,t), \ldots,$$

$$V(x,t), V_{,k}(x,t), \dot{V}(x,t), \ldots)$$

5.2 What Does Scale Invariance Mean for Non-Relativistic Quantum Mechanics?

In (relativistic) gravitation, there is a direct connection between units of space and time. In non-relativistic case, there is no such direct connection, so we can independently change the unit for space $x^i \to x'^i = \lambda \cdot x^i$ and a unit of time $t \to t' = \mu \cdot t$. It is reasonable to require that the optimality criterion on the set of all Lagrange functions for non-relativistic quantum mechanics be invariant with respect to both scaling transformations.

How do L, $\psi(x,t)$, and $V(x,t)$ change under these transformations? A specific feature of quantum measurements is that simple experiments enable us to obtain a unit of action $\hbar$; therefore action $S = \int L(x,t)\, d^3x\, dt$ must be invariant with respect to scale transformations. Hence, $L(x,t)$ (which is action/(volume×time)) must transform as $L \to L' = \lambda^{-3} \cdot \mu^{-1} \cdot L$.

Similarly, since action is energy×time, and action is invariant, the potential energy $V(x,t)$ must transform as $V \to V' = \mu^{-1} \cdot V$.

Energy is mass×velocity^2; we know how energy is transformed and how velocity is transformed; therefore, for mass, we get $m \to m' = \lambda^{-2} \cdot \mu \cdot m$.

The transformation law for the wave function $\psi(x,t)$ can be deduced from its physical meaning: the integral $\int |\psi|^2\, dV$ is a probability and is therefore independent (invariant) on the choice of length or time units, i.e. invariant. So, $|\psi|^2 \sim 1/\text{length}^3$, hence, $|\psi|^2 \to \lambda^{-3} \cdot |\psi|^2$, and $\psi \to \psi' = \lambda^{-3/2} \cdot \psi$.

Therefore, the expression

$$L'(x,t) = \lambda^{-3} \cdot \mu^{-1} \cdot L(m, m^{-1}, \psi(x,t), \psi_{,k}(x,t), \dot\psi(x,t), \ldots,$$

$$\psi^*(x,t), \psi^*_{,k}(x,t), \dot\psi^*(x,t), \ldots, V(x,t), V_{,k}(x,t), \dot V(x,t), \ldots)$$

leads to

$$L' = g_{\lambda,\mu}(L) = \lambda^{-3} \cdot \mu^{-1} \cdot L(\lambda^2 \cdot \mu^{-1} \cdot m, \lambda^{-2} \cdot \mu \cdot m^{-1},$$

$$\lambda^{3/2} \cdot \psi, \lambda^{5/2} \cdot \psi_{,k}, \lambda^{3/2} \cdot \mu \cdot \dot\psi, \ldots, \lambda^{3/2} \cdot \psi^*, \lambda^{5/2} \cdot \psi^*_{,k}, \lambda^{3/2} \cdot \mu \cdot \dot\psi^*, \ldots,$$

$$\mu \cdot V, \lambda \cdot \mu \cdot V_{,k}, \mu^2 \cdot \dot V, \ldots).$$

5.3 Main Result for Non-Relativistic Quantum Mechanics

Now, we are ready to present the main result of this section.

Theorem 3. *For every scale-invariant final optimal criterion on the set of all Lagrange functions for non-relativistic quantum mechanics, the optimal Lagrange function has the form*

$$L = \mathrm{i} \cdot b \cdot \left(\psi \cdot \frac{\partial \psi^*}{\partial t} - \psi^* \cdot \frac{\partial \psi}{\partial t} \right) + \frac{c}{m} \cdot (\nabla \psi \cdot \nabla \psi^*) + d \cdot V \cdot \psi \cdot \psi^* + L_0,$$

where b, c, and d are real constants, and L_0 is an expression which does not contribute to variational equations.

This Lagrange function leads to Schrödinger equation which is, thus, optimal.

Proof. Let us first fix m and consider only transformations which preserve m, i.e., transformations for which $\mu = \lambda^2$. For these transformations,

$$L' = g_\lambda(L) = \lambda^{-5} \cdot L(\lambda^{3/2} \cdot \psi, \lambda^{5/2} \cdot \psi_{,k}, \lambda^{7/2} \cdot \dot\psi, \ldots,$$

$$\lambda^{3/2} \cdot \psi^*, \lambda^{5/2} \cdot \psi^*_{,k}, \lambda^{7/2} \cdot \dot\psi^*, \ldots, \lambda^2 \cdot V, \lambda^3 \cdot V_{,k}, \lambda^4 \cdot \dot V, \ldots).$$

Similarly to the proof of Theorem 1, we conclude that the optimal Lagrange function is scale-invariant, and therefore, that it can only contain monomials which do not depend on λ. On the other hand, each monomial is proportional to $\lambda^{(3/2) \cdot n_\psi + 2n_V + n_S + 2n_T - 5}$, where n_ψ is the total number of terms ψ, ψ^*, and their derivatives, n_V is the total number of V and its derivatives, n_S is the total number of spatial differentiations, and d_T is the total number of differentiations with respect to time. Thus, we must have $(3/2) \cdot n_\psi + 2n_V + n_S + 2n_T = 5$. Since all four numbers n_ψ, n_V, n_S, and n_T are integers, we must have n_ψ even. Since all these integers are non-negative, we have the following options:

- $n_\psi = 2, n_V = 1, n_S = n_T = 0$;
- $n_\psi = 2, n_V = 0, n_S = 2, n_T = 0$;
- $n_\psi = 2, n_V = 0, n_S = 0, n_T = 1$;
- $n_\psi = 0$ and $2n_V + n_S + 2n_T = 5$.

In the first case, we get a product of V and two terms of type ψ and ψ^*; the only way to make it real-valued is to have $V \cdot \psi \cdot \psi^*$. Another possibility would be

$$V \cdot (\psi^2 + (\psi^*)^2),$$

but the corresponding variational equations are not phase-invariant.

In the second case, we have two derivatives of two functions ψ. Due to the requirement that L is real-valued, one of them must be ψ, and another one ψ^*. Due to rotation-invariance, we have two possibilities: $\psi_{,i} \cdot \psi^*_i$ and $\psi \cdot \Delta \psi^*$; the second term differs from the first one by a full derivative, so we can assume that we get the first term, and add the full derivative to L_0.

In the third case, we have two functions ψ and ψ^* and one time derivative. This leads to the corresponding term in L.

In the fourth case, the monomial does not depend on ψ at all, so it does not contribute to the variational equations at all; so all terms of these type go directly to L_0.

We have almost proved the theorem, except for the dependence on m. To do that, we can take the expression that we have obtained so far, substitute the dependence on m, and explicitly require that the result be invariant with respect to all scaling transformation. This will enable us to find the exact dependence on m. The theorem is proven.

Comments.

1. If in the formulation of Theorem 3, we allow L to depend also on the cosmological field Λ and on its derivatives, then we'll obtain the Lagrange function which can be obtained from that of Theorem 3 by a nonessential change $V \to V + \text{const}$. This result implies that *the cosmological lambda term does not influence non-relativistic effects.*

2. The wave function ψ is not directly observable. Therefore, it may seem natural, instead of using $\psi(x,t)$, to use a directly observable probability density $\rho(x,t)$. We can repeat the same arguments as above and try to get a Lagrange function depending on m, m^{-1}, $\rho(x,t)$, $V(x,t)$, and their derivatives of different orders that is optimal with respect to some scale-invariant optimality criterion. A similar proof can describe the corresponding Lagrange functions; it turns out that they do not lead to any dynamics at all, because the only possible term containing time derivative is $\dot\rho$, which is a full derivative. Therefore, *our approach explains why we cannot restrict ourselves to directly observable quantities in the formulation of quantum mechanics.*

6 First Auxiliary Result: Gravitation With a Λ-Term

6.1 Lagrange Function for Gravitation with a Λ-Term: General Definition

Gravitation theory with a Λ-term is not invariant with respect to scale transformations, because it contains a fixed unit of length $\sqrt{\Lambda^{-1}}$. But if we consider Λ not as a constant, but as a new field, transforming according to the law

$$\Lambda \to \Lambda' = \lambda^{-2} \cdot \Lambda, \tag{11}$$

then we get a possibly scale-invariant situation. So, we arrive at the following definition:

Definition 7. *By a* Lagrange function for gravitation with a Λ-term L, *we mean a generally covariant analytical function of the differences $g_{ij}(x) - \eta_{ij}$, field $\Lambda(x)$, and of the derivatives $g_{ij,k}(x), \ldots, g_{ij,k\ldots l}(x), \ldots, \Lambda_{,k}(x), \ldots, \Lambda_{,k\ldots l}(x), \ldots$ at the same point x:*

$$L(x) = L(g_{ij}(x) - \eta_{ij}, g_{ij,k}(x), g_{ij,kl}(x), \ldots, \Lambda(x), \Lambda_{,k}(x), \Lambda_{,kl}(x), \ldots).$$

6.2 What Does Scale Invariance Mean for Gravitation with a Λ-Term?

Under scale transformations, the new field Λ gets transformed according to the formula (11). Therefore, $\Lambda = \lambda^2 \cdot \Lambda'$, and hence, the expression

$$L'(x) = \lambda^{-2} \cdot L(g_{ij}(x) - \eta_{ij}, g_{ij,k}(x), g_{ij,kl}(x), \ldots, \Lambda(x), \Lambda_{,k}(x), \Lambda_{,kl}(x), \ldots)$$

leads to

$$L' = g_\lambda(L) =$$
$$\lambda^{-2} \cdot L(g_{ij} - \eta_{ij}, \lambda \cdot g_{ij,k}, \lambda^2 \cdot g_{ij,kl}, \ldots, \lambda^2 \cdot \Lambda, \lambda^3 \cdot \Lambda_{,k}, \lambda^4 \cdot \Lambda_{,kl}, \ldots). \tag{12}$$

6.3 Main Result for Gravitation with a Λ-Term

Now, we are ready for the main result.

Theorem 4. *For every scale-invariant final optimal criterion on the set of all Lagrange functions for gravitation with a Λ-term, the optimal Lagrange function has the form $L = b \cdot R + a \cdot \Lambda$ for some constants a and b.*

If we rename $\Lambda' = (a/b)\cdot\Lambda$, we get the standard Einstein's theory $L = b(R+\Lambda')$, which is, thus, optimal.

Proof. Similarly to the proof of Theorem 1, we conclude that the optimal Lagrange function is scale-invariant, that it is an analytical function of the metric field, of the curvature, of the field Λ, and of their covariant derivatives, and, therefore, that it can only contain monomials which do not depend on λ. On the other hand, each monomial is proportional to $\lambda^{2n_R+n_D+2n_\Lambda-2}$, where n_R and n_D are defined as in the proof of Theorem 1, and n_Λ is the total number of Λ and its derivatives in this monomial. Thus, we must have $2 = 2n_R + n_D + 2n_\Lambda$. Since all three numbers n_R, n_D, and n_Λ are non-negative integers, we have either $n_\Lambda = 0$ (then $P = b \cdot R$), or $n_\Lambda = 1$, in which case $n_R = n_D = 0$ and $P = a \cdot \Lambda$. The theorem is proven.

7   Second Auxiliary Result: Scalar-Tensor Gravitation

7.1   Lagrange Function for Scalar-Tensor Gravitation: General Definition

The main idea of the scalar-tensor theory is that the gravitational constant G which relates the gravitational force F to masses ($F = G \cdot m \cdot M/r^2$) is not necessarily a constant, it may change with time, i.e., in other words, it represent a new physical field. Traditionally, the inverse value $\varphi = 1/G$ is used in such theories; to make a comparison with the existing theories easier, we will use this requirement.

Since φ is not necessarily a small number, we can assume that the Lagrange function is analytically depending not only on φ, and on the derivatives of φ, but also on φ^{-1}. So, we arrive at the following definition:

Definition 8. *By a* Lagrange function for scalar-tensor gravitation L, *we mean a generally covariant analytical function of the differences* $g_{ij}(x) - \eta_{ij}$, *field* $\varphi(x)$, *its inverse* $\varphi^{-1}(x)$, *and of the derivatives* $g_{ij,k}(x)$, $g_{ij,kl}(x)$, $\ldots$, $\varphi_{,k}(x)$, $\varphi_{,kl}(x)$, $\ldots$, *in the same point* x:

$$L(x) = L(g_{ij}(x) - \eta_{ij}, g_{ij,k}(x), g_{ij,kl}(x), \ldots, \varphi(x), \varphi^{-1}(x), \varphi_{,k}(x), \varphi_{,kl}(x), \ldots).$$

7.2   What Does Scale Invariance Mean for Scalar-Tensor Gravitation?

In metric-only gravitation, G was a constant and therefore, when the unit of length changes, the unit of mass must change accordingly. In the scalar-tensor gravitation, G is no longer a constant, and therefore, we can independently change a unit of length $x^i \to x'^i = \lambda \cdot x^i$ *and* a unit of mass $m \to m' = \mu \cdot m$. In this case, the Lagrange function, whose physical meaning is energy$\times$time/length^4, transforms as $L \to L' = \mu \cdot \lambda^{-3} \cdot L$. Due to the definition of φ as $1/G$, where $m \cdot a = G \cdot m \cdot M/r^2$ and $G = a \cdot r^2/M$, where $a = r/t^2$, we have $G \to G' = \lambda \cdot \mu^{-1} \cdot G$, and $\varphi \to \varphi' = \lambda^{-1} \cdot \mu \cdot \varphi$.

Therefore, the expression

$$L'(x) = \lambda^{-3} \cdot \mu \cdot L(g_{ij}(x) - \eta_{ij}, g_{ij,k}(x), g_{ij,kl}(x), \ldots, \varphi(x), \varphi^{-1}(x), \varphi_{,k}(x), \ldots)$$

leads to

$$L' = g_{\lambda,\mu}(L) = \lambda^{-3} \cdot \mu \cdot L(g_{ij} - \eta_{ij}, \lambda \cdot g_{ij,k}, \lambda^2 \cdot g_{ij,kl}, \ldots,$$
$$\lambda \cdot \mu^{-1}\varphi, \lambda^{-1} \cdot \mu\varphi, \lambda^2 \cdot \mu^{-1} \cdot \varphi_{,k}, \lambda^3 \cdot \mu^{-1} \cdot \varphi_{,kl}, \ldots). \tag{13}$$

7.3 Main Result for Scalar-Tensor Gravitation

Now, we are ready for the main result of this section.

Theorem 5. *For every scale-invariant final optimal criterion on the set of all Lagrange functions for scalar-tensor gravitation, the optimal Lagrange function has the form*

$$L = a \cdot \varphi \cdot \left(R - \omega \cdot \frac{\varphi_{,i} \cdot \varphi^{,i}}{\varphi^2}\right) + L_0,$$

where a and ω are constants, and L_0 is an expression which does not contribute to variational equations.

Thus, we get Brans-Dicke scalar-tensor theory (see, e.g., (Misner, Thorne and Wheeler 1973), which is, therefore, optimal.

Proof. Similarly to the proof of Theorem 1, we conclude that the optimal Lagrange function is scale-invariant, that it depends only on the metric, curvature, scalar field, and their covariant derivatives, and, therefore, that it can only contain monomials which do not depend on λ and μ. On the other hand, each monomial is proportional to $\lambda^{2n_R+n_D+n_\varphi-n_{-\varphi}-3} \cdot \mu^{n-\varphi-n_\varphi+1}$, where n_R and n_D are defined as in the proof of Theorem 1, n_φ is the total number of φ and its derivatives in this monomial, and $n_{-\varphi}$ is the total number of terms φ^{-1}.

Thus, we must have $3 = 2n_R + n_D + n_\varphi - n_{-\varphi}$ and $-1 = n_{-\varphi} - n_\varphi$. Adding these two equalities, we get $2 = 2n_R + n_D$, hence either $n_R = 1$ and $n_D = 0$, or $n_R = 0$ and $n_D = 2$. In both cases, we have $n_\varphi - n_{-\varphi} = 1$.

In the first case, the monomial can only contain φ, R_{ijkl}, and no derivatives. The only covariant term of this type is $a \cdot \varphi \cdot R$.

In the second case, we do not have any curvature terms, and we have two derivatives which can be only applied to φ. Thus, we have two options:

$$\left(\varphi_{,i} \cdot \varphi^{,i}\right)/\varphi \text{ and } \varphi^{,i}_{;i}.$$

The term corresponding to the second option differs from the term corresponding to the first option by a full derivative; therefore we can replace this term by the term of the first option without changing the variational equations. The theorem is proven.

8 Conclusion

Inspired by Leibniz's idea that our world is the best possible world, physicists showed that many equations of fundamental physics can be equivalently formulated as the requirement that the value of a certain functional (called action) is the largest possible. For each physical equation, there is a corresponding expression for the action.

This approach have been very successful in physics applications. However, from the fundamental viewpoint, there is still a challenge: in principle, we can consider many different action functionals, but only a few of them correspond to fundamental physical equations. In this paper, we use the same Leibniz-inspired idea to answer this question. Namely, it turns out that it is possible to set up natural optimality criteria on the set of all possible action functionals. With respect to this criterion, optimal action expressions are the ones corresponding to the known fundamental physical equations:

- Equations of general relativity that describe gravity (and their Λ-term and scalar-tensor modifications),

- Maxwell's equations that describe electrodynamics,

- Schrödinger's equations that describe quantum phenomena, etc.

Acknowledgments

This work was supported in part by the Chinese Scholarship Council (CSC) and by the US National Science Foundation grants HRD-0734825 and HRD-1242122 (Cyber-ShARE Center of Excellence) and DUE-0926721.

The authors are greatly thankful to Prof. Raffaele Pisano for his encouragement and to the anonymous referees for their valuable suggestions.

References

Feynman, R. P., Leighton, R. B., Sands, M. L. 1989. *The Feynman Lectures On Physics*. Redwood City, California: Addison-Wesley.

Finkelstein, D. R. 1997. *Quantum Relativity: A Synthesis of the Ideas of Einstein and Heisenberg*. Berlin-Heidelberg: Springer-Verlag.

Finkelstein, A., Kosheleva, O., and Kreinovich, V. 1996. Astrogeometry, error estimation, and other applications of set-valued analysis. *ACM SIGNUM Newsletter* 31(4):3–25.

Finkelstein, A., Kosheleva, O., and Kreinovich, V. 1997. Astrogeometry: towards mathematical foundations. *International Journal of Theoretical Physics* 36(4): 1009–1020.

Finkelstein, A., Kosheleva, O., and Kreinovich, V. 1997a. Astrogeometry: geometry explains shapes of celestial bodies. *Geombinatorics* 6(4):125–139.

Finkelstein, A. M., Kreinovich, V. 1985. Derivation of Einstein's, Brans-Dicke and other equations from group considerations. In: Choque-Bruhat, Y., Karade, T. M. (eds) *On Relativity Theory. Proceedings of the Sir Arthur Eddington Centenary Symposium, Nagpur India 1984*, Vol. 2, pp. 138–146.

Finkelstein, A. M., Kreinovich, V., Zapatrin R. R. 1986. Fundamental physical equations uniquely determined by their symmetry groups. *Lecture Notes in Mathematics*, Berlin-Heidelberg-N.Y.: Springer-Verlag, 1214:159–170.

Kreinovich, V. 1976. Derivation of the Schroedinger equations from scale invariance. *Theoretical and Mathematical Physics* 8(3):282–285.

Landau, L. D., Lifschitz, E. M. 1987. *The Classical Theory of Fields*. Oxford, UK: Butterworth-Heinemanm.

Misner, W. Thorne, K., Wheeler, J. A. 1973. *Gravitation*, San Francisco: Freeman Co.

Moshinsky, M., Wolf, K. B., Frank, A. 1992. *Symmetries in Physics: Proceedings of the International Symposium Held in Honor of Prof. Marcos Moshinsky, Cocoyoc, Morelos, Mexico, 1991*. Berlin, New York: Springer-Verlag.

Nguyen, H. T., Kreinovich, V. 1997. *Applications of Continuous Mathematics to Computer Science*. Dordrecht: Kluwer

Olver, P. G. 1995. *Equivalence, Invariants, and Symmetry*. Cambridge: Cambridge University Press.

Remnant, P. 2009. *New Essays on Human Understanding*. Cambridge, UK: Cambridge University Press.

Wolf, K. B., Seligman, T. H., Frank, A., Moshinsky, M. 1992. *Group Theory in Physics: Proceedings of the International Symposium Held in Honor of Prof. Marcos Moshinsky, Cocoyoc, Morelos, Mexico, 1991*. New York: American Institute of Physics.

Vladik Kreinovich, University of Texas at El Paso, USA
vladik@utep.edu
Guoqing Liu, Nanjing University of Technology, China
guoqing@njut.edu.cn

A Background Condition for Analysis

Montgomery Link

Abstract. While Gottfried Wilhelm von Leibniz (1646–1716) often contends against a non-mental infinite mathematical actuality, his analysis of mathematics presupposes exactly that. Leibniz seems to have roughed out a way that this tension can be resolved. The resolution is not entirely satisfactory but does throw light on the background for analysis that he presupposes.

Keywords: Leibniz, calculus, infinite analysis, infinite sequences, relation, syncategorematic, foundation, labyrinth, *analysis situs*, synthesis, conceptual containment (*praedicatium inest subjecto*), *saltus* (leap), osculations, conceptualism, realism, actual infinite, *to apeiron,* mind.

1 Introduction

How do we organize our knowledge? In concluding a discussion of reason in the *New Essays on Human Understanding* (Book IV, Chapter 21) Leibniz specifies three systems for classifying the "totality of doctrinal truths".[1] His three forms of classification are "synthesis", "analysis", and "classification by terms" (G V 506–507). The first is "synthetic" and "theoretical". It "involves setting out truths according to the order in which they are proved, as the mathematicians do, so that each proposition comes after those on which it depends". Under this procedure, "truths are set out according to their origins".

The second form is "analytic and practical". This is the "art" of analysis. These classifications are applicable to the whole realm of knowledge but also to a particular science. Euclid's work, for example, was synthetic but others have treated geometry as an art.

The third form, the classification by terms, is a "kind of inventory". It covers the realm of logic. The advantage of the third form over the other two is that the truths that concern one term can occur all together. What has come to be called the paradigmatically Leibnizian analysis is the first or theoretical sense.

---

[1] G V 506. These are the words of Theophilus. Translated in Leibniz 1981, IV.21. The capital letter 'G' refers to Leibniz 1875–1890. In what follows, 'C' stands for Leibniz 1903, 'L' denotes Leibniz 1969, and '*Math. Schr.*' refers to Leibniz 1849–1863.

Montgomery Link (2017) Conditions for Analysis in Leibniz's Thought. In: Pisano R, Fichant M, Bussotti P, Oliveira ARE (eds.), *The Dialogue between Sciences, Philosophy and Engineering. New Historical and Epistemological insights. Homage to Gottfried W. Leibniz 1646-1716.* College Publications, London, pp. 227-253
© 2017 College Publications Ltd | ISBN: 978-1-84890-227-5 www.collegepublications.co.uk

Looking ahead, section 2 (below) sets out the building blocks of the "synthetic" approach. Section 3 incorporates the mathematical aspect. Section 4 considers the philosophical consequences of the theoretical form of analysis for our assessment of Leibniz. We will see that it has certain limitations that prevent the theoretical framework from being grounded in a definite way. To compensate, one might try an appeal to the third form of classification, but this form of classification does not take the organization of knowledge outside or beyond the theoretical, but only sorts it from within. We will see that in either case the background remains infinitely divisible, unbounded, and indefinite, but even this clarification will not prevent one reservation concerning the investigation of the nature of concepts.

Leibniz designed his analysis to overcome logical and metaphysical constraints, barriers he saw on the one hand in the Scholastic understanding of the underlying subject, and on the other hand in the Cartesian construal of geometric figures in terms of algebraic equations. The paradigmatic analysis is constrained by the philosophical tradition in which Leibniz was working, within an understanding of reality as composed of substances and attributes. Propositions reflect that reality insofar as they are true predications of subjects (G VII 43–44; Mates 1986, p. 53). The challenge was to formulate the analysis of the proposition in order to incorporate relations.

The difficulty in the reduction arises because relations, unlike accidents, had "one leg in one [subject] and the other in the other" (Mugnai 1992, p. 30). They followed a "third way". Therefore, the traditional logic constrained analysis because asymmetric relations between subjects technically cannot be reduced to equivalent statements about just the predicates of individual subjects without losing the logical interconnections between the *relata*. To maintain a nominalist position on relations, Leibniz should in principle accomplish a reduction equivalent to the original, but he was able to carry out the analysis only in one direction, and may have for his own purposes seen that as sufficient to provide a foundation (Mates 1986, pp. 215–218). Frege's later more comprehensive analysis, of course, keyed off the function concept using multiple quantified variables instead of the concept of subject. In this way the *Begriffsschrift* extended logic to incorporate the analysis of relations in a new light. As we will see, although for Leibniz relations were "mental" things, that makes them no less important to the project (Mugnai 1992, ch. V).

The second principle influence on the Leibnizian analysis is the clarity gained by the Cartesian analysis of geometry (Knobloch 2006); however, the Cartesian picture of algebraic exactness had finite parameters. Leibniz realized that it is no less clear to allow for other curvilinear shapes that can be accessed only by infinite analysis, under the condition that the form is preserved. Leibniz himself extended algebra to incorporate the analysis of these objects. Not just the objects of geometry but also the rigor of geometric deduction and logical reasoning are essential to the picture. These inferences are based on the inherence of the concept of the predicate in the concept of the subject, and on replacement by definition.[2]

The paradigmatic Leibnizian analysis of mathematics, once augmented by the *altitudinem divitarum,* the depth of divine wisdom, is far-reaching but not completely comprehensive. A counter-proposal we have already mentioned is that

---

[2] Rescher 1967, p. 23. The latter constitutes a platonic influence.

this limitation can be overcome by appealing to analysis in the third sense. But this third type of analysis, which Leibniz thinks of as bookkeeping, does not seem to escape the technical difficulties.

2 Leibnizian Analysis

In their groundbreaking interpretations, Russell and Couturat emphasized the logical aspects of the Leibnizian analysis, almost to the exclusion of the metaphysical.[3] Yet, Leibniz was not only looking into the formal nature of the proposition but also investigating what exists.[4] Leibniz himself anticipated general formulae in speaking of the outcome of his synthetic form of analysis, by which I mean an analysis described as follows: "Synthesis is when, beginning from principles and running through truths in order, we discover certain progressions and form tables, as it were, or sometimes even general formulae, in which the answers to what arises can later be discovered" (Leibniz 1683?, p. 16).

One might, of course, argue in response that Leibniz featured the intensional, not the extensional. But then one could respond in kind that the intensional can be given an equivalent formulation extensionally, in terms of set theory, for example, where the axiomatic structure gives us a plain way to assess what we are up against if we take the Leibnizian approach. These are dual approaches that share a common structure (Swoyer 1995, p. 96). Because of this common structure, extensions "express" intensions. But then, because of Leibniz's own account of truth, his view on intensions requires concepts to be extensional. We tend to group the Leibnizian with the intensional and separated from the extensional point of view. The guiding thread is that the extensional and the intensional are two sides of the same coin.[5]

2.1 The Subject

Leibniz took for granted the doctrine about the subject being the ultimate essential substance and assumed an ancient distinction between concrete and abstract entities (*abstracta*). In the Aristotelian *Categories,* the most basic things

---

[3] For which see Russell 1937 and 1972, and Couturat 1901 and 1972.

[4] Parkinson 1965, p. 3, for example, notes that the metaphysics is not derived from logic alone: "What he derives are necessary, but conditional, propositions about what the world must be like, if there is a world; these may be called consequences which relate to metaphysics, rather than metaphysical without qualification".

[5] These debates involve the contentious issue of the presence in Leibniz's philosophy of science of both realism and idealism. Garber, whose interpretation is properly at the center of the debate, says (2011) that Broad and others had earlier noted the presence of the physical realism, for which see, e.g., Broad 1975, pp. 87–90. Garber's view is that this constitutes a tension in Leibniz's thought. While 'infinity' has different meanings for Leibniz, one crucial distinction in his thought, which runs in tandem with his view of relations (Mugnai 1992, ch. VII), is that it is at once among both the "real objective things" and the "merely mental things".

are individual substances, like Socrates, that underlie all the rest, even the universal forms that Plato had emphasized. This is an attractive picture for Leibniz, as it is in keeping with his nominalist predilections, such as his attachment to Occam's Razor: "From this principle they [the *Nominales*] have deduced that everything in the world can be explained without any reference to universals and real forms. Nothing is truer than this opinion, and nothing is more worthy of a philosopher of our own time", he wrote early in his career.[6]

The picture in the *Metaphysics,* however, is slightly different. In Z3 Aristotle distinguishes two senses of the *hypokeimenon.* It is in one sense the basic stuff, *hyle,* prime matter, that underlies all the rest; in another sense, the properly metaphysical sense, it is the essential form that is the subject. There is no material component to account for in the mathematical case, just the essential form that can be given by definition. The mathematical case is at a distance from the contingent case. While the idea of the complete individual concept[7] can be compared to the Aristotelian subject, that is not all, for Leibniz makes a conceptual innovation. The foundation of predication lies in the concept of the individual subject, not in the individual form, a position that clarifies Leibniz's conceptualism and separates him from the Aristotelian tradition. His conceptualism is a kind of nominalism, with an increased emphasis on the relation. We should look at some details to make this vague claim more precise.

The complete concept of an individual uniquely distinguishes and completely characterizes any individual falling under it.[8] The mark of a true subject-predicate proposition is the inherence or conceptual containment of the concept of the predicate in the foundation of the concept of the subject (*praedicatum inest subjecto*):

> It is of course true that when a number of predicates are attributed to a single subject while the subject is not attributed to any other, it is called an individual substance. But this is not enough, and such a definition is merely nominal. We must consider, then, what it means to be truly attributed to a certain subject. Now it is certain that every true predication has some basis in the nature of things, and when a proposition is not an identity, that is to say, when the predicate is not expressly contained in the subject, it must be included in it virtually.[9] This is what the philosophers call *in-esse,* when they say that the predicate *is in* the subject. This being premised, we can say it is the nature of an individual substance or complete being to have a concept

---

[6] G IV 158; L 128. From the 1670 preface to an edition of Marius Nizolius.

[7] For which see sections eight and nine of the *Discourse on Metaphysics:* Leibniz 1686. For further discussion, see the correspondence with Arnauld (G II 1–138) and Sleigh 1990.

[8] Mates 1986, pp. 62–64. Mugnai (1992) agrees with Mates on Leibniz's nominalism, while clarifying and emphasizing the conceptual component in Leibniz's point of view.

[9] Cf. at this juncture the Kantian account in the *First Critique* (A 6–7/B 10); also the logical interpretation in Anderson 2015, chs. 1–4, and specifically p. 7.

so complete that it is sufficient to make us understand and deduce from it all the predicates of the subject to which the concept is attributed (Leibniz 1686, sec. 8; L 307).

A concept for Leibniz is a "simple thinkable" (*simplex cogitabile;* C 512), whereas a proposition is a complex thinkable (Parkinson 1965, p. 11). Leibniz gives many different formulations of the principle of *praedicatium inest subjecto*, but the main idea is that truth is grounded *in summa rerum,* "in the nature of things" (Rescher 1967, pp. 25–26).

This is not an arbitrary truth assignment, in contrast with the view of Hobbes and more subtly with the Cartesian understanding of the will of God. Leibniz at first glance would seem to agree with Hobbes that truths "can be demonstrated from definitions"; however,

> Hobbes saw that all truths can be demonstrated from definitions, but he believed that all definitions are arbitrary and nominal, since the imposition of names on things is arbitrary. He therefore wanted truths to consist in names, and to be arbitrary. But it must be known that concepts cannot be combined in an arbitrary fashion, but a possible concept must be formed from them, so that one has a real definition (Leibniz 1683?, p. 13).

The analysis *praedicatum inest subjecto* is a reduction by definitional replacement that resolves finally into a series of identities so as to achieve a real definition. Leibniz says, "if the characters can be used for ratiocination, there is in them a kind of complex mutual relation [*situs*] or order which fits the things [...] at least in their combination [...]. Though it varies, this order somehow corresponds in all languages"; indeed, "the analytic or arithmetical calculus confirms this view" (L 184).

For Leibniz, every true proposition is analytic, a statement which itself expresses one form of the principle of sufficient reason. For a proposition to be analytic is for it to be analyzed by substituting definiens for definiendum until it is reduced to a logical truth in terms of either an identity or a series of identities in combination. Not all analyses result in explicit identities, but there are also virtual identities that can only be realized in an infinite number of steps. Leibniz puts the matter as follows:

> In contingent truths, however, though the predicate inheres in the subject, we can never demonstrate this, nor can the proposition ever be reduced to an equation or an identity, but the analysis proceeds to infinity, only God being able to see, not the end of the analysis, since there is no end, but the nexus of terms, or the inclusion of the predicate in the subject, since he sees everything which is in the series (Rescher 1967, p. 26).

Due to human limitations, we cannot treat the process so directly, and require a notation to make things exact and preserve the appearances, for such a concept would have to be infinitely complex. Let us come back to that later (section 4.3).

2.2 The Predicate

A Leibnizian analysis is a two-step procedure. First, taking the complete individual concept of the subject of the proposition, list the properties of the substance included in the complete individual concept. This is the "definitional replacement" (Rescher 1967, pp. 23–24). If, for example, a property is not on the list, but can be derived from properties that are on the list, then it is to be included in the complete individual concept. Second, we examine whether the list includes the properties that the proposition is predicating of the subject. This is the "determination of predicational containment".

This procedure has two outcomes. Even if the analysis of a true proposition does not resolve into explicit identities, it still approaches that outcome without ever actually yielding actual identities. These are virtually identical, the main idea being that reductions of contingent propositions concerning matters of fact are being said to have an infinite number of intermediate stages. In these cases, finite intellects come closer and closer to the truth, without ever attaining truth *qua praecisio veritatis*.

The idea is to determine whether or not the concept of the predicate is *founded on* the concept of the subject. Leibniz says that this procedure may not terminate for certain propositions, particularly complex ones, such that the explicit identity is not revealed after a finite number of steps. Leibniz, like many of his contemporaries, might have agreed with Stifel that the properly contingent analysis "lies hidden in a cloud of infinity". That remark reflects Leibniz's modesty about human knowledge, but he is no skeptic. One way to disperse the cloud is through a mathematical analogy: "Nicholas even gives the mathematical analogue of the increasingly many-sided regular polygons inscribed in a circle whose content comes closer and closer to that of the circular disk in question without ever reaching it".[10] This will prove to be a useful analogy.

2.3 The Relation

Leibniz stands within the Scholastic tradition on relations in distinguishing relations from both individuals and attributes of individuals. Thomas, in keeping with the Aristotelian ontology, had distinguished between the relation in the accidental sense of being inherent in something else, and relation in the proper sense, not of inhering in, but of referring to something else.[11] Leibniz stands outside this tradition in his view that relations also in some sense inhere in two subjects at the same time, which means they cannot be accidents per se (Mugnai 1992, ch. I, secs. 2–3). So, they follow a third way; yet, they are based on an objective foundation (*fundamentum*) in things, for once given the individuals, and so given the

---

[10] Rescher 1967, p. 24. See Nicholas of Cusa, *De docta ignorantia,* chs. I–II.

[11] In *Quaestio disputata de potentia*, q. 7, a. 9, he writes: '[…] ipsa relatio […] aliud habet in quantum est accidens et aliud in quantum est relatio […]' (*Opera,* 3, 1980, p. 247). For more on the decisive influence of Thomas' doctrine of the relation, see Mugnai 1992, ch. II, and p. 97, n. 19.

modifications thereof, the relations are an immediate outcome or result. Leibniz said:

> I believe that a mode is properly an accident which determines, or adds certain limits to what is perpetual and undergoes modification. But I would not attribute this property to the relation and indiscriminately to all accidents. The relation results in fact from the substance and the modes, without producing any change of itself but only in virtue of a *consequential*. In a sense, the relation may be defined as an *ens rationis*, which yet is real at the same time; all things in fact are constituted by virtue of the divine intellect [...] relation and order are not therefore imaginary [...] since they are founded in truths.[12]

As an *ens rationis,* the relation requires the perceiving mind. As Mugnai explains, a relation is "an automatic inference made by the intellect but which has its basis in given states of things" (Mugnai 1992, pp. 26–27). The mechanism requires first, two individuals to which we can refer (e.g. "Socrates" and "Theaetetus") that must be given in existence; second, they must have some mode or accident in common; third, they must be compared in a single act of thought. In the exchange with Clarke, Leibniz wrote:

> The ratio or proportion between two lines L and M may be conceived three several ways: as a ratio of the greater L to the lesser M, as a ratio of the lesser M to the greater L, and, lastly, as something abstracted from both, that is, the ratio between L and M without considering which is the antecedent or which the consequent, which the subject and which the object. And thus it is that proportions are considered in music. In the first way of considering them, L the greater, in the second, M the lesser, is the subject of that accident which philosophers call 'relation'. But which of them will be the subject in the third way of considering them? It cannot be said that both of them, L and M together, are the subject of such an accident; for, if so, we should have an accident in two subjects, with one leg in one and the other in the other, which is contrary to the notion of accidents. Therefore we must say that this relation, in this third way of considering it, is indeed out of the subjects; but being neither a substance nor an accident, it must be a mere ideal thing, the consideration of which is nevertheless useful.[13]

Leibniz said that numbers are like relations in this third way. Mathematical relations, then, for Leibniz, are among the *entia rationis*. They are ideals that are realized in the rational thinking subject. That is where the connection is forged. The takeaway in connection with our present concern is that order is an attribute "inherent in several subjects at once", in one sense (Mugnai 1992, p. 30), and self-standing in another sense.

---

[12] Translation by Mugnai (1992, p. 26) of a passage from *Notationes "Ad schedam Hamaxariam"*.

[13] Fifth paper; L 704. Cf. Russell 1937, p. 13.

A detail in this reduction is the problem of relations and relational predicates.[14] One wants to say that the relation of Alexander the Great to his father Philip of Macedonia could be reduced such that we predicate of Alexander the concept of *son of Philip,* while predicating of Philip the concept of *father of Alexander,* and then combine these two propositions by some operation like conjunction.[15]

Leibniz spends much time and energy in his attempt to reduce the relations between two substances to monadic predicates of each substance, following Thomas in reducing the relation to two sentences expressing a relation and its converse (Mugnai 1992, p. 57). Take the sentence 'Helen loves Paris', which can be split into the following: 'Helen is a lover' and 'Paris is beloved', and nothing else. The problem is the unity of the proposition, for what we want is to be able to infer the truth of each from the truth of the other. He uses, for example, the *quatenus* locution, *eo ipso,* and others, to forge a logical connection across the conjunction, e.g. *et eo ipso*: Helen is a lover insofar as Paris is beloved. Leibniz is successful with symmetric relations, which were his focus.

Relations also have a distinctive nature as they are in the 'subject-relation-subject' form: Philip is the father of Alexander. In this distinctive sense the relation is not a proper accident but is thought of as a relationship "*hors des sujets*" (Mugnai, loc. cit.). 'Theaetetus is taller than Socrates' could be reduced roughly to something like a three-part conjunction of 'Theaetetus is taller than someone', 'someone is taller than someone else', and 'someone is taller than Socrates'. Still, although that helps clarify a traditional worry about the Leibnizian analysis of relations, it does not seem to match the historical Leibniz. In particular, what is the standing of the middle conjunct, the 'xRy', for some x and y? It looks to be self-standing, which would be in keeping with the position that Leibniz was committed to the reality of certain relations,[16] a position difficult to endorse without major qualifications. Instead, the relation as *hors du sujets* is a "purely ideal thing" and syncategorematic for Leibniz. This is the third way, the sense in which a relation stands with one leg in one subject and the other in the other. That allows the relation to be an absolutely precise construct adequate to the analysis of mathematics.

Socrates, as he remarks in the *Theaetetus,* while not changing in terms of his own size, grows smaller in relation to Theaetetus. The new widower in India is changed by the death of his wife back in Europe, even without his knowledge of it. And that interconnectedness comes to the assumption of the pre-established harmony. It involves the hypothesis of concomitance. In "A Specimen of Discoveries of the Admirable Secrets of Nature in General", Leibniz gives a picture of that hypothesis:

> [...] everything that happens in the soul can be explained by means of the laws of perception alone [...]. It is also clear what perception should be if it

---

[14] See Hintikka 1972 and Ishiguro 1972, and Mugnai's response (1992, ch. VI).

[15] This shows the influence to a certain extent of the Hobbesian account of mental processes as operational, except that for Leibniz the symbolism is not a mere convention but is somehow in accord with reality. See sec. 4.3 below.

[16] That is Hintikka's position.

is to be applicable to all forms, namely, the expression of many in one [...]. In the mind, however, apart from the expression of objects, we also find consciousness or reflexion, in which there is a kind of expression or image of God himself (Leibniz 2001, p. 321).

2.4 The *Analysis Situs* as a "Mechanical Thread of Meditation"

The *analysis situs* is the idea of specifying, for example, a geometric object like a triangle by situating the points in respect to each other, an idea Leibniz discusses in the 10 August 1679 letter to Huygens, and returns to later in life. In the *Dissertatio de arte combinatoria* he had defined '*situs*' as "the location of parts" (L 77–78). It is absolute or relative: the former is expressed by a line; the latter is expressed by the circle. Within a decade after the publication of his dissertation, Leibniz will come to understand the tangent by giving an absolute (linear) but infinitary form of expression to what expresses the relative (circular). Indeed, the very contention of the truth being a matter of conceptual inherence of predicates in the foundation of the concept of subject might be said to be a situational interrelation between subject and predicate components. Leibniz appears to have thought that the situational interrelationship is self-standing, imbuing this "new type of calculus" with potential power not possessed by any other. Like Huygens in his correspondence with Leibniz, we pass over that claim in silence.

What cannot be avoided is the optimism that a combinatorial symbolic analysis, a "mechanical thread of meditation", can somehow clarify reality (Mates 1986, ch. XIV). Leibniz wrote to Tchirnhaus in a letter of May 1678: "No one should fear that the contemplation of signs will lead us away from the things in themselves; on the contrary, it leads us into the interior of things". But Leibniz did not believe that the human mind is infinite: that is the distinguishing feature of God's mind. We are conscious only of a partial mirroring of that infinity. That distinguishes us as humans and prevents us from recognizing such things as the entirety of a complete individual concept. It seems, then, that we have an interesting modification of Spinoza's *sub specie aeternitatis*, for Leibniz does not allow humans that access entirely, and yet does allow humans useful fictions, such as infinitary analysis, as we shall see, to mirror the real actions going on in the infinite mind. In Leibniz's own words:

> The[r]e is something which had perplexed me for a long time—how is it possible for the predicate of a proposition to be contained in (*inesse*) the subject without making the proposition necessary. But the knowledge of Geometrical matters, and especially of infinit-esimal analysis, lit the lamp for me, so that I came to see that notion too can be resolvable in infinitum. (C 18; Rescher 1981, p. 113)

In Leibniz's philosophy, therefore, out of respect for human limitations, the real actualization completing and complementing the conceptual and nominal happens *in mente Dei*.[17] But how, and how far, does this security extend? To answer that

---

[17] See L 627. Cf. Mugnai 1992 and Mates 1986, pp. 245–6.

question, we might first remind ourselves of Leibniz's well-known Law of Continuity, and its dual, the principle forbidding "leaps" or jumps.

3 Technical Aspects of the *Saltus*

There are "no leaps in nature", Leibniz reminds us again and again. Nature is *smoothly continuous.* Yet, what of the successor operation, the idea of moving from one step to the next in an operation: does this not constitute a mini-leap?

Leibniz was not oblivious to this question; still, for him a succession of terms in a sequence need not constitute a jump. The conceptual transition from operations on finite sequences of differences and sums contributes to the broader view. In 1687 he wrote: "Jumps are forbidden not only in motions, but also in every order of things and of truths". These are all to be "continuous", such that their "parts are indeterminate and can be assumed in infinite ways", he will say years later (Leibniz 1981, p. 700).

Leaps, then, can be understood by contrast with a numbering, which involves a discrete repetition and no skips. Still, this principle turns out to be more puzzling than it at first appears, for, as Russell brilliantly quips: "In spite of the law of continuity, Leibniz's philosophy may be described as a complete denial of the continuous" (Russell 1937, sec. 59). Leibniz himself says that the confusion of the actual and the ideal leads to a mistaken understanding of analysis: "In actuals, single terms are prior to aggregates, in ideals the whole is prior to the part. The neglect of this consideration has brought forth the labyrinth of the continuum" (G II 379; Russell 1937, p. 245). In 3.2 and 3.3 we will see that what the law of continuity comes to is the infinite mathematical series, although Leibniz's first step in that direction is combinatorial.

3.1 "No Leaps in Nature"

In Paris already Leibniz had drawn a clear mathematical distinction, which he came to identify as the *saltus,* or leap.[18] "Leap" is an unfortunately tinny translation of '*saltus*', but it is the standard one in use in Leibniz studies. The concept has also been picked out by the translation "jump". A *saltus* occurs when, "as in algebra", an analysis "reduces a given problem to what is simpler", but not "by intermediate problems".[19] The simpler replaces the more cumbrous, providing a new aspect. Leibniz's three main mathematical contributions—the combinatorial, the ordered series, and the calculus—each represents a step in the same sustained project by which the Cartesian algebra is expanded to the infinite case, and each constitutes a leap.

---

[18] See Levey 2003.

[19] C 557; again, in the *Elementa nova matheseos universalis* he wrote: "Methodus solvendi problema est vel synthetica vel analytica. Utraque vel per saltum vel per gradum (C 350).

We can see a development in Leibniz's thought from the pre-Paris combinatorial vision to the restructuring of these investigations in terms of the ordered series during the Paris years. During these years, he extended the concept of mathematical series to the infinite case. He then applied that extension to the analysis of geometry such that any curvilinear line can be given an exact algebraic expression, thus solving an old problem about tangents.

The *Dissertatio de arte combinatoria* marks the outset of Leibniz's publication history (Leibniz 1666). His overarching view seems to be that the combination itself is self-standing and integrated. Given the principle of sufficient reason, such a combinatorial operation in and of itself, however ideal, suggests to a thinker like Leibniz a real basis. A leading example of the combinatorial idea is the operator Leibniz uses for "real addition".[20] Swoyer raises the question of whether the operation is adequate, given Leibniz's account of the logical structure of concepts, to provide for the infinite complexity of concepts (Swoyer 1994, pp. 27–28). Leibniz said that "infinite things can be compounded out of the combination of a few" (C 430; Leibniz 1973, p. 2), which seems to allow that an operation could be infinitary. He mentioned infinite conjunctions in the paper on real addition (G VII 245; Leibniz 1685–7, p. 142); however, strictly speaking he never provides the technical apparatus. The analysis also requires an infinite number of primitive concepts, for which there is some faint textual evidence (Leibniz 1683?, p. 10).

3.2 The Mathematical Series

It is an insight to recognize, in any such combinatorial scheme, a single ordered series. The second and accompanying insight was to extend that ordered series to infinity. That was the maneuver that allowed Leibniz to break through the parameters Descartes had imposed on the analysis of geometry: "Descartes's mind was the limit of science", Leibniz said (Knobloch 2006, p. 113). Leibniz opened the door to geometric objects that cannot be determined by finite algebraic equations, first by developing linear algebra, and then, having witnessed the inability of algebra to handle the transcendental, by introducing infinite equations. In this way, he conceived of a "science of the infinite", or in Leibniz's own words, "a veritable complement of algebra for the transcendentals". Knobloch calls it a "geometry of the transcendentals" (Knobloch 2006, p. 113, 127, 118, 115).

Leibniz achieves his breakthrough by pursuing geometric exactness. So, while for Descartes geometry dealt with a realm that was static, fixed, and closed, for Leibniz geometry is wide open and dynamic in the sense that it is in flux along with current knowledge of the subject. This progress in algebra Leibniz saw as combinatorial. In Proposition 49 of *De quadratura arithmetica*, written in 1676, Leibniz established a criterion for the convergence of an infinite alternating series. He had already deduced such a series for $\pi/4$ in Proposition 32. This, he pointed out in his 51st and final proposition, is the ultimate arithmetical quadrature; it could not be more geometrical in the Cartesian sense without being finite, but there is no such

---

[20] Leibniz 1685–1687. Scholars once thought that this item was composed after August 1690.

finite algebraic equation (Leibniz 1993, pp. 128, 17). That is how he marked off the limit of what can be done under the aspect of what is human and finite.

What philosophical commitments are involved in extending the realm of quantification to the infinite? Two of the main convictions that shape Leibniz's view were, first, that the whole is greater than the part, and, second, that the infinite is to be treated by the same rules as the finite, a version of the law of continuity (Knobloch 2006, pp. 122–124). The first conviction was based in Euclid and widely shared by the intellectual community in Leibniz's time. The working out of the second conviction led him to a position in which the existence of the actual infinite would violate the first conviction, a conviction obvious already in the Paris period.[21]

It is in this respect that for Leibniz the actual infinite is a fiction. It is an ideal of thought, a connection of order that is made real in a single combinatorial act of a rational mind. So, he wisely insists that part-whole relation is not entirely applicable to the relation between the collection of even natural numbers and the entire collection of the natural numbers.

De quadratura arithmetica does not include the *saltus* per se; however, in Proposition 32, just after the highpoint of the paper, he considers the equation:

$$\text{area of circle : circumscribed square} = \text{arc of the quadrant : diameter}$$
$$= 1 - \frac{1}{3} + \frac{1}{5} - \frac{1}{7} etc. : 1$$

This is like the *saltus*. He said it "is an exact equation. Yet, it is readily perceived because the mind rushes through it so to speak by a single stroke (*unus ictus*)" (Knobloch 2006, p. 128). The *uno velut ictus* should be attended to as central to the philosophy: "*uno velut ictu mens pervadit*". This quote from Leibniz follows hard on the major technical development in Proposition 31, in which he proposed the general analytic relationship between the arc and the tangent of the circle (Leibniz 1993, p. 17). In the Scholium to Proposition 31, Leibniz directly connects the infinite mathematical series to the solution to problems in the analysis of geometry.[22] An infinite mathematical series consists of ordered elements, or even, it

---

[21] See "Notes on Galileo's *Two New Sciences*" (Leibniz 2001, pp. 4–9), dated 1672, and "On Minimum and Maximum; On Bodies and Minds" (Leibniz 2001, pp. 8–19) dated 1672/3.

[22] Hoc theorema totius tractationis nostrae palmarium est: ejusque causa reliqua scripsimus. Series longitudine infinitas, magnitudine finitas, esse quantitates veras, multis exemplis ostendi potest: imprimis vero manifeste exemplo progressionum Geometricarum, de quibus supra. Progressionibus autem geometricis et nostrae nituntur. At inquies magnitudo quaesita sic non potest exhiberi, quoniam in nostra potestate non est progredi in infinitum. Fateor: neque enim eam constructione quadam geometrica exhibere promitto, sed expressione Arithmetica sive analytica. Seriei enim, licet infinitae, natura intelligi potest, paucis licet terminis tantum intellectis, donec progressionis ratio appareat. Qua semel inventa frustra progredimur, quoties de mente potius illustranda, quam de operatione quadam mechanica perficienda agitur.

might be said, of self-ordering elements, which Leibniz conceived as in combination, linked together by an operation like conjunction.

3.3 The Calculus

The most formidable of all Leibniz's Renaissance achievements was the calculus (Leibniz 1684 and 1686a). What was essential to its development is in a sense the jump (*saltus*) from finite to infinite.

Leibniz must have appealed to the infinitesimal rather than the derivative in the early development of the calculus, since the derivative depends on the function concept, while that very concept of the function had not been clarified at that time. The notion of the infinitely small itself is not completely clear at the outset (Ishiguro 1990, p. 81). It seems clear that an interpretation of Leibniz that relies on the function concept is anachronistic, for he is at most contributing to the process of the formation of that concept, in particular in his formalization in mathematics of formulae apart from geometry (Bos 1974, sec. 1.9).

That said, from another perspective, things are not so easy, for Leibniz had some ideas that resemble the concepts of function and derivative. The function concept, although not clarified, was coming into view in his thought and in the thoughts of other mathematicians, as well. Leibniz will use something like a mathematical function to describe a process of synthesis as a life of change (Rescher 1981, p. 113). Likewise, the derivative, while nascent, is not completely hidden from view in Leibniz's accounts. 'Function' originally had a geometric connotation. For Leibniz, "functiones" of a curve "include variable geometric quantities such as coordinates, tangents, radii of curvature, *etc.*" All that data can be exactly described through algebraic equations.

As we will see, it is the convergence of the infinite series that will allow Leibniz to challenge the view that an infinite analysis presents a vicious regress. Once again, as with the function concept, one wants to say that he is using the limit concept here; and indeed, like the function concept the limit concept in nascent form was on the mathematical scene already, so that that these ideas would have been available to him in Paris. But, strictly speaking, he did not in the early work on the calculus use a limit concept like we would formalize it.

In extrapolating to the actual infinite, a difference sequence transforms into an actually infinite sequence of infinitely small terms, called *differentials* (Bos 1974, sec. 2.5). Leibniz wrote "*continuus*" for a variable that ranges over an infinite sequence. He talked about growth and motion, writing about "increasing by minima" ("*per minima crescentes*"), "continually increasing by inassignables" ("*continue crescentes per inassignabilia*"), and "momentaneously increasing" ("*momentanee crescentes*"). In these examples, Bos writes, "'minima' and 'inassignables' stand for the differentials as differences between successive terms of the sequence. If these differences are equal, Leibniz sometimes used the term 'uniformly increasing' ('*aequabiliter crescere*')" (Bos 1974, sec. 2.4). Recalling Russell's apt quip, continuity for Leibniz is the idea of a variable going proxy for a mathematical series.

A finite sum sequence transforms into an actually infinite sequence of infinitely large terms, called *sums*. Differentials and sums form sequences. The finite sequence operators transform as well. d, the symbol for the differentiation operator, assigns to a finite variable an infinitely small variable: e.g. to y it assigns dy. $\int$, the symbol for the summation operator, assigns to a finite variable an infinitely large variable: e.g. to y it assigns $\int y$. "Therefore", Bos writes:

> [...] the fundamental concepts of the Leibnizian infinitesimal calculus can best be understood as extrapolations to the actually infinite of concepts of the calculus of finite sequences. I use the term "extrapolation" here to preclude any idea of taking a limit. The differences of the terms of the sequences were not considered each to approach zero. They were supposed fixed, but infinitely small (Bos 1974, sec. 2.1).

In defining the "infinitely small", Leibniz wrote that the increment that results from "the addition of an incomparably smaller line to a finite line"

> [...] cannot be exhibited by any construction. For I agree with Euclid Book V Definition V that only those homogeneous quantities are comparable, of which the one can become larger than the other if multiplied by a number, that is, a finite number. I assert that entities, whose difference is not such a quantity, are equal. ... This is precisely what is meant by saying that the difference is smaller than any given quantity (Leibniz 1695, p. 322; Bos 1974, sec. 2.1).

Finally, the operators d and $\int$ are reciprocal:

> *Foundation of the calculus*: Differences and sums are reciprocal to each other, that is, the sum of the differences of a sequence is the term of the sequences, and the difference of the sums of a sequence is also the term of the sequence. The former I denote thus: $\int dx = x$; the latter thus: $d\int x = x$.[23]

The Cartesian algebraic analysis preserved geometric exactitude, and Leibniz preserved the form of that analysis, thereby preserving the exactness, while extending the objects that can be treated to include those curves that can be accessed algebraically only in the actually infinite case. That is a jump.

His strategy is based on the concept of a series or numeric sequence: "The consideration of differences and sums in number sequences had given me my first insight, when I realized that differences correspond to tangents and sums to

---

[23] Leibniz, from a manuscript, "Elementa calculi novi pro differentiis et summis, tangentibus et quadraturis, maximis et minimis, dimensionibus linearum, superficierum, solidorum, aliisque communem calculum transcendentibus", partly translated in Bos 1974, sec. 2.9.

quadratures".[24] The basic insight, then, in his analysis of tangents was to use the curvilinear figure to approximate the curve by treating the curve as a geometric object with infinite angles, an infinitangular polygon (Leibniz 1684a, p. 126).

We have seen aspects of the Leibnizian infinite that bear rough similarities to our post-Cantorian understanding. Philosophically, it also has a most ancient aspect of the indefinite infinite, perhaps like *to apeiron* in Aristotle and even Anaximander.[25] Of course, putting it like that hardly sheds any light, but one of the features of the indefinite is that it is the unlimited source of genesis and decay. We might, as others have, contrast it metaphorically with entropy. In his debate with Clarke (and Newton), Leibniz specifically disagrees with the position that God would have to intervene at various points in universal history to make sure that energy loss was countered by winding the universe back up.

Aristotle said that it is "the principle of other things", "held to encompass all and steer all".[26] Leibniz said: "This infinite would more correctly be called the immeasurable".[27] It is like an indefinite unstructured background condition. It bears repeating the Aristotelian distinction mentioned above. The subject can be conceived in two ways. In the second—formal—sense, the foundation of the concept of the subject is to contain the concepts of all its accidents. The indefinite background is presupposed as the place for the formal analysis, although Leibniz never puts it that way. Leibniz himself will give this autobiographical assessment later in life:

> At length some new and unexpected light appeared from a direction in which my hopes were smallest—from mathematical considerations regarding the nature of the infinite. In truth there are two labyrinths in the human mind, one concerning the composition of the continuum, the other concerning the nature of freedom. And both of these spring from exactly the same source—the infinite.[28]

The purpose of the next section is to give a philosophical account of that "same source".

---

[24] Leibniz in a letter to Wallis of 28 May 1697; *Math. Schr.* IV, p. 25.

[25] See the fragment that Simplicius, in his commentary on Aristotle's *Physica*, attributes to Anaximander: "Whence things have their origin, there their destruction happens as it is ordained. For they give *justice* and *compensation* to one another for their *injustice* according to the ordering of time".

[26] *Physica* 203b, Hardie and Gaye, trans.

[27] Leibniz 1992, p. 27, quoted from the 1676 "On the Secrets of the Sublime, or on the Supreme Being".

[28] Couturat 1901, p. 210, notes; quoted by Rescher 1981, p. 112. See the "Guilielmus Pacidius on the Secrets of Things" (Leibniz 1992, pp. 88–91).

4 The *Saltus* Constitutes a Philosophical Assumption that Transcends Nominalism

Leibniz gave a more philosophical presentation of the mathematical developments identified in sections 3.1–3.3. He did this for example in essays on the philosophical questions of infinite number and infinite degrees of the soul (Leibniz 1676a and 1686?). Leibniz himself ties the leap to the curvilinear line to the problem of the human connection to the divine. He was thinking about Spinoza quite a bit in the 1670s and even went to meet him, one of his concerns being the question of human freedom given the divine will. In the "*Infiniti possunt gradus esse inter animas*", Leibniz used the curvilinear line to schematize a series of degrees of knowing and being, in an interesting twist on Plato's Divided Line. But the geometrical metaphor for the Soul that Leibniz gravitates to is not the circle, but the infinitangular polygon.

4.1 Osculations

It is not entirely clear when "There Can Be Infinite Degrees of Souls" was written, but it could have been 1686.[29] Our main quote from this text is as follows:

> But mind corresponds to an osculation of infinitieth degree; and expresses the entire curvature of a line in a given point, in other words, whatever is not assignably missing from the given point. Whence it is clear that minds are to simple souls as the infinite is to the finite, or as the finite is to the infinitely small. But an infinite mind corresponds to the whole progression of motion through the line, to the transition from the given point to any other point assignably distant from it, that is to say, a leap. (Leibniz 1686?, pp. 298–301).

From the standpoint of substance, Leibniz will give the operative or recursive stipulation of an infinity of degrees of souls based on the initial state, which he identifies with body, a lifeless aggregate.[30]

In building up the concept of substance, Leibniz draws an analogy with a curvilinear line conceived as a polygon of infinite sides. The zero state is body.

---

[29] Arthur placed it just before "A Specimen of Discoveries of the Admirable Secrets of Nature in General" and just after 1686?a.

[30] During the Paris years Leibniz is thinking of sphere-like bodies such as "globules" in a way reminiscent of corpuscles, for which see Leibniz 1676. These globules for Leibniz, as with the corpuscles for Locke, solve an important problem about action at a distance. Leibniz attributes the globules to Democritus. Bodies are essentially in motion. What separates them from mass itself is "endeavor" ("*l'effort*" in Leibniz 1686?a, p. 298); without the *conatus*, "all variety in bodies ceases". The globules fill all of space, with infinitely smaller ones fitting in nicely between larger ones. Since each globule moves around its own center, the smaller ones are moving at a faster rate.

This corresponds to a point in space, a present state without change. Add a point and the direction of motion is expressed. Likewise, the first iteration takes us from body, which is in itself a lifeless aggregate, to life and the lowest degree in the soul. If "we connect three points, we have not only the direction but also the bending, or change of direction; and so we have the osculating circle" (Leibniz 1686?, p. 299). Likewise, the second iteration is to a soul with sensation of the first degree. In general, the $n+1^{th}$ point added to the curvilinear line presents an osculation of the $n-1^{th}$ degree and corresponds to the nth iteration of substance within non-rational soul with sensation of the $n-1^{th}$ degree.

The osculations are an interesting inversion of the Cartesian picture. Biologically, the glandular connection between mind and body disappoints almost every undergraduate, yet Descartes knew from his study of line segments that a two-dimensional grid can be projected onto a single axis or line segment, and this great dimensional reduction was no doubt deep in his mind when he wrote the *Meditations,* for his Archimedean point is the *cogito,* and not the "general rule" of the third Meditation. The body, for all its clarity as *res extensa,* and its distinctness as an unthinking thing, is second in the Cartesian *ordo cognoscendi.* It is a real second dimension. When we realize that second dimension through a wax argument, the underlying substantial reality of the persisting wax is known by the mind alone. Descartes said that the duality of body and soul is unlike the duality of ship and pilot; instead, the body complements the soul and completes an integrated human nature. So, for Descartes, the body is the added dimension that arises in meditating as a clarification to distinguish by contrast, like the hill and the valley, what it is for one substance to think and not extend, and for another to extend and not think. In reducing Cartesian dimensions one uses characteristic functions so that the dimensionality may be recovered; to persist in our analogy, then, human nature is for Descartes at its barest foundation not merely soul alone but the union or unity of body and soul.[31] But Leibniz, by direct contrast, in this text sets the initial state to be body, with the mind being a further degree, or rather, a further series of degrees.

An apparent shortcoming of this component of Leibniz's position would be that, in responding to the Cartesian account of soul and the problem that it separates the human soul from nature, Leibniz goes too far in the other direction, for on his account there really is nothing to distinguish human souls from other animal souls, on one way of looking at this. That is to say, a traditional complaint has been that from this position there is only a difference of degree between the human and animal, but not a difference of a fundamental kind. As is evident in the text we have been looking at, for Leibniz by this analogy the recursion, in this case a concept drawn from Leibniz's *lineation* operation, is essential to our commencing out of our physical nature. For us, from the perspective of 21st-century biology, the basic recursive ability, that infinite capacity of human language, is genetic and evolutionary.[32] That is to say, the complaint does not really hold of our text. For we saw that what distinguishes rational soul from non-rational soul was that non-

---

[31] See also "On the Union of Soul and Body" (Leibniz 1992, pp. 34–37), written in 1676.

[32] See Chomsky 2009, p. 200: "What concerned the Cartesians was something different: the creative use of language, what Humboldt later called 'the infinite use of finite means,' stressing *use*".

rational soul can handle one certain degree or another but that the rational soul does lineation to any degree, which is an entirely different level. Indeed, this is essential to a further step Leibniz takes in this text, for there is a fundamental difference between finite and infinite mind. The fundamental break in this text is between God, who incorporates all leaps, and humans, who cannot.

We can now return to the osculations and the concept of a jump to all distances. An "infinite mind corresponds to the whole progression of motion through the line, to the transition from the given point to any other point assignably distant from it, that is to say a leap" (Leibniz 1686?, p. 301). The first part is uncontroversial. But the second part now sounds like a return to the concept of lineation, which is characteristic of a finite mind. Characteristic of the infinite mind is not merely lineation but the actual line, however extended, and actually making the leap. Leibniz said that himself, just following the main quote:

> Now it might be thought possible here to make a further distinction among minds making a leap through different finite distances, which all infinitely exceed a mind lacking a leap, and yet are still infinitely inferior to a mind whose leap is to all distances: but to this it must be said that whenever it is a mere continuity, or only a difference through greater or less, it is not possible for different species of mind to be assigned. And so a mind that makes a leap makes it to however great a distance, since it is evident that osculations differ numerically, i.e. are discrete (*Ibidem*).

This infinite mind leaps "to all distances"; but then Leibniz seems to reverse himself by saying that this is the mind that when it makes a leap makes it to however great a distance, because we are talking about discrete numeric osculations.

There is some conflict here, it seems to me. For Leibniz has rightly emphasized that there is a real conceptual difference between the lineation of which humans are capable and the concept of all lines however extended. That certainly is a real step. Is it a jump? If so, then, it could not be in nature itself. But Leibniz was committed to the actual infinite in nature.[33] There is a second degree of tension. Leibniz has a continuum between animals and humans; yet, no species except the human is capable of lineation. For all others, it is one degree or another. Other animals can be taught to work with individual numbers but they cannot do the recursion. So, there seem to be three levels at work, which we can distinguish by kisses (osculations). The first is the level of a certain limited number of osculations, the second is the level of stepping from any osculation to the next, and the third is the level of all osculations.

On this construal a conflict arises because there are two jumps depicted right there, on Leibniz's own terms. How do we resolve this? But again the difference between going from two to three and going from any number to the next represents to some degree increased consciousness, at least in terms of the problem of

---

[33] "[T]here exist some globules smaller than others to infinity" (Leibniz 1676, p. 60); see also Leibniz 1686?a, p. 296; 2001, p. 234; and Levey 2012, p. 34.

generality. For it represents at the minimum a different order in conceptuality, but Leibniz braced against that being of a different order.

Leibniz wrote: "There are two kinds of analysis: one is the common type proceeding by leaps, which is used in algebra, and the other is a special kind which I shall call 'reductive'" (Leibniz 1683?, p. 16). He explained this distinction further in *Elementa nova matheseos universalis*. Analysis, Leibniz said, happens "through a leap" when

> [...] we begin to solve the problem itself, making no other presuppositions. [...] Analysis by degrees (per gradum) is when we reduce the given problem to one which is easier, and this in turn to one which is easier still, and so on until we come to one which is within our power.[34]

Again he wrote: "Analysis reduces a given problem to what is simpler; this occurs either by a leap, as in algebra, or by intermediate problems, in topics or reduction".[35] He presented the synthetic analysis in the *Nouveaux essais* as one of three forms of classification (Leibniz 1981, IV.21).

A separate issue is that a particular fact or memorable anecdote can be classified under multiple headings, as "for instance, the very apt story that is told in the biography of Cardinal Ximénez, of how a Moorish woman cured him of an almost hopeless bout of hectic fever merely by rubbing" has multiple categorizations. This form of classification, bookkeeping, could be counted as a third form of analysis.

> But now let us speak only of general doctrines, setting aside particular facts, history, and languages. I know of two main ways of organizing the totality of doctrinal truths. Each has its merits, and is worth bringing in. One is *synthetic* and *theoretical*: it involves setting out truths according to the order in which they are proved, as the mathematicians do, so that each proposition comes after those on which it depends (G V 506).

The theorems of Geometry and other demonstrative reasoning are synthetic on Leibniz's account. The theoretical division, with its synthetic arrangement, is "natural philosophy". It is the combinatorial and operational approach to knowledge. Leibniz, in a 1683 distinction drawn between analytic and synthetic, specifies that the synthetic is the combinatorial (Leibniz 1683?, p. 17). He specifically sets aside the approach by bookkeeping when speaking of general doctrines.

---

[34] C 351, translated in Leibniz 1683?, note m.
[35] C 557, translated in Leibniz 1683?, note m.

4.2 Pure Mathematics

By "pure mathematics", Leibniz means "mathematics which contains only numbers, figures, and motions".[36] This is an algebraic conception; however:

> Algebra is but one example of it. I have a new kind of geometry as different from Mr. Descartes's as his is from the geometry of the ancients. For as Descartes has added supersolid lines and [...] even problems of the fifth, sixth, seventh, and higher degrees to the plane and solid lines and problems of the ancients, so I have added transcendental problems which are of no degree, or rather of all degrees at once, to his (L 274).

We have seen a reduction of the operating notation to an infinite series of conjunctions. The propositional reduction through definitions and predicate containment is to instances of the law of identity. This quintessentially Leibnizian analysis is like our idea of an infinitary logical language.[37] The paradigmatic Leibniznian mathematical analysis is an infinite reduction of contingent propositions with intermediate stages of propositions of infinite length. That is synthetic overall, and has the odd effect of giving the propositions of mathematics an air of contingency. After all, Leibniz's viewpoint is that only through an infinite analysis can they be clarified, which means their truth is contingent upon the existence of a certain number of things. That calls to mind Knobloch's apt and succinct description; a science of the infinite would amount to a contingent investigation of the infinite, for it would presuppose facts about the world such as how many things there are.

The clue is to separate the categorematic from the syncategorematic.[38] What is infinite is not really the subject or predicate of a proposition for Leibniz in the sense that there is no real meaning. Leibniz, in the famous early paper on minima and maxima, has what looks to me something like an argument establishing that the points on the diagonal are not in a one-one correspondence with the coordinates on the axis. We have already witnessed in his discussion of the degrees of soul that the Leibnizian position is that it is the human ability for any curvilinear line to add a point thereby extending the old line to a new one. That would seem to presuppose

---

[36] L 274. From a letter to La Chaise in which he claims to retain the philosophy of Aristotle and Thomas.

[37] Quine 1960. In that sense even the second edition of the *Principia Mathematica* is circular. See Gödel 1944.

[38] The comments on Galileo's *Discorsi* give a clue to Leibniz's take on the infinite. Galileo had demonstrated that there is a one-one map of the natural numbers onto the set of squares. See Knobloch 2015. Also, Knobloch 2012, p. 19, 1999, p. 89, and cf. the interpretation of Levey 2015, pp. 184–185.

the actually completed line, but, as Leibniz thought, there are some things that can be done without assuming infinity.[39]

It is possible that a confusion of the actual and the ideal leads to a mistaken understanding of analysis. Maybe it is not too much of an historical stretch to urge that for Leibniz the difference between the actual and the potential infinite comes to the same thing as the difference between the world viewed under the aspect of eternity and the world viewed under the aspect of the human. One subtle example of the infinite in terms of a series or progression was written 10 April 1676:

> Thus if you say that in an unbounded [series] there exists no last number that be written in, although there can exist an infinite one: I reply, not even this can exist, if there is no last number. The only other thing I would consider replying to this reasoning is that the number of terms is not always the last number of the series. That is, it is clear that even if finite numbers are increased to infinity, they never—unless eternity is finite, i.e. never—reach infinity. This consideration is extremely subtle (Leibniz 1676a, p. 101).

This text is suggestive of a shift from the strictly finite to the finite in a more ideal sense and then another shift to the actual infinite.[40] Leibniz seems to have thought that the potential and the actual infinite are two different ideas, appropriate to different locations. He located the potential infinite internally in the human mind and the actual infinite *in mente Dei,* but here we have two conceptual shifts. Another way to look at the same is from the perspective that for Leibniz a unity constitutes a leap, so that he is a pluralist about the natural numbers, but recognizes that the leap to the unity of the natural numbers is one that must be made, for he is ready to quantify over well-founded sequences of infinite length.

4.3 *Altitudinem Divitarum*

I have mentioned the limitless, unbounded, and unstructured nature of *to apeiron* as the background but did not include Aristotle's remark that it is "the Divine" (*Physica* 203b). An "essential feature", Mates tells us, of Leibniz's "reasoning on matters metaphysical is its dependence on a background condition of assumed truths about the existence and nature of God" (Mates 1986, p. 244). This condition might be something like an infinite ordering of ideas, including ideas of concepts, into the one unity by "God, who grasps the infinite in one intuition" (Leibniz 2001, p. 303). Since numbers, relations, and the actual infinite are all ultimately realized in God's mind, his view although primarily idealistic recalls ontological realism, as well (Mugnai 1992, p. 134).

We saw at the outset that the Aristotelian premise about substances and substantial reality was essential to Leibniz's view of the analysis of propositions and

---

[39] Leibniz 1976. There are obvious modern examples of analyses that do not assume the axiom of infinity, as for example in Zermelo 1909. For set theory without infinity, see Parsons 1987.

[40] For those jumps, see Bernays 1935.

to his metaphysics in general. A thoroughgoing realism cannot be ginned up from that Aristotelian premise. In other words, I don't see how—on his own terms—Leibniz achieves the "whole philosophical apparatus of concepts, propositions, ideas [...]" (Mates 1986, p. 246; also pp. 247–250). Just because a concept is realized as an idea *in mente Dei*, that is still not adequate, by the principle of sufficient reason, to establish the relational interplay of the ideas as expressed in an axiomatic structure such as set theory, for example. If one responds that the actual infinite is real in the nature of things, I would say that was missing the point. The Leibnizian analysis in itself, for all its power, is limitative. If one responds by asking what sort of special features are there that distinguish the limit from the other side, I would answer that the Leibnizian analysis can stay within itself by rejecting what is outside, but we can determine the exact limit by providing extensions of the analysis in terms of sets, for example.

If one responds that God's mind might have sufficient structure to be identified with an elaborate mathematical model, and obviously this as such would constitute the maximal case for a Leibnizian ontological realism, I would say we should draw the limit. The classic twentieth-century Leibnizian analysis is, of course, that of the *Principia Mathematica*.[41] That analysis reaches its absolute limit in extension as the constructible universe.[42] Let us be expansive in this anachronism but follow Leibnizian instincts, and so emphasize the definable. Still, that sort of transfinite order only provides a backbone to certain levels of the infinite hierarchy. That much structure, given modern presentations of the Leibnizian analysis, can be mapped into set theory as the hypothesis that the entire set theoretic hierarchy is identical with the constructible universe, namely, $V=L$. What would be required for a strongly dedicated realism is not merely an infinite order, but something equivalent to an additional axiom stronger than a measurable cardinal. But do we not, indeed, also identify categories of sets and do the algebra of sets? Such classifications certainly could be thought of as Leibnizian in the third sense, and perhaps could constitute stronger axioms. Let us leave behind these far-flung musings, and conclude with the point that these days 'Leibnizian' can refer to quite different things.

We have seen that at times Leibniz thought of infinity in terms of a series, for example, in the expression for π. Infinity, in that sense, Kant will revisit in his antinomies of pure reason. But we have also seen (in section 3) another sense of infinity in the idea of the infinitesimal.

These two senses of infinity, both plainly in Leibniz, closely correspond to Aristotle's classic distinction between the infinity of a progression and the infinitely divisible.[43] Does one sense weigh heavier with Leibniz than the other? In terms of the technical developments in mathematics that Leibniz achieved, his advance was from the formal series to the infinitesimal. The historian suggests that this is evidence more for the infinitely divisible conception. But it is the comparison on the other side of the distinction that is liable to be misleading, so I should clarify the

---

[41] Actually, the type theory changes so the analyses are not identical in the first and second editions; they are both, however, Leibnizian.

[42] Gödel 1938, p. 556; see also the introduction by Robert M. Solovay (Gödel 1986, p. 8).

[43] For which see Kanamori 2009.

analogy between Leibniz and Aristotle with regard to the potential infinite. Both Leibniz and Aristotle endorse the concept of a mathematical sequence of any finite length. Leibniz extrapolated from this idea and committed to well-founded sequences of infinite length. That extrapolation is fundamental to Leibniz's thought about analysis. Whether or not Aristotle has the same commitments remains an open question.

Any further question about the choice of which theory to use is a matter of bookkeeping. As such, it would fall under the third form of analysis identified in the *Nouveaux essais*. One might argue that this third sort of analysis qualifies as a conceptual analysis; however, the idea of a realist conceptual analysis is the idea of an exploration of the meaning of the concepts that surround mathematics and all the background structure that goes along with that, and that is not entirely like bookkeeping. We are faced with a background lacking an independent objective order embedded in the conceptual structure. So, the whole conceit—which I accept—that Leibniz's conceptualism, while inadequate on its own for mathematical analysis, has God's mind in the background, would be mistaken if it led to the belief that this made his version of conceptualism adequate for all of mathematics as we understand it, given the "mind whose leap is to all distances". Although a conceptualist and a realist, Leibniz's own form of analysis fails to provide sufficient reason unless the infinite mind is conceived in a more determinate fashion than Leibniz would have realized or might have thought. Here again, the aspect of infinite divisibility is in keeping with the weighted historical perspective.

The trouble remains. The analysis of mathematics requires at its foundation the unity of the natural numbers for an infinite reduction with intermediate steps consisting of infinitely long sentences.[44] What is lacking is an analysis of the meaning of the concepts that occur in the propositions.[45] We have seen something of a conceptual array, or at least that seemed to have been suggested. But an adequate conceptual analysis would fix the array by definition, whereas Leibniz never fixed the conceptual apparatus in that background condition, the source of the two labyrinths. His overall view remains *in situ*, this concept containing that concept; each of these asymmetrical containment relations stands and unfolds within an endlessly divisible and indefinite background.

5 Concluding Remarks

Having sketched the containment of predicate in subject, we saw the relation not quite fit into that picture. Relations follow a third way. Like other pure mathematical notions such as the infinite series, and the infinitesimal, the asymmetric relation is for Leibniz an ideal. Having admitted to a slight disagreement with Hintikka on that point, I acknowledged that, even if the idealism is complemented by a realism actualized by God's mind, still there remain regions of the higher infinite that are ruled out on this account.

The indefinite background for the Leibnizian analysis has enough structure for a nice combinatorial reduction, up to the level of the conceptual. Indeed, the

---

[44] See the introduction by Parsons to Gödel 1944, p. 115.
[45] See Gödel 1944, p. 151.

analysis must presuppose that much structure, which leads to one concern. My impression is that the Leibnizan analysis is empty of content. While the Leibnizian analysis can provide a formal treatment of the relation and other formal concepts, it is not adequate to an investigation of their nature. The broader position is in keeping with Leibniz's idea that there are no jumps in nature.

One can defend Leibniz against the charge of Russell that the analysis of asymmetric relations was inadequate by urging that Leibniz only intended the foundation as a sufficient but not a necessary condition. Only if, that is, Leibniz himself gives up the quest for the deeper meanings of the concepts involved. That consequence, as well, would be in keeping with an ultimately indefinite and unlimited background condition.

Acknowledgements

Thanks first and foremost to Eberhard Knobloch, Akihiro Kanamori, Charles Parsons, an anonymous reviewer, and to Brian Kiniry, who supervised the project in the early going. Thanks to Alison Simmons in regard to the third paragraph of 4.1 (4.1,3). Thanks also to Kelly Warwick, David Rollow, T. S. Link, Lewis Schneider, S. Balash, Richard H. Jandovitz, Alex Taylor, and Joanne Montgomery Link, and to Kathryn D. Schinabeck, Andres Barragan, and Michelle Lim. Jeffrey K. McDonough's interpretation of Leibnizian optics, although nowhere mentioned above, has been a model for this paper. Thanks to the organizers of the Northern New England Philosophical Association for allowing me to present this paper at Bates on 26 September 2015, and to the small audience for their excellent questions. Thanks very much to the editors, in particular to Raffaele Pisano. Errors are mine.

References

Anderson, R. L. 2015. *The Poverty of Conceptual Truth. Kant's Analytic/Synthetic Distinction and the Limits of Metaphysics.* Oxford: Oxford University Press.

Bernays, P. 1935. Sur le platonisme dans les mathématiques. *L'enseignement mathématique* 34:52–69.

Bos, H. J. M. 1974. Differentials, Higher-Order Differentials and the Derivative in the Leibnizian Calculus. *Archive for History of Exact Sciences* 14:1–90.

Broad, C. D. 1975. *Leibniz. An Introduction.* Levy, C. (Ed.). Cambridge; New York: Cambridge University Press.

Chomsky, N. 2009. The Mysteries of Nature: How Deeply Hidden? *The Journal of Philosophy* 106:167–200.

Couturat, L. 1901. *La Logique de Leibniz d'après des documents inédits.* Par Louis Couturat, clargé de cours à l'université de Toulouse. Paris: Félix Alcan.

Couturat, L. 1902. Sur la Métaphysique de Leibniz (avec un opuscule inédit). *Revue de Métaphysique et de Morale* 10:1–25.

Couturat, L. 1972. "On Leibniz's Metaphysics". Translation of Couturat 1902 by R. Allison Ryan. In Frankfurt, H. G. (Ed.). *Leibniz. A Collection of Critical Essays.* Garden City, New York: Anchor Books, pp. 19–45.

Garber, D. 2011. *Leibniz: Body, Substance, Monad.* Oxford; New York: Oxford University Press.

Gödel, K. 1944. Russell's Mathematical Logic. In Schilpp, P. A., (Ed.). *The Philosophy of Bertrand Russell.* Evanston: Northwestern University, pp. 123–153. Reprinted with an additional note of 1964 expanded in 1972 in Gödel, *Collected Works. Vol. II. Publications 1938–1974.* Feferman, S., et al. (Eds.). (Oxford; New York: Oxford University Press, 1990), pp. 102–141.

Hintikka, K. J. J. 1972. Leibniz on Plenitude, Relations, and the 'Reign of Law'". In Frankfurt, H. G. (Ed.). *Leibniz. A Collection of Critical Essays.* Garden City, New York: Anchor Books, pp. 155–190.

Ishiguro, H. 1972. Leibniz's Theory of the Ideality of Relations. In Frankfurt, H. G. (Ed.). *Leibniz. A Collection of Critical Essays.* Garden City, New York: Anchor Books, pp. 191–213.

Ishiguro, H. 1990. *Leibniz's Philosophy of Logic and Language.* Cambridge: Cambridge University Press. First ed. (1972).

Kanamori, A. 2009. The Infinite as Method in Set Theory and Mathematics. *Ontology Studies* 9:31–41.

Knobloch, E. 1999. Galileo and Leibniz: Different Approaches to Infinity. *Archive for History of Exact Sciences* 54:87–99.

Knobloch, E. 2006. Beyond Cartesian Limits: Leibniz's Passage from Algebraic to "Transcendental" Mathematics. *Historia Mathematica* 33:113–131.

Knobloch, E. 2012. Leibniz and the Infinite. *Documenta Mathematica.* Extra Volume ISMP:19–23.

Knobloch, E. 2015. Analyticité, équipollence et théorie des courbes chez Leibniz. In Goethe, N. B., Beeley, P., and Rabouin, D. (Eds.). *G. W. Leibniz, Interrelations between Mathematics and Philosophy.* Dordrecht: Springer, pp. 89–110.

Levey, S. 2003. The Interval of Motion in Leibniz's *Pacidius Philalethi. Noûs* 37: 371–416.

Levey, S. 2012. On Time and Dichotomy in Leibniz. *Studia Leibnitiana* 44:33–59.

Levey, S. 2015. Comparability of Infinities and Infinite Multitude in Galileo and Leibniz. In Goethe, N. B., Beeley, P., and Rabouin, D. (Eds.). *G. W. Leibniz, Interrelations between Mathematics and Philosophy.* Dordrecht: Springer, pp. 157–187.

Mates, B. 1986. *The Philosophy of Leibniz. Metaphysics and Language.* New York; Oxford: Oxford University Press.

McDonough, J. K. 2009. Leibniz on Natural Teleology and the Laws of Optics. *Philosophy and Phenomenological Research* 78 (May):505–544.

Mugnai, M. 1992. *Leibniz' Theory of Relations.* Studia Leibnitiana Supplementa XXVIII. Stuttgart: Franz Steiner.

Parkinson, G. H. R. 1965. *Logic and Reality in Leibniz's Metaphysics.* Oxford: Clarendon.

Parsons, C. 1987. "Developing Arithmetic in Set Theory without Infinity: Some Historical Remarks". *History and Philosophy of Logic* 8:201–213.

Quine, W. V. 1960. "Carnap and Logical Truth". *Synthese* 12:350–374.

Rescher, N. 1967. *The Philosophy of Leibniz.* Englewood Cliffs, N.J.: Prentice-Hall.

Rescher, N. 1981. *Leibniz's Metaphysics of Nature: A Group of Essays.* Dordrecht; Boston: D. Reidel.

Russell, B. 1937. *A Critical Exposition of the Philosophy of Leibniz.* With an appendix of leading passages. Second ed. Reprinted (London: Routledge, 1992). First ed. (Museum Street, London: George Allen & Unwin, 1900).

Russell, B. 1972. Recent Work on the Philosophy of Leibniz. In Frankfurt, H. G. (Ed.). *Leibniz. A Collection of Critical Essays.* Garden City, New York: Anchor Books, pp. 365–400. Originally published in *Mind* 12 (Apr., 1903): 177–201.

Sleigh, R. C., Jr. 1990. *Leibniz and Arnauld: A Commentary on Their Correspondence.* New Haven; London: Yale University Press.

Swoyer, C. 1994. Leibniz's Calculus of Real Addition. *Studia Leibnitiana* 26:1–30.

Swoyer, C. 1995. Leibniz on Intension and Extension. *Noûs* 29:96–114.

Thomae Aquinatis, S. 1980. *Opera omnia: ut sunt in indice thomistico, additis 61 scriptis ex aliis medii aevi auctoribus.* Busa, R. (Ed.). 7 vols. Stuttgart-Bad Cannstatt: Frommann-Holzboog.

Zermelo, E. 1909. Sur les ensembles finis et le principe de l'induction complète. *Acta Mathematica* 32:185–193.

Leibniz, Selected Works

1666. *Dissertatio de Arte Combinatoria* [...]. *Math. Schr.* V, pp. 1–87. Originally published (Lipsiae). Partially translated in Leibniz 1966, pp. 1–11, and in L 73–84.

1676. De plenitudine mundi. Translated as "On the Plenitude of the World". In Leibniz 2001, pp. 58–63.

1676a. Numeri infiniti. Translated in Leibniz 2001, pp. 82–101.

1683?. Of Universal Synthesis and Analysis; or, of the Art of Discovery and of Judgement. Translation of G VII 292–298 in Leibniz 1973, pp. 10–17. Also translated in L 229–234, where it is dated 1679(?).

1684. Nova Methodus pro Maximis et Minimis, itemque Tangentibus, quae nec fractas nec irrationales quantitates moratur, et singulare pro illis calculi genus. *Math. Schr.* V, pp. 220–226. Originally published (*Acta Erud.*).

1684a. Additio [...] de dimensionibus curvilineorum. *Math. Schr.* V, pp. 226–233. Originally published (*Acta Erud.*).

1685–7. A Study in the Calculus of Real Addition. In Leibniz 1966, pp. 131–144.

1686. Discourse on Metaphysics. Translated from G IV 427–463 and also from the Lestienne ed. in Leibniz 1989, pp. 35–68, and in L 303–330.

1686a. De Geometria recondita et Analysi Indivisibilium atque infinitorum [...]. *Math. Schr.* V, pp. 226–233. Originally published (*Acta Erud.*).

1686?. Infiniti possunt gradus esse inter animas. Translated as "There Can Be Infinite Degrees of Souls". In Leibniz 2001, pp. 298–303.

1686?a. Dans les corps il n'y a point de figure parfaite. Translated as "There Is No Perfect Shape in Bodies". In Leibniz 2001, pp. 296–299.

1695. Responsio ad nonnullas difficultates a dn. Bernardo Nieuwentijt circa methodum differentialem seu infinitesimalem motas. In *Math. Schr.* V, 320–326. Originally published (*Acta Erud.*).

1849–1863. *Mathematische Schriften.* Gerhardt, C. I. (Ed.). 7 vols. Berlin: Halle. Reprinted (Hildesheim: Georg Olms, 1962).

1875–1890. *Die philosophischen Schriften von G.W. Leibniz.* Gerhardt, C. I. (Ed.). 7 vols. Berlin: Weidmann. Reprinted (Hildesheim: Georg Olms, 1960–1961).
1903. *Opuscules et fragments inédits de Leibniz.* Couturat, L. (Ed.). Paris: Félix Alcan.
1966. *Logical Papers. A Selection.* Parkinson, G. H. R. (Ed.). Oxford: Clarendon.
1969. *Philosophical Papers and Letters.* Loemker, L. E. (Ed.). Second ed. Dordrecht: Reidel. Reprinted (Kluwer Academic, 1989). First ed. (Chicago University Press, 1956).
1973. *Philosophical Writings.* Parkinson, G. H. R. (Ed.). Translated by Mary Morris and G. H. R. Parkinson. London: Dent; Totowa: Rowman and Littlefield. First ed. (Everyman's Library, 1934).
1976. *Ein Dialog zur Einführung in die Arithmetik und Algebra.* Knobloch, E. (Ed.). Stuttgart: Frommann-Holzboog.
1981. *New Essays on Human Understanding.* Translated by Peter Remnant and Jonathan Bennett. Cambridge; New York: Cambridge University Press.
1989. *Philosophical Essays.* Ariew, R. and Garber, D. (Eds.). Cambridge, Mass.; Indianapolis: Hackett.
1992. *De Summa Rerum. Metaphysical Papers, 1675–1676.* Parkinson, G. H. R. (Ed.). New Haven and London: Yale University Press.
1993. *De quadratura arithmetica circuli ellipseos et hyperbolae cujus corollarium est trigonometria sine tabulis.* Knobloch, E. (Ed.). Göttingen: Vandenhoeck & Ruprecht.
2001. *The Labyrinth of the Continuum. Writings on the Continuum Problem, 1672–1686.* Arthur, R. T. W. (Ed.). New Haven: Yale University Press.

Montgomery Link, Suffolk University Boston, USA
mlink@suffolk.edu

From Leibniz to the Information Age. Leibniz's deep Footprints in Wiener's Scientific Path

Leone Montagnini

Abstract. In his 1948's book *Cybernetics*, Norbert Wiener chose Leibniz as the "patron saint" of cybernetics. Why did he do that, being Wiener a 20th century's mathematician working at MIT, an engineering institution? On the base of the last three decades studies on Wiener, we can now say that Leibniz was important throughout his intellectual itinerary, articulated in three periods. 1) *Philosophy*: from his childhood (he was born in 1894), as a prodigy, until he got to MIT Mathematics Department, in 1919. He obtained a B.S. in 1908. In 1913, he obtained under Royce a Ph.D. in philosophy, followed by a post–doctorate period under Russell, Husserl and Dewey. In all these years Wiener was mainly attracted by the Leibnizian rigour. 2) *Mathematics*: from 1919 until 1939, Wiener became a renowned mathematician. He felt that his mathematics, strongly concerned with the developments of the variational calculus and the Gibbsian physics fields where minimising and maximising procedures are used, was very much in debt to the Leibnizian attitude of thought, in particular by the "principle of sufficient reason" combined with Leibniz's optimism. In addition, during the Thirties, Wiener began to consider Leibniz as a thinker who was free from the Newtonian consolidated paradigms and invoked to "go back to Leibniz" to be able to find new concepts for the emerging new physics. At the same time Leibniz became for him the basis of reflections together with physiologist J.B.S. Haldane, who had proposed a reduction of the Platonic ideas via the quantum mechanics supposed to be operating in the brain. Wiener considered that a neo–leibnizian approach. 3) *Cybernetics*: from 1940 until his death, in 1964. Throughout his war work, Wiener achieved the vision of cybernetics, an interdisciplinary field studying animals and machines as systems elaborating information. In this phase Wiener assimilates Leibniz, Newton and Huygens, as scientists of an age of clocks. He kept having a great esteem for Leibniz's interdisciplinarity and assures us that Leibnitz was the "patron saint" of cybernetics having combined logics with mechanical calculation.

Keywords: Interdisciplinarity, Mind–Body Problem, Calculus of Variations, Optimism, Analysis Situs, Quantum Mechanical and Brain, Monadology, Continuum and Corpuscular approaches, Computers, Logics, Determinism, Consciousness, Panpsychism, Materialism.

Leone Montagnini (2017) From Leibniz to the Information Age. Leibniz's deep Footprints in Wiener's Scientific Path. In: Pisano R, Fichant M, Bussotti P, Oliveira ARE (eds.), *The Dialogue between Sciences, Philosophy and Engineering. New Historical and Epistemological insights. Homage to Gottfried W. Leibniz 1646-1716*. The College's Publications, London, pp. 255-286
© 2017 College Publications Ltd | ISBN: 978-1-84890-227-5 www.collegepublications.co.uk

1 Introduction

In his 1948's book *Cybernetics*, Norbert Wiener (1894–1964) states: "If I were to choose a patron saint for cybernetics out of the history of science, I should have to choose Leibniz" (Wiener 1961 [1948], 12).[1] This election is neither a *boutade*, nor a fleeting inspiration of the moment. As early as 1932 Wiener was very convinced that, "as few people realize – the Leibnizian work [...] possesses a startling modernity" (Wiener 1932, 201).

Even the book *Cybernetics* had a deep value for the 54 years old author at that time, who states in his autobiography:

> Cybernetics, or the theory of communication and control, wherever it may be found, whether in the machine or in the living being [...] has not been merely the aperçu of a moment. It has its deep roots both in my personal development and in the history of science (Wiener 1964 [1953], 8).

Today "cybernetics" has become an esoteric disciplinary designation. However more recent historical research points out at it like at a bold attempt to draw a unitary generalized science of "control and communication in the animal and the machine", *i.e.* of today's leading disciplines and technologies called automata theory, computer science, mathematical theory of information, control theory, neurosciences, bio–engineering, biomathematics etc. (cf. Segal 2003; Montagnini 2017a). A scientific set often referred to as "information sciences". It should appear clear to the reader that the locution covers at the moment only a juxtaposed set of things sharing a reference to information, but they do not enjoy the epistemological unitariness dreamed by Wiener in the book *Cybernetics* (cf. De Luca 2006; Termini 2006a and 2006b; Montagnini, Tabacchi and Termini 2015).

Nevertheless it should appear also evident that, choosing Leibniz as the patron saint of the cybernetics, Wiener elected him as the scientist "for" and "of" our age: an age called "information age" or "information society", in a non–banal sense a "cyber–society", an age envisaged by Wiener in *Cybernetics* (Breton 1992; Montagnini 2001–2002; 2017a and 2014a).

2 At Least Some "half Leibniz"

In Wiener's opinion, the first characteristic denoting Leibniz was interdisciplinarity.[2] Speaking of his early philosophical studies, at the age of eleven, he tells us of his "admiration for him as the last great universal genius of philosophy" (Wiener 1964 [1953], 109). In 1932 he adds that the "German mathematician–physicist–philosopher–statesman, Gottfried Wilhelm Leibniz" "stands unique in intellectual history", "there is no aspect of scholarship which he did not touch and adorn" (Wiener 1932, 201). Speaking of the need of

---

[1] In some English quotations one finds "Leibnitz", in other "Leibniz", even in the same author, I have always adopted the latter form.

[2] About interdisciplinarity in Wiener see Montagnini 2008, 2013 and 2017b.

interdisciplinarity in today's science, precisely in the second page of *Cybernetics* he states:

> Since Leibniz there has perhaps been no man who has had a full command of all the intellectual activity of his day. Since that time, science has been increasingly the task of specialists, in fields which show a tendency to grow progressively narrower. (Wiener 1961 [1948], 2).

Wiener himself was an interdisciplinary scientist, as recognized by his contemporaries. In 1963 he was awarded with the US National medal of science "for marvellously versatile contributions, profoundly original, ranging within pure and applied mathematics, and penetrating boldly into engineering and biological sciences" (Rosenblith and Wiesner 1965, 3), and after his death, in 1964, his colleagues of the American Mathematical Society published a special number of their Bulletin *in memoriam*, "[...] in recognition of his towering stature in American and world mathematics, his remarkably many–sided genius, and the originality and depth of his pioneering contributions to science." (Browder, Spanier and Gerstenhabler 1966, dedicatory section)

This acknowledged versatility is rooted in Wiener's educational and scientific history.[3] He was born in 1894 in the United Stated by a Jewish Russian professor of Slavic studies at Harvard University. He was a prodigy. Educated at home until nine by his eclectic father, from 1903 to 1906 he attended the Ayer High School, and from 1906 to 1909 the Tuft College, where, at fourteen, he got his Bachelor of Science, majoring in Mathematics. After that, he enrolled at Harvard University to study biology. But because of the clumsiness in laboratory showed by the boy, his father prompted him once again towards philosophy. Therefore he attended the Sage School of Philosophy at Cornell University, from 1910 to 1911, and the Department of Philosophy at Harvard University, from 1911 to 1913, where he took his Master of Arts and, in 1913, his Ph.D., both in Philosophy. Later he went to Europe for a postdoctoral period, from 1913 until 1915, at the Trinity College of Cambridge University, to study under Bertrand Russell, with an interlude at Göttingen to attend Edmund Husserl's courses. The period ended at Columbia University, New York, attending John Dewey, because of the increasing difficulties to cross the Atlantic caused by WWI.

During the academic year 1915–1916 Wiener worked as an Assistant at the Department of Philosophy of Harvard. The University refused to hire him for the following year, and Leo advised the son to switch and teach mathematics at the University of Main. A three–year period of uncertainties began, in which Norbert was in doubt about his career: was he to become a philosopher or a mathematician? The engagement as an Instructor at the Department of Mathematics of Massachusetts Institute of Technology in 1919 finally solved the riddle. For Norbert

---

[3] For the Wiener's intellectual itinerary I follow here, besides the autobiography Wiener 1964 [1953] and 1964 [1956], the pioneering Heims 1984 [1980], then Masani 1990 a very important book for Wiener's mathematical aspects, and Montagnini 2017a, that investigates philosophical, sociological and various historical aspects left out by Heims and Masani.

Wiener a successful lifelong career as a mathematician began, ending only with his death.

Wiener's zigzagging from philosophy to biology to mathematics "did not result from any particular plan on my part or on the part of my father" (Wiener 1964 [1953], 295), he explains. In fact Leo wanted only to try and find the field which best fitted his son. Even Norbert did not cherish the ideal to be a universal man. He wanted to become a specialized man, and was very happy to embrace the identity of the mathematician, so much so that he entitled *I am a Mathematician* the second part of his autobiography, beginning from his hiring by MIT.

Nolens volens from this itinerary emerged a singular many–sided mathematician. Norbert Wiener says:

> Everything is grist to his mill. Indeed, the peculiar advantage of the ex–infant prodigy in science [...] is that he has had a chance to absorb something of the richness of many fields of scientific effort before he has become definitely committed to any one or two of these. Leibniz was an infant prodigy, and in fact the work of Leibniz is precisely the sort of work for which the training of the infant prodigy is peculiarly suitable. The scientist must remember and he must reflect and he must correlate. It does not change the situation in any fundamental way that the field of science has so grown that the scholar of the present day must perforce be nothing more than a half–Leibniz. The task of scientists is even more essential than it was in Leibniz's time; and if it cannot be fulfilled with the completeness which seemed possible in the seventeenth century, that part of it which can be carried out is more demanding and less avoidable (Wiener 1964 [1953], 295).

Here Wiener considers the condition of being a *child prodigy* almost as a precondition to be an interdisciplinary man: not a man provided with a universal knowledge, but a specialist with a broad basic knowledge of other sciences, a "half–Leibniz", which he considered more necessary in his time — and I would add for today as well — than Leibniz's universalism in '600 and '700. In 1948 he explains his formula for an interdisciplinary scientific work for our age:

> The mathematician needs not have the skill to conduct a physiological experiment, but he must have the skill to understand one, to criticize one, and to suggest one. The physiologist need not be able to prove a certain mathematical theorem, but he must be able to grasp its physiological significance and to tell the mathematician for what he should look (Wiener 1961 [1948], 1–3).

In his opinion, interdisciplinarity provides the way to fertilize the scientific creativity. It is clear that this approach needed men trained in a particular way. In a 1962's conference he specifies:

> We want people who will be able to face yet unknown situations, by as yet unknown combinations of ideas from different fields of work. For this, a broad basic training is necessary. So, too, are crossing the boundaries of scientific specialization, interdisciplinary thinking, and a

willingness to take all that one has acquired as part of one's available assets. [...] I believe that is extremely important to have a broad basis in very different sciences for one's intellectual work so that one can follow the problem wherever it leads, even though it crosses boundaries. [...] If a problem leads us into a new field in which we have no knowledge, we should acquire such knowledge. It is no excuse, when working on a problem, to say 'but that's not my field'. At some stage or other one must decide to learn what is needed about the field; those who do not are "stooges" who are not serving their social function for science (Wiener 1962, 20).

Today it is hoped to solve complex problems through big laboratories, a massive attack, producing a huge amount of scientific literature. Wiener disliked this mechanical approach. In his opinion a massive attack can be useful to systematize and implement new ideas, but new ideas could be yielded only by first class minds, coming from different fields, and working together in small groups. In 1950 he raised his voice:

[...] we need a range of thought that will really unite the different sciences, shared among a group of men who are thoroughly trained, each in his own field, but who also possess a competent knowledge of adjoining field. No, size is not enough. We need to cultivate fertility of thought as we have cultivated efficiency in administration (Wiener 1950, 57–58).

In addition, the way the big science works clogs the libraries with a huge amount of books and papers. How can we separate good work from trash? The solution suggested by one of Wiener's friends, the engineer Vannevar Bush, consisted in coping with the amount of information by means of mechanical aids, to search automatically through the vast bodies of material produced. Since 1936, Bush had begun to work — for civil and military goals — on a machine then known as "Memex", to deal with big amounts of text, through microfilm, photocells and analogical computers (Colin Burke, 1994); someone has seen in Memex the embryo of the present hypertexts (Nyce and Kahn, 1991). We could therefore consider in some way Bush's solution as an early example of what is today's normal practice in bibliometrics. But Wiener was skeptical about it. These systems

[...] probably — Wiener argued — have their uses, but they are limited by the impossibility of classifying a book under an unfamiliar heading unless some particular person has already recognized the relevance of that heading for that particular book. In the case where two subjects have the same techniques and intellectual content but belong to widely separated fields, *this still requires some individual with an almost Leibnizian catholicity of interest* (Wiener 1961 [1948], p. 158. Italics added).

All in all, in a way or in another, even the methods of Big Science in the end needed "an almost Leibnizian catholicity".[4]

3 From Leibniz to the Modern Computers

In Wiener's education and career, Leibniz figure played a special role even concerning the contents. Let's read a passage in Wiener's autobiography:

> Cybernetics [...] has its deep roots both in my personal development and in the history of science. Historically it stems from *Leibniz*, from Babbage, from Maxwell, and from Gibbs. Within me, it stems from the little I know of these masters and from the way this knowledge has fermented in my mind. [...] Therefore, perhaps, an account of the origins of my predisposition towards these ideas, and of how I came to take them as significant, may be of interest to the others who are as yet to tread my road (Wiener 1964 [1953], 8. *Italics* added).

In some sense Wiener looked at his own whole intellectual path, at least partly, as an ontogenetic recapitulation of a phylogenetic scientific development starting precisely from Leibniz.[5] In the set of names quoted in the passage we can single out two threads. The first being the Leibniz–Babbage thread, i.e. a path conducing from Leibniz to the modern computers, considering that Babbage can correctly be seen as the early one who proposed a project for a multipurpose calculator provided with a program. The second being the Leibniz–Maxwell–Gibbs thread, i.e. a path arriving exactly to the very personal kind of Wienerian mathematics.

Let's start from the first path. According to Wiener, Leibniz was the first to open the long road leading to the modern computer, conceived by Wiener as a thinking machine. The point is a subtle one to understand. As we know Leibniz designed arithmetical machines (cf. Leibniz 1929 [about 1671]). As Wiener himself acknowledges, he "like his predecessor Pascal, was interested in the construction of computing machines in the metal" (Wiener 1961 [1948], 12). Actually, Leibniz seems overall, to have only taken part to a long tradition, just as Wiener says: "mechanization progressing through the abacus and the desk computing machine to the ultra–rapid computing machines of the present day", i.e. the modern computer in Wiener's language. At the same time Leibniz was looking for a *Characteristica Universalis*, a universal symbolic language, to reduce the reasoning to a type of calculus. Here, again, Leibniz was only taking part to a long tradition getting to the "mathematical notation and the symbolic logic of the present day" (Wiener 1961 [1948], 12) going back in the West at least to Raymond Lully.[6]

---

[4] The various possible sources of Wiener's scientific creativity have been investigated by Montagnini 2015 and 2017c.

[5] Martin Davis (2000) as well argues for a priority of Leibniz in this context, but it seems to me that this great logician does not catch the shades pointed out by Wiener.

[6] Lully wrote several books about his "Art", the main of them being Ars demonstrativa (1275), Ars brevis (1308), Ars magna generalis et ultima (1305–1308). The first two works were translated in Lully [1985]. About Raymond Lully

Therefore in what sense Wiener could consider Leibniz as the initiator of those two traditions? His originality, in Wiener's view, does consist in having opened the road for a synthesis between logics and engineering. A road that has led to the conception of a logical machine, [7] and therefore of a thinking machine. "The *machina ratiocinatrix* — writes Wiener, that is the computer as a thinking machine — is nothing but the *calculus ratiocinator* of Leibniz with an engine in it" (Wiener 1961 [1948], 125). Leibniz did not pioneer the modern computer, if not in the sense — which was fundamental for Wiener — that he was able to give the first hints that opened the road.

In addition Leibniz did it, according to Wiener, through a process of merging different traditions in science, a fundamental way for Wiener to produce new pieces of science. We could also give a wide variety of examples taken from his long experience as a creative scientist (cf. Montagnini 2015).

But here there is another biographic aspect to catch. Wiener had been a young philosopher who had grown up in the world of the most abstract ideas, in an environment such as the Harvard's Departments of Philosophy, the Trinity College of Cambridge, the Göttingen's Faculty of Philosophy. In that period he had been one of the best–loved pupils of Russell, attending his restricted reader seminar on the *Principia Mathematica* (Russell and Whitehead. 1910–1913) with only two other students. It was the same seminar that Ludwig Wittgenstein had been attended until the summer of 1913 (cf. Monk 1990, 92 ff.).

In those years, symbolic logic was indeed considered such an esoteric chapter of philosophy and mathematics (cf. also the testimony of Couturat 1901. VII). In a letter written in the summer 1914 from Göttingen, Wiener writes to Russell:

> Symbolic logic stands in little favor in Göttingen. As usual, the Mathematicians will have nothing to do with anything so philosophical as logic, while the philosophers will have nothing to do with anything so mathematical as symbols (Wiener to Russell, June or July 1914, quot. by Russell 1968, p. 41).

The period was antecedent to the metamathematical turn of David Hilbert, who actually in 1914 taught precisely at the Faculty of Philosophy of Göttingen, which housed the teaching of mathematics as well. During his stay in Germany, Wiener found that the only person really interested in symbolic logic there was Gottlob Frege, who he went to meet in Brunnshaupten, Mecklenburg (cf. Wiener to Russell, June or July 1914; quoted. by Russell 1968, 41).

(c. 1232–1315) — Ramon Llull in his mother tongue, the Catalan — see Bonner 2007; about the path from Llull to Leibniz, through Nicholas of Kues and Giordano Bruno, see Rossi 2000 [1983].

[7] However we have to recall here that the argument for Leibniz being the real initiator of the modern symbolic logic has been stated by one of Wiener's main master, Bertrand Russell: "It is from the recognition of asyllogistic inferences that modern Symbolic Logic, from Leibniz onward, has derived the motive to progress" (Russell 1903, 10). An almost identical sentence can be found in his contemporary Couturat 1901, XII.

In any case, at this stage, Wiener appears to be a subtle, well-educated philosopher, specialized in logic. In 1916 the refusal opposed by the Department of Philosophy was a real wound for him. An injury burning even more when he had to accept a position as an instructor at MIT, a school for engineering. Subjectively for him the passage was a traumatic one. But it was just through it, that he was enabled to constantly create, for 45 years, a unique synthesis between the pure theoretical thought and the practical technology. From this standpoint a machine like the modern computer, a "logic machine", a "thinking machine", an "electronic brain" appeared to Wiener as one of the most mature outcomes of his personal story. Therefore he loved to think that a high theoretical and unique philosopher and mathematician like Leibniz had wanted to design a calculating "machine".

Wiener was one of the protagonist of the creation of the modern computer. Today his contribution to the creation of the electronic digital computer is scarcely recognized. Most computer science historians recognized him as being one of the creators of analogical computing machines, neglecting his contributions to the electronic digital computer.[8] As a matter of fact, Wiener played a fundamental part in the creation of the general high speed computer with von Neumann's architecture, in particular from 1940 to 1945.

It is certainly not here the right place to establish Wiener's credits in the matter. I will be pleased telling the reader just one episode of his life. Warren Weaver, the mathematician of the Rockefeller Foundation who supervised the whole research of Wiener during WWII, in 1946–1947, supported with 100.000 dollars from the Foundation a project for building a general purpose electronic digital computer at MIT. He wanted at all costs Wiener to participate in it. But Wiener appeared to be reluctant, after what had happened in Hiroshima and Nagasaki, in engaging himself in the project, that, he thought, could have had military aims (Montagnini 2017a, 2014a, 2014b). After various attempts to try and involve him, MIT suggested to replace Wiener with a group of mathematicians. Warren Weaver curtly replied:

> I think the question, 'What does one want a computer to do?' requires great knowledge of mathematics, great imagination, great sweep and depth of mind. One could get some inspiration, doubtless, by talking to a dozen leaders in applied mathematics. But this is where I had hoped and expected we would have the genius of Norbert Wiener (Weaver to J. A. Stratton, February 1947, quoted by Wildes and Lindgren 1985, 234–5).

At the end MIT returned the fund to the Rockefeller Foundation, and bet all its chances on another project which was arising in the meantime: the Whirlwind by the MIT Servo–Lab. The prototype of a large production of computers that actually became the core of the entire antiaircraft shield built by USA over North America during the early years of the Cold War.

Getting back to our talk, Wiener considers the implications in the construction of computers as the outcome of the synthesis we pointed out above, between the pure world of speculation and the more terrestrial field of engineering. In *Cybernetics* he draws on a tradition that begins precisely with Leibniz: whose

---

[8] Philippe Breton (1987) assumes an upstream position I agree with, see Montagnini 2017a.

calculus ratiocinator "contains the germs of the *machina ratiocinatrix*" (Wiener 1961 [1948], 12). Then he quotes himself, leaving out his war contribution to computers, and says: "I am myself a former student of Russell and owe much to his influence" (Wiener 1961 [1948], 13). After that, he quotes Claude Shannon, who, in 1938, had discovered the way to implement the Boolean logic by means of switching circuits. It is interesting to remark the satisfaction with which, in the autobiography, Wiener speaks about Shannon:

> It is through his work that a training in symbolic logic, that most formal of all disciplines, has come to be on recognized mode of introduction into the great complex of scientific work of the Bell Telephone Laboratories (Wiener 1964 [1956], 179).

Wiener enjoys seeing such "esoteric symbolic logic", as once was considered, now taught at the Bell Laboratories, i.e. the Department of Research and Development of the AT&T, the colossal international corporation of telegraph and telephone. In that image Wiener could contemplate the outcome of his distressing association as a former philosopher who had become mathematician within the tree of engineers.

After Shannon, *Cybernetics* recalls the name of Alan Turing, he had known in 1947 in Teddington, UK. Turing, Wiener says,

> [...] perhaps the first among those who have studied the logical possibilities of the machine as an intellectual experiment, served the British government during the war as a worker in electronics, and is now in charge of the program which the National Physical Laboratory at Teddington has undertaken for the development of computing machines of the modern type (Wiener 1961 [1948], 13).

The last name Wiener quotes is that of Walter Pitts, who, with Warren McCulloch

> [...] began to work quite early on problems concerning the union of nerve fibers by synapses into systems with given overall properties. Independently of Shannon, they had used the technique of mathematical logic for the discussion of what were after all switching problems (Wiener 1961 [1948], 13).

Pitts, a young man, gifted with a deep logical mind, in 1943 had become Wiener's assistant at MIT. Wiener remarks that, when he met him for the first time, he lacked of a technological training and that the first thing he made him do was showing

> [...] him examples of modern vacuum tubes and explained to him that these were ideal means for realizing in the metal the equivalents of his neuronic circuits and systems. From that time, it became clear to us that the ultra-rapid computing machine, depending as it does on consecutive switching devices, must represent almost an ideal model of the problems arising in the nervous system (Wiener 1961 [1948], 14).

Leibniz, Wiener, Shannon, Turing and Pitts. We have drawn here the first steps of cybernetics, albeit at its very beginning, if we consider the brain as a logical machine and the computer as an electronic brain. Wiener does not quote von Neumann, but I think, in this case, it has to be taken as a venial oversight. Anyway, and not in a trivial sense, Wiener considers Leibniz as the first one to have paved the way for the cybernetic tradition of thinking machines: the "*calculus ratiocinator* of Leibniz contains the germs of the *machina ratiocinatrix*, the reasoning machine" (Wiener 1961 [1948], 12).

4   The "Leibnizian Frame" Taking Shape

The other thread, the Leibniz–Maxwell–Gibbs one, leads us even more in depth in Wiener's thought, at the very sources of his worldview and of his way of making mathematics and science.

Firstly we have to go back again to his philosophical years. The triggers for Wiener's interest for logic had come from Spinoza and Leibniz, the philosophers who influenced him the most during his second year at Tufts College (1907–1908) (cf. Wiener 1964 [1953], 109). "As a very young man — he confesses — I appreciated the help and discipline of a rigid logic and a mathematical symbolism" (Wiener 1964 [1953], 222–3). That he had derived his enthusiasm for Logic from the rationalist philosophers is easily understandable reflecting on the fact that the majority of his studies in philosophy had been of historic character.

In the same period, while staying at Tufts College, he came in contact for the first time with Pragmatism, the prevailing philosophy in America at that time. Wiener knew personally William James, a friend of his father, and devoured James' book on *Pragmatism* (James 1907) "almost as much as literary tidbits as for their serious content" (Wiener 1964 [1953], 109). Later at Harvard he attended the last courses of Josiah Royce, through whom he also got acquainted with the thought of Charles Sanders Peirce (who died in 1914, while Royce in 1916).

In order to understand the intellectual atmosphere Wiener breathed in those years, it is useful to read a Richard Bernstein description, explaining that Pragmatists abandoned

> [...] the presupposition of much of modern philosophy that the rationality and legitimacy of knowledge require necessary foundations. Inquiry has nor needs any such foundations. The pragmatists did not think that abandoning all foundational claims and metaphors leads to skepticism (or relativism). They stressed the fallibility of all inquiry [...]. The classical pragmatists shares a cosmological vision of an open universe in which there is irreducible novelty, chance, and contingency (Bernstein 1992, 813–814).

From the Pragmatists Wiener received an imprint as strong as the one he got by the Rationalists. The latter pushed him to dislike the lack of logical rigor shown by the former. For example, about James, he writes that

> [...] there was more than the style of the novelist in William James — writes Wiener — and perhaps less of the philosopher than one might

have thought, for his ability to evoke the concrete was to my mind many times greater than his ability to organize it in a cogent logical form (Wiener 1964 [1953], 110).

Similar words he used about John Dewey (cf. Wiener 1964 [1953], 222–223). As a consequence of the double and opposite Wiener's fascination for Rationalism and Pragmatism, precisely when he was at Tufts he tried "to combine in some manner the logical standards of formal consistency" with the ideas of Pragmatism (Wiener's letter for doctoral submission, quot. by Grattan–Guiness 1975, p. 106). One can imagine, that he hoped to create a sort of "Pragmatismus Ordine Geometrico Demonstratum", paraphrasing Spinoza's famous book on Ethica. A goal that appears intrinsically paradoxical: is it possible to put in a logical form a philosophy denying the need, or better, the same possibility of epistemological foundations? During the years at Tuft the results for the young Wiener were disappointing: "finding little success in his undertaking — he states — my view again shifted towards an agnosticism, and my main interests turned towards the natural sciences" (Wiener's letter for doctoral submission, quoted by Grattan–Guiness 1975, 106).

It was for this reason that he majored in mathematics, and that he chose to study biology after the Bachelor. Forced to come back into the world of the philosophers, he preferred however philosophical technical work. His Ph.D. thesis was "A comparison between the treatment of the algebra of relatives by Schröder and that by Whitehead and Russell"; he wrote other very original papers in mathematics and logic as well even during all the years before the hiring at MIT. Science for him proved to be a safe shelter.

However he did not dislike to write philosophical papers; articles revealing his need to conciliate rigor and Pragmatist instances. In particular this is very clear in Relativism (Wiener 1914) and *Is mathematical certainty absolute?* (Wiener 1915).

In both cases he maintained, pragmatistically, the idea of the impossibility of an absolute foundation of the scientific knowledge, including also mathematics. Therefore he considered seriously and effectively the criticism against the science of Bergson, for example. But in general, Wiener excludes to solve the difficulties that one could meet in science to build some sort of parallel sciences, better than the traditional one. He dislikes Henri Bergson's suggestions to use the "intuition" or those by Husserl using, in Wiener's words, that "intellectual contortions through which one must go before one finds oneself in the true Phenomenological attitude" (Wiener to Russell, June or July 1914, quoted by Russell 1968, 41). In Wiener's opinion the old traditional methods of science are, with all the faults one could find, the only possible tools for science.

The result of the impossible synthesis sought by him was the acceptance of the limitations of science, an ontology of contingency and a moderate fallibilism in epistemology, that he carried on into his scientific and mathematical research. The outcome must have pushed him necessarily towards Leibniz, the theorist of a contingent universe rather than towards Spinoza. In 1932 Wiener resumed the Leibnizian optimism together with the principle of sufficient reason in this way:

> For Leibniz, all contingent truths, that is, all truths of particular fact, are determined not merely by the principle of contradiction, but by an additional principle which he calls the principle of sufficient reason,

which asserts that some particular perfection must be realized by each phenomenon in this world and by this world as a whole to distinguish it from all other possible worlds. This principle is closely connected with Leibnizian optimism (Wiener 1932, p. 203).

Wiener was very impressed by this way of thinking: on one hand God had introduced the set of all the possible worlds, as the result of the divine logical rigor; on the other, he had chosen only one world based on the best compromise. If we reflect on this argument, it appears to be, in some sense, exactly as the best synthesis between Rationalism and Pragmatism.

Leibniz's way of thinking here formed in Wiener's mind a sort of cognitive frame, a "Leibnizian frame" that will become an abstract skeleton we nearly always find in his mathematics to come. With the injection of two others ingredients — the "chance" and the "self–similarity" — we will obtain the typical way to Wiener's thought in mathematics.

5 The Random Nature of Reality

Let's consider the first ingredient, that is "chance". Its injection into the "Leibnizian Frame" came about through several phases. Chance has been a pivotal notion in Charles S. Peirce's thought. Karl Popper writes:

> Peirce conjectured that the world was not only ruled by the strict Newtonian laws, but that it was also at the same time ruled by laws of chance, or of randomness, or of disorder: by laws of statistical probability. This made the world an interlocking system of clouds and clocks, so that even the best clock would, in its molecular structure, show some degree of cloudiness. This made the world an interlocking system of clouds and clocks, so that even the best clock would, in its molecular structure, show some degree of cloudiness. So far as I know Peirce was the first post–Newtonian physicist and philosopher who thus dared to adopt the view that to some degree all clocks are clouds; or in other words, that only clouds exist, though clouds of very different degrees of cloudiness (Popper 1972, 212).

The probabilistic approach Peirce introduced in science is contained in various papers that Royce read and also proposed to read and comment to the members of his interdisciplinary seminar on the scientific method he had led for decades at Harvard University (about it, see Royce 1963).[9]

In the first chapter of *Cybernetics*, entitled "Newtonian and Bergsonian Time" (Wiener 1961 [1948], 30–43) every science is included between two extremes, astronomy and meteorology as archetypes of a "science of clocks" and a "science of clouds". The chapter emphasizes how gradually all sciences have adopted the statistical approach of the latter. Wiener had definitely absorbed Peirce's lesson through Josiah Royce, whose seminar he attended during his stay at Harvard in

---

[9] See the several papers by Peirce from 1891 to 1892 quoted in our bibliography and Peirce 1923.

1911–'12 and 1912–'13, gave him "some of the most valuable training I have ever had" (Wiener 1964 [1953], 165–6; we can also find a reference to the seminar in the first page of *Cybernetics* (Wiener 1961 [1948], 1). Exactly in that period, following Peirce, Royce had been reflecting on the superiority of the statistical method in science, arguing that:

> [...] our mechanical theories are in their essence too exact for precise verification. They are verifiable only approximately. Hence, since they demand precise verification, we never know them to be literally true. But statistical theories, just because they are deliberate approximations, are often as verifiable as their own logical structure permits. They often can be known to be literally, although only approximately, true (Royce 1914, 55).

Wiener would translate these ideas into his mathematics, as he tells us:

> [...] what we put into our problem not only consists of precise data which we later have to ease off in accordance with the inaccuracy of the equations and the initial conditions but contains intrinsically the very inaccuracy which hinders our work. [...] If this recognition of the statistical nature of all science is already proving to be valuable in the most Newtonian type of mechanical–engineering computation, how much more must it then be the natural method of computation in those fields in which our errors of observation are naturally very large! (Wiener 1964 [1956], 255–229)

However, in order to became real mathematics, Wiener's predilection for the statistical and probabilistic method had to assume particular characteristics, passing through the statistical mechanics of Willard Gibbs (cf. Gibbs 1906) and the mathematics of Henri Lebesgue, a kind of approach in which Wiener could easily found material for mathematizing the "Leibnizian frame".

As Wiener explains, the Leibnizian way of thinking had penetrated into physics from Maupertuis. The latter had discovered, in the Newtonian celestial mechanics, a quantity called "action", "the integral of the energy with respect to the time is smaller over the actual path of a particle than over any possible alternative path" (Wiener 1932, 222). Therefore Maupertuis had proposed to consider all the possible trajectories and chose that in which the quantity of "action" is assumed into a minimum. In this way he obtained an alternative method to get to the same results Newton had achieved. Maupertuis thought that the fact was a sign "that God in creating the world has done it with the greatest possible economy of effort". Wiener remarks: "From one standpoint, this is utter anthropomorphic trash, but it leads to sound physics and originates from a very profound principle, which has modified the entire course of science" (Wiener 1932, 222).

In such a way the "Leibnizian frame" penetrated into mathematics and physics, introducing the variational calculus and more in general an approach that we find also in the functional analysis and statistical mechanics. But how does the Leibniz's mental attitude manage to shape the way to do mathematics by Wiener? It is himself to explain the path:

> In the form of the principle of least action, the other possible worlds are introduced merely to be rejected, for they do not satisfy the desired principle of minimization. On the other hand, in a statistical mechanics — the quantum theory reduces the whole of physics to a form of statistical mechanics — these other possible worlds are considered from the standpoint of probability. To put it crudely, the propositions of statistical mechanics assert nothing about any individual possible world, but rather about the overwhelming majority of all of them. This concept, without which modern physics could not have assumed anything like its present form, is a definite part of the philosophy of Leibniz (Wiener 1932, 224).

When Wiener arrives at MIT, he comes into contact with the physics of Willard Gibbs, pervading various Departments of engineering and physics. Gibbs had introduced a particular way to deal with the statistical phenomena in physics, distinguishing him from the one introduced by Clerk Maxwell. As Wiener says:

> The intuition on which he [Gibbs] based his work was that, in general, a physical system belonging to a class of physical systems, which continues to retain its identity as a class, eventually reproduces in almost all cases the distribution which it shows at any given time over the whole class of systems. In other words, under certain circumstances a system runs through all the distributions of position and momentum which are compatible with its energy, if it keeps running long enough. This last proposition, however, is neither true nor possible in anything but trivial systems. [...] Gibbs' innovation was to consider not one world, but all the worlds which are possible answers to a limited set of questions concerning our environment" (Wiener 1989 [1954], 9 and 12).

Wiener had the clear notion that, at least from his standpoint, the cognitive frame derived from Leibniz had been, or at least could be, subtended with such a view. Gibbs' approach required a second aspect: a rational method to operate on the set of the possible worlds. At the same time, in France, independently, Wiener remarked, "Borel and Lebesgue in Paris were devising the theory of integration which was to prove apposite to the Gibbsian ideas" (Wiener 1989 [1954], 9).

In particular it was the Lebesgue's method to reveal itself as the best tool to solve the Gibbsian problems, even if Lebesgue had not suspected the physical usefulness of it. Wiener had known well the methods of Lebesgue attending the lessons of Hardy (Wiener 1964 [1953], p. 190).

It is noticeable, in passing, the reappearance of the typical Wiener's idea of the importance of confluence among two of more different independent traditions in science, we have already, earlier on, come across speaking of the confluence of pure symbolic logic and the practice of engineers, to discover the computer as a logical machine.

6 Self–Similarity in Royce

Let's consider now the second additional ingredient needed to obtain Wiener's "Leibnizian frame": "self–similarity". To best understand the origins of this aspect in the Wiener's path, we have to illustrate, at least summarily, one of the main philosophical currents in the Anglo–American panorama between the 19th and the 20th century, that is Idealism or "Absolutism". United Kingdom has been the homeland of Empiricism, Evolutionism, and in the last quarter of the 19th century has been dominated by the Max Spencer's Positivism. But, over the centuries, it has also housed a variously Platonizing line of thought, intended to counter those trends, as well as to give a speculative support to the religious thought. The case of Bishop George Berkeley was emblematic. In the 17th century he denied the existence of things reduced to a mere appearance, abiding by the principle: *"esse est percipi"* ["To be is to be perceived"]. God and men were for Berkeley the only existing beings because of their spiritual nature. Following this long tradition, between 19th and the 20th century we can trace the appearance of English Absolutism.[10] Among the others we need to quote here the Hegelian John McTaggart (1866–1925), and, above all, the powerful theoretical thinker Francis Herbert Bradley (1846–1924). It is not a banal consideration the fact that, as for Berkeley, these philosophers were very interested in mathematics and logic, especially when dealing with their ontological interests.

In his masterpiece *Appearance and Reality* (1893) Bradley analyzes the logic relations among the attributes of individuals, things or men. He discovers contradictions in this net of relations, and concludes that the individuals constituting the world as we experience it do not exist. This world is only "appearance", while the Absolute alone, which is God, unique, simple and devoid of internal relations is real. In particular one of the main contradictions that Bradley discovers is a *Regressio ad infinitum* concerning the number of attributes assumed by the individuals once their existence was admitted.

The true early Norbert Wiener's master, Josiah Royce, who represented the American version of English Absolutism, did not want to deny the existence of the Absolute. Nevertheless he wanted to save the existence of individuals and their relations. In his own masterpiece, The world and the individual (Royce 1900), he devoted the last long part to confute Bradley's stands.

In the last part of his book, entitled Supplementary Essay. The One, the Many, and the Infinite (Royce 1900. 473–588), he does not avoid the *regressio ad infinitum* found by Bradley. He looks instead for a conceptual tool to save the One and at the same time the possibility to admit the existence of an infinite number of individuals and attributes. And he finds it in an "iterative or recurrent process of thought". Royce argues:

> Now [...] the observation that reflection makes upon *the general nature of any iterative or recurrent process of thinking*, becomes at once of great interest for the comprehension of the question about the One and

---

[10] For this approach, connecting the Idealism of Berkeley to that of the Absolutists, I owe to the Italian Idealist, writing in that time, De Ruggiero 1912.

the Many. We want to find some case of an unity which develops its own differences out of itself (Royce 1900. 496. *Italics* are original).

Royce draws his inspiration from two mathematicians of his time, Georg Cantor (1845–1918) and Richard Dedekind (1831–1916). In particular Royce focuses on the concept of self–representation explained by an example. He imagines to put an exact map of England on some place of England. Representing with absolute exactness England it would include of course a representation of itself as well, and so on without limits (cf. Royce 1900, 504–505).

Wiener attended Royce's courses and his interdisciplinary seminar during his stay at Harvard Department of Philosophy (1911–'12 and 1912–'13) to obtain his M.A. and Ph.D. in philosophy. Having already gained a B.S in Mathematics, he became interested in logic and philosophy of science through Royce. As a declared Atheist since he was 6 years old, Wiener never followed Royce's theological and metaphysics goals. Nevertheless he always held in great consideration his master, whose views in the last period of his life (Royce died in 1916) had become wider and more open to the philosophical novelties of the new century.

Wiener earned his doctorate under Royce's advisement discussing a Doctoral Dissertation on the logic of relations, which was precisely the hot topic in Bradley philosophy. The Thesis dealt in a very technical way with Schröder's approach to this topic — derived in its turn from Peirce — compared with that of Whitehead and Russell's *Principia*.[11] Later, Wiener went to study to Trinity college of Cambridge for two post–doctoral years (1913–'15), under Bertrand Russell, his new mentor.

Wiener learnt the rigorous approach to Russell' symbolic logic. He was able to move freely and with great expertise into the arising field of mathematical logic. Nevertheless his American Pragmatist imprinting prevented him from sharing Russell's theoretical goal: the foundation of mathematics on absolute logical basis. Ideas then well expressed in *Is mathematical certainty absolute?* (Wiener 1915). As he tells us in the autobiography.

> I already then felt that an attempt to state all the assumptions of a logical system, including the assumptions by which these could be put together to produce new conclusions, was bound to be incomplete. It appeared to me that any attempt to form a complete logic had to fall back on unstated but real human habits of manipulation (Wiener 1964 [1953]. 192).

To understand Wiener's stance on the proposal of Royce to confute Bradley it is very helpful to read the entry "Infinity" that Wiener wrote for the *Encyclopedia Americana* approximately in 1917–'18 (Wiener 1919a). He synthesizes the Roycean proposal in this way: "no multiplicity is reducible to unity except through processes involving self–representation", and that "the Reality is such a self–represented and infinite system" (Wiener 1919a, 121). An interesting criticism follows:

---

[11] The complete title of the Thesis is Wiener, N., *A comparison between the treatment of the algebra of relatives by Schröder and that by Whitehead and Russell*. Never published. A copy is in the Wiener's Archives of M.I.T.

The Royceian theory of the infinite is based on the analogy of cardinal infinitude, and presupposes that there is such a thing as a complete universe. Certain paradoxes discovered by Russell and Burali–Forti tell very strongly against the existence of a complete unity embracing all lesser unities. Royce's work possesses value rather as an account of potential infinity of a universe capable of indefinite enlargement than as a description of a complete infinite. *Furthermore, it is given, systems of much less extent than the Royceian infinite may possess the self–reflecting property of cardinal infinity* (Wiener 1964 [1953]. 192. Italics added).

Clearly here we are reading the point of view of an expert by now. Considering the advancements of mathematical logic, in his opinion, we have to exclude "the existence of a complete unity embracing all lesser unities". The speech of Royce is valid therefore only for a potential infinity; nevertheless the existence of "subsystems" possessing "the self–reflecting property of cardinal infinity" cannot be excluded. And the "Wiener process", the mathematical model of the Brownian motion he would discover between 1919 and 1920, provided exactly with self–reflecting proprieties, with the supplement of chance, and dealt with mathematical tools derived by Lebesgue and Gibbs, would be just one of those "subsystems". But before we talk about this discovery we need to go back to Leibniz.

7   Self–Similarity in Leibniz

Curiously enough, Royce, in *The world and the individual,* never mentions Leibniz and, symptomatically, even in the same abovementioned Wiener's article *Infinity* does not think important to deal with Leibniz. And yet Leibniz who had been an advocate of the actual infinity was one of the philosophers who were much closer to Royce's point of view. Consequently in his work, the notion of self–similarity comes out tenaciously. However Wiener is well aware of this and will often linger on that concept when talking about Leibniz.

Wiener always retained that the Leibnizian monadology in philosophy was a consequence of the discovery of the microscope made by Leeuwenhoek: "behind Leibniz's philosophical views of the monads there lie some very interesting biological speculations. It was in Leibniz's time that Leeuwenhoek first applied the simple microscope to the study of very minute animals and plants" (Wiener 1950, 106). Wiener writes:

> *The discovery of a new instrument often leads immediately to a new insight.* [...] Leeuwenhoek's microscope showed by direct observation that a drop of pond water was a teeming mass of life suggestive of a crowded city. The new power lent to the eye engendered a new range of imagination, and everyone's thoughts turned to the fine structure of the world and to the philosophical implications suggested by the process of magnification.

One of the results of this experience, perhaps, was Swift's famous jingle:

> So, naturalists observe, a flea Hath smaller fleas that on him prey; And these have smaller still to bite 'em; And so proceed ad infinitum [Swift 1733, *my note*]. [...]
> If the spermatozoon was itself an early stage of the fetus, it was natural to think that it was a human being in miniature, with all the organs of the human being on a smaller scale, distorted but still essentially there. By this token, it should contain smaller spermatozoa, much as Swift's flea carried lesser fleas on a scale far smaller than the microscope of the day could show. These in turn could be thought to contain still smaller spermatozoa, and so on ad infinitum, so that the whole future of the human race actually lies preformed within the bodies of those now existing (Wiener 1964 [1956], 98–100. Italics added).

Moreover, Wiener supposed that Leibniz's Infinitesimal Calculus itself had to be integrated in this point of view:

> This opinion, which was generated as we have seen by the microscopic observations of his day as well as by the inner workings of his own philosophy, *led Leibniz eventually to a new interpretation of mathematics*. He was, we must remember, one of the co–inventors of the calculus, and he originated the notation which we use even now. For him not only are time and space infinitely subdivisible, but quantities distributed in time and space may have rates of change in all their dimensions (Wiener 1964 [1956], 101. Italics added).

Wiener will often use this argument. "The thought of every age is reflected in its technique", he writes in *Cybernetics*. Which causal direction this influence assumes for him is scarcely important: it might even be the other way round. In the case of Leibniz actually we know that his ideas on the Infinitesimal calculus can be roughly ascribed to 1675 and prepared by two or three years of reflections earlier on (see in particular Leibniz 1920). He met Antony van Leeuwenhoek in the city of Delft during his zigzagging journey from Paris to Hannover, between the end of 1676 and the very beginning of 1677, via London, Delft and The Hague (where he met Spinoza). Justin Smith points out that the interest of Leibniz in microscopes arose before his meeting with Leeuwenhoek; therefore Wiener's supposition could be reasonable. But, as Smith adds, "Leibniz explicitly claims on more than one occasion that his commitment to the doctrine of preformation comes not from any particular microscopical discovery but from a decidedly philosophical source" (Smith 2011, 266).

Therefore Wiener's hypothesis is not supported by Leibniz himself. Be that as it may, what fascinated Wiener was the idea of self–similarity on which Leibniz insisted and which actually arose even in a mathematical text, where Leibniz stated: "I have diverse definitions for the straight line. The straight line is a curve, any part of which is similar to the whole, and it alone has this property, not only among curves, but also among quantities" (Leibniz 1858, 185). [12] Wiener explains:

---

[12] IV (2) "Ego varias lineae rectae definitiones habeo: veluti Recta est linea, cujus pars quaevis est similes toti, quanquam Recta non solum inter lineas, sed etiam inter magnitudines hoc sola habeat." (Leibniz 1858, 185). I found the English quotation in

Leibniz conceived the world after the analogy of the drop of water and the similarly teeming drop of blood as a plenum. That is, he conceived that all the apparent spaces between living beings and within living beings are themselves filled with living beings on a smaller scale. This theory led Leibniz to postulate the infinite subdivisibility of life and, consequently, the continuity of matter (Wiener 1964 [1956], 101).

Actually in the *Monadology* Leibniz states:

> 65. [...] Each portion of matter is not only infinitely divisible, as the ancients observed, but is also actually subdivided without end, each part into further parts, of which each has some motion of its own; otherwise it would be impossible for each portion of matter to express the whole universe. 66. Whence it appears that in the smallest particle of matter there is a world of creatures, living beings, animals, entelechies, souls. 67. Each portion of matter may be conceived as like a garden full of plants and like a pond full of fishes. 67. But each branch of every plant, each member of every animal, each drop of its liquid parts is also some such garden or pond. 68. And though the earth and the air which are between the plants of the garden, or the water which is between the fish of the pond, be neither plant nor fish; yet they also contain plants and fishes, but mostly so minute as to be imperceptible to us (Leibniz 1898 [1714], 255–256).

The theory of monads belongs to Leibniz's late period: the *Monadology*, published posthumously since most of the other fundamental Leibniz's works had been written in 1714, however for Wiener it was crucial to understand his thought; and this was also true for Russell (1900).

8   The "Leibnizian frame" Becomes New Mathematics

In the article *Infinity* (Wiener 1919), which appeared in the same year he was hired by the M.I.T. Department of Mathematics, Wiener states that we cannot exclude the existence of "subsystems" possessing "the self–reflecting property of cardinal infinity". And he began at once to study a kind of self–reflecting process, subjected to chance and treated by means of those conceptual tools derived by Lebesgue and Gibbs: the Wiener process.

Wiener asked to be credited as the one who, in 1920, had been, "the first person to apply the Lebesgue integral to a specific physical problem — that of the Brownian motion" (Wiener 1989 [1954], 9–10). As he just arrived at MIT, from 1919 until 1920, he made one of his most important discoveries, the so called "Wiener process", a mathematical description of the Brownian motion. In this research, all the ingredients we have pointed out above appear clearly. In this early

---

Mandelbrot 1983, p. 419. The English translation from Latin I'm proposing here is slightly different (*e.g.* I do not think that "magnitudines" could be translated as "sets").

period at MIT Wiener was very interested in processes such like the flow of the rivers. Looking at the River Charles from the window of his MIT room he used to reflect:

> How could one bring to a mathematical regularity the study of the mass of ever shifting ripples and waves, for was not the highest destiny of mathematics the discovery of order among disorder? [...] What descriptive language could I use that would portray these clearly visible facts without involving me in the inextricable complexity of a complete description of the water surface? (Wiener 1964 [1956], 33).

From the end of '800 and the beginning of '900, mathematicians had introduced examples of continuous non differentiable curves. Speaking for the non–mathematicians, these curves can be drawn without ever lifting the hand from the sheet of paper (this is, put simply, a definition for "continuity"); but they are characterized at the same time by peaks or cliffs in each of their points, so that one cannot trace a tangent anywhere (this is, put simply, a definition for "non differentiability"). We can quote Koch's curves — known also as Koch Snowflake — or the Peano curve, that are obtained as the limits of iterative procedures. At any scale of enlargement these curves keep their geometrical proprieties.

In the book *Atoms* by the physicist Jean Perrin, that Wiener had read, the author had sustained that this kind of curves are not rare but very frequent in nature. Only apparently, at a first approximation, nature showed smooth continuous differentiable curves. He wrote:

> Though derived functions are the simplest and the easiest to deal with, they are nevertheless exceptional [...]. And often those who hear of curves without tangents, or underived functions, think at first that Nature presents no such complications, nor even offers any suggestion of them. The contrary, however, is true [...]. Consider, for instance, one of the white flakes that are obtained by salting a soap solution. At a distance its contour may appear sharply defined, but as soon as we draw nearer its sharpness disappears. The eye no longer succeeds in drawing a tangent at any point on it [...]. So that if we were to take a steel ball as giving a useful illustration of classical continuity, our flake could just as logically be used to suggest the more general notion of a continuous underived function (Perrin 1916 [1913], IX).

Perrin added a familiar consideration for such a pupil of Royce like Wiener, because of the reference to the map of Brittany, even if not of Great Britain, but with reference to different degrees of magnification. He stated:

> We must bear in mind that the uncertainty as to the position of the tangent plane at a point on the contour is by no means of the same order as the uncertainty involved, according to the scale of the map used, in fixing a tangent at a point on the coast line of Brittany. The tangent would be different according to the scale, but a tangent could always be found, for a map is a conventional diagram in which, by construction, every line has a tangent. An essential characteristic of our flake and,

indeed, of the coast line also when, instead of studying it as a map, we observe the line itself at various distances from it is, on the contrary, that on any scale we suspect, without seeing them clearly, details that absolutely prohibit the fixing of a tangent (Perrin 1916 [1913], IX–X).

The passage was immediately followed by the consideration of Brownian motion in which he stated:

> [...] in order to be able to fix a tangent to the trajectory of such a particle, we should expect to be able to establish, within at least approximate limits, the direction of the straight line joining the positions occupied by a particle at two very close successive instants. Now, no matter how many experiments are made, that direction is found to vary absolutely irregularly as the time between the two instants is decreased. An unprejudiced observer would therefore come to the conclusion that he was dealing with an underived function, instead of a curve to which a tangent could be drawn (Perrin 1916 [1913], X).

It was in this way that Wiener began to work on the Brownian motion, arriving to the "Wiener process". The Brownian motion is an irregular and causal motion of very small particles suspended in a fluid, caused by the collision with the molecules, discovered by the botanist Robert Brown in 1827. It had been studied in physics by Albert Einstein, in one of his famous articles of 1905, and by Marian Smoluchowski in 1906.

The main difference between the Einstein and Smoluchowski approaches and Wiener's one lies in the fact that the models of the formers looked at the particles motion considered as material points, while Wiener considered the trajectories described by the points. Basically he used here the "Leibnizian frame": the set of the Brownian trajectories represented "every possible worlds", he needed just a method to operate on them finding very useful a mathematical method derived from Lebesgue. This way he was successful in demonstrating that, analogously to the Koch Snowflakes and Peano curves, the trajectories of a Wiener process are "all" continuous not differentiable curves, denoting self–similarity at any level of magnification.

In the specific case of the Wiener process this self–similarity was not geometrical as for the Koch and the Peano curves, but merely statistical: for any degree of enlargement, they kept their statistical propriety, that is, they contained the same degree of chance. And via this method Wiener could reach the same numerical results found by Einstein.

In passing, some reflections about the relation between Wiener and the Mandelbrot's fractals are useful. In his last book, God & Golem, Inc., published the year of his death, Wiener quotes enthusiastically one of the first pieces of research made by the young Mandelbrot (1963), concerning time series of the prices of commodities markets and showing a kind of self–similarities, akin to the Wiener Process. Wiener writes:

> He [Mandelbrot] has shown that the intimate way in which the commodity market is both theoretically and practically subject to

random fluctuations *arriving from the very contemplation of its own irregularities* is something much wilder and much deeper than has been supposed, and that the usual continuous approximations to the dynamics of the market must be applied with much more caution than has usually been the case, or not at all. (Wiener 1964a, 92. Italic added).

The passage is not very easy to understand. In the phrase in Italic, Wiener emphasizes the fact that the casualness discovered by Mandelbrot is made volatile and wayward in such a way because of "autocatalysis" in which the curve becomes more irregular and casual, "contemplating" its own irregularities.[13]

Four years later Mandelbrot was to discover the "Fractals". He introduced them in the article: *How Long Is the Coast of Britain? Statistical Self–Similarity and Fractional Dimension* (Mandelbrot 1967), whose central idea is that the "seacoast shapes are examples of highly involved curves with the property that — in a statistical sense — each portion can be considered a reduced–scale image of the whole. This property will be referred to as 'statistical self–similarity' " (Mandelbrot 1967, 636). Mandelbrot used Perrin's image, but he read it through, so to speak, Wiener's glasses.

The young Benoît Mandelbrot (1924–2010), between 1947 and 1948, had worked as the editor of the first issue of *Cybernetics*, by the publisher Hermann & Cie. (cf. Segal 2003). In fact the book had been published before in France, although in English. Considering the two earlier works he had published, it would appear clearly Wiener's influence on him, and actually in the first French edition of his masterpiece on Fractals (Mandelbrot 1975), he acknowledged Wiener's work as his main source of inspiration, seeing — maybe excessively so — that the "Wiener process" was the "first fractal". In the second French edition Wiener's credits are reduced, to be eventually confined in a small corner in the big English edition (Mandelbrot 1985).

9 From Leibniz to the Information Theory

The discovery of Wiener's process was not confined to the breakthrough in itself. Wiener had obtained a new and very personal mathematical method forming in some sense new intellectual "glasses" to watch, mathematically, the physical reality. It had convinced him "that a significant idea of organization cannot be obtained in a world in which everything is necessary and nothing is contingent" (Wiener 1964 [1956], 323). A way of seeing, in the final analysis, based on his cognitive "Leibnizian frame".

The mathematical method of Wiener became a deep and powerful tool in his hands. As the mathematician and physicist Mark Kac will state:.

> In retrospect one can have nothing but admiration for the vision which Wiener had shown when, almost half a century ago, he had chosen Brownian motion as a subject of study from the point of view of the theory of integration. To have foreseen, at that time, that an impressive

---

[13] I owe a special thank–you to Giuseppe O. Longo that more than ten years ago suggest me the complex meaning hidden in the Wiener's sentence.

edifice could be erected in such an esoteric corner of mathematics was a feat of intuition not easily equaled now or ever (Kac 1966).

Wiener used several variants of his mathematical method in various fields. I cannot go here over all the mathematical discoveries he made afterwards based on this method. But I think that the most useful thing, in order to corroborate my point as well, could be to introduce in simple words his Prediction theory and the mathematical theory of communication directly deriving from it, usually called information theory and credited to Claude Shannon alone.

From 1940 to 1942 Wiener worked for the US government on a military project about antiaircraft prediction. He developed a statistical mathematical theory to predict the future position of the cruising altitude planes to shoot down by means of a gun, connected to a radar and usually to an analogical computer, implementing Wiener's theory of prediction that automatically controlled the aim of the gun. The deep mathematical conclusions to which Wiener arrived flowed into the classified book *The Extrapolation, Interpolation and Smoothing of Stationary Time Series, with Engineering Applications*, nicknamed as *Yellow Peril*, (1st February 1942).

In the book Wiener speaks of the method used by the modern statistician dealing with the so called time series or historical series (i.e. graphs in which quantities are a function of time: number of inhabitants per year; prices per day etc.). He explains:

> Behind all statistical work lies the theory of probabilities. The events which actually happen in a single instance are always referred to a collection of events which might have happened; and to different subcollections of such events, weights or probabilities are assigned […]. The strictly mathematical theory corresponding to this theory of probability is the theory of measure, particularly in the form given by Lebesgue. […] In other words, the statistical theory of time series does not consider the individual time series by itself, but a distribution or ensemble of time series. Thus the mathematical operations to which a time series is subjected are judged, not by their effect in a particular case, but by their average effect (Wiener 1949 [1942], 3–4).

It should appear also here the usual Wiener's "Leibnizian frame". And we should note also that the way in which he speaks of the methods of the modern statisticians is very much tailored on his way of interpreting statistics in terms of mechanical statistics.

Wiener adds in the book that communication engineers do the same work of the statisticians. Communication engineering works on wired telegraphs, telephones, radio, television, but also — in his opinion — on servomechanisms (using sensors to operate), and analogical or digital computers. All these systems convoy, record, elaborate, use messages, and in his view a message is a time series (one can think e.g. to a speech as a sequel of sounds in time). Consequently messages could be treated as any other time series. Therefore Wiener could state:

> While one does not ordinarily think of communication engineering in the same terms, this statistical point of view is equally valid there. No

> apparatus for conveying information is useful unless it is designed to operate, not on a particular message, but on a set of messages, and its effectiveness is to be judged by the way in which it performs on the average on messages of this set. "On the average" means that we have a way of estimating which messages are frequent and which rare or, in other words, that we have a measure or probability of possible messages. The apparatus to be used for a particular purpose is that which gives the best result "on the average" in an appropriate sense of the word "average" (Wiener 1949 [1942], 4).

Once again we found here the "Leibnizian frame", in the form that Wiener had given it from the Brownian motion on, and that now we see arriving up to the field of communication engineering.

In the quoted passage there is the fundamental intuition bringing to the mathematical theory of communication, now best known as Shannon's Information Theory. In the first page of his masterly article, Shannon writes:

> [...] semantic aspects of communication are irrelevant to the engineering problem. The significant aspect is that the actual message is one *selected from a set* of possible messages. The system must be designed to operate for each possible selection, not just the one which will actually be chosen since this is unknown at the time of design (Shannon 1948, 379).

Shannon neither acknowledged here he was in debt to the Wiener's Yellow peril, nor quoted its author. Shannon could not see that the "Leibnizian frame" itself underpins clearly in his sentence. We will find a Shannon's acknowledgment only in a passage of the second part of the paper, were he wrote:

> An ensemble of functions is the appropriate mathematical representation of the messages produced by a continuous source (for example speech), of the signals produced by a transmitter, and of the perturbing noise. Communication theory is properly concerned, as has been emphasized by Wiener, not with operations on particular functions, but with operations on ensembles of functions. A communication system is designed not for a particular speech function and still less for a sine wave, but for the ensemble of speech functions (Shannon 1948, 627).

Wiener had had the earlier fundamental ideas, Shannon systematized the matter and the contribution of the former was forgotten. In any case we cannot but be admired by the fact that even in this passage the "Leibnizian frame" shines like a star, shaping a large conceptual set of technical tools which is still useful to communication engineers.

At the end of our "walk", the reasons why Wiener chose Leibniz as the "patron saint for cybernetics" appears certainly clearer. Not only Leibniz is the perfect icon of the interdisciplinary man of science Wiener and the other cyberneticists longed for, an asymptotic ideal proposed to the hyper–specialized post–war science; not only Leibniz had glimpsed from afar the modern computer; but in his metaphysical thought Wiener had found the seed from which his powerful mathematical methods had bloomed.

10 The dissolution of the Vitalism–Mechanism opposition

One of the themes Wiener dealt with in his youth were the disputes among Vitalism and Mechanism. The former sustained a clear cut distinction between the mental and vital phenomena on one hand, and the mechanical ones on the other, asking for a different method to study the first ones. The latter instead considered living and psychical phenomena explainable by the usual method of Newtonian mechanics.

Henri Bergson had been a famous vitalist in Wiener's youth. In *Cybernetics* Wiener had shown to agree with him about "the difference between the reversible time of physics, in which nothing new happens, and the irreversible time of evolution and biology, in which there is always something new". But Wiener considers:

> [...] the many automata of the present age are coupled to the outside world both for the reception of impressions and for the performance of actions. [...] What is perhaps not so clear is that the theory of the sensitive automata is a statistical one. We are scarcely ever interested in the performance of a communication–engineering machine for a single input. To function adequately, it must give a satisfactory performance for a whole class of inputs, and this means a statistically satisfactory performance for the class of input which it is statistically expected to receive. Thus its theory belongs to the Gibbsian statistical mechanics rather than to the classical Newtonian mechanics (Wiener 1961 [1948], 43–44).

And concludes that:

> [...] the modern automaton exists in the same sort of Bergsonian time as the living organism; and hence there is no reason in Bergson's considerations why the essential mode of functioning of the living organism should not be the same as that of the automaton of this type. *Vitalism has won to the extent that even mechanisms correspond to the time–structure of vitalism; but as we have said, this victory is a complete defeat [...], the new mechanics is fully as mechanistic as the old* (Wiener 1961 [1948], 44).

You can consider that the world of Leibniz was really different from the one drawn here. It was a determinist one like that of Newton. Certainly, Wiener notes, "Leibniz is as dynamically minded as Spinoza is geometrically minded" (Wiener 1961 [1948], 41) but, recalling the theory of the pre–established harmony, he retains that

> Leibniz considers a world of automata, which, as is natural in a disciple of Huyghens, he constructs after the model of clockwork. Though the

monads reflect one another, the reflection does not consist in a transfer of the causal chain from one to another. [...]. As he says, they have no windows. [...] The monad is a Newtonian solar system writ small (Wiener 1961 [1948], 41).[14]

Wiener does not share the Leibniz's Panpsychism. Leibniz, he knows, "sup-poses an infinity of substances, the monads, all partaking to some extent of a mental nature. [...] regarded the monads as possessing ideas of various grades of clearness and distinctness, and believed that matter was made up of those with the vaguest ideas" (Wiener 1919b, p. 939).

Cybernetics is clear in this regard: "instead of building a wall between the claims of life and those of physics, the wall has been erected to surround so wide a compass that both matter and life find themselves inside it" (Wiener 1961 [1948], 38).

About Panpsychism we cannot neglect here a friendly dispute which arose in 1934 between Wiener and physiologist J. B. S. Haldane. In 1932 at Cavendish Laboratory, Cockcroft e Walton had made an experiment known as the first case of fission of an atom. Among other things, it appeared clearly that some alpha and beta particles, because of their wave–like nature, could pass through the potential barriers of atoms, as supposed by George Gamow. A phenomenon now established and known as "tunnel effect" or "Quantum tunnelling".

Now, in 1934, J. B. S Haldane, a friend of Wiener, published a paper on *Quantum Mechanics as a Basis for Philosophy* (Haldane 1934). He was a professed materialist and his speech actually seems more an apology of materialism against the vitalists denying materialism by opposing the mental phenomena as irreducible to matter. Haldane argued that the tunnelling effect was similar to the behaviour of an animal able to see something beyond a barrier with his imagination.

He concluded that the human mind could be considered as a wave–like phenomenon of the brain, underling quantum phenomena, responsible of the existence of the universals, that is the Platonic ideas, as well.

Wiener, engaged in the dispute, published a response on the same journal entitled *Quantum mechanics, Haldane and Leibniz* (Wiener 1934). He showed the remarkable similarities between Haldane's and Leibniz's thinking. In Wiener's opinion, Haldane appeared substantially Leibnizian, because if we speak of a continuous from psychical monads becoming more and more material, as Leibniz does, the situation is nearly similar to a continuous formed by material particles assuming more and more the mental aspect, as Haldane does. The difference between Leibniz's pluralistic spiritualism and Haldane's pluralistic materialism appears to Wiener very thin, if not nominalistic at all.

Returning on the theme in successive paper on The role of the observer (Wiener 1936), Wiener considered the question of a brain underlying Quantum mechanical phenomena as an empirical problem solvable therefore only through experiments.[15]

---

[14] About the relations between Huygens and Leibniz see Bos (Bos 1978), Hofmann (Hofmann 1974) and Chareix (Chareix 2010).

[15] For a commentary about the dispute between Wiener and Haldane see Gale 1997. About the status of the idea to consider scientifically useful the hypothesis

In *Cybernetics* Wiener, however, claims: "it is true that the matter of the newer physics is not the matter of Newton, but it is something quite as remote from the anthropomorphizing desires of the vitalists" (Wiener 1961 [1948], 38). The friendly dispute with Haldane had the merit to push Wiener to think to the mental phenomena and to resume his youth reflection about Vitalism and Mechanism. In *Cybernetics* Wiener shows to have arrived to suggestive answers regarding the ancient questions he had met when he was young.

Introducing the notion of information, derived from his long reflections on the mathematical theory of communication and of the message, he believes it possible to bypass the old opposition vitalism–mechanism. He retains that the human brain, as well as an electronic brain like ENIAC, has to be considered not as a heat engine. They are both elaborators of information, and the power they absorb has no correlation with their potential to elaborate information. Concluding, he states *"Information is information, not matter or energy. No materialism which does not admit this can survive at the present day"* (Wiener 1961 [1948], 132).

11 Conclusion

From the last chapter it appears clearly that Wiener was not Leibnizian. However he was very convinced of the importance of studying the ancient scientists and philosophers, in particular Leibniz. But the question for Wiener was not to assume mechanically pieces of information from Leibniz. He had always maintained a dialogue with him, often finding the best insights in his own way of thinking and in his metaphysics *strictu sensu*.

In 1932 he published a paper entitled *Back to Leibniz! Physics reoccupies an abandoned position* (Wiener 1932). The main idea he sustained was that in a time in which the ancient Newtonian certainties in physics had fallen, one needed to go back to "times when Newtonian physics was itself but one alternative theory struggling for recognition" (Wiener 1932, 201). Leibniz's thought was "unprejudiced by the success of a single theory", that is the Newtonian one. And therefore it was apt to a new period of chaos in which "we again find it necessary to escape from the blinding glare of a theory to see our physical universe face to face and free from prejudice" (Wiener 1932, 224).

In some sense, reading Leibniz, Wiener was able to breathe new air. He was aware of the criticisms the proposal would have had to deal with. He claims:

> This historical attitude in science, it is true, is suspect to many of our more "tough–minded" contemporaries. What does the stupid repetition of the blunders of the past, they will say, have to do with so live and growing a subject as science? Science is — or at least I imagine the up-to-date reader of outlines, symposia, and digests will claim that it is — the utter antithesis of such intellectual activities as metaphysics, which spirals around in a continual reoccupation of abandoned positions, and which has an interesting history only because it has a dead future.

---

that brain underlying Quantum mechanics phenomena, see Stapp 1997. See also Montagnini 2017a.

Science is progressive, we are told, and when it once outlives a primitive stage, that stage is past for all time (Wiener 1932, 201).

On the contrary, this going back to study the philosophers and scientists of the past proved to be extremely fruitful, as we have seen here. Twenty–four years later, concluding his epilogue of his autobiography, Wiener had not changed his mind. Criticizing the science of his (and our) days, as too much bounded to the needs of the moment, in order to corporations' interests. He spurs it to widen its temporal horizons:

> Like a tradition of scholarship, a grove of sequoias may exist for thousands of years […]. There are scientific ideas which we can trace clearly to the time of Leibniz, a quarter of a millennium ago, which are just beginning to find their applications in industry. Can a business firm or a government department, moved primarily by the immediate needs for new weapons, compass this period of time in its backward glance? (Wiener 1964 [1956], 362).

References

Bernstein, J. R. 1992. The Resurgence of Pragmatism. *Social research* 59(4): 813–840.
Bonner, A. 2007. *The Art and Logic of Ramon Llull: A User's Guide*. Leiden: Brill.
Bos, H. J. M. 1978. The influence of Huygens on the formation of Leibniz' ideas. In: Heinekamp, A. (Ed.). 1978 *Leibniz à Paris. 1672–1676*. Wiesbaden: Steiner.
Bradley, F. H. 1902 [1893]. *Appearance and reality. A metaphysical essay*. London: Sonnenschein & co.
Breton, P. 1987. *Histoire de l'informatique*. Paris: La Découverte.
Breton, P. 1992. *L'utopie de la communication. L'emergence de l'homme sans interieur*. Paris: La Découverte.
Browder, F., Spanier E. H., Gerstenhaber M. (Eds.) 1966. Norbert Wiener 1894–1964. *Special Edition of the Bulletin of the American Mathematical Society* 72(1/2).
Chareix, F. 2010. Geometrization or mathematization: Christiaan Huygens's critiques of infinitesimal analysis in his correspondence with Leibniz. Dascal, M. (ed) *The practice of reason: Leibniz and his controversies*. Philadelphia: John Benjamins Pub. Co. 2010:33–49.
Colin B. 1994. *Information and Secrecy: Vannevar Bush, Ultra, and the Other Memex*. Metuchen, N.J.: Scarecrow Press.
Couturat, L. 1901. *La Logique de Leibniz: d'après des documents inédits*. Paris: Félix Alcan Editeur.
Davis, M.. 2000. *The universal computer. The road from Leibniz to Turing*. New York and London: W. W. Norton & C.
De Luca, A. 2006. Some Reflections on Cybernetics and its Scientific Heritage. *Scientiae Mathematicae Japonicae* 64 (2):243–253.
De Ruggiero, G. 1912. *La filosofia contemporanea*. Bari: Laterza.

Gale, G. 1997. The role of Leibniz and Haldane in Wiener's cybernetics. In Mandrekar, V. and Masani, P. (eds.), *Proceedings of Norbert Wiener Centenary Congress*, Providence, Rhode Island: The American Mathematical Society:247–262.

Gibbs, W. J. 1906. *The scientific papers. I: Thermodynamics.* London, New York, and Bombay: Longmans, Green, and Co.

Grattan–Guiness, I. 1975. Wiener on the Logics of Russell and Schröder. An Account of His Doctoral Thesis, and His Discussion of it with Russell. *Annals of Science* 32: 103–132.

Haldane, J. B. S. 1934. Quantum Mechanics as a Basis for Philosophy. *Philosophy of Science* 1(1): 78–98.

Heims, S. J. [1980] 1984 *John von Neumann and Norbert Wiener. From mathematics to the technologies of life and death.* Cambridge, MA: The MIT Press.

Hofmann, J. E. 1974. *Leibniz in Paris, 1672–1676. His growth to mathematical maturity.* London and New York: The Cambridge University Press.

James, William. 1907. Pragmatism. A new name for some old ways of thinking. New York: Longman Green and Co.

Kac, M. 1966. Wiener and integration in function spaces. *Bulletin of American Mathematical Society* 72. Part 2:52–68.

Leibniz, G. W. 1858. *Euclidis ΠΡΩΤΑ, Mathematische Schriften*, Halle: Druck und Verlag von H. W. Schmidt. V:183–211.

Leibniz, G. W. 1898. *The Monadology and other philosophical writings.* Transl. with introduction and notes by Robert Latta. Oxford: The Clarendon Press.

Leibniz, G. W. 1920. *The Early Mathematical Manuscripts of Leibniz.* Transl. Carl Immanuel Gerhardt. Chicago and London: The Open Court Publishing company.

Leibniz, G. W. 1923 ff. *Sämtliche Schriften und Briefe* [All writings and letters. Berlin. Berlin: Akademie Verlag [Academy edition].

Leibniz, G. W. 1929 [ca. 1671]. Leibniz on His Calculating Machine Smith, in David Eugene (ed). A Source Book in Mathematics. New York and London: McGraw–Hill Book Company, Inc.

Leibniz, G. W. 2002. *The Labyrinth of the Continuum: Writings on the Continuum Problem, 1672–1686.* Translated and edited by Richard T. W. Arthur. New Haven, CT: The Yale University Press.

Lully, R. 1985. *Selected works of Ramón Llull (1232–1316).* Edited and Translated by Anthony Bonner. Princeton: The Princeton University Press. Vol. 1: The book of the Gentile and the three wise men. Ars demonstrativa. Ars brevis.

Mandelbrot, B. 1963. The Variation of Certain Speculative Prices. *The Journal of Business* 36(4):394–419.

Mandelbrot, B. 1967. How Long Is the Coast of Britain? Statistical Self–Similarity and Fractional Dimension. *Science* 3775(156):636–638.

Mandelbrot, B. 1975. *Les objets fractals. Forme, hazard et dimension.* Paris: Flammarion.

Mandelbrot, B. 1983. *The Fractal Geometry of Nature.* San Francisco: W. H. Freeman.

Masani Pesi, R. 1989, Leibnitz, Quantum Mechanics, and Cybernetics. Comments on [32c], [34c], [36g]. In Wiener, Norbert. 1989. *Collected works*. With commentaries, Masani Pesi. R. (Ed.). Cybernetics, science, and society; ethics, aesthetics, and literary criticism; book reviews and obituaries, Cambridge, MA: The MIT Press, Vol. 4, pp. 98–104.

Masani Pesi, R. 1990. *Norbert Wiener, 1894 – 1964*. Basel, Boston, Berlin: Birkhäuser Verlag.

Monk, R. 1990. *Ludwig Wittgenstein. The duty of genius*. London: Jonathan Cape.

Montagnini, L., Tabacchi M. E., Termini S. 2015, Out of a creative jumble of ideas in the middle of last Century: Wiener, interdisciplinarity, and all that. *Biophysical Chemistry*, June.

Montagnini, L. 2001–2002. La rivoluzione cibernetica. L'evoluzione delle idee di Norbert Wiener sulla scienza e la tecnica. *Atti e memorie dell'Accademia Galileiana di Scienze, Lettere ed Arti* II(114):109–135.

Montagnini, L. 2005. *Le Armonie del Disordine. Norbert Wiener Matematico–Filosofo del Novecento*. Venezia: Istituto Veneto di Scienze, Lettere ed Arti.

Montagnini, L. 2008. Philosophical Approaches towards Sciences of Life in Early Cybernetics. In eds. Ricciardi, L. M., Buonocore, A. and Pirozzi, E., Collective Dynamics on Competition and Cooperation in Biosciences, pp. 11–17. Melville–New York: American Institute of physics. DOI: 10.1063/1.2965081

Montagnini, L. 2013. Interdisciplinary issues in Early Cybernetics. In: Lilia G., László R. and C. Pléh (Eds.). *New Perspectives on the history of cognitive science*. Budapest: Akadémiai Kiadò, pp. 81–89.

Montagnini, L. 2014a. Norbert Wiener. Il matematico che avvistò il nostro tempo. *Scienza in rete*. 1st May 2014.

Montagnini, L. 2014b. Come Norbert Wiener divenne l'icona di una scienza pacifica e John von Neumann del suo opposto. *Scienza e Pace* 26th November 2014.

Montagnini, L. 2015. The mathematical art of Norbert Wiener. *Lettera Matematica. International edition*. September 3(3):129–134.

Montagnini, L. 2017a. *Harmonies of Disorder. Norbert Wiener a Mathematician–Philosopher of our time*. Berlin: Springer. DOI:/10.1007/978-3-319-50657-9

Montagnini, L. 2017b. Interdisciplinarity in Norbert Wiener, a mathematician-philosopher of our time. *Biophysical Chemistry*. Online: 28 June 2017. DOI:/10.1016/j.bpc.2017.06.009

Montagnini, L. 2017c. W for Wiener. Consistency is no longer a virtue. *Lettera Matematica International edition*. Online: 10 July 2017. DOI:/10.1007/s40329-017-0186-0

Nyce, J. M., Kahn, P. (Eds). 1991. *From Memex to hypertext. Vannevar Bush and the mind's machine*. Boston: Academic Press.

Peirce, C. S. 1891. The architecture of theories. *The Monist*. January 1:161–176.

Peirce, C. S. 1891–1892. The Law of Mind, *The Monist*. February 2: 533–559.

Peirce, C. S. 1892a. Man's Glassy Essence. *The Monist*. October 3: 1–22.

Peirce, C. S. 1892b. On slaught on the doctrine of necessity. *The Monist*. July 2: 560–582.

Peirce, C. S. 1892c. The doctrine of necessity examined. *The Monist*. April 2: 321–337.

Peirce, C. S. 1923. *Chance, Love and Logic: Philosophical Essays*. New York: Harcourt, Brace & Co., Inc.; London: Kegan Paul Trench, Trubner & Co., Ltd.

Perrin, J. [1913] 1916. *Atoms*. New York: D. van Nostrand Co.

Popper, K. 1972. *Objective knowledge. An evolutionary approach*. Oxford, The Clarendon Press.

Rosenblith, W., Wiesner, J. 1965. From Philosophy to Mathematics to Biology. *The Journal of Nervous and Mental Disease* 140(1):3–8.

Rossi, P. [1983] 2000. *Logic and the art of memory. The quest for a universal language*. Chicago: The University of Chicago Press.

Royce, J., et. al. 1963. *Josiah Royce's seminar, 1913–1914: as recorded in the notebooks of Harry T. Costello*. Ed. by Grover Smith. New Brunswick, N.J.: Rutgers University Press.

Royce, J. 1900. *The world and the individual*. Gifford Lectures delivered before the of University of Aberdeen. [With a supplementary essay: The one, the many, and the infinite]. London: The MacMillan Co.

Royce, J. 1914. The Mechanical, the Historical and the Statistical. *Science* New Series, 1007(39):551–566.

Russell, B. 1900. *The philosophy of Leibniz*. London: George Allen and Unwin.

Russell, B. 1903. *Principles of Mathematics*. Cambridge: Cambridge University Press.

Russell, B. 1968. *The autobiography. II: 1914–1944*. London: George Allen and Unwin.

Segal, J. 2003. *Le Zéro et le Un. Histoire de la notion scientifique d'information au 20 a siècle*. Paris: Syllepse.

Shannon, C. 1948. A Mathematical Theory of Communication. Bell System Technical Journal 27(3):379–423 [see also: Bell System Technical Journal 27(4):623–656].

Shannon, C. E. 1938. A Symbolic Analysis of Relay and Switching Circuits. *Transactions of the American Institute of Electrical Engineers* 57(12):713–723.

Smith, J. E. H. 2011. Leibniz and the Life Sciences Justin. In Look, Brandon C. ed. The Continuum Companion to Leibniz. London and New York: Continuum Press.

Stapp, H. P. 1997. Quantum mechanical coherence, resonance, and mind. In: Mandrekar, V., Masani Pesi R. (Eds.) *Proceedings of Norbert Wiener Centenary Congress*. Providence, RI. The American Mathematical Society Press, pp. 263–300.

Swift, J. 1733. *On Poetry: A Rhapsody*. London:

Termini, S. 2006a. Remarks on the development of cybernetics. *Scientiae Mathematicae Japonicae* 64(2):461–468.

Termini, S. 2006b. Imagination and Rigor: their interaction along the way to measuring fuzziness and doing other strange things. In Termini S. (ed.). 2006. *Imagination and Rigor: essays on Eduardo R. Caianiello's Scientific Heritage*. Milano: Springer, pp. 157–173.

Whitehead, A. N., Russell, B. 1910–1913. Principia Mathematica. Cambridge: The Cambridge University Press.

Wiener, N. 1914. Relativism. *Journal of philosophy, psychology and scientific method* 11:561–577.
Wiener, N. 1915. Is mathematical certainty absolute? *Journal of philosophy, psychology and scientific method* 12:568–574.
Wiener, N. 1919a. Infinity. In *Encyclopedia Americana* 15:120–122.
Wiener, N. 1919b. Mechanism and vitalism. In *Encyclopedia Americana* 18:527–528.
Wiener, N. 1919c. Metaphysics. In: *Encyclopedia Americana* 18:707–710.
Wiener, N. 1920. The mean of a functional of arbitrary elements. *Annals of Mathematics* 22(2):66–72.
Wiener, N. 1921. The average of an analytical functional and the Brownian movement. *Proceedings of [US] National Academy of Sciences* 7: 294–298.
Wiener, N. 1932. Back to Leibniz! Physics reoccupies an abandoned position. *The Technology Review* 34:201–203, 222, 224.
Wiener, N. 1934. Quantum mechanics, Haldane and Leibniz. *Philosophy of Science* 1(4):479–482.
Wiener, N. 1936. The role of the observer. *Philosophy of Science* 3(3):307–319.
Wiener, Norbert. 1949 [1942]. *Extrapolation, interpolation, and smoothing of stationary time series, with engineering applications.* Cambridge–MA: The MIT Press.
Wiener, N. 1950. *The Human Use of Human Beings.* London: Eyre and Spottiswoode.
Wiener, N. 1958. My connection with cybernetics. Its origin and its future. *Cybernetica* 1–14.
Wiener, N. 1961 [1948]. *Cybernetics. Or control and communication in the animal and the machine.* New York and London: Wiley & Sons.
Wiener, N. 1962. The mathematics of self–organizing system, Recent developments in information and decision processes. In: *On the function of science in society*, Appendix III. New York and London: Macmillan.
Wiener, N. 1964 [1953]. *Ex–prodigy. My childhood and youth.* Cambridge–MA: The MIT Press.
Wiener, N. 1964 [1956]. *I am a mathematician. The later life of a prodigy. An autobiographical account of the mature years and career of childhood in Ex–prodigy.* Cambridge–MA: The MIT Press.
Wiener, N. 1964a. Intellectual Honesty and the Contemporary Scientist. *The Technology Review* 66:17–18, 44–45, 47.
Wiener, N. 1964b. God & Golem, inc. A comment on certain points where Cybernetics impinges on religion. Cambridge–MA: The MIT Press.
Wiener, N. 1989 [1954]. *The Human Use of Human Beings.* Revised edition. London: Free Association.
Wildes, K. L., Lindgren, N. A. 1985. *A century of electrical engineering and computer science at MIT. 1882–1982.* Cambridge–MA and London: The MIT Press.
Woolhouse, R. S. 1993. *Descartes, Spinoza, Leibniz. The Concept of Substance in Seventeenth Century Metaphysics.* Routledge: New York.

Leone Montagnini, Biblioteche di Roma, Italy
leonemontagnini@katamail.com

Leibniz and the Sciences of Engineering

Agamenon R. E. Oliveira

Abstract. Leibniz (1646-1716) is one of the most prominent figures of the Scientific Revolution of the seventeenth century, an important precursor of the Enlightenment, and a genius of modern thought. Nevertheless, his name does not appear in the history of science literature to the same extent as Galileo (1564-1642), Descartes (1596-1650), or Newton (1642-1727). In this paper the role of his scientific work is emphasized, as well as his contribution to the development of engineering sciences. The intellectual context where Leibniz lived is described, as well as his participation in the seventeenth century Scientific Revolution. However, the basis for a true revolution in the scientific conceptual framework was laid by Leibniz with the development of differential and integral calculus and dynamics. It is worth emphasizing the establishment of the conservation of living forces principle postulating the expression mv^2 as the quantity that remains constant in motion. Not only did this concept have great importance in theoretical mechanics but it also marked the birth of applied mechanics, as shown in the text by means of Lazare Carnot's (1753-1823) contribution to a new theory of machines. The principle of conservation formulated by Leibniz was an important contribution to the principle of conservation of energy which emerged in the nineteenth century.

Keywords: Scientific Revolution, Mechanics Foundation, Engineering Disciplines, Leibnizian Studies.

1 Introduction

If we look at specialized literature dedicated to the Scientific Revolution, Leibniz's contribution is frequently underestimated or even omitted. In the third chapter of Bernard Cohen's book *Revolution in Science* (Cohen 2001), entitled *Scientific Revolutionaries of the Seventeenth Century*, Leibniz's name only appears in the context of the development of infinitesimal calculus and its intellectual disputes. Another book by Bernard Cohen, *The Birth of a New Physics* (Cohen 1985), does not even mention Leibniz. Even in the works of Alexandre Koyré (1892-1964), Leibniz occupies a secondary or marginal position (Koyré 1973).

On the other hand, if we look at Floris Cohen's book, *The Scientific Revolution: A Historiographic Inquiry* (Cohen 1994), Leibniz appears in several contexts, mainly in the history of the development of the concept of force as follows:

Agamenon R. E. Oliveira (2017) Leibniz and the Sciences of Engineering In: Pisano R, Fichant M, Bussoti P, Oliveira ARE (eds.), *The Dialogue between Sciences, Philosophy and Engineering. New Historical and Epistemological insights. Homage to Gottfried W. Leibniz 1646-1716.* College Publication, London, pp. 287-308
© 2017 College Publications Ltd | ISBN: 978-1-84890-227-5 www.collegepublications.co.uk

In Leibniz, the conflict between the two [currents] began to resolve itself by the modification of the mechanical philosophy... The development of a conception of force as action on a body to change its state of motion, a conception that contributed greatly to the further elaboration of mathematical mechanics, was inhibited by the mechanical philosophy during the century... It remained for Isaac Newton to pick up that conception again and to use it both to extended mechanics and to revise mechanical philosophy[1].

Leibniz became a victim of publication problems as noted by Jurgen Lawrenz:

The only books he published under his name were 'News from China' and 'Theodicy'. So, for nearly 300 years his posthumous reputation suffered from the twofold malaise that his invention of the calculus was disputed by followers of Newton and that his philosophy was seen as pure metaphysics without effective connections to the problems of real life. I mean that historically we had to cope with the notion that Leibniz fell between two stools. As a scientist he was far advanced and proposed many brilliant new ideas especially in mathematics; by as a philosopher he was (supposedly) superseded by Kant.

Leibniz's work remained unknown for a long time. Only at the beginning of the twentieth century would some important books and documents finally appear. Practically at the same time two fundamental books were published. Bertrand Russell's *A Critical Exposition of the Philosophy of Leibniz* (Russel 1900), which appeared in London in 1900, and Louis Couturat's work *Logique de Leibniz*, published in Paris in 1901. Other important books include E. Cassirer, *Leibniz' System* (Cassirer 1902), published in Marbourg in 1902 as well as the M. Gueroult book, *Dynamique et Métaphysique Leibnizièenne* (Gueroult 1967), published in Strasbourg in 1934.

The above mentioned books by B. Russell and L. Couturat argued for a remarkably similar interpretation of Leibniz's philosophy. Couturat supported the argument of his book by publishing in 1903 a collection of Leibniz manuscripts. Independently, later authors corroborated Russell and Couturat's analyses.

The seventeenth century was a period of deep social and scientific changes. The Aristotelian science that had dominated the Middle Ages and the Renaissance schools was replaced by a new physics (Westfall 1977). To understand this transformation from the Aristotelian framework to the new physical world vision, it is fundamental to study the role that Leibniz played (Jolley 1995). According to Aristotelian physics the basic explanatory principles are matter and form. Together these two principles compose the idea of body. However, according to Leibniz's doctrine *force and body are directed solidly against the mechanist doctrines of Descartes, Hobbes and their followers. There is every reason to think that it was largely in place by 1686 when, seemly out of nowhere, Newton published his* Philosophiae naturalis principia mathematica, *which came to be known simply as the* Principia, *and developed a conception of force that was very different from what Leibniz had developed. The ultimate success of the Newtonian program has all but driven Leibniz's concept of force off the playing field* (Garber 2009, pp. 99-179).

One of the central interests of Leibniz's philosophy was understanding the physical world (Garber 2009). He was one of the most important physicists of the late seventeenth century. Apart from Newton there is no other physicist of his

---

[1] This citation of Westfall (1977) book: The Construction of Modern Science appears in p. 141, Part Two: The Great Tradition, of Cohen's book.

generation who contributed more to mathematical physics. Thus, to understand the history of sciences in this crucial period, we must understand Leibniz's theories of physics (Leibniz 1694).

2 Intellectual and Scientific Renewal in the Seventeenth Century

In 1660 - 1687, the apogee of Louis XIV's reign, a deep intellectual change occurred in Europe, mainly in France (Léger 1935). This involved the victory of the rationalist philosophy of Descartes (Alquié 1963) and Spinoza (1632-1677) at the same time that the experimental science of Galileo, Kepler (1571-1630), and William Harvey (1578-1657) was established (Schuster 1990), with all these changes being crowned by the publication of Isaac Newton's Principia. The great scientific discoveries transformed the vision of the physical world and significantly changed research methods, improving and increasing knowledge in all fields. In addition, in the seventeenth century, the great thinkers frequently appeared as great philosophers, as Descartes, Pascal (1623-1662) and Leibniz.

For the first time, metaphysical principles seemed to contradict the scientific spirit. However, we cannot forget that in this period a fruitful relationship was established between science and philosophy (Allard 1963). The idea of infinity and other concepts of God appeared. Consequent to this was the idea of the infinity of the world and the idea of perfection associated with the immutability of natural laws leading to idea of order and harmony, without a rejection of change. Added to these simple and *a priori* principles were the scientific principles: the idea that phenomena could only be known by means of observation and that experiments obeyed laws derived from geometry and mechanics. As a result, both of these sciences underwent fundamental development, not to mention infinitesimal calculus. By induction there arose the idea of a universe as a mechanism governed by laws discovered by the human spirit.

The development of mathematics created new and better conditions for other sciences to emerge. Between the end of the sixteenth century and the first half of the seventeenth, algebra and geometry were renewed, with the former being transformed by its use in geometry. At the hands of Descartes, and later Fermat (1601-1665) and Pascal, there emerged a new discipline, analytic geometry (Gadoffre 1961), as well as important contributions to infinitesimal calculus.

Although infinitesimal calculus was created independently by Newton and Leibniz there were differences between the two, which need to be explained. Newton used notions of limits and introduced in his method 'fluxions' or 'flowing quantities,' i.e., velocities (Newton 1994). He applied geometry to mechanics using a very complicated notation in order to express variations. Leibniz found the principles of infinitesimal calculus and simplified the notations of differential quantities (Leibniz 1694). By means of the solution of difficult problems, he showed the power and potentiality of the new method.

It is also important to emphasize the previous discoveries of Fermat and Pascal, especially the theory of numbers and probability calculus. In this context, the study of exact sciences became the objective of the young scientists in England where John Wallis (1616-1703) and Isaac Barrow (1630-1677), professors in Oxford and Cambridge, propagated the methods of Descartes and Fermat (Lagrange 1788).

Progress in mathematical science resulted in the development of mechanics, astronomy, and in physics as a whole. The mechanical principles, inertial force, equilibrium and the composition of forces, were established by Simon Stevin (1548-1620) and Galileo. The last, based on well-known experiments in inclined planes, arrived at the law of falling bodies (Hall 1981).

On the other hand, the problem of planetary motion in the solar system was eventually solved by Kepler's three laws. However, the force which maintained each planet in its orbit and the law which governed this movement was still unknown (Westfall 1971). After Kepler and Galileo this question drew the attention of several mathematicians and physicists. Huygens (1629-1695), with good mathematical knowledge followed in the footsteps of Stevin, Descartes, and Galileo, but criticizing Descartes' theories of motion and using the experimental findings of Galileo, discovered how to calculate the centrifugal force acting on the planets, thereby coming close a universal gravitational law. His *Horologium Oscillatorium* (1673) is a natural introduction to Newton's work (Huygens 1673, 163-167).

Newton, using the conception of centrifugal force proposed by Huygens (Huygens 1673), in which the model of attractive force is inversely proportional to the square of the distance and the measurement of the earth's radius made by Picart (1673-1733), arrived at the conclusion that the unknown force is equal to the product of the planet's mass (in this case the moon) multiplied by an already known quantity, gravitational acceleration. This force is also equal to weight. By knowing the force it is possible to calculate impressed motion. Thus, Newton verified the validity of Kepler's laws. By following the method of utilization of experimental results and mechanical laws Newton discovered the universal gravitational law, generalizing and expanding the application of his theory of motion to the planets (the solar system or the known universe).

Progress in science thereby completely renewed philosophical thought, encouraging discussions about the world system, and showing the method to be followed to discover the truth.

3   Leibniz the Reformer of Rationalism

Leibniz commenced his investigations with a profound knowledge of mathematics, law, scholastics, as well as a special preference for logic and theological meditation which he did not separate from science (Ross 2001). To the contrary, he tried to unify science and Christian faith. In this context it is important to emphasize the question of the existence of God, the most central problem of Catholic theology, for Leibniz and Descartes. The central contrast between the two philosophers is the degree of theocentrism in their respective philosophies. Leibniz's theocentrism was absolute, while Descartes was only theocentric as far as his philosophy of Self was concerned. Both philosophers, with their encyclopedic spirits, used scientific and empirical methods to prove the existence of God, notwithstanding their rationalism.

Leibniz reacted against Locke's (1632-1704) empiricism (Feyrabend 1970) and reinforced Cartesian rationalism. However, he recognized the importance of experience for knowledge, but what was essential to him was the reasoning power which organized thought.

His passion for logic was expressed in his desire to understand everything. Everything could be understood. He believed that there was perfect agreement between thought and reality. As a consequence the method to be followed in order to investigate reality was logical deduction and geometrical reasoning. The logic of necessity and according to him another logic, the logic of the probable, were the same as the logic of the truth in the moral sciences and in history. All this reasoning about logic and methodology led him to a kind of hierarchy in science. Initially it was necessary to study the nature of man and medicine. Second, the history of mankind, and finally the technics of arts, the connection between theory and application. Poetry did not appear in his considerations.

Nature for him was the first objective of study (Leibniz 1994). The world could not be explained by a fatal or arbitrary mechanism. It was submitted to a need which surpassed logic and geometry, this need being of a metaphysical order: it was the result of choosing wisdom.

In his critique of Bacon (1561-1626), Descartes, and the mechanists in general, in his *Principles of the Nature and Grace*, he reestablished the idea of final causes. Neither being nor the world were reduced, as in Descartes' doctrine, to an extension, rather they were force, energy, mind, perception and life. From inanimate things to animal species, on to man, nature produced a gradual and continuous effort to achieve consciousness. But the world is not only matter and motion. It is mainly energy which belongs to it and is conserved. In fact, in his vision these forces are spiritual forces which are in 'pre-established harmony,' according to the will of God, and build the best of the possible worlds (Schuster 1980).

4   Leibniz and the Scientific Revolution

Seventeenth century thought is characterized by the reaction against Aristotle (384-322 BC), not the Aristotle of antiquity, but Aristotle as seen through the eyes of the medieval scholastics. In this reaction there appeared the mark of what has come to be called empiricism, the attitude of the new science which appeared in the Renaissance and a characteristic which persists to the present among a certain school of thinkers. However, not every reaction against medievalism was of an empirical nature. Descartes' philosophy, for instance, was a typical product of the age and a direct result of this fight against scholasticism. Indeed, in some aspects it was the direct antithesis of empiricism.

The seventeenth century Scientific Revolution is a theme that continues to attract the attention of many historians of science. These portray this process as occurring roughly in the following sequence: Copernicus' (1473-1543) reformulation of Ptolemy's (100-170) solution of the problem of planets with the need to restore their lost harmony; Kepler and Galileo's acceptance of its realistic proposition; based on this perspective the development of mathematical tools to study the heavens; the mathematization of free fall and projectile motion to confirm the realistic basis of Copernicanism (Cohen 2001); and the development of a new inertial conception of motion, associating an abstract idealized concept of nature, linked to empirical and artificial means of experiment (Cohen 1994, 21-147).

This revolution was mainly a revolution in scientific method which took place during the Renaissance culminating with Newton's *Principia* (Cohen 1978). It had a double aspect: 1) first, scholastic notions of essence and final causes were abandoned and a new explanation of phenomena was sought in efficient causation and the assessment of quantitative change. Changes were to be explained solely in terms of matter and motion and the mechanical character of natural phenomena was expressed in quantitative terms. Measurement was the instrument of scientific discovery. 2) Second, it became clear that the discovery of efficient causes and the assessment of quantitative change could be made only by means of observation and experiment and not any more by argument from first principles. This was the only fruitful method for the discovery of the structure of the natural world and the interaction of its parts, the details of which could not be explained by any general principles.

Descartes' metaphysics sets out the first of these two assumptions of Renaissance science, while Locke's epistemology is the theoretical expression of the second presupposition.

In relation to the assessment of quantitative change, as is well known, the progress in infinitesimal calculus developed independently by Newton and Leibniz during the Scientific Revolution was a fundamental part of this and was crucial for the transformations towards a new physics and the sciences of engineering (Elster 1975), as we will see later (Leibniz 1995). The capacity to solve old and new problems as well as to create modern new branches of knowledge, also demonstrated the power of the scientific method.

With respect to the development of infinitesimal calculus, there emerged in 1682 the great scientific review *Acta Eruditorum*, in which Leibniz would publish the results of his scientific investigations of the foundation of this new branch of mathematics. 1682 saw the first publication of his work in *Acta Eruditorum*. In 1684 what is considered his first work on differential calculus was published: *A new method for maximum, minimum and tangents*. In this paper Leibniz used for the first time the symbol of differential d and listed the rules for the differentiation of addition, subtraction, product and quotient, the chain rule, the second differentiation, the method of separation of variables for solving differential equations. Furthermore, Leibniz abandoned the expression *methodus tangentium directa* and adopted the term *differential calculus*.

Two years later, in his essay *De geometria recondita*, Leibniz presented the fundamental rules of integral calculus, indicating the inverse character of the integration and differentiation operations and also for the first time using the symbol $\int$, although he used the expression *methodus tangentium inversa o calculus summatorius*. The term integral was first used in 1690 by Jacob Bernoulli (1655-1705), while in 1698 Leibniz and Johann Bernoulli (1667-1748) agreed with the denomination of *Calculus Integralis*.

The second fundamental step, not in chronological order, taken by Leibniz to produce a revolution in scientific thought in the seventeenth century was his dynamics (Leibniz 1991). His theory of motion can be considered to have been occurred before his period in Paris, when he developed an abstract mechanics in his *Theoria motus abstracti* (1671), as well as his *Hypothesis physica nova*, which also appeared in 1671. These works implied a hypothetical physics consisting of complex structures and a very creative preparation. In addition to these theoretical questions, Leibniz took into account the empirical laws governing collisions between bodies proposed by Huygens, Wallis, Wren (1632-1723) and Mariotte (1620-1684).

Notwithstanding the above mentioned works from 1671, it seems that Leibniz had adopted mechanical philosophy some years before, around 1661. In a letter to Nicolas Remond, he wrote:

After having finished the trivial schools, I fell upon the moderns, and I recall walking in a grove on the outskirts of Leipzig called the Rosental at the age of fifteen and deliberating whether to preserve substantial forms or not. Mechanism finally prevailed and led me to apply myself to mathematics.

Although none of Leibniz's writings exist from the years immediately following his adoption of mechanism, there is much evidence that in the late 1660s, he studied a wide range of mechanist philosophers. The early influence of natural philosophers such as Pierre Gassendi (1592-1655) and Thomas Hobbes (1588-1679) appears in his *Theoria motus abstracti*, where he introduced the Hobbesian notion of *conatus* which he describes as the beginning and the end of motion that seems a tendency to motion in a particular direction.

Leibniz, in his period in Paris from 1672 to 1676, worked hard to reconcile the above mentioned laws with theorems about the forces of percussion and with the principle of momentum conservation postulated by Descartes in 1644 (Descartes 1997). With this step, trying a new synthesis of physics, Leibniz announced a new principle of conservation (*vis viva* conservation) as shown by Michel Fichant, involving a systematic reform of dynamics. This was done in 1678 in his *De corporum concursu* (Leibniz 1994, 71-171).

These results only appeared and became known in 1686 with the publication of *Brevis demonstratio erroris memorabilis Cartesii*. Thus, a complex theoretical structure was built to reinforce the emerging new science. His further works *Dynamica de potentia* (1689), *Specimen Dynamicum* (1695), and *Essay on tardive dynamics* (1700) meant a new step in Leibnizian dynamics. These works as a whole represented one of the more impressive accomplishments of modern science.

5 The New Conceptual Framework of Dynamics

Leibniz's initial scientific project, which appears in his 1671 publications, *Theoria motus abstracti* and *Hypothesis physica nova*, was characterized by many imperfections. Frequently these are used as demonstrations of the new theories. However, it was in his period in Paris, from 1672 to 1676, mainly under the influence of Huygens (CNRS 1982), that he begun to develop modern mathematics, thereby founding infinitesimal calculus. In this period he was also influenced by the experimental philosophy of Bacon which implied the adoption of empiricist methodologies, strongly inspired by Boyle (1627-1691). Following the same empirical approach, he adopted the inertia principle of Hobbes and his notion of *conatus* as a kind of embryonic motion and a concept closely related to acceleration, as we will see below.

Also in 1671, Huygens, Wallis and Wren presented mathematical models representing the laws of collision, drawing on empirical investigations contradicting the abstract laws presented by Descartes in his famous *Principia philosophia* (1644). Clearly it indicated that geometrical models represent natural laws only if based on experience. Consequently, Leibniz produced *Phoronomia elementalis* starting from *a priori* definitions and abstract concepts but by means of a combinatory approach which could result in a demonstrative mechanics. It was in this context that he adopted the notion of *conatus*, a Hobbesian inheritance which he modified with the help of the Cavalieri's (1598-1647) indivisibles. In his concept, *conatus* is an indivisible and can even behave with a plurality of internal determinations for a given instant. *Conatus* can be algebraically added or subtracted but cannot be applied to circular motion. According to Martial Gueroult (Gueroult 1967), *conatus* represents a difference of velocities, that is the increment of velocity, during an infinitesimal instant. Thus, algebraically *conatus*, the differential of velocity, is the product of the infinitesimal variation of velocity per unit of time multiplied by time element (*adt*), where *a* is the acceleration.

In the Leibnizian view, *conatus* is a postulation very close to the point of forces intrinsically discernible in its series effect. In order to take into account the cohesion and the elasticity of bodies, Leibniz developed another theory as an auxiliary postulation which is called now an *ad hoc* theory. This approach avoids a metaphysical solution to reconcile the *Phoronomia elementalis* with the physical models. Hence, Leibniz conducted his investigation using a system of postulates which led to a unique hypothesis with the capacity to explain the new order of phenomena.

It is this compensatory strategy which is used in *Hypothesis physica nova* (1671) in order to try to integrate the elements of his physics into the *Theoria motus abstracti*, in which the following propositions are denied: the Cartesian principle of the conservation of the quantity of motion, the empirical laws of collision, the laws of angular reflections, as well as the relative cohesion of bodies and their determinations in circular motion.

Until this stage of his investigations, Leibniz also considered the formulation of hypotheses as an essential part of his theoretical constructions. These hypotheses were to be replaced by the building of analogic models in complete coherence with geometry and which did not contradict the physical properties which the models tried to explain. In spite of many difficulties and some explanatory weakness that can be found in *Hypothesis physica nova*, these methodological characteristics need to be emphasized.

Leibniz made his first scientific synthesis in 1671, which would be radically reviewed when he also carried out what he called his *reformatio* of mechanics (Leibniz 1994). At the beginning of 1676, towards the end of his time in Paris, this was done in his *De corporum concursu*, when Leibniz tried to examine the problem of the laws of motion in depth, especially collisions between bodies, in an attempt to move beyond the relativistic point of view expressed by Huygens and Mariotte with respect to theoretical mechanics based on geometry.

Leibniz believed that he knew the key to reconciling empirical laws and the *a priori* principle of the quantity of motion conservation similar to Descartes' work. This key was the principle of equivalence between total cause and complete effect. Leibniz then intended to combine both equivalents in terms of both definitions of cause and effect which would permit a single uniform measure of driven force for all possible cases to be obtained. He believed that this measure would be the product of mass by velocity.

In this sense *De corporum concursu* is a unique work from the epistemological point of view. Leibniz then adopts a systematic 'deduction' of the laws of collision in order to establish conformity with the Cartesian principle of quantity of motion conservation; he compared the theoretical calculations with the deduction of results from an experience based on pendulum properties to measure their displacements effects under the collision of unequal masses.

Initially Leibniz reformulated mechanics using a hypothetical principle of conservation, which considered that there is a fall in height equivalent to the product of the mass multiplied by the velocities of the bodies previously in collision. He also believed that if the explanation was constructed according to the equivalence postulation between total cause and complete effect, as well as the methodological rule of continuity, the theory would work without a proper conceptualization of the underlying forces. This aspect guided Leibniz to reform mechanics by developing a theoretical system of concepts and arguments which maximized the coherence and functionality of this new mechanics. However, some people only see it as a subtle differentiation between concepts on the frontiers of science and metaphysics (Garber 2009).

According to Duchesneau (Duchesneau 1994, 147-258) when Leibniz produced his first global formulation of metaphysics, this represented the official birth of reformed mechanics. This can be seen in some articles in his *Brevis demonstratio erroris memorabilis Cartesii et aliorum circa legem naturalem* (1686), while the same arguments appear in *Discourse of metaphysics*, written in the same year. The context is given by concerns with the laws of nature.

Brevis demonstratio is characterized by the opposition of Leibniz's reformed dynamics and Cartesian science to its argumentative structure. The turning point towards dynamics appears in *Phoranomous* (1689), where Leibniz applies the equivalence between total cause and complete effect. He is also mainly concerned with a dynamic science explained by action and force. This dynamic structuration organized as a true science is completely revealed in his *Dynamica de potentia* (1689-1690). In this great demonstrative synthesis, definitions and heuristic principles allow the conception of multiple theoretical constructions with the capacity to take into account empirical laws. Thus, Leibniz follows an *a priori* way to demonstrate the theorem of the conservation of driven action.

In 1695, *Specimen dynamicum* revealed another aspect of the Leibnizian method, the theoretical construction of his dynamics. In appearance this text follows an *a posteriori* path for demonstrations. In addition the typology of primitive and derivative forces allows the integration of formal components of force: *conatus*, *impetus*, *vis viva*.

In *Essay de dynamique tardive* (1700) we can see the last phase of the theoretical structuration of Leibniz's dynamics. He shows the integration of the relative conservation principles in the absolute principles of conservation. The synthesis of these models is based on architectural principles.

During 1690-1700 and 1700-1710 the question of a system of *a priori* proofs was emphasized within Leibnizian dynamics.

6 Leibnizian Origins of the Sciences of Engineering

6.1 Differential and Integral Calculus

The first work where Leibniz presents the rules for the differentiation of functions is *Nova methodus pro maximis et minimis, itemque tangentibus, quae nec fractas nec irrationals quantitates moratur et singular pro illis calculi genius* which can be translated into English as *A new method for searching for the maxima and minima including tangents, a method which does not impede fractional expressions, following original applied calculus* (Leibniz 1684, 4-9). It was published in *Acta Eruditorum* in October 1684.

These rules are as follows:

Multiplication $\quad d(xv) = xdv + vdx$

Division $\quad d(v/y) = \frac{vdy - ydv}{yy}$

Power $\quad d(x^a) = ax^{a-1}dx$

Root $\quad d(\sqrt{x}) = 1/2\, x^{-1/2}\, dx$

In relation to integral calculus, Leibniz also published in June 1686 in *Acta Eruditorum* his *De geometria recondita et analysi indivisibilium atque infinitorum*. Its main objective was to provide, using his new calculus, the way to calculate areas of figures, which in fact is the inverse problem of tangents. In modern terms the tangent of a function *f(x)* is defined by a new function $f'(x) = dy/dx$. Then, the area under this new function is $y = \int f'(x)\, dx$.

6.2 The Principle of *Vis Viva* Conservation

If we look at the reformation of dynamics, developed by Leibniz in January 1678, we are mainly concerned with how we can characterize his use of the expression mv^2, which Michel Fichant designates as "the nucleus of a complete doctrinal set" and its consequences for metaphysics. After this the text *De corporum concursu* written in January 1678 gives a definitive answer to Leibnizian dynamics.

Eight years before the *Brevis demonstratio erroris mirabilis Cartesii* and the *Discourse of metaphysics,* Leibniz arrived at his canonic definition of force, as well as the formulation of his conservation principle. As we know, his main motivation was to establish the rules governing the collision of two bodies which implied the substitution of the product mv by mv^2 as a measurement of force (Westfall 1971).

De corporum concursu is a manuscript whose pages are numbered by Leibniz from one to ten (Scheda prima to Scheda decima). The first nine pages date from January 1678, while the tenth bears the date of January and February 1678, thus suggesting that the manuscript was rewritten. In addition two sheets have appendices which are designated as Scheda secundo-secunda and Scheda secundo-sexta, respectively. These additions show decisive points of rupture in his argumentative chain, while the insertions in fact prepare a decisive turning at the beginning of Scheda octava.

Schedae 8 and 9 explain how a coherent solution for the collision problem was obtained, and emphasized that this problem was previously involved in conceptual contradictions. According to Leibniz the problem was completely solved by starting from the three principles of conservation: adding the quantities mv^2, total direction, and appearance or relative velocity. In addition, the tenth sheet from January-

February 1678 increases the field of investigation, including non-frontal collision.

Nonetheless, according to Fichant, it is possible to describe the logical propositional structure underlying the process of *vis viva* construction as follows: if the combination of the rule of translation of the center of gravity with the constancy of relative velocity is true, then, it is not (always) true that $\sum m|v|$ is constant. Although the identification of the quantity effect which measures force by means of the height of the lifting or falling thereby provides a new formula for the forces of conservation according to mv^2. Distance conservation, or the relative velocity after collision, provides the second formula as a starting point according to which the theory of elastic direct collision can be developed.

Leibniz uses several other arguments to contradict the Cartesian quantity of motion that is conserved. Using for instance the principle of the equality of cause and effect, he postulates that if the quantity of motion is conserved, one could build a perpetual motion machine, a machine that would create the ability to do work out of nothing at all. Leibniz is concerned not only to show that quantity of motion differs from force but, that quantity of motion is not conserved. This, it is the quantity mv^2 that correctly measures force and mv^2 is conserved in the world.

Using Figure 1 below, Leibniz argue that if two bodies A and B have different quantities of motion, their size multiplied by the square of their speeds will be equal. It is easy to generalize this and show whenever they have equal force, the size multiplied by the square of their speed will be equal and that whenever this is violated, the ability to do work will either be gained or lost, violating the principle of the equality of cause and effect. (See S. D., part I, par. 16, G. M. VI 244-45: A. G. 128)

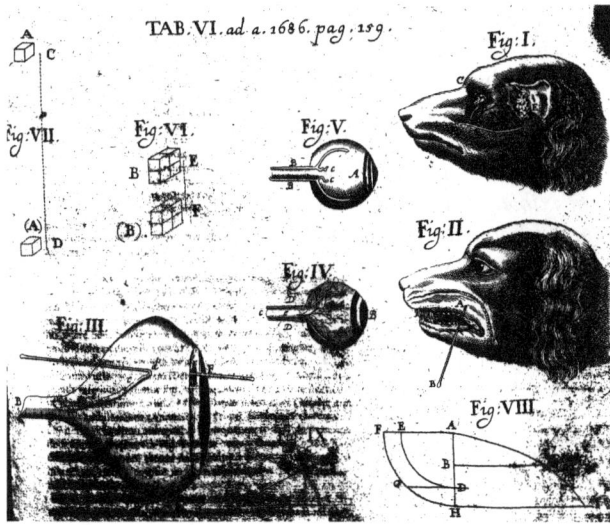

Fig. 1. Leibniz's drawing on *vis viva* proof (Leibniz 1686 [*Brevis demonstratio erroris memorabilis Cartesii, et aliorum circ a legem naturalem, secundum quam volunt a Deo eamdem semper quantitatem motus conservari; qua et in re mechanica abutuntur*, Acta Eruditorum, pp. 161-163], Table. VI, p. 159, Figs. VI-VII, on the left). Image source: Google books – Public domain.

A more complete discussion of this problem can be found in the article *Leibniz: physics and philosophy* (Jolley 1995).

6.3 Mechanical Properties of Materials

Leibniz, in his *New Demonstration on Strength of Solids* (Leibniz 1684, 15-20) published in 1684, wrote:

The mechanical science seems to have two parts: one regarding the power of action or motion, and the other regarding the power of resisting, with respect to the solidity of bodies. The last one was studied by a few number of people. As we know, the same postulation is made by Galileo in his *Discorsi* (1638), where the two new sciences are the dynamics and strength of materials, corresponding to part one and part two mentioned by Leibniz, respectively (Galileu 1988).

With regard to Leibniz's ideas about the mechanical properties of materials, it can be noted that initially he was in agreement with Pierre Gassendi and the atomists, but his mature writings clearly reject atomism. In spite of this, he presented a variety of arguments against the existence of atoms. Sometimes arguing a kind of principle that two things in world cannot be perfectly similar.

In 1690, Leibniz presented a different kind of argument, using reduction to absurdity to analyze the idea of atom. These ideas against atoms began to be developed in *Confessio naturae contra atheistas* (1699) and in the letter to Jacob Thomasius (1622-1684) in defence of Christian faith also published in the same year. He argued that atoms can arise in any form and proposed the existence of a cubic atom as well as two prismatic triangular atoms, which when combined form a undistinguishable cubic atom. This argument was called the principle of sufficient reason. In addition to this reasoning, Leibniz used the principle of continuity to refute the idea of atom. According to this principle, changes in nature cannot happen by means of jumps. In Part II of the SD he wrote:

If we were to imagine that there are atoms, that is, bodies of maximal hardness and therefore inflexible, it would follow that there would be a change through a leap, that is, an instantaneous change. For at the very moment of collision the direction of the motion reverses itself.

In addition to the property of cohesion and solidity of bodies, Leibniz also discussed the elasticity of matter saying that "no body is so small that it is without elasticity". (See SD, pt II, par.3 GM VI 249: AG 32, cf Dynamica, GM VI 491). However, he also argued that if all bodies were elastic, then they all must be made up of small parts. For him and his contemporaries, elasticity is not a basic and fundamental property of matter, but a characteristic that must be explained mechanically by means of the configuration of parts that compose a given body.

Leibniz wrote in Specimen Dynamicus:

Elasticity ought always to derive from a more subtle and penetrating fluid, whose movement is distributed by the tension or by the change of the elastic body.

He also stated:

And since this fluid itself ought to be composed, in turn, of small solid bodies, themselves elastic, one well sees that this replication of solids and fluids goes to infinity.

Using the principle of continuity, we can conclude that the possibility of division of matter can continue to infinity. Some important consequences of the property of elasticity can be emphasized. One interesting implication is the case of two elastic bodies in collision. Initially both bodies become compressed and deformed, by virtue of their elasticity, and then, they return to their original shapes, and thus push themselves off from another.

In *Specimen Dynamicus*, Leibniz argued:

The repercussion and bursting apart [of a body in impact] *arises from the elasticity it contains, that is, from the motion of the fluid ethereal matter permeating it, and thus it arises from an internal force or a force existing within itself.*

Finally, the properties and the elastic behavior of solid bodies in Leibniz's vision support and guarantee his metaphysical principle of continuity. It is only satisfied due to the physical properties of the world, sometimes appearing like a ghost from his early thought.

6.4 Calculating Machines

In 1671, Leibniz designed a calculating machine called the Stepped Reckoner. It was actually first built in 1673, based on Pascal's ideas and did multiplication by repeated addition and division by repeated subtraction. Two prototypes were built; today only one survives in the National Library of Lower Saxony in Hanover, Germany.

Hardware: The machine has the following characteristics:

- Structure: Two attached parallel parts, an accumulator section to the rear and an input section to the front. There is also an indicator and a control to reset the machine.
- Dimensions and materials: 67 cm long, 27 cm wide and 17 cm high. Polished brass and steel mounted in a big oak case with dimensions: 97 cm x 30 cm x 25.
- Operations: Add and subtract an 8 digit number to/from a 16 digit result; multiply two 8 digit numbers to obtain a 16 digit number; divide a 16 digit number by an 8 digit divisor.

Software: Leibniz was one of the first mathematicians, after Ramon Llull (1232-1316) and almost at the same time with Athanasius Kircher (1602-1680), who conceived a logical device for calculations. In 1666, he published his first book, in fact his qualifying thesis in philosophy: *On the Art of Combinations*, inspired by the *Ars Magna* of Ramon Llull.

Leibniz was also one of the first logicians who realized the importance of the binary system. As we know, the inventor of this was an obscure Indian author who wrote in about 300 BC the *Chandahsastra*, or *Science of Meters*. Leibniz discovered that computing processes can be done much easier with binary number coding in his treatises: *De Progressione Dyadica* of 1679 and *Explication de l'Arithmetique Binaire*, of 1703. In this sense, Leibniz was a precursor of Charles Babbage (1791-1871; Rosenberg 1994, 24-46).

The modern binary system was fully documented by Leibniz in the seventeenth century in the above cited 1703 article. He used 0 and 1 in a similar manner to the system used today. The binary arithmetic developed by Leibniz looked like another numeral system. Summation, subtraction, multiplication and division could be performed on binary numerals. All arithmetic operations in the binary system are reduced to binary numeration. The summation of two single-digit binary numbers is relatively simple:

$$0 + 0 = 0$$

$$0 + 1 = 1$$

$$1 + 0 = 1$$

$$1 + 1 = 10$$

In Leibnizian logic the ones and zeros also represent true and false values or on and off stages. It took more than a century before George Boole (1815-1864) published his Boolean algebra in 1854 with a complete system that allowed computational processes to be mathematically modeled.

7 Sciences of Engineering and their Relations with Leibnizian Sciences

7.1 Formalization of the Laws of Dynamics

An important consequence of the development of differential and integral calculus was the formalization of mechanical laws. This formalization was made in two different ways, with Euler and Lagrange (1736-1813) being the most important figures for both of these. The first manner was the different form of Newton's second law presented by Euler (1707-1783) in 1752. According to accepeted historical accounts, before the first formalization, the pioneering work of Pierre Varignon (1654-1722) represented an important step towards the formalization of Newton's laws. Starting with Leibnizian algorithms Varignon built the concepts of instantaneous velocity and acceleratory force. These new results appeared in two memoirs to the Royal Academy. The first on July 5 1698, and the second on September 6 of the same year:

a) Règles générales pour toutes sorte de mouvements de vitesses quelconques variées à discrétion.
b) Application de la règle générale des vitesses variées, comme on voudra, aux mouvements par toutes sortes de courbes, tant mécaniques que géométriques, d'où l'on déduit encore une nouvelle manière de démontrer les chutes isochrones dans la cycloide renversée.

In 1750, Euler finally realized that the principle of linear momentum applied to mechanical systems of all kinds, whether discrete or continuous. His paper is entitled: *Discovery of a new principle of mechanics*, published in 1752 as mentioned above, in which he presents the equations:

$$F_x = Ma_x; \qquad F_y = Ma_y; \qquad F_z = Ma_z$$

where the mass M may be either finite or infinitesimal.

The second manner was Lagrange's formalization of Analytical Mechanics, published in 1788 (Lagrange 1788). Newtonian mechanics underwent a profound change using the formalization carried out by Lagrange. With Lagrange's work rational mechanics reached what the Cartesians had long desired, becoming a branch of pure mathematics. Some years before, at the beginning of the eighteenth century, differential and integral calculus had been sufficiently developed, becoming a useful tool in the resolution of a series of problems in physics and mathematics. Hence, the conditions for the application of the new analytical and algorithmic procedures to the science of motion were provided.

In 1736, Euler wrote the first treatise on the mechanics of the material point, known as *Analytice exposita*. Some years later d'Alembert explained his mechanical philosophy in a preliminary discourse to his famous 1743 *Traité de Dynamique*. D'Alembert (1717-1783) had a different conception of force, as a derived notion, in complete disagreement with Newton (D´Alembert 1921). In addition, he attributed the fundamental importance to the concept of mass, as well as all pure kinematical elements.

Lagrange's reading of Euler's treatise *Methodus inveniendi lineas curvas maximi minimine proprietate gaudentes*, published in 1744, led Lagrange in 1755 to discover his 'method of variations' which Euler had sought in vain. Lagrange communicated his method to Euler who recognized its enormous importance. At the same time, Lagrange drew his attention to the principle of least action, initially formulated in a vague way by Maupertuis (1698-1759) in 1744. In 1759 he informed Euler that he had almost finished a treatise concerning his method of variations and deduction of mechanics starting with this principle. Euler apparently showed no interest in the treatise.

Before the publication of *Analytical Mechanics*, Lagrange published two works in Turin in 1760: *Essai d'une nouvelle method pour determiner les maxima et les minima des formules integrales indefinies* and *Application de la methode exposée dans le mémoire precedent a la solution de differents problems de Dynamique*. Also in 1760 Euler published *Theoria motus corporum solidorum rigidorum*, which was reviewed and expanded by his son Johann Albrecht Euler (1734-1800).

Lagrange's *Analytical Mechanics* (Lagrange 1778, 1-24) was published in 1788, crowning a series of his writings and other important contributions previously developed by d'Alembert and Euler. In relation to the new mechanics, Lagrange stated: *We already have various treatises on mechanics, but the plan of this one is completely new. I propose to reduce the theory of this science, and the art of solving problems related through to formulae, by means of general formulae, and in this way simple development gives all the equations necessary for the solution of each problem. I hope that the method which I have just developed will achieve this objective, not leaving anything wanting.* And he continued: *Figures cannot be found in any part of this book. The methods outlined here do not need constructions, nor geometrical or mechanical reasoning, but rather only algebraic operations, subject to a regular and uniform pace. Those who love analysis will, with pleasure, see mechanics as a new branch, and will be grateful to me for thus having extended its domain.*

7.2 Mechanics of Elastic Bodies

The ideas of elasticity of bodies associated with the development of infinitesimal calculus provided the necessary foundations for several disciplines of engineering: strength of materials, theory of elasticity, machine elements, etc. One important development using infinitesimal calculus was made by Jacob Bernoulli, it concerned the shape of the deflection curve of an elastic bar and in this way he began a remarkable chapter in mechanics of elastic bodies. As is well known, Galileo and Mariotte investigated the strength and solidity of beams, but it was Jacob Bernoulli who calculated their deflections.

John Bernoulli, the younger brother of Jacob, was the first to formulate the principle of virtual velocities in a letter addressed to Varignon (Oliveira 2013). This principle became a fundamental tool for studying the equilibrium of systems, even those with elastic properties. Another decisive step in this new field of mechanics was taken by Daniel Bernoulli (1700-1782), son of John Bernoulli, and his pupil Leonhard Euler. Daniel is best known for his famous book *Hydrodynamics* (Bernoulli 1738, 1-17), but it was he who suggested to Euler to apply variational calculus to derive the equations of elastic curves (Bernoulli 1738).

Another important contribution to the mechanics of elastic bodies was made by Coulomb (1736-1806). In 1784 he published a memoir on torsion. He studied this problem from the theoretical and experimental point of view. What he did was determine the torsional rigidity of a wire by observing the torsional oscillations of a metal cylinder suspended by it. He assumed that the resisting torque or a twisted wire was proportional to the angle of twist, obtaining a linear second order differential equation as a torsional oscillator where the period of oscillation was $T = 2\pi\sqrt{I/n}$ and the torsional moment of inertia of the cylinder was I.

At the time of the publication of his masterpiece *Analytical Dynamics* (1788), Lagrange became interested in elastic curves. Looking at the correspondence between him and Daniel Bernoulli, we can see that the problem of lateral vibrations of elastic beams drew his attention. He also investigated the corresponding differential equations. The most important and best known finding for the theory of elastic bodies appeared in his memoir *Sur la Figure des Colonnes*, where he discussed the problem of an elastic bar with hinges at the ends under the effect of an axial compressive force P. He proposed then a solution in the form:

$$C\frac{d^2x}{dx^2} = -Py$$

He showed that the solution $y = f\sin\sqrt{P/C}\,x$ satisfied the end conditions only if $\sqrt{P/C}\,I = m\pi$. It followed that for a very small bending the value of $P = m^2\pi^2 C/l^2$ for m an integer is the solution for an infinite number of buckling curves.

These differential equations for studying the elastic deflections of bars under lateral loads were again used and his application generalized by Navier (1785-1836). In 1826, the first edition of Navier's book on the strength of materials was published. Introducing the moment of bending M, the Young modulus E, and the moment of inertia I of the cross sectional area with respect to neutral axis, he generalized the Euler equation:

$$EI\frac{d^2y}{dx^2} = M$$

The above equation can be used for any kind of lateral load. Its integration to obtain the deflection curve $y = f(x)$ can be done analytically or graphically. Before Navier this equation was used only for simply supported beams.

7.3 Application of the Concept of Living Forces (mv^2) to Machines

One application of the concept of living forces (*vis viva* or kinetic energy) was an important step towards the development of the machines sciences. According to Navier, the first study where we find the principle of the conservation of living forces applied to machines was *Hydrodynamics*, published by Daniel Bernoulli in 1738. This important achievement was ignored by physicians and engineers for several decades. Only after the memoir of Claude Borda (1733-1799) entitled *Memoir on the Hydraulic Wheels* appeared in 1767, did the principle of the conservation of living forces start to be applied to machines. He was the first to apply this principle to hydraulic wheels. Some years later, in 1781, Coulomb published his *Theoretical and Experimental Considerations on the Effect of Windmills* using the same principle (Coulomb 1821).

According to Navier, Borda and Coulomb's contributions were fundamental steps and showing remarkable progress in regard to Bernoulli's *Hydrodynamics*. Navier also emphasizes that there was a need for the creation of a general theory involving that principle with the capacity to calculate machine efficiency. It is exactly in this context that he affirms that this theory was created by Lazare Carnot (1753-1823; Gillispie 1978, 5-138; Gillispie and Pisano 2014) in his *Fundamental Principles of Equilibrium and Motion*, in 1803. He also attributes to Carnot (Charnay 1990) the general demonstration of the theorem which calculates the loss of living forces due to collisions between non-elastic bodies (hard bodies).

Carnot then studied the problem of the transformation of work in motion by considering all the parameters involved. From this viewpoint, this meant established convenient variations of the terms of the quantity FVT, i.e., the moment of activity, later denominated in Coriolis' work (1792-1843; Coriolis 1829, 1-34). If time is the most important parameter and it should be minimized, the effect must be produced in a very short time. It is possible to generalize this reasoning for the case of a system of forces, for instance; if we have the forces F, F', F'' with the velocities V, V', V'', acting during the times T, T', T'', respectively, then one reads:

$$FVT = F'V'T' = F''V''T'' = PH$$

If the motion of each one of the forces is variable, we will take the quantity: $\int (FVdt + F'V'dt' + F''V''dt'')''$, or, if we have the forces directions with respect to velocities, one has:

$$\int [FVdt \cos(F \wedge V) + F'V'dt' \cos(F' \wedge V') + F''V''dt'' \cos(F'' \wedge V'')]$$

This is the definition of work done by all forces.

The quantity *PH*, the effect to be produced by a machine, is, by Carnot called latent living force. Obviously, it presupposes a transformation of living forces from latent to actual one. If we call *M* the mass of the weight *P*, and *V* the velocity correspondent to a height *H*, one reads:

$$PH = {1}/{2} MV^2$$

The relation above is always valid for any variation of the effect. When Carnot presented the above equation he mentions Leibniz as being its author and says that only after Leibniz were the forces acting in bodies in motion calculated in a different form than the equilibrium situation.

After these achievements, which made an important contribution to the development and application of the conservation of living forces principle, Navier made some remarks about and additions to Belidor's (1698-1761) *Hydraulic Architecture*. He demonstrated the principle of the conservation of living forces in a single mass but also generalized this result to a system of n particles using the d'Alembert principle.

However, a great evolution in the living forces principle occurred with the publication of Coriolis' book *Du Calcul de l'Effet des Machines* in 1829. This work is considered one of the most important nineteenth century works in mechanical engineering. The term *work* was coined and the constant ½ was incorporated in the expression of living forces in this book. A great advance in machines science was achieved with this book which did not consider the machine as a conservative system. The balance of living forces throughout the entire system was equal to global work also spent or produced by the system.

7.4 Computing Sciences

Boolean algebra, the basis for computer operations was introduced by George Boole in his first book *The Mathematical Analysis of Logic* (1847), and later more fully described in *An Investigation of the Laws of Thought* (1854). In this algebra the variables are the truth values true and false, usually denoted by 1 and 0, respectively. Boolean algebra has been fundamental in the development of digital electronics and is provided for all modern programming languages, but also is used in set theory and statistics. Logic sentences that can be expressed in classical propositional calculus have an equivalent expression in Boolean algebra.

Hardware: Charles Babbage, an English mechanical engineer, was the first to conceive a programmable computer. He conceptualized and invented the first mechanical computer in the early nineteenth century, a difference engine, designed to aid navigational calculations. In 1833 he invented a more general machine, an analytical engine. The input of programming and data was designed to be provided to the machine using punched cards, an influence of the Jacquard machine. Babbage's engine incorporated an arithmetic logic unit, a control flow system, and an integrated memory, in such a way that this machine could be considered a general-purpose computer that could be described in modern terms as a Turing-complete machine.

Software: The principles of the modern computer were first described by Alan

Turing (1912-1954), an English scientist, in 1936 in his paper *On Computable Numbers* (Petzold 2008). He used the results of Kurt Gödel's (1906-1978) 1931 theorem, replacing Gödel's formal language with formal and simple hypothetical devices, known as Turing machines. He proved that these machines could perform any conceivable mathematical construction if it were represented by an algorithm. However, there remained the so-called halting problem for Turing machines which was undecidable in general, because it is not possible to decide algorithmically whether a given Turing machine will ever halt. Another fundamental work to computer sciences was the manuscript: *A Mathematical Theory of Communication*, published by Claude Shannon (1916-2001) in which it was found how binary logic could be used to program a computer.

8 Concluding Remarks

Our main purpose here was to show the great importance of Leibniz's work for the sciences of engineering, with the starting point that dominant Newtonian ideas since the hegemony of Newtonian mechanics has left Leibniz in a secondary position as a founder of mechanics. Furthermore, his modernity is emphasized in the text, while it is also shown that engineering disciplines have been derived from Leibnizian concepts.

Also shown is that interest in Leibniz is relatively recent. Only in the twentieth century did there appear some important studies showing unknown aspects of his work. Some of these were discussed in this text.

The intellectual and scientific context where Leibniz worked is discussed, emphasizing his prominent role in the seventeenth century Scientific Revolution. In this sense, the basis for a true transformation in the scientific conceptual framework was left by Leibniz with the development of differential and integral calculus and dynamics. In addition, the establishment of the conservation of living forces principle, postulating the quantity mv^2 such as the quantity that remains constant in motion instead of the quantity of motion mv as proposed by Descartes. Not only did this concept have great importance in theoretical mechanics with the introduction of ideas which led to energy methods in engineering but it also marked the birth of applied mechanics, as shown with the contribution of Lazare Carnot to a new theory of machines. The vis viva principle of conservation as postulated by Leibniz was an important anticipation to the principle of conservation of energy which emerged in the middle of nineteenth century.

In addition to all these theoretical constructions we can add the pioneering work of Leibniz in binary logic, which was associated with the construction of a calculating machine, which really does make Leibniz an important precursor of modern computing sciences. In the text both contributions are studied, separated in terms of hardware and software.

The developments of infinitesimal calculus made by Newton and Leibniz, provided the necessary mathematical tools which propelled the majority of engineering fields like the formalization of motion laws by Euler and Varignon as well as the algorithm form proposed by Leibniz used to study elastic bodies deformations, as the text emphasizes. These are fundamental contributions to engineering. Finally, Leibniz's ideas for explaining the mechanical properties of materials have influenced modern theories that support methods for solving complex problems of solid deformations that can be also associated with the concept of potential energy in the context of conservative processes.

References

Allard, J. L. 1963. *Le mathématicisme de Descartes*. Ottawa: Éditions de l'Université d'Ottawa.
Alquié, F. 1963. *Oeuvres philosophiques de Descartes*, t.1. Paris: Garnier Frères.
Bernoulli, D. 1738. *Hydrodynamics*. New York: Dover Publications.
Cassirer, E. 1902. *Leibniz System in seinem wissenschaftlichen Grundlagen*. Marburg: Elwert.
Charnay, J. P. 1990. *Lazare Carnot ou le Savant-Citoyen*. Paris: Presses de l'Université de Paris-Sorbonne.
CNRS 1982 (Aa. Vv). *Table Ronde: Huygens et la France*. Paris: Vrin.
Cohen, I. B. 1978. *Introduction to Newton's Principia*. Cambridge–MA: The Harvard University Press.
Cohen, I. B. 1985. *El nacimiento de una nueva física*. Madrid: Alianza Editorial.
Cohen, I. B. 2001. *Revolution in Science*. Boston: The Havard University Press.
Cohen, I. B. 1980. *The Newtonian Revolution*. Cambridge: Cambridge University Press.
Cohen, H. F. 1994. *The Scientific Revolution: A Historiographical Inquiry*. Chicago: The University of Chicago Press.
Coriolis, G. G. 1829. *Du Calcul de l'Effet des Machines*. Paris: Carillan-Goeury.
Coulomb, C. A. 1821. *Théorie des Machines Simples*. Paris: Bachelier Librairie.
D'Alembert, J. L. 1921. *Traité de Dynamique*. Vol.2. Paris: Gauthier-Villars et Cie Éditions.
Descartes, R. 1964-1974. *Principes de la Philosophie*. Vol. IX, II, In: *Oeuvres*. Adam J et Tannery A. Nouvelle présentation par Rochet E, et Costabel P, 11 Vols. Paris: Vrin.
Dilthey, N. 1947. *Leibniz e sua Época*. S. Paulo: Saraiva Editores.
Duchesneau, F. 1994. *La Dynamique de Leibniz*. Paris: Vrin.
Elster, J. 1975. *Leibniz et la Formation de l'Esprit Capitaliste*. Paris: Aubier Montaigne.
Feyerabend, P.K. 1970. 'Classical Empiricism'. In: *The Newtonian Heritage*. R.E. Butts and J.W. Davis Editors. London: Blackwell, pp. 150-170.
Gadoffre, G. 1961. *Descartes' Discours de la méthode*. 2nd Ed., Manchester: The Manchester University Press.
Galilei, G. 1988. *Duas Novas Ciências*. Translated by Pablo Mariconda. S. Paulo, Brazil: Nova Stella.
Garber, D. 2009. *Leibniz: Body, Substance, Monad*. Oxford: The Oxford University Press.
Gilson, E. 1947. *René Descartes, Discours de la Méthode: Texte et Commentaire*. Paris: Vrin.
Gillispie, C. C. 1979. *Lazare Carnot et sa Contribuition a la Théorie de l'Infini Mathematique*. Paris: Vrin.
Gillispie, C. C. 2014. *Lazare and Sadi Carnot. A Filial and Scientific Relationship*. 2nd edition. Dordrecht: Springer.

Gueroult, M. 1967. *Leibniz: Dynamique et Metaphysique.* Paris: Éditions Aubier Montaigne.
Hall, A. R. 1981. *From Galileo to Newton.* New York: Dover Publications.
Huygens, C. 1673. *Horologium Oscillatorium.* Paris: Librairie A. Blanchard.
Jolley, N. 1995. *The Cambridge Companion to Leibniz.* New York: The Cambridge University Press.
Koyré, A. 1973. *The Astronomical Revolution.* London: Methuen & Co.
Lagrange, J. L. 1788. *Mécanique Analytique.* Paris: Éditions Jacques Gabay.
Léger, A. & Sagnac, P. 1935. *La Préponderance Française: Louis XIV (1661-1715).* Paris: Librairie Félix Alcan.
Leibniz, G. W. 1686. Brevis demonstratio erroris memorabilis Cartesii, et aliorum circ a legem naturalem, secundum quam volunt a Deo eamdem semper quantitatem motus conservari; qua et in re mechanica abutuntur [pp. 161-163]. *Acta Eruditorum.* Lipsiæ, prostant apud J. Grossi hæredes & J. F. Gleditschium [...] Anno MDCLXXXII. [-MDCCXXXI.] Table. VI, p. 159, Figs. VI-VII.
Leibniz, G. W. 1694. *Oeuvres concernant le Calcul Infinitesimal.* Traduit du Latin en Français par Jean Peyroux. Paris: Blanchard.
Leibniz, G. W. 1694. *Oeuvre concernant la Physique.* Traduit du Latin em Français par Jean Peyroux. Paris: Blanchard.
Leibniz, G.W. 1991. *Escritos de Dinâmica.* Madrid: Editorial Tecnos.
Leibniz, G.W. 1994. *La Reforme de la Dynamique.* Paris: Vrin.
Leibniz, G.W. 1995. *La Naissance du Calcul Differentiel.* Paris: Vrin.
Newton, I. 1994. *La Méthode des Fluxions et des Suites Infinies.* Translated to French by M. Buffon. Paris: Blanchard.
Oliveira, A. R. E. 2013 *A History of the Work Concept: from physics to economics.* Dordrecht: Springer.
Petzold, C. 2008. *The Annotated Turing.* Indianapolis: Wiley.
Rosenberg, N. 1994. *Exploring the black box.* Cambridge: The Cambridge University Press.
Ross, G. M. 2001. *Leibniz.* S. Paulo: Edições Loyola.
Russel, B. 1900. *The Philosophy of Leibniz.* London: Redwood Books.
Schuster, J. A. 1980. 'Descartes' Mathesis Universalis: 1619-28'. In: Gaukroger S. (ed). *Descartes: Philosophy, Mathematics and Physics.* Sussex: Harvester, pp. 41-96.
Schuster, J. A. 1990. 'The Scientific Revolution'. In: Olby R. C., Cantor G. N., Christie J. R. R. and Hodge M. J. S. (Eds.).*The Companion to the History of Modern Science.* London: Routledge, pp. 217-242.
Schuster, J. 2000c. 'René Descartes'. In: *Encyclopedia of the Scientific Revolution.* Applebaum W., (Ed.). New York: Garland Publishing.
Westfall, R. S. 1971. *Force in Newton's Physics.* New York: Neale Watson Academic Publications.
Westfall, R. S. 1977. *The Construction of Modern Science: Mechanisms and Mechanics.* Cambridge University Press.

Agamenon R. E. Oliveira, Federal University of Rio de Janeiro, Brazil
agamenon.oliveira@globo.com

Logica Mathematica: Mathematics as Logic in Leibniz

Anne Michel-Pajus and David Rabouin

Abstract. On several occasions in the 1690s, Leibniz mentions that he considers algebra and more generally 'universal mathematics' as a kind of 'mathematical logic' (*Logica Mathematica*). While this idea of a 'mathematical logic' has given rise to many interpretations, it is rare for anyone to focus on the texts in which he sets out this 'logic', or even to contextualize them. This paper is an attempt to address this lack. The thread of *logica mathematica* will allow us to revisit a number of classic chapters of Leibnizian exegesis such as the status of the *mathesis universalis,* the role of the *ars combinatoria* and of formal axiomatics in Leibniz's conception of mathematics.

Keywords: Mathematics, Logic, Mathematical logic, *Mathesis universalis, Ars combinatoria, Caracteristica universalis, Speciosa generalis,* Universal Mathematics, Universal algebra, Axiomatics, Logical calculus, Algebra, Fermat little theorem, Whole-parts relationship, Mereology.

1 Introduction

On several occasions in the 1690s, Leibniz mentions that he considers algebra and more generally 'universal mathematics' (*mathesis universalis*) as a 'mathematical logic' (*Logica Mathematica*). It is no surprise to come across this expression under his pen. After all, isn't Leibniz regularly credited as the precursor of mathematical logic in a century where it seems, rather, to have been dormant[1]? Furthermore, isn't he the first known writer to have proposed formal axiomatic systems for various types of calculation? However, this perception, far from being confirmed by a close examination of the texts, seems in fact invalidated by the opening of one of the most elaborated projects dealing with this issue[2]:

> Many people have tried to illustrate Logic by comparing it to a Computation, and Aristotle expressed himself in a mathematical manner in the Analytics. Conversely Arithmetic and Algebra, but most of all the *Mathesis* which is truly *universalis,* could be treated

---

[1] To use the expression of Robert Blanché in his chapter dedicated to the early modern period (Blanché 1970, p. 169).

[2] *Matheseos universalis pars prior* (GM VII, 53-76; circa 1699), quoted as MU.

Anne Michel-Pajus et David Rabouin (2017) *Logica Mathematica*: Mathematics as Logic in Leibniz. In: Pisano R, Fichant M, Bussotti P, Oliveira ARE (eds.), *The Dialogue between Sciences, Philosophy and Engineering. New Historical and Epistemological insights. Homage to Gottfried W. Leibniz 1646-1716. College Publications, London,* pp. 309-330
© 2017 College Publications Ltd | ISBN: 978-1-84890-227-5 www.collegepublications.co.uk

in a logical manner, as if they were Mathematical Logic, so that in fact *Mathesis universalis* or Logistic would coincide with the Logic of Mathematicians[3].

Leibniz clearly mentions here that conceiving logic as a calculation was a relatively common phenomenon in his time—and one which, in his eyes, derived quite naturally from Aristotle's approach to logic. His own motivation, which consisted on the contrary of conceiving calculation as a form of logic, seems to him more original. What follows in the text makes it clear that his aim was to construct parallels between the ingredients of logical reasoning (notions, judgments, proofs, methods) and the ingredients of algebraic reasoning (quantities, relations, operations, methods). Moreover, Leibniz did not equate 'mathematical logic' with *ars combinatoria,* since he argues a few pages on that "so far the subordination of Algebra to the *Ars Combinatoria* or of Specious Algebra to the *Speciosa generalis* has been ignored or neglected" (MU, 61).

These questions are certainly not simple to tackle as the respective positions of algebra, *mathesis universalis* and *ars combinatoria* fluctuate from one text to another. But it is precisely this which seems interesting for us to study: not so much a certain organization of knowledge assumed to be fixed once and for all, but an exploration of the complex relationships between mathematics and logic that Leibniz could bring together under the umbrella of a *logica mathematica*. Here we are especially interested in how calculation can be understood by Leibniz as logic rather than the reverse path on which commentators have usually focused (logic conceived as a form of computation).

Thus the thread of *logica mathematica* will allow us to revisit a number of classic chapters of Leibnizian exegesis such as the status of the *mathesis universalis,* the role of the *ars combinatoria*, the extension of the domain of mathematics from quantity to quality (or 'form'), and the precise role of formal axiomatics in his conception of mathematics. As many mathematical texts of Leibniz are still unpublished or only imperfectly edited, this can only be an exploratory approach. To do this, we shall first briefly recall how Leibniz's *logica mathematica* has been interpreted to date (section 1), before moving on to study the precise occurrences of the term in the corpus (section 2). This will offer us a privileged point of view from which to study two classic themes of this 'logical' approach: the role of *ars combinatoria* (section 3) and the axiomatization of algebra (section 4).

---

[3] MU, 54: *Et quemadmodum multi Logicam illustrare tentaverunt similitudine computi, ipseque Aristoteles in Analyticis Mathematico more locutus est, ita vicissim et multo quidem rectius Mathesis praesertim universalis, adeoque Arithmetica et Algebra tractari possunt per modum Logicae, tanquam si essent Logica Mathematica, ut ita in effectu coincidat Mathesis universalis sive Logistica et Logica Mathematicorum.* Unless otherwise stated, all translations are ours.

2 Mathematical Logic, *Mathesis Universalis* and *Ars Combinatoria*

The project of a logical treatment of *mathesis universalis* or 'mathematical logic' would seem completely natural to an author like Leibniz who did so much to merge Logic and Mathematics. For some commentators, beginning with Louis Couturat, this merger could even be seen as a form of identification:

> In sum, Leibniz had the merit of perceiving (well before the modern discoveries and advances which have shown this to be an obvious truth) that there is a Universal Mathematics upon which all mathematical sciences build their principles and their most general theorems, *and that this Mathematics merges with Logic itself, or, at least, is an integral part of it*. No longer is there only a formal analogy between Logic and Mathematics, but an identity, which is at least partial. It is that, on the one hand as we have seen, Universal Mathematics constitutes the general science of relationships [...]. On the other hand, formal logic extends so far as to coincide with Mathematics. Indeed, it is the formal nature of reasoning that guarantees the universal, necessary worth of deduction (Couturat 1901, pp. 317–318. Our emphasis).

A famous passage from the *Nouveaux Essais* seems to support this form of identification:

> I hold that the invention of the syllogistic form is one of the finest, and indeed one of the most important, to have been made by the human mind. It is a kind of universal mathematics whose importance is too little known. It can be said to include an art of infallibility, provided that one knows how to use it and gets the chance to do so – which sometimes one does not. But it must be grasped that by ''formal arguments' I mean not only the scholastic manner of arguing but also any reasoning in which the conclusion is reached by virtue of the form, with no need for anything to be added. So: a sorites, some other sequence of syllogisms in which repetition is avoided, even well drawn-up statements of accounts, an algebraic calculation, an infinitesimal analysis – I shall count all of these as formal arguments, more or less because in each of them the form of reasoning has been demonstrated in advance so that one is sure of not going wrong with it (NEEH, IV, Chap. 17, § 4)[4].

In addition, several lines further on, Théodore, *alias* Leibniz, mentions the fact that "any syllogistic argument could be demonstrated by that of *de continente et contento*, of the containing and the contained, which is different from that of the

---

[4] GP V, 460-461; A VI, 6, 478, transl. Peter Remnant and Jonathan Bennett (Leibniz 1981, p. 478). This text appears on the very first page of Couturat 1901, Book I, chapter 1. Amongst many others, the text is quoted by Russell 1900, p. 170; Husserl 1900, § 60, Weyl 1926, p. 12.

whole and the part; for the whole always exceeds the part, but the containing and the contained are sometimes equal, as happens in reciprocal propositions". If Syllogism belongs to 'universal mathematics' and if it has to be included in the much greater calculus *de continente et contento,* it is tempting to consider that it is this general logical calculus which merits the name of 'universal mathematics'. This was the conclusion reached by Couturat: "Leibniz thus conceived, more exactly, of his logic as a mathematics of thought, and following his wording a 'universal Algebra' applying to all objects capable of precise determination, and comprehending as many special algebras as there are genres of relationships to consider between objects" (Couturat 1901, pp. 319-320)[5]. In fact, that is what Philalèthe seems to understand when he replies: "I'm starting to form an entirely different idea of logic from my former one. I took it to be a game for schoolboys, but I now see that, in your conception of it, it involves a sort of universal mathematics" (NEEH IV, 17, § 8).

At first glance, these texts seem to set out a *mathesis universalis* which would function as a logical framework and under which particular mathematical theories would fall. By juggling neighbouring concepts, as is common in Leibnizian commentary, one could then bring this logical framework together with *ars combinatoria* conceived as a 'general science of forms and formulae'. In fact, Leibniz never ceases to insist on the fact that universal mathematics, in the narrow sense mentioned in the passage quoted above (through examples such as 'drawn up statements of accounts', 'algebraic calculations' or 'infinitesimal analysis') must be part of a much vaster theory:

> The art of combinations in particular, as I take it (it can also be called a general characteristic or *speciosa*), is that science in which are treated the forms or formulas of things in general, that is, *quality* in general or similarity and dissimilarity; in the same way that ever new formulas arise from the elements a, b, c themselves when combined with each other, whether these elements represent quantities or something else. This art is distinct from common algebra, which deals with formulas applied to *quantity* only or to equality and inequality. This algebra is thus subordinate to the art of combinations and constantly uses its rules. But these rules of combination are far more general and find application not only in algebra but in the art of deciphering, in various games, in geometry itself when it is treated linearly in the manner of the ancients, and finally, in all matters involving relations of similarity (*De synthesi et analysi universali seu Arte inveniendi et judicandi* 1683-1686?, A VI, 4 545; GP VII, 292-298; translated by Loemker 1989, p. 233)

Once again, this interpretative strategy was that followed by Louis Couturat and since repeated by many commentators. Indeed, the beginning of the chapter on the combinatorial art of *La Logique de Leibniz*, straight after mentioning the passage from the *Nouveaux Essais*, reminds us: "It is to this higher logic and not to that of

---

[5] See also Schneider 1988, p. 165, where logic, of which 'mathematical logic' is presented as an application, receives three names: universal characteristic, rational grammar and *universal calculus*.

Aristotle that he [Leibniz] gives the title 'universal mathematics.' And it is this that we must now investigate and explain." Even if the 'logicist' interpretation of Couturat has by now been discussed and criticized, this characterization of Leibnizian 'mathematical logic' as a general theory of relationships remains the most common description of his *mathesis universalis*[6].

What is striking about these descriptions is that they rely on statements, often programmatic, taken from different contexts and placed end to end[7]. It is rare for anyone to focus directly on the texts in which Leibniz intends to set out this *logica mathematica*, or even to contextualize them. When and where did this notion appear? What exactly does it mean? How is it implemented? These are the questions we want to address in the sections that follow.

3 The Logic of Mathematicians

In his exchange with Augustin Vaget in 1696, Leibniz described his desire for a systematic treatment of algebra in the following way:

> [...] For what you ask of me on the subject of some Arithmetical compendium, which could serve you for teaching, I have nothing of use for you. And would I wish that it be written as one in which Algebra and Arithmetic are treated παραλλήλως. They are, in truth, the same science and differ in nothing except that in Arithmetic it is a question of determinate numbers and, in Algebra, of indeterminate numbers. From this also comes the fact that Algebra or Arithmetic is the *Mathesis Universalis*, that is to say the science of quantity in general, because magnitude is nothing other than the multitude of parts. [...] As for me, I am accustomed to conceive of the science of magnitudes, sometimes called Logistics, as a Mathematical logic (to Vagetius, 5(15) Juin 1696; A III, 6, 781).

This description matches the beginning of *De ortu, progressu et natura algebrae* (GM VII, 203-216). Similar expressions are found in the so-called *Praefatio*[8] and in the correspondence with Johann Andreas Schmidt at the end of 1698—with the difference that Leibniz then seems to have put the scripts in order himself. In fact, the exchanges show evidence of the sending of a draft treatise, of which Schmidt

---

[6] See Risse 1970, p. 175, Serres 1968, p. 5, Arndt 1970, p. 110, Cassirer 1970, p. 175, Mittelstrass 1979, p. 603; Burkhardt 1980, p. 395; Schneider 1988, p. 163.

[7] The situation is quite similar to that which reigns in the interpretation of the no less celebrated *scientia generalis* which, up to now, as A. Pelletier has rightly pointed out, has functioned essentially by anadiplosis (Pelletier 2013, p. 288).

[8] GM VII, 50. *Praefatio* was the title given by Gerhardt, but is not to be found in the manuscript. Moreover, it cannot be the preface of the *Matheseos universalis pars prior* edited by Gerhardt just after it (the dating from watermark indicates an earlier period. It moreover conforms with a content whose context appears to be, above all, in confrontation with Cartesian physics).

subsequently made a copy that Leibniz revised[9]. This corresponds to the handwritten state of the text of *Matheseos universalis pars prior* edited by Gerhardt – without making mention of the existence of two manuscripts (GM VII, 53-76). It includes the desire for both a 'parallel' treatment of arithmetic and algebra, and for a 'logical' approach to mathematics. Finally, the expression reappears in a group of texts from the same period (ca. 1700) again devoted to the *scientia mathematica generalis,* which seems to have been the guiding theme of this 'logical' approach. These texts are conserved in the LH XXXV Series I, 9, and will be of special interest to us in Section 4.

Before addressing the details of these documents, some preliminary remarks are necessary. In view of the somewhat late dating of these texts, we could wonder why Leibniz expresses himself as if the 'logical' approach to mathematics was a long-standing method for him. The other remark, which immediately arises out of these texts is the unfixed character of the *mathesis universalis*. Sometimes it is identified with algebra, itself considered as the universal science of quantity[10]. Sometimes it is the general science of quantity, but containing algebra as well as differential calculus (*scientia infiniti*)[11]. Sometimes, finally, it does not limit itself to quantity and consists of two parts: one dedicated to quantity (algebra or logistic) and the other to quality or form (*ars combinatoria*)[12].

To the first question we can reply that the idea of a 'mathematical logic' is actually very old. At the start of the letter to Conring of 9/19 April 1670, Leibniz compliments him for having developed the 'art of judging' as a form of logic applied to moral issues. He then evokes the idea, of scholastic origin, of a *logica utens* (or *serviens*), and already refers to the new algebra as an example[13]. The same

---

[9] Schmidt mentions in November 1698, that he is in the process of reading the part of the *Logica analytica* that Leibniz had sent to him (A I, 16, 295). As noted by the editors (A I, 16, 295, note to line 7), this allusion refers to a manuscript about the *mathesis universalis* explicitly mentioned in the letters of 8 December (A I, 16, 341) and of 28 December (A I, 16, 393) – where Schmidt says very clearly that he made one of his servants recopy the text.

[10] This is the case, as we have seen, in the letter to Vaget of 1696, as well as in those to Schmidt at the end of 1698 and again in the letter to Ludewig in April 1700 (A I, 18, 611). This narrow definition is equally utilised by the *Specimen Geometriae Luciferae* (GM V, 261) and the *De Calculo situum* (C 550).

[11] This is the case in the *Matheseos universalis pars prior* (MU), the fragments of LH XXXV as well as in the numerous texts in which Leibniz interests himself in questions of physics (*De legibus naturae et vera aestimatione virium Motricium, Acta eruditorum,* 1691; GM VI, 211; *Praefatio,* GM VII, 50).

[12] This is the case in *De Ortu* (GM VII, 205), but also already in (*Idea Libri cui titulus erit*) *Elementa nova mathesis universalis* (A VI, 4, A, 513–524); *Initia Scientiae Generalis* (1679, A VI, 4, 362); *Guilielmi Pacidii Plus Ultra* (1686, A VI 4, 675).

[13] *Illud igitur prorsus assentior: prudentiam dicasticam seu artem judicandi in genere paucissimis regulis absolvi, esse enim nihil aliud quam Logicam ad moralia applicatam. Porro ut Ars Experimenta faciendi alia est a physica, ita ars quaestiones juris definiendi alia a Jurisprudentia; tantum enim hae distant inter se quantum* Logica serviens *(ita enim appellare malo quam cum Scholasticis Logicam utentem) a* Scientia utente. *Nihil enim aliud est haec generalis prudentia dicastica,*

expression is repeated in a text contemporary with our corpus, which is equally interesting for the description of the relationship between universal mathematics and combinatorial art:

> What concerns the rational (*Rationalia*) is distinguished in the general and the particular. The general belongs to logic. And surely finding particulars is nothing other than putting logic to use (*Logica utentes*). Thus *mathesis universalis*, which is a certain kind of the rational, is nothing else than Mathematical Logic (*Logica Mathematica*). If logic is the doctrine of relationships, isn't it more likely to belong to combinatorics? Particulars are only the intelligible—with which Metaphysics deals, the imaginable—with which Mathematics deals, and those composed of the intelligible and the imaginable (*De ordine cognitionum*, 1695, A IV 6, 497).

Such a conception crops up from time to time in Leibniz's work, especially in encyclopedic presentations, many of which already mention a *Logica mathematica* corresponding to the universal theory of mathematics[14].

This already makes it possible to reconsider the interpretation arising from the passage of the *Nouveaux Essais* mentioned in Section 1—one of very few texts in which Leibniz makes logical forms (syllogisms, sorites) figure *inside* universal mathematics. As we have seen, this description has fed the interpretation that Leibniz would have therefore meant a kind of universal logical calculus, of a mathematical type. But this anachronistic interpretation of the term 'Mathematical logic' does not hold up when we recall the other examples mentioned by Leibniz: "a well-trained account, an algebraic computation, analysis of infinitesimal"— presented along with syllogisms and sorites as 'arguments *in forma*'. It is not a matter of putting mathematics in a logical framework, but of considering their *intrinsic* logical nature (which does not depend on a prior formulation in a calculus). This is what the following, very rarely cited passage also makes clear:

> And it is commonly so that Euclid's proofs are formal arguments most of the time; for when he gives what appear to be enthymemes, the suppressed proposition and what seems to be missing is supplemented by a marginal quote which gives the way to find what has been already demonstrated. This allows for considerable abbreviation without affecting the force of the argument. These inversions, compositions and divisions of ratios which he uses, are only species of argumentation particular and proper to mathematicians and to the matters that they treat, and they demonstrate these formal arguments with the aid of universal forms taken from logic (NEEH IV, 17 § 4).

---

quam vera dialectica juris, prorsus ut Logicam Theologicam non pauci, Logicam Medicam Methodistae, Logicam Mathematicam Algebraici Vieta, Ougthredus, Cartesius dedere (A II, 1 2006, 68. Our emphasis).

[14] *Pacidius Philalethi* (1676, A VI, 3, 533); *Usus geometriae* (printemps-été 1676, A VI, 3, 439 and A VI, 4, 1197 (*logica logistica*).

If the universal logical forms are the ultimate framework of our reasoning, we see that Leibniz also considers portions of *existing* mathematical theories to be logical (and not their reformulation in a logical formalism). Hence the other name of the *logica mathematica*: the 'logic of mathematicians' (*Logica mathematicorum*). This point was well noted by Louis Couturat, who, to Bertrand Russell's relative surprise, saw in universal mathematics something similar to what he was advocating under the name of 'universal algebra' (an axiomatic theory of algebraic structures which stood with one foot in logic and the other in mathematics)[15].

As for the fuzzy characterization of *mathesis universalis*, contrary to the majority of commentators, it does not seem necessary to us to rectify it. Indeed most commentators who are interested in the 'universal mathematics' have proposed reconstructions that tend to present a coherent organization that is simply not provided by Leibniz's texts. Making use of descriptions like that of the *De Ortu*, they consider that Leibniz envisaged a narrow and a broad characterization of *mathesis universalis*. According to this interpretation, he preferred the second, which was ultimately identified with *ars combinatoria*, itself seen as a general theory of relationships. But this perception is not confirmed by our corpus. No sign of a preference for the broad definition can be detected. The *Matheseos universalis pars prior*, which rests on the narrow definition, is later than the *De Ortu*. Moreover the fragments from the 1700's *all* involve the narrow conception of universal mathematics as the general science of quantity, and even more strictly, of magnitude in general (*scientia magnitudinis in universum*)[16]. Furthermore, these texts are concerned with a set of axioms for *elementary* algebraic operations on numbers (LH XXXV I, 9, 7; 8 et 9-14) or with foundational discussions about the concept of magnitude (LH XXXV I, 9, 1-4), in an apparent regression from the program of treating the whole of mathematics in the context of a general science 'of forms and formulae'. But in fact, there is only a regression for those who have initially projected a vision of a progressive ascent towards generality, borne by logical formalisation. The true challenge, for those who wish to simultaneously grasp the idea of *mathesis universalis* in Leibniz and his proposal for a 'logical' treatment of mathematics, is to understand precisely why the late texts return to what, in the teleological interpretation, should have been the *beginning* of the investigation. The next sections (3 and 4) will be dedicated to these aspects (the role of the *ars combinatoria* in the structuring of mathematics and the axiomatization of elementary algebraic calculations).

As soon as we enter into the content of texts such as *Matheseos Universalis pars prior* or *De Ortu*, another remark imposes itself on us. The texts are explicit

---

[15] On these intepretations of *mathesis universalis* in Leibniz, see Rabouin 2011. However, Couturat's interpretation is no more tenable than that which would place mathematical logic as a universal language, as Russell thought. We establish this in more detail later, by looking at the exact role of the axiomatic in Leibniz's thinking.

[16] The letter of April 1700 to Johann Peter Ludewig could equally be added to this corpus. In it Leibniz writes: *Algebra revera Logica Mathematica est; et avidit ut eadem sit etiam mathesis universalis, quid aliud enim sunt Generales scientiae, quam Logicae applicationes quas vocant Logicas mentes.* (A I, 18, 611). M. Schneider 1988 has already remarked that the corpus does not permit a division in favour of an evolution of the concept of *mathesis universalis* towards greater breadth. A better knowledge of the corpus shows that quite the contrary.

and consistent regarding the nature of the 'logical' structuring that Leibniz has in mind:

> In Logic there are Notions, Propositions, Arguments and Methods. It is the same in Mathematical Analysis, where there are quantities, truth statements involving these quantities (equations, setting of upper and lower bounds, proportions etc.), arguments (i.e. operations in a computation) and finally the methods, that is to say, the process that we use to find what is sought (MU, 54)[17].

In other words, what Leibniz has in mind under the heading of 'mathematical logic' is essentially a conceptual analysis of theories (at least of their "universal" part). Few commentators have addressed the details of this analysis[18]. True, it is somewhat tedious and rather confused (for example Leibniz takes three pages to describe the notation of powers in detail!). But its apparent disorder is precisely what should interest those who want to understand what *logica Mathematica* is. Space will not allow us to go into too much detail. We will thus limit ourselves to a few remarks for directing our investigation.

First and foremost, it is notable that although Leibniz regularly announces a 'parallel' treatment of notions and quantities on the one hand, and relationships and truths on the other, he will not follow that approach. In *Matheseos universalis pars prior*, he distinguishes quantities and relationships *within the notions* (separated into 'categorematic' and 'syncategorematic'). Surprisingly, he starts with relations (similarity, ratio, coincidence, etc.), without much apparent concern about being systematic. A second surprise comes from the treatment of 'categorematic' notions. As Leibniz discusses primitive terms through their notation, he is led to distinguish between two very different types of representations: one in which the symbols of numbers refer to objects and those in which they designate places (what he calls 'fictive numbers')[19]. This dual presentation of the basic concepts introduces an element of uncertainty indicated at the end of the passage:

> Where it also appears how Algebra has so far remained in a state of imperfection: indeed, a way to properly express even simple terms has not been established, to say nothing of the many defects in the Connotations which were remedied here, and those to which remain to be remedied (MU, 61).

We will return to this aspect in the next section since it occurs under the heading of *ars combinatoria*. However, it should be noted at the outset that *ars combinatoria*

---

[17] The same as in the letter to Vaget 1696 and *De Ortu* (GM VII, 207-208). This description explains the link with analysis whose structure is above all logical (*analysis notionum et veritatum*).

[18] The most developed study from this point of view is assuredly Schneider (1988), but it is limited to the study of fundamental relationships in parallel with the texts on *analysis situs*.

[19] GM VII, 59-60. On the notion of "fictive numbers" see section 3 below. Same development at the beginning of the *De Ortu*.

here *disturbs* the 'logical' order—to the point of casting doubt on the expression of simple concepts themselves.

The other notable element in this development comes from Leibniz being led to distinguish two ways of dealing with quantities that he called 'progressive' and 'regressive'.[20] By the latter he means the inverse operations (subtraction, division and square root extraction), which "are not always successful" and result in "substitute" (*succedanea*) quantities[21]. Contrary to a widespread reading of symbolic (or 'blind') thought considered as evidence of an inspired 'formalist' philosophy[22], we see here that Leibniz does not see a description of the operations as being sufficient to ensure the existence of the objects on which they operate. This is made clear by the many passages where he then explains, using detailed examples, that entities such as negative or imaginary quantities "indicate that the question was badly posed" (*quaestionem fuisse male conceptam*: MU, 70 and 73-74).

Thus, when Leibniz resolves to make a 'logical' presentation of universal mathematics, even considered under the narrow form of the universal science of quantity, he runs into a number of difficulties. These difficulties appear clearly when they are compared to some previous attempts, such as the *Elementa nova Matheseos Universalis* (the content of which is closer to the work on *analysis situs*: A VI, 4, 513-524; 1683?). In fact it is striking that the presentation in the earlier texts is much more ordered, Leibniz setting out different types of relationships under the rubric of a 'logic of the imagination' (*logica imaginationis*: A VI, 4, 514-515). In a presentation of this type, the relationships are the basic terms, structured in a logical

---

[20] Echoing a similar description proposed by John Wallis. In chapter LXXXIII of his *Treatise of Algebra* devoted to demonstrating that there is no algebraic expression for the quadrature of the circle, Wallis indicated, in fact, that this phenomenon was more general and concerned all of the 'resolutive' part of arithmetic: "*Nor is it strange, that such impossibility should arise; for the same happens in all the resolutive parts of arithmetick. And for most of them, we have already provided notations to express that impossibility. As for instance, addition is genetical, (or synthetical) and to any positive numbers, any positive numbers may be added, without coming to any impossibility. But subduction, is analytical, or resolutive: and here the case is sometimes possible ; as if a lesser be to be subducted from a greater: (3 − 2 = 1). But sometimes impossible; as if a greater to be taken from a lesser: (2 − 3), in which case we are provided of a notation, to express that impossibility (and the measure of that impossibility) by a negative quantity, (− 1 = 2 − 3) imparting somewhat less than nothing*" (Wallis 1685, p. 316).

[21] "We have so far used only addition and equal terms, that is to say multiplication, and the latter with equal terms, that is to say the power, and combinations of these; this is directly progressive calculation, which is always successful. Now is the time to mention regressive calculation, which is not always in our power, and this calculation applies to the inverse of addition, of multiplication and the elevating to a power, that is to say to subtraction, division and extracting square roots." (GM VII, 68).

[22] See the classic study of Belaval 1960, also the article of Granger 1981, or more recently Sasaki 2004, whose last chapter is significantly entitled: "Leibniz's 'logicist-formalist' philosophy of mathematics".

order and the indicative/explicative duality is much easier to grasp[23]. It seem to us that the disorder apparent in our group of texts is of particular interest, especially since it comes after the ordered presentation of the 1680s. It indicates that an initial project, where quantity and quality were presented as two domains for dealing with the imaginable ('form' and 'magnitude'), was more difficult to characterize than expected—enabling us to gain a better understanding of the many hesitations in the characterization of 'universal mathematics' (See Rabouin 2016). In particular, we have seen that the combinatorial approach permeates algebra to the point where it *disturbs* the organization of its primitive notions. Above all, these difficulties allow for a better understanding of what reconstructive approaches must carefully hide: that Leibniz ultimately *arrives* at questions such as the axiomatization of elementary algebra and integers, when they should have been at the beginning of a logical approach to mathematics. In the following sections, we will focus on these two aspects, on the one hand examining the role of *ars combinatoria*, and on the other the precise form taken by axiomatics.

4 Ars Combinatoria

On many occasions, Leibniz insists that algebra, although the first model for a 'mathematical logic', is nevertheless insufficient. The mere fact that it uses formulae makes it dependent on the 'general science of forms and formulae':

> Up to now the subordination of Algebra to the Combinatorial Art, or of specious Algebra to the *Speciosa generalis*, or again, of the science of formulae representing quantity to the theory of formulae—i.e. of expressions of order, of similarity, of relations in general—or again, of the general science of quantity to the general science of quality, has been ignored or neglected to the point that our Symbolic Mathematics is none other than a specific sample of the Combinatorial Art or of *Speciosa Generalis* (MU, 61).

The same dissatisfaction led him, in the *De Ortu*, to refuse the title of authentic 'universal mathematics' for Algebra:

> So there are two parts, if I am not mistaken, in *Mathesis generalis*: Combinatorial Art [which deals] with the variety of things and forms or qualities in general, in so far as they are subject to distinct reasoning, with similarity and dissimilarity, and Logistic or Algebra [which deals] with quantity in general. Certainly the art of deciphering, of playing the game of brigands (*ludus Latrunculorum*) and similar arts which are considered to relate to the *Mathesis*, have more need of Combinatorics than of Algebra, and Algebra itself, in so far as it expresses quantities by precise formulae that indicate the

---

[23] This is the model followed by Schneider 1988 to elucidate the notion of *mathesis universalis* in Leibniz, which he read in close connection with the work on *analysis situs*.

various relationships between quantities, is subordinated to the combinatorial art, and can progress thanks to it (GM VII, 205-206).

Thus, one interesting aspect of our texts is in providing a way to explore the role of *ars combinatoria* in 'mathematical logic'. The fact that the two passages assign different positions to this art is not necessarily a problem. As Louis Couturat has already noted, we can build on the classification of the *Guilielmi Pacidii Plus Ultra* which distinguishes between 'general' and 'special' Combinatorics, the first allied to Logic and the second to Mathematics[24]. Indeed, one can understand *ars combinatoria* in the narrow sense ('special' combinatorics) as the mathematical theory of combinations. Leibniz uses these results to calculate the chances of winning at gambling, for example. These enumerations are also important in establishing formulae (products of polynomials and powers of multinomials. Cf. De Mora-Charles 1992). Their determination as well as their use is aided by the choice of a well suited 'characteristic'. In this more general setting, *ars combinatoria* help to detect regularities (order, similarity, harmony), governed by architectonic laws such as the law of justice or the law of homogeneity[25]. These laws allow not only for the testing, but also for the organization of the calculations. They have a structural role within algebra itself, and more generally in any calculation. Hence a methodological role of 'general combinatorics' in which it is very close to the project of a *characteristica universalis*. If *mathesis universalis* in the narrow sense (algebra and differential calculus) is subject to *ars combinatoria* in the broad sense, it is itself able to adopt a broad sense, as is presented in *De Ortu*, where *ars combinatoria* (in the narrow sense this time) is subordinate to it. However, one should note that this broad sense *disappears* in the known texts from the 1690s.

For commentators like Louis Couturat, the identification between logic and mathematics derived directly from the unifying role of the combinatorial art conceived of as a meeting place between the 'general science of relationships' and the development of various axiomatic systems for calculation. This orientation is found in a number of commentators and summed up by M. Schneider as follows: *mathesis universalis* holds the same position vis-à-vis *ars combinatoria* as *mathesis specialis* bears to it. In this sense, *mathesis universalis* appears as an *application* within the domain of the imagination of universal logical structures abstractly treated by *ars combinatoria*[26].

---

[24] *Guilielmi Pacidii PLUS ULTRA sive initia et specimina SCIENTIAE GENERALIS de instauratione et augmentis scientiarum, ac de perficienda mente, rerumque inventionibus ad publicam felicitatem* (G VII, 50, A VI, 4, A, 675).

[25] MU, 65-68 and *Initia mathematicarum rerum metaphysica* (GM VII, 24-25).

[26] "The relationship of *mathesis universalis* to *mathesis specialis* is similar to that of logic (in the general sense where it is called *Scientia generalis*, not in the specific sense of the syllogistic) to *mathesis universalis*. Both are characterized, not by differing principles in their fundamental concepts and relationships, but rather only in this that *mathesis universalis*, whose relationships also find their use in logic, is limited to the specific domain of the imagination, while logic is abstracted from this particular area as in other possible areas of application" (Schneider 1988, p. 165). A few lines below, Schneider points out that the most common designation of this 'general science' is, according to him, *ars combinatoria*.

Here again, a more precise study of the documents proves to be very interesting. In fact, we will see that *ars combinatoria* does not intervene in *logica mathematica* as the application of a logical theory to specific areas. Still, it governs an approach to the (hidden) structure of symbolic systems. Let us follow one of Leibniz's examples to make this point clear, the product of polynomials:

> The letters a, b, etc., do not naturally signify the order and the relationship of quantities among themselves enough; so that in the progress of the calculation these beautiful harmonies do not appear, as well as the laws and theorems that are immediately manifested to view when a certain regular order is preserved.
> For example, if we multiply $cx^3 + bx^2 + qx + r$ by $gx^2 + px + e$ in the usual way, the product will be:
>
> $$\begin{cases} cgx^5 + bgx^4 + qgx^3 + rgx^2 \\ \quad + cpx^4 + bpx^3 + qpx^2 + rpx \\ \quad\quad + cex^3 + bex^2 + qex + re \end{cases}$$
>
> But if we multiply $10x^3 + 11x^2 + 12x + 13$ by $20x^2 + 21x + 22$, where no notation has been adopted without reason, and where nothing has been adopted that might not be expressed and distinguished in the notation, then the expansion appears perfectly in the product (MU, 59).

Leibniz calls this type of notation, that we would call a double index (except that the letter indicating the "content" does not appear), 'fictive' (*ficti*) or 'supposed' (*suppositi*) numbers. It is also mentioned at the beginning of the *De Ortu* and it plays a central role in the theory of elimination (or what in modern terms is called 'determinants')[27]. Three aspects are then particularly emphasized: how the *ars combinatoria* helps us to develop notations revealing the deep structures of the field under study[28]; the way in which it allows us to avoid calculation and work directly on the 'forms' (here of polynomials) governing the relationships between quantities; and finally, the way it allows us both to reorganize the theory rationally and to provide methods of verification:

> Where it is clear that, in the product, all possible combinations are made according to a certain law and in order. From whence it follows that we can write this product without calculation, once we have established theorems of this type. For example, for the term x^2 we first write:
>
> | 13. | then | 13.2 | and finally | 13.20 x^2 |
> | 12. | | 12.2 | | 12.21 ... |

---

[27] Cf. Knobloch 1976; Serfati 2001.

[28] See in particular the letter to Tschirnhaus of November 1678 where Leibniz already defends the primacy of the combinatorial art over algebra and argues that the study of symbols, far from leading us away from things, leads us to what is deeper in them. (*Nemo autem vereri debet, ne Characterum contemplatio nos a rebus abducat, imo contra ad intima rerum ducet*, A II, 1 (2006), 623).

11.	11.2	11.22 …
10.	10.2	10.23 …

[...] These things are of the very greatest utility when we have to multiply three formulae or more; indeed, if we apply a regular and precise notation, and not an arbitrary notation as is usually done, we can often predict the result and often bring some indubitable and appropriate theorems to light, and just as easily avoid errors or correct them (MU, 60).

What is remarkable here is that the *ars combinatoria* is not a general theory involved in *applications* to particular cases. It works, as Leibniz often repeats, as an *ars inveniendi*, that is to say, a way to *explore* theories and, from that, to *reorganize* concepts and truths. This gives a completely different meaning to the term 'logic' from that of a form of theoretical subordination (or worse of an application of a 'universal calculus').

In fact, we've already seen that Leibniz himself insists on the fact that when the *ars combinatoria* is taken into consideration, the conceptual organization of existing mathematics is disturbed more than confirmed, particularly as regards primitive terms and relationships (MU, 61). This immediately extends to other branches of mathematics: "Likewise I also show that the notation of arithmetic has so far been poorly established, from a theoretical point of view, so that naturally enough, neither the relationship between numbers nor their order is apparent. It is for this reason that much of arithmetic remains unknown right up to the present, which might seem surprising for a very easy and useful science" (MU, 61).

It is in the *Nova Algebrae Promotio*, another text in which Leibniz presents some of his algebraic results as well as epistemological reflections, that one of the finest examples of a successful treatment of difficult questions of arithmetic by combinatorial means is found. This is a demonstration of what we now call 'Fermat's little theorem': if e is a prime number and x any number, e divides $x^e - x$[29]. Leibniz stresses the importance of combinatorics in the discovery of this result: "How many transpositions of the letters in the form (*Forma*) there are, still does not emerge from what I know, even though these issues are at the forefront of the Combinatorial Art. I have obtained this, with the powers of polynomials and other expressions of this kind, during a time of inaction on a boat trip" (GM VII, 179).

The result used is known as the 'expansion of a multinomial'. Leibniz's idea is to first write x as a sum $a+b+c+d+\ldots$ (finite) and, at the very end, to replace each letter by 1, thereby obtaining any integer. He also writes $e = k+l+m+n+\ldots$ and the calculation is used for all possible integer decompositions of e. Indeed, when the power is expanded, it is the sum of terms of the form $a^k b^m c^l d^n \ldots$ multiplied by a coefficient M, equal to the number of all possible ways of finding this product in an expansion[30].

---

[29] For French translation and mathematical commentaries see Bühler and Michel-Pajus 2007.

[30] Note that the terms in the expansion observe the law of homogeneity: they all have the same overall degree, e.

Leibniz determines *M* in illustrating his explanations by graphs. He then gives the calculation for the generic example $e = 19 = 5 + 4 + 3 + 3 + 2 + 1 + 1$ of the coefficient *M*, of $a^5 b^4 c^3 d^3 e^2 f^1 g^1$. In modern notations:

$$M = \frac{e!}{k!\, l!\, m!\, n!}$$

We then see that if *e* is prime, it divides all the coefficients *M*, except when it is equal to one of *k*, *m*, *l*, *n*... and so *e* divides $(a+b+c+d...)^e - a^e - b^e - c^e - d^e + ...$ It only remains to set $a = b = c = d = ... = 1$, i.e. for any integer *x*, in order to reach the conclusion.

Through this example, we can clearly see to what point arithmetic, in the sense of number theory, can be 'subordinated' to *ars combinatoria*: it allows for the *discovery* of profound laws, without this subordination having anything to do with the application of logical structures or the interpretation of logical relationships in a particular domain.

5   Axiomatizations

We have just seen that *ars combinatoria* is not involved in our texts as an abstract theory, a 'general science of relationships' that is then *applied* to objects of the imagination. What of the other main line of interpretation of Leibniz's *logica Mathematica*, that of a universal calculus, in which mathematical theories would appear as specifications[31]? Isn't this what Philalèthe seems to describe, just after mentioning a calculus of 'the containing and the contained', in speaking of *mathesis universalis* as an 'entirely different idea of logic'? Again, our corpus offers a privileged field of study by providing examples of the axiomatization of algebraic calculations. We will now pay particular attention to the manuscript LH XXXV, 1, 9 fol. 9-14 (circa 1700), in parallel with a similar text entitled *Prima calculi magnitudinum elementa*[32].

The two texts are set out with numbered items enunciating axioms, definitions, notations, demonstrations, explanations and/or applications. The second is a more refined work, but it is in the abortive attempts, the erasures and the comments on the manuscript that we find clarification of Leibniz's approach. Furthermore, MS[33] presents multiplication, while PC is limited to addition/subtraction, and MS treats only integers, whereas PC applies to all numbers. Both attempts are set in an axiomatic form consistent with a declaration often repeated by Leibniz[34]:

---

[31] See the quote from Couturat 1901, pp. 319-320.

[32] *Prima calculi magnitudinum elementa demonstrata in additione et subtractione, ususque pro ipsis signorum + et –* (GM VII, 77-82).

[33] MS designates the LH XXXV I, 9 manuscript and PC, the *Prima calculi*. Their items will be coded MS[N] or PC[N], the figure indicating the number of the item in the text.

[34] NEEH I, 2 § 22; to Clarke, GP VII, 355; *Historia et origo calculi differentialis*, GM V, 395-396.

[...] demonstrations are finally resolvable into two kinds of indemonstrables: definitions or ideas, and primitive propositions or identities, such as B is B, anything whatever is equal to itself, and a great many others of this kind[35].

The first item of MS is very similar to the opening of MU : "[1] *Mathesis generalis* is the science of magnitude in general and has two parts: the science of the finite, that is to say, Algebra, that we will first set out, and the science of the infinite, which has just been established". In an alternative formulation, presented in the previous folio[36], Leibniz wrote: "*Mathesis Generalis* is the science of magnitude, also called *protomathesis* or even Logistic, i.e. regarding calculation, since it deals with indefinite numbers; but equally well [called] Mathematical Analysis, since it is like the logic used by Mathematicians." As in MU, number is then associated with magnitude through measure: "[2] Magnitude is what is designated by the number of congruent parts. Thus the magnitude of the *orgia* is designated as six feet, the feet being congruent indeed with each other, or it can be designated by the number of inches, 72, if we take an inch to measure the twelfth part of a foot. Thus, the general number designating magnitude varies with the action to be repeated, which can take on a variety of different forms."[37] In PC, these inaugural definitions are followed by that of equality: "Equal are those things which can be substituted one for the other *salva magnitudine*, which is designated by $a = b$, i.e. *a* may be substituted for *b* throughout the calculation dealing with magnitudes". In MS, however, this definition appears only at item [8], following a definition of integer (MS [3]), after which the main part of the development follows. It consists of a construction of the integers (MS [4]), their nomenclature (MS [5]), and the famous demonstration that "two and two are four" (MS [6]), providing an interesting context for reading the passage of the *Nouveaux Essais* dedicated to this demonstration (NEEH IV, 7, § 10).

A whole number is defined as follows: *Numerus integer est totum ex unitatibus tanquam partibus collectum*. Another attempt proposed the same definition but without the *tanquam partibus*[38]. Emily Grosholz writes: "This definition adds the geometrical word 'part' alongside the arithmetical word 'unit' stressing the analogy by the word *tanquam*, 'just as'" (Grosholz and Yakira 1998, p. 81). We could equally say this definition adds the *mereologic* word 'part' alongside the words 'unit of measurement'. Indeed '*unum*' is never defined, and the term '*unitas*' appears only in PC [2], clearly designating the unit of measurement. If Leibniz introduces 'parts'

---

[35] *Initia Rerum Mathematicarum Metaphysica* (GM VII, 20; transl. Loemker (1969), 668).

[36] LH XXXV, 1, 9, fol. 8. MS uses in part the same paper.

[37] PC [1] replaces "congruent" with "determined" and gives more explicit examples, in a scholium in particular: "If *l* is the side of a square, the diagonal will be $l\sqrt{2}$, or if *l* is 1, *l* is $\sqrt{2}$." So it really involves all the numbers, even the irrationals. Following on, he sets out notation, then discusses the notation and the term 'unit' (*Unitas*): "[2] *a*, *b* and similar notations mean numbers of things, [...] something being posited, to which the unit we call Measure is associated", accompanied by the definition of homogeneous: "Homogeneous things are those whose values can be expressed by numbers, the same measure being taken for the unit."

[38] LH XXXV, 1, 9 fol. 7 (Leibniz used for this folio a paper corresponding to certain sheets of MS).

here, it is because the whole/part relationship plays a fundamental role in his definitions. The presence of parts is characteristic of the domain of magnitudes, which can be geometric but not solely so.

Within the same bundle of documents, there are four folios devoted to the *scientia mathematica generalis* which bear more specifically on the concept of magnitude. Here is the beginning of this text (LH XXXV I, 9 fol. 1): "The General Mathematical Science is concerned with quantity in general, that is to say with way of estimating, which takes place not only in the sciences, commonly called pure or mixed *Mathesis*, but also wherever [struck out: greater, lesser or equal are involved] the whole and the parts are involved, and where the one is said to be greater than, or less than, or equal to." The correction shows the importance of whole and parts. It is this, in fact, that permits inequality to be defined: "The lesser (*minus*) is what is equal to a part of something else, called the greater (*major*)" (MS [9]).

Then comes the axiom enunciated in PC [4]: "a = a", and in MS [10]: "This axiom is unprovable: any thing is equal to itself". This is the version in the field of magnitudes of the fundamental "B is B" (by means of the definition MS [8] "are *equal* things whose magnitude is the same [...] and thus equal things can be substituted one for the other *salva magnitudine*"). It allows the proof of Euclid's axiom: "the part is smaller than the whole" (MS [11])[39]. Leibniz, however, adds a decisive clarification here: "The magnitudes of all parts having nothing in common together constitute the magnitude of the whole. Indeed, the whole is the same as all of its parts. Thus the magnitude of the whole coincides with the magnitudes of all the parts taken simultaneously, provided they do not involve any new magnitude: do not involve parts which have a common part whose magnitude would be used twice. As a consequence, in this calculus, we will always understand that the parts have no part in common."[40]

The condition "having no common part" is fundamental here. It differentiates these axioms from another axiomatization, concerning what Leibniz calls "real addition" and which is characterized by the axiom: 'A + A = A'. Leibniz clearly points to the difference between these two axiomatizations[41]:

> When two and two are said to make four, the latter two must be different from the former two. If they were the same, they would produce nothing new. This would be like trying, in jest, to make six eggs out of three, by counting first the three eggs, then taking away one and counting the remaining two, and again taking away one and counting the remaining one. But in the calculus of numbers and magnitudes, A or B or any other signs do not signify a definite thing, but any cases whatever of the same number of congruent parts. (GP VII, 237).

---

[39] This proof is often repeated by Leibniz since the time of the *Demonstratio propositionum primarum* (A VI, 2, 482), and plays a central role for him from this time on.

[40] Note that the premise: "the whole is the same as all its parts" is not taken as an axiom here, and, moreover, becomes an object of discussion cf. NEEH IV, 7, § 10.

[41] This is a scholium subsequently added to the original wording; translation by Loemker 1969, p. 380.

This is the key point regarding the relationship between the two calculus and is precisely what Philalèthe stated: "all syllogistic doctrine could be demonstrated by that of *continente et contento,* the containing and the contained, *which is different from the whole and the part*; because the whole always exceeds the part, but the containing and the contained are sometimes equal, as happens in reciprocal propositions" (Our emphasis).

It is possible that the mereological basis of the axiomatization of 'real addition' indicated the need to make the axiomatization of algebraic operations more precise. This would explain why these developments come rather late in the corpus—such as the difficulties found in the analysis of simple notions in *mathesis universalis* would explain the need for axiomatizing the integers. The point to note is that in no case does the calculus *de continente et contento* enter here as a general logic to which *universal mathematics* could be *subordinated*[42]. On the contrary, it is a matter of *opposing* two forms of calculation depending on how they treat the composition of objects. It is moreover significant that the manuscripts of the group LH XXXV I, 9 all then return to the 'narrow' definition of *mathesis universalis* as the universal science of quantity.

MS then proceeds with the definition of addition: "[12] Addition occurs if several magnitudes are in fact one, in such a way that nothing else is needed than to put them down simultaneously, and then of course, we will put before each of them the + sign, i.e. 'plus', so that if a, b and c are conjoined to make m, $+ a + b + c = m$, as $2 + 3 + 4 = 9$". An interesting point is that the addition does not appear here as a binary operator. This could explain the absence of the associative law in the list of items and the absence of parenthesis in the calculations—a lack that some commentators, including Lenzen, attribute to Leibniz's carelessness[43]. The use of 'simultaneity' reinforces this hypothesis, at the same time justifying commutativity (as represented in MS [13] by the fact that "the order is not important").

We then find in MS a long series of items preparing the introduction of subtraction and negative numbers. Addition is presented as a 'progression' (*progressus*) of increasing magnitudes, supported by the image of displacement on a straight line (MS [16]). Subtraction is presented as a regression (*regressus*) of decreasing magnitudes, associated with a movement of displacement in the opposite direction (MS [17])[44]. If the movement regresses more than it progresses, there is a false progression, which will actually be a regression. Such an increase will be less than zero (*minus nullo*), and the number designating this will be less than nothing (*minus nihilo*). Other examples (the speed of a boat carried by the wind and the current, debts) precede the presentation of positive (*affirmativa*) or negative

---

[42] Contrary to what Berlioz and Drapeau-Contim put forward: "Leibniz proposes here a common syntax not only for many logical calculations (extensional logic of aggregates, intensional logic of properties, propositional logic and modal logic), but also for several formal disciplines (logic, geometry, *arithmetic*). In doing so he clearly opts for the existence of other types of arguments like those of logic [NDE: the note refers here, without any surprise, to NEEH IV, 17, 4]. The formal calculus thus constructed is interpreted inside and outside of the field of logic in the strict sense" (Berlioz and Drapeau-Contim 1998, p. 40. Our emphasis).

[43] Lenzen 2004, p. 19: "Leibniz never took very much care about bracketing".

[44] Same development in MU, 70. The number line had already been used by John Wallis to make sense of the negative quantities, see Wallis 1685, p. 267.

(*negativa*) quantities, hence the property: "a decrease of a negative quantity amounts to an increase" (MS [18 and 19]). Leibniz concludes: "[20] and these things are certainly visualizations useful to assist the progress of the imagination, but now we will set out specific proposals that can be the foundation of proofs." In PC, these preliminary examples are found at the end of the text, grouped under the heading "Application of the calculation to the thing, where it is a matter of the whole, the part, larger, smaller, positive (*positivo*) and negative (*privativo*)". We can therefore assume that Leibniz introduced these examples as much to show the usefulness of his calculus as for the intuitive illumination they provide.

Certain proofs are deduced by the principle of substitution of equals. This is the case with MS [22]: "If equals are added to equals, equals will be obtained. Let $a = b$ and $l = m$, we have $a + l = b + m$". Same with MS [23]: "If we remove equals from equals, the remainders are equal". Others are based on the relationship of whole/part and refer to the "notation", as in MS [24]: "If $e = a + b$, we will have $-e = -a - b$. If we were to remove 5 and remove 3 and similarly 2, this is like saying $-5 = -2 -3$. This is evident from Article 11.2". Similarly MS [25]: "$a - a = 0$. If we add as much as we remove the rest is nothing; or if the progression is equal to the regression, in short, there is neither progress nor regression; this arises from the very meaning of the notation."[45]

Certain items are very rigorously justified. For example, MS [27] uses previous items that characterize what we would call symmetry (MS [25]) or neutrality (MS [26]): "[27] if $h = -g$ we will have $-h = g$. Indeed, $h = -g$ by hypothesis, therefore $g + h = g - g$ (by item 22). So $g + h = 0$ (Article 25) So (by 23) $g + h - h = 0 - h$. Therefore (by 22) $g = 0 - h$, that is to say (by 26) $g = -h$". The following items demonstrate the manipulation of signs in various cases: "$-(+a - b)$ means $-(+a) - (-b)$" (PC [22])[46]. It is also interesting to note that Leibniz insists on the reciprocity of operations: "[30.2] Addition and subtraction resolve each other (*resolvunt*) and serve mutually as a proof (*comprobant*)".

While PC stops at addition, MS tries to pursue with the multiplication of integers. The definition proves difficult: there are a lot of erasures, returns and aborted passages in the manuscript. The final version appears to be MS [31]: "Multiplication is the addition of equals, and if the same number is placed as many times as there are units in another, the first number is said to be multiplied (*multiplicatus*) and the following the multiplier (*multiplicans*); but the sum (*summa*) will be the product (*productus*)". This definition does not allow Leibniz to demonstrate the properties in a way that would satisfy him. He enunciates associativity law in MS [34], justified in the following manner: "since nothing matters in what is observed in the order of producing a multiplication" and "nothing changes when multiplying by unity" (MS [36] "*1. a = a*"). The distributivity of multiplication over addition is given as a method without justification (MS [40] and [41]).

By refusing to increase the number of axioms and always seeking for support both intuitive and unifying for the definitions, Leibniz imposed an impossible task on himself, despite his astonishing intuitions. He thought perhaps to have found this

---

[45] See also MS [26]: "$a = a + 0 = -0 + a$, that is to say, 0 can be added to or removed from a member with impunity".

[46] Note that here Leibniz does not neglect parentheses.

support in the general theory of the whole and the part, adapting the procedure of axiomatization used in philosophical logic to mathematicians' logic, but it could not be used for multiplication.

6 Conclusion

Since the end of the nineteenth century, many studies have examined Leibniz's works for the relationship between his philosophy and his mathematics. These, rather naturally, have given priority to the mathematical fields where the philosopher opened up new areas: differential calculus, *analysis situs*, combinatorics (with its extensions into what is today called linear algebra), the arithmetic of series, or the various attempts to develop a logical calculus. However, when we concentrate on those areas where Leibniz developed his deepest philosophical insights into mathematics, we find to our surprise that they are concerned with another area altogether, barely explored by commentators, that of algebra and its intrinsic 'logic'. The main thread of this approach is given by the idea of a 'mathematical logic': *Algebra revera Logica Mathematica est.*

In this article, we have proposed to take up this thread of a 'Mathematical Logic', in order to draw out a certain number of themes from Leibniz's philosophy of mathematics which, it seems to us, is too often approached from the viewpoint of reconstruction. This approach allows a fresh look at the subtle affinity between algebra and logical analysis, but also at the programme of a 'universal mathematics', at the status and the place of *ars combinatoria*, or at Leibniz's study into axiomatic foundations. What we have shown is that we do not find in our texts the simple picture that dominates those reconstructions. Mathematical logic is not presented as a general science, more or less identified with *ars combinatoria* in its broadest sense. 'Universal mathematics' is not presented as an application of this 'logic' to the field of imagination. The *Logica Mathematica* is neither a general theory of abstract relations nor a universal logical calculus. It concerns the conceptual analysis of mathematical concepts within their complex architecture.

Acknowledgements

The authors would like to warmly thank Pam and Stuart Laird for their invaluable help in translating this paper into English.

References

[GM] Leibniz, G. W. 1849–63. *Leibnizens Mathematische Schriften*, ed. C. I. Gerhardt, 7 vols. Berlin: A. Asher and Halle. H. W. Schmidt.
[GP] Leibniz, G. W. 1875–90. *Die Philosophischen Schriften und Briefe von Gottfried Wilhelm Leibniz*, ed. C. I. Gerhardt, 7 vols. Berlin: Weidman.
[A] Leibniz, G. W. 1923– *Sämtliche Schriften und Briefe*, Berlin-Brandenburg Academy of Sciences and the Academy of Sciences in Göttingen. 8 series. Berlin: Akademie Verlag.

Arndt, H. W. 1971. *Methodo Scientifica Pertractatum. Mos Geometricus und Kalkülbegriff in der Philosophischen Theorienbildung des 17. Und 18. Jahrhunderts.* Berlin: De Gruyter.
Belaval, Y. 1960. *Leibniz, critique de Descartes.* Paris: Gallimard.
Berlioz, D., Drapeau-Contim, F. 1998. Un essai logique de Leibniz, "Le calcul des ingrédients ». *Revue d'Histoire des Sciences* 51(1):35-64.
Blanché, R. 1970. *La Logique et son histoire.* Paris: Armand Colin.
Bühler, M., Michel-Pajus, A. 2007. Sur différents types de démonstrations rencontrées spécifiquement en arithmétique. *Mnémosyne* 19:61-66.
Burkhardt, H. 1980. *Logik und Semiotik in der Philosophie von Leibniz.* München: Philosophia.
Cassirer, E. 1970. *Die Logik der Neuzeit.* Vol. 2. Stuttgart: Frommann.
Couturat, L. 1901. *La logique de Leibniz d'après des documents inédits.* Paris: Alcan.
De Mora-Charles, M. 1992. Quelques jeux de hazard selon Leibniz. *Historia Mathematica* 19:125-157.
Husserl, E. 1900. *Logische Untersuchungen I.* Halle: Niemeyer.
Knobloch, E. 1976. *Die mathematischen Studien von G.W. Leibniz zur Kombinatorik.* Wiesbaden. Steiner. Studia Leibnitiana Supplementa XVI.
Granger, G. G. 1981. Philosophie et mathématique leibniziennes. *Revue de Métaphysique et de Morale* 86(1):1-37.
Grosholz, E., and Yakira, E. 1998. *Leibniz's science of the rational.* Stuttgart. Franz Steiner Verlag. *Studia leibnitiana Sonderhefte* 26.
Leibniz, G. W. 1981. *New essays on human understanding.* Translated by Peter Remnant and Jonathan Bennett. Cambridge: The Cambridge University Press.
Leibniz, G. W. 1989. *Philosophical essays.* Translated by Garber D., Ariew, R. Indianapolis: Hackett.
Leibniz, G. W. 1989. *Philosophical papers and letters.* 2nd ed. Dordrecht: Reidel.
Lenzen, W. 2004. Leibniz's Logic. In: Gabbay, D. M., Woods, J. and Kanamori, A. (Eds). *Handbook of the History of Logic, volume 3: The Rise of Modern Logic: From Leibniz to Frege.* Amsterdam: Elsevier-North-Holland, pp. 1–83.
Mittelstrass, J. 1979. The Philosopher's Conception of Mathesis Universalis from Descartes to Leibniz. *Annals of Science* 36:593-610.
Pelletier, A. 2013. Logica est scientia generalis. Leibniz et l'unité de la logique. *Archives de Philosophie* 76(2):271-294.
Rabouin, D. 2011. Interpretations of Leibniz's Mathesis universalis at the Beginning of the XXth Century. In: Krömer, R., Chin-Drian, Y (Eds.). *New essays on Leibniz reception in philosophy of science 1800-2000.* Basel: Birkhäuser. 187-201.

Rabouin, D. 2016. A fresh look at Leibniz' Mathesis universalis. In: Li, W. (Ed.), *Ad felicitatem nostram alienamve. "Für unser Glück oder das Glück anderer": Vorträge des X. Internationalen Leibniz-Congress*, Olms, Bd 4, pp. 505-519.
Russell, B. 1900. *A Critical Exposition of the Philosophy of Leibniz*. Cambridge. The Cambridge University press.
Sasaki, C. 2004. *Descartes's Mathematical Thought*. Dordrecht: Kluwer.
Schneider, M. 1988. Funktion und Grundlegung der Mathesis Universalis in Leibnizschen Wissenschaftsystem. *Studia Leibnitiana Sonderheft* 15:162-182.
Serfati, M. 2001. Mathématiques et pensée symbolique chez Leibniz, Mathématiques et physique leibniziennes (1ère partie). *Revue d'Histoire des Sciences* 54(2):165-222.
Serres, M. 1968. *Le système de Leibniz et ses modèles mathématiques*. Paris: P.U.F.
Wallis, J. 1685. *Treatise of Algebra both historical and practical, shewing the original, progress and advancement thereof*. London: Playford.
Weyl, H. 1926. *Philosophie der Mathematik und Naturwissenschaft*. München u. Berlin: Oldenbourg.

Anne Michel-Pajus, Université Paris Diderot, France
annie.pajus@sfr.fr
David Rabouin, CNRS-Université Paris Diderot, France
david.rabouin@wanadoo.fr

Describing Reality: Bernoulli's Challenge of the Catenary Curve and its Mathematical Description by Leibniz and Huygens

Miguel Palomo

Abstract. In 1690 Bernoulli proposed to Leibniz to describe mathematically the catenary curve and the later opened this challenge to other scientists. Leibniz, Huygens and Bernoulli presented their solutions to be published in the *Acta Eruditorum* one year later. However, more than a simple challenge, for Leibniz this was a demonstration of the superiority of his calculus in contrast to Huygens' geometrical methods. With this paper we aim to examine the role of the problem of the catenary curve on Leibniz's thought by addressing the following questions: How did the mathematical description of the catenary determine Leibniz's scientific vision? What is the connection between the catenary and Leibniz's metaphysics? We argue that through the development of the catenary challenge it is shown that even though Huygens had a main role in Leibniz's scientific thought, he qualitatively surpassed his former master creating a new method and searching not only for the immediate results of the problem but for the implications of those results; and that mathematical descriptions of phenomena were, for Leibniz, a way to reach a profound knowledge of reality in connection with his metaphysics.

Keywords: Leibniz, Catenary, Mechanical Curves, Infinitesimal Calculus, Huygens, Bernoulli, History of Mathematics, Metaphysics, Acta Eruditorum, Geometry.

1 Introduction

Mathematics has greatly evolved since the years of the Greeks, who shaped how the discipline would develop in the future. However, some issues remained unsolved for centuries, and that is the case regarding geometrical and mechanical curves. Greek mathematicians had already realised that not every curve had the same characteristics. For example, the use of a ruler and compass was not enough to represent certain curves such as the trisectrix or the quadratrix of Dinostratus. Let us take the case of the spiral: mechanical forces like gravity generate this kind of curve and therefore it could not be drawn with a ruler and compass. Consequently, they called them "mechanical curves" to distinguish them from the "geometrical curves", which were those that could be easily drawn and represented with a simple equation.

Miguel Palomo (2017) Describing Reality: Bernoulli's Challenge of the Catenary Curve and its Mathematical Description by Leibniz and Huygens. In: Pisano R, Fichant M, Bussotti P, Oliveira ARE (eds.), *The Dialogue between Sciences, Philosophy and Engineering. New Historical and Epistemological insights. Homage to Gottfried W. Leibniz 1646-1716. New Historical and Epistemological Insights.* College Publications, London, pp. 331-340
© 2017 College Publications Ltd | ISBN: 978-1-84890-227-5 www.collegepublications.co.uk

On the one hand, we talk of a geometrical curve when one line moves away from a straight line forming a continuous trajectory without any angle. On the other hand, mechanical curves are generated by one or more superimposed mechanical forces, a characteristic that makes every mechanical curve unique.

This distinction between curves was inherited by modern mathematicians. While they did not have any problem mathematically representing geometrical curves using simple equations, the mechanical curves could not be described using the same methods due to the variety of forces that came into play. However, Descartes claimed that if Geometry was the science that studied and showed the measure of every object, then there was not a reason to exclude the more complex curves from the simple ones (Descartes 1925, p. 316). The problem of the mathematical representation of the mechanical curves was still unresolved nevertheless, as Crippa explains in his analysis of curves (Crippa 2014, pp. 155-167).

The application of those mathematical descriptions was clear: through them the movement could be illustrated. Thus, the search for a general method that could lead to the description of these mechanical curves was required because it would eliminate the necessity of having to resolve each case individually.

The XVIIth Century saw in Descartes the beginning and the culmination of his search for a final systematisation of philosophy. He was aware of the need for general methods, and with his philosophical method he became the father of modern philosophy. Moreover, he invented analytic geometry, which showed how to interpret geometrically algebraic operations. This eased the path to the creation of a general method for the mathematical description of the mechanical curves. However, until the apparition of the infinitesimal calculus, there was not a rule that could offer any general procedure to reduce these curves to algebraic equations as Leibniz claims in his "Historia et Origo Calculi Differentialis":

> Now it certainly never entered the mind of any one else before Leibniz to institute the notation peculiar to the new calculus by which the imagination is freed from a perpetual reference to diagrams, as was made by Vieta and Descartes in their ordinary or Apollonian geometry; moreover, the more advanced parts pertaining to Archimedean geometry, and to lines which were called "mechanical" by Descartes, were excluded by the latter in his calculus. But now by the calculus of Leibniz the whole of geometry is subjected to analytical computation, and those transcendent lines that Descartes called mechanical are also reduced to equations chosen to suit them, by considering the differences dx, ddx, etc., and the sums that are the inverses of these differences, as functions of the x's; and this, by merely introducing the calculus, whereas before this no other functions were admissible but x, xx, x^3, $\sqrt{x}$, etc., that is to say, powers and roots (Leibniz 1920, pp. 25-26).

2 The Catenary Curve

The catenary curve is formed when a string or chain with a uniformly distributed mass is suspended by its ends. Since its form is caused by mechanical forces, it is a mechanical curve. Huygens gave it the name "catenary" in a letter to Leibniz on the 18th of November, 1690 (AA III, 4, pp. 654-658). This name comes from the Latin word *catena* which means "chain". Before this letter, the catenary was known as *the corde pendante* or the *chaine pendant*, and it would play an important role in the consolidation of the infinitesimal calculus.

The story of its mathematical description is peculiar both in the sense that it is distinctive and determining. Galileo had mistakenly claimed that the *chaine pendante* took the shape of a parabola since both curves are quite similar.

> The other method of drawing the desired curve upon the face of the prism is the following: Drive two nails into a wall at a convenient height and at the same level; make the distance between these nails twice the width of the rectangle upon which it is desired to trace the semiparabola. Over these two nails hang a light chain of such a length that the depth of its sag is equal to the length of the prism. This chain will assume the form of a parabola, so that if this form be marked by points on the wall we shall have described a complete parabola which can be divided into two equal parts by drawing a vertical line through a point midway between the two nails (Galilei 1914, p. 149; see also Galilei 1898, p. 186).

He came to this conclusion by studying the movement of the projectiles. And because he thought demonstrated that that curve was a parabola (that is, a geometrical curve), Galileo failed to see that the *chaine pendante* was definitely formed by mechanical forces.

Huygens, when he was only 17[1], noticed the mistake Galileo had made, which he pointed out in a letter to Leibniz on the 9th of October 1690 (AA III, 4, p. 585; OC 9, p. 497). The *chaine pendante* could not take the form of a parabola since it could not be described with a simple equation as established in a letter Huygens wrote on 28th October 1646 for Mersenne[2], friend of his father Constantijn. This fact undoubtedly showed the high capabilities young Huygens had, bringing to light a mistake made by one of the greatest minds of all time.

Even though it had become clear that the catenary was not the same curve as the parabola, several years passed before the issue was properly studied. Joaquim Jungius also proved Galileo's mistake in 1669, while the first published work on the catenary was *La statique, ou la science des forces mouvantes*, written by Ignace Gaston Pardies in 1673. However, although both proved Galileo to be wrong, neither of them offered a mathematical description of the catenary. The issue to

---

[1] Huygens claimed he sent his discovery to Mersenne when he was 15 years old. However, this should be seen as a mistake since the letters he is referring to were written between 28th October and November 1646 (OC 1, 24-28; 34-44), when he was 17. Another explanation is that he may have kept the discovery for himself for two years.

[2] Several references mistakenly claim Joaquim Jungius was the first who pointed out Galileo's mistake: Leslie 1821, 391; Nahin 2004, 241; Salmon 1879, 220 (footnote).

answer seemed easy: if the *chaine pendante* is not a parabola, what form does it take and how can it mathematically be described?

3   Bernoulli's Challenge in the *Acta Eruditorum*

At the end of the 17th century the problem of the description of the catenary was well known by the mathematicians. With the idea of settling this issue and simultaneously proving the utility of the Leibnizian calculus, Jacob Bernoulli[3] proposed (in the article "Analysis problematis antehac propositi, de inventione linea descensus a corpore gravi precurrende uniformiter, fic ut temporibus equalibus equales altitudines emetitur: & alterius cujusdam Problematis Propositio", AE May 1690, pp. 217-219) a challenge to Leibniz: to describe the catenary curve[4].

Even though Leibniz had not cultivated his mathematical skills in his youth as Huygens had (Leibniz had instead obtained a doctorate in law), he soon became an expert while studying mathematics with Huygens in Paris between 1672 and 1676. In his paper "Ad ea, quae vir clarissimus J.B. mense Majo nupero in bis Actis publicavit, Responsio" (AE July 1690, pp. 358-360; Leibniz 1989, pp. 166-172)[5] Leibniz accepted Jacob Bernoulli's challenge. By that time Bernoulli had been converted to the Leibnizian calculus after the calculation of the isochrone curve in May 1690 (Leibniz 1989, p. 192), and he certainly knew Leibniz was going to use it in this proposition. Moreover, because Leibniz had hardly any doubt about the utility of his infinitesimal calculus, he opened the challenge up to the contribution of other mathematicians. The idea was that during a period of six months (which was later extended six additional months) anyone could present his own solution to the mathematical description of the catenary. All of the results would be published at the same time in the *Acta Eruditorum*, and three other intellectuals decided to contribute their approaches; Huygens, Tschirnhaus and John Bernoulli[6]. Each one of them would test their mathematical skills in public, although in the end Tschirnhaus did not submit his solution.

In the words of Huygens, this challenge would be a unique opportunity to show the efficacy of Leibniz's new calculus. In the letters they exchanged during 1690-1691 both discussed the utility of the infinitesimal calculus. Leibniz's former teacher was not sure of the real applications of this method and it was not the first time Leibniz was facing this situation; years before he had tried in vain to convince Huygens of the utility of his *anaylisis situs*. Huygens was now eager to discuss and compare their methods to describe the catenary and see which method was more elegant and simple in case both arrived at a viable solution. Due to his impatience Huygens sent Leibniz his partial solution before the publication in the *Acta Eruditorum* and was expecting Leibniz to send him back his own description of the catenary. In a letter to Huygens on the 13th of October 1690 Leibniz claimed:

---

[3] Also known as Jacobus or James Bernoulli.

[4] "[P]roblema vicissim proponendum hoc esto: Invenire, quam curvam refer at funis laxus et inter duo puncta fixa libere suspensus. Sumo autem funem esse lineam in omnibus suis partibus facillime flexilem" (AE May 1690, 219).

[5] Not included in GM.

[6] Also known as Johan o Johannes.

En considerant vostre chiffre de la chaine pendante, j'y trouve quelque rapport à mon calcul, mais aussi quelque difference, car au lieu de l'equation xxyy= a^4 - aayy, je voy dans mon calcul reduit à certain termes xxyy = a^4 + aayy, qui sert à arriver à la ligne de question, et quoyque cette ligne soit du nombre des transcendantes, je ne laisse pas (supposita ejus constructione) d'en pouvoir donner non seulement les touchantes, mais encor la dimension de la courbe, la surface du solide de sa rotation et la dimension de l'espace compris de la courbe et de l'axe; et le calcul m'offre tout cela comme de soy meme (AA III, 4, p. 622).

Even though Leibniz claimed that both results were identical except for one sign, he did not send his demonstration to Huygens. In a letter on the 18th of November 1690 (where Huygens gave name to the catenary), the Dutch mathematician claimed that since the only difference between their results was a sign, it was surely caused by their different methods. He insisted again that Leibniz send him his complete solution (AA III, 4, p. 654). However, Leibniz maintained silence until a letter on the 2nd of March 1691 (AA III, 5, pp. 58-64), in which he confirmed to Huygens that John Bernoulli had also found his own solution and that the infinitesimal calculus certainly should have aided him (although for a mind like Leibniz's this issue was not especially complex, as he said). Huygens wrote Leibniz two more letters insisting on seeing his solution, but Leibniz did not agree. For that reason, Huygens had to wait until the final publication in the *Acta Eruditorum* on June of 1691.

The articles published as part of Jacob Bernoulli's challenge were:

- Leibniz, "De linea in quam flexile se pondere propio curvat, ejusque usu insigni ad inveniendas quotcunque medias proportionales et logarithmos" (AE June 1691, pp. 277-281; GM 5, pp. 243-247; Leibniz 1989, pp. 186-199; Leibniz 2001, pp. pp. 55-58).
- John Bernoulli, "Solutio Problematis Funicularis, exhibita a Johanne Bernoulli" (AE June 1691, pp. 274-276; GM 5, pp. 248-250).
- Huygens, "Dynastae in Züchelem, solutio ejusdemproblematis" (AE June 1691, pp. 281-282; OC 10, pp. 95-98; GM 5, pp. 251-252).

Added to these articles are four more pieces:

1) The first one, written by Jacob Bernoulli, that acted as an epilogue closing the challenge and the solutions presented by Leibniz, Huygens and his brother John, "Specimen alterum calculi differentialis in dimetienda spirali logarithmica, loxodromiis nautarum, et arcis triangulorum sphaericorum; una cum aditamento quodam ad problema funicularum alisquee" (AE June 1691, pp. 288-290; GM 5, pp. 252-254).
2) A short article by Leibniz, "De solutionibus problematis Catenarii vel Funicularis in Actis Junii A. 1691, aliisque a Dn. Jac. Bernoullio propositis" in which he provided a more detailed description of the catenary published in the *Acta Eruditorum* of September 1691 (AE September 1691, pp. 435-439; GM 5, pp. 255-258; Leibniz 1989, pp. 200-205; Leibniz 2001, pp. 58-61).

3) Finally, Leibniz recapitulated the challenge of the catenary in two articles written in 1692: "De la chainette, ou solution d'un problème fameux proposé par Galilei, pour servir d'essai d'une nouvelle analyse des infinis, avec son usage pour les logarithmes, et une application a l'avancement de la navigation" (GM 5, pp. 258-263; JS 1692, pp. 147-153), written in French and in which Leibniz notes the uses of the mathematical description of the catenary for navigation; and,
4) "Solutio illustris problematis a Galilaeo primun propositi de figura chordae aut catenae e duobus extremis pendentis, pro specimine novae analyseos circa infinitum" (GM 5, pp. 263-266; GLM 1692, pp. 128-132).

Once all the solutions were published Leibniz gave Huygens his opinion about the similarities and differences of each of the mathematical descriptions (it seems Leibniz had access to the *Acta Eruditorum* of June before Huygens). In a letter to Huygens on the 24th of July 1691 he made reference to the use of his new method by John Bernoulli, reducing the problem to the quadrature of the hyperbola[7] due to their mathematical similarity:

> [C]omme les Equations communes pures se resolvent par la seule extraction des racines ; de même les Equations pures Transcendentes sont reduites aux sommes ou quadratures. Et comme on n'a pas encor perfectionné l'Analyse des Equations ; de même n'ay je pas encor tellement perfectionnée ma methode que je puisse tousjours reduire tous ces problemes Transcendans aux seules quadratures. Cependant j'en ay ouvert le chemin, et même je suis assez avancé. Aussi les deux problemes, que vous m'avés proposés, Monsieur, qui donnent d'abord une Equation Transcendente Affectée, se peuvent tous deux reduire aux quadratures ; at même je trouve qu'ils ne supposent que la quadrature de l'Hyperbole pour estre construits. Et quand je puis reduire ces problemes aux quadratures, je crois alors d'avoir surmonté la plus grande difficulté. Cependant pour perfectionner cette Methode il faut aussi achever la doctrine des quadratures (AA III, 4, p. 596).

Huygens' response was delayed until the 1st of September 1691 (AA III, 5, pp. 157-164) when he admitted he had not related the catenary with the quadrature of the hyperbola. The truth is that he suspected that Leibniz and John Bernoulli could have shared their solutions before the publication in the *Acta Eruditorum*. That could explain the enormous similarity between their procedures as he expressed to Leibniz in a not very affable tone (AA III, 5, p. 168). The truth is that even though John Bernoulli realised that the quadrature of the hyperbola could be used to draw the

---

[7] "Luy [Bernoulli] et moy nous avons reduit le probleme à la quadrature de l'Hyperbole, nous avons donné tous deux non seulement les tangentes et l'extension de la courbe, mais aussi le centre de gravité de la courbe, et moy j'y ay adjouté le centre de gravité de l'espace. Nous avons donné tous trois les tangentes et l'entendue de la courbe. Mons. Bernoulli s'est rencontré avec Vous Mosieur à penser à la courbe dont l'evolution sert à descrire la ligne catenaire, et il a remarqué là dessus de fort jolies choses" (AA III, 5, p. 133).

catenary, Leibniz went further reducing everything to logarithms, and in doing so determining how future mathematics would be used:

> I have even added to my solution the center of gravity of this last figure, that is, of its area. M. Huyghens gives the construction of the curve by supposing the following quadrature: $xxyy = a^4 - aayy$, while M. Jean Bernoulli, and myself, have related the catenary to the quadrature of the hyperbola; this last one makes an absolutely judicious use of the quadrature of a parabolic curve, while for my part, I have reduced everything to logarithms; I have determined in this way the type of expression, as well as the best of all possible constructions, for transcendentals. Indeed, all you need to know is a unique constant proportion, which will enable you to discover an infinity of points, using only ordinary geometry, and without any more need of quadrature or rectification (AE September 1691, p. 436; GM 5, p. 255; Leibniz 1989, p. 204; Leibniz 2001, p. 59).

The calculus created by Leibniz was so precise that he was suspicious of plagiarism. However, Leibniz defended himself in a letter on the 21st of September 1691 (AA III, 5, pp. 171-179) claiming that the editors of the *Acta Eruditorum* had dealt with John Bernoulli's contribution with strict confidentiality. In addition, Leibniz pointed out various mistakes made by Huygens. He had published in the *Acta Eruditorum* June 1688 an article called "De Angulo contactus et osculi" in which he had reduced the catenary to the sum of the secants of the arc; therefore he should have related the catenary with the quadrature of the hyperbola. Huygens' answer in a letter on the 16th of November 1691 (AA III, 5, pp. 196-202) served as the reconciliation of this dispute. He stated it was his desire to study, in deep, his new infinitesimal calculus to finally understand the relation between the catenary and the quadrature of the hyperbola.

According to Yoder even L'Hôpital, who was not as good of a mathematician as Huygens was, managed to modify his calculations to rectify the exponential curve that Huygens had introduced in his solution (Yoder 1988, p. 176). In spite of this, and even after the finalization of Jacob Bernoulli's challenge, Huygens still insisted on the adequacy of his method in respect to Leibniz's (OC 10, pp. 305-306). This clearly shows that Huygens had not been able to follow the path created by his former student; and that even though Leibniz's new calculus would become the standard in mathematics, he did not understand its utility.

4 Philosophical Implications of the Catenary Challenge

Following Leibniz's article in June of 1691 (*De linea in quam flexile se pondere propio curvat...*), the mathematical description of the catenary has two applications: first, to understand and extend the utility of *analysis* (or calculus) with which Leibniz wanted to solve the most important problems of geometry; and second, to make progress on the art of construction, or problem solving.

Leibniz's solution to the problem of the catenary served as a justification for his contemporaries, especially Huygens, of the use and utility of his infinitesimal calculus. With this solution, he was able to present a general method to find the mathematical description of every mechanical curve. This method provided a way to overcome the mathematical infinity problem.

> [J]e crois de voir qu'on ne sçauroit reduire toutes le quadratures generales ou indefinies à celles du cercle et de l'Hyperbole et lors qu'on aura reduit les problemes auz quadratures, et les quadratures à de certains chefs, comme je le projette, cette Espece d'Analyse sera arrivée à sa perfection. Et vous sçavés, Monsieur, que c'est là plus sublime partie de la Geometrie, et même la plus importante, car ordinairement lors qu'on applique la Geometrie à quelques Problemes difficiles de la nature ou de la Mecanique, on vient à ces Equations Trascendentes ; dont la raison est, que la nature va ordinairement par des changemens continuels, ou instantanées, qui ne sont autre chose que mon dx ou dy (AA III, 4, p. 597).

Leibniz's new method also justified his vision of the empirical and metaphysical reality; that every possible problem could be solved by starting from the empirical cases and the solution for these empirical cases would indicate how to solve metaphysical problems (so it is a departure from philosophy in order to arrive at a philosophical solution). The new calculus was successfully applied by John Bernoulli, who had arrived to the same conclusion in the catenary challenge. This was for Leibniz a clear indication of the impartiality of his analysis which produced truth in science. Therefore, in a certain way, one could apply a philosophical method in order to solve all philosophical problems as if it was actually an algebraic philosophy.

Moreover, Leibniz's calculus is closely connected with metaphysics through the use of the infinitesimals and the notion of infinite. However, its results are clear and evident, in contrast with the results that any metaphysical method could offer in principle. That does not mean that during the process of its creation it did not have any philosophical nature or major implication. First, metaphysics actually works as a kind of universal mathematics (AA VI, 6, p. 478); and second, it was precisely while studying a philosopher, Hobbes, when Leibniz began to be interested in infinitesimals. Nevertheless, because Hobbes had a bad reputation as a mathematician, many researchers have rejected the idea of him truly influencing Leibniz's mathematics (Goldenbaum and Jesseph 2008, pp. 55-56). More specifically, we can state that one of the origins of his interest in infinitesimals was the Hobbesian *conatus*.

In his *analysis situs* Leibniz tried to systematise the study of space; with mathematics, as we have seen, the problems generated by empirical phenomena; with his mechanics, problems over the nature and movement of bodies; and with his theodicy, issues concerning evil, suffering and the necessity of God. For Leibniz, each of these disciplines had its own role in order to explain reality's phenomena. The project of a universal science lies beneath his works and it would later unify all these small efforts to form a final system, a universal science. And even before he studied with Huygens, Leibniz already believed that mathematics had a central point in this corpus of sciences together with theology, metaphysics, moral philosophy and physics because it deals with the form or idea of things[8].

Leibniz's goal and vision greatly contrasted with the praxis of other modern scientists such as Newton and Huygens. In a letter to Remond on the 10th of January 1714 Leibniz claimed that Huygens had not cultivated any taste for metaphysics (GP III, p. 607)[9], and that he was not prepared to deduce the truly philosophical consequences from his works. In the context of the letter Descartes' philosophy was considered a precursor of the truth, which would be reached little by little. This shows us that, in a certain way, Leibniz was thinking about the possibility of reaching a final explanatory system that although not built following the Cartesian method, it could take its explanatory spirit and make it real. The case of the study of the catenary confirmed Leibniz's idea about this.

What's more, Leibniz's claim that Huygens did not cultivate metaphysics was misguided; Huygens knew his works had philosophical implications. As Dugas said, Huygens was aware of the necessity of adding principles and hypothesis to the experience since the experience alone could never produce any law of motion in mechanics (Dugas 1954, p. 286). It is difficult to claim that Huygens was a metaphysician per se due to the mathematical nature of the vast majority of his works; but this claim shows that the necessity of a philosophical base for mechanics was not unknown to him[10]. However, Huygens' approach to mathematical and mechanical problems, like the description of the catenary, never allowed him to transcend to metaphysical implications as Leibniz did.

5 Concluding Remarks

The success of Leibniz's infinitesimal calculus in the challenge of the *Acta Eruditorum* would consolidate its utility. Moreover, Leibniz's vision of a universal science would also be strengthened.

---

[8] "So there occurs to me, as I write this, a beautiful harmony among the sciences; namely, that under careful examination it appears that theology or metaphysics deals with the efficient cause of things, or mind; moral philosophy, whether ethics or law (for as I learned from you, these are one and the same science), deals with the final cause of things, or the good; mathematics (I mean pure mathematics, for the rest is a part of physics) deals with the form or idea of things, or figure; physics deals with the matter of things and the unique affection resulting from the combination of matter with the other causes, or motion. For mind supplies motion to matter in order to achieve a good and pleasing figure and state of things for itself. Matter in itself is devoid of motion. Mind is the principle of all motion, as Aristotle rightly saw" (Leibniz 1956, pp. 98-99; AA, II, 1, 31).

[9] Heinekamp mistakenly claimed in "Huygens vu par Leibniz" (Taton 1982, p. 106) that the reference is GP, III, 611.

[10] Moreover, Huygens wrote a piece at the end of his life called *Cosmotheoros*, posthumously published, which could be seen as a philosophical exercise. In this work Huygens recognized the necessity of certain phenomena, such as the existence of valleys, mountains and trees on other planets, based on their existence on the Earth. This speculative methodology marked the whole book.

Huygens, as his former master, played an essential role in Leibniz's academic training. However, it should be noted that he was reluctant to accept new methods, especially if it were as innovative as Leibniz's. Thus, at first he did not understand neither its advantages nor Leibniz's promises of its great usefulness. In spite of that, Huygens was responsible for guiding Leibniz in the path that would lead to the infinitesimal calculus, and it appears that in the end he convinced Huygens as evidenced it in his "Historia et Origo Calculi Differentialis":

> But Huygens, who as a matter of fact had some knowledge of the method of fluxions as far as they are known and used, had the fairness to acknowledge that a new light was shed upon geometry by this calculus, and that knowledge of things beyond the province of that science was wonderfully advanced by its use (Leibniz 1920, p. 25).

The fact that an empirical phenomenon like the form of the catenary can be described through geometry and mathematics supports the idea of mathematics describing all phenomena and presenting a model to describe reality through the infinitesimal calculus[11]. This means that the description of the catenary curve serves as a schema on how we should solve other problems to reach "the perfection of science"[12]. In principle the meaning of the word *description* does not have an explanatory sense by itself, but this word has another sense that include a reproduction of categories of phenomena. To describe the curve of the catenary is not to find a mere mathematical representation but to find a unitary equation in which all infinite points of the catenary are given. This description is the unification of those infinite points; and it entails a biunivocal correspondence between the points, which is the unity of the dispersed. Leibniz exports this schema to other disciplines, as philosophy.

This could lead to various consequences. First, that for Leibniz every phenomenon is translatable to scientific mathematical language; secondly, that every scientific explanation points at a truth; and finally, that philosophical problems like

---

[11] "While Leibniz in his mathematical practice generally sought to steer clear of metaphysical disputes and instead placed emphasis on effectiveness and reliability of procedures, particularly when working with the infinite, he ultimately produced a philosophical system which in remarkable fashion was able to account for the successes of mathematical science in contributing to our understanding of nature and over and above this to the increase in human wellbeing. In a very true sense his philosophical deliberations on mathematics were deliberations pertaining to life" (Beeley 2014, pp. 46-47).

[12] "Itaque quod primun attinet, constat iis, quibus nota est Historia nostri temporis literaria, magnam incrementorum scientiae partem problematum propositioni deberi, idemque in futurum licet augurari, si scilicet problemata nondum sint in potestate receptae Analyseos. Ita enim discitur, quae *desiderata* ad perfectionem artis supersint, simulque ingenia *ad augmento scientiarum* animantur. Certe ut olim Cycloidem, ita nuper Catenarium plurimum profuisse constat. Neque ego Catenariam ipse delegi, sed ab alio mihi propositam et confestim solvi et proposni aliis porro. Nec *Dn. Johannes Bernoullius* in suo problemate Lineae brevissimi descensus diu laboravit: nempe non casui, sed methodo succesum debuimus" (GM V, 341-2; OFC 7A, 373).

the existence of infinites could be resolved in practice (or skipped in the case of the infinitesimal calculus).

The mathematical description of the catenary supported Leibniz's idea of an explanatory system of reality. His hypothesis about how to decipher, use and understand the world would be supported by the success of his calculus in Jacob Bernoulli's challenge. In this sense, Leibniz rapidly overtook his master Huygens regarding the implications of their studies on geometry and mathematics. While Huygens was mainly focused on the solution of mathematical problems and the creation of proofs, Leibniz was more interested in the philosophical implications of those results and proofs reached through the use of mathematics. This determined not only his methodology but also his vision in regards to the reason why mathematics is essential to understand reality.

Acknowledgements

This work has been financed with the funds of the program FPU (Formación del Profesorado Universitario) of the Spanish Government, Ministerio de Educación, Cultura y Deporte, with reference FPU13/00725. I am also indebted to the anonymous referees who contributed helpful commentaries and suggestions.

References

Abbreviations

>AA = Leibniz 1923 ff.
>AE = Acta Eruditorum.
>GLM = Giornale de Letterati di Modena.
>GM = Leibniz 1849-1863.
>GP = Leibniz 1875-1890.
>JS = Journal des Sçavans.
>OC = Huygens 1888-1950.
>OFC = Leibniz 2007 ff.

Beeley, P. 2014. Leibniz, Philosopher Mathematician and Mathematical Philosopher. In: Goethe, N. B., Beeley, P., Rabouin, D. (Eds.). *G.W. Leibniz, Interrelations Between Mathematics and Philosophy*. Dordrecht: Springer.

Crippa, D. 2014. *Impossibility Results: From Geometry to analysis*. Doctoral Thesis. Paris: Université Paris Diderot (Paris 7).

Descartes, R. 1925. *The Geometry of René Descartes*. Chicago: The Open Court Publishing Company.

Dugas, R. 1954. *La mécanique au XVIIe siècle : des antécédents a la pensée classique*, Neuchatel: Griffon.

Galilei, G. 1898. Discorsi e Dimostrazioni Matematiche Intorno a Due Nuove Scienze. In: Favaro, A. (Ed.). *Edizione nazionale delle opere di Galileo Galilei*. Vol. 8. Florence: G. Barbèra.

Galilei, G. 1914. *Dialogues Concerning Two New Sciences*. New York: The McMillan Company.

Goldenbaum, U., Jesseph, D. (Eds.). 2008. *Infinitesimal Differences: Controversies Between Leibniz and His Contemporaries*. Berlin–New York: Walter de Gruyter.

Huygens, C. 1888-1950. *Œuvres complètes de Christiaan Huygens*. 22 vols. The Hague: Martinus Nijhoff.

Leibniz, G. W. 1849-1863. *Mathematische Schriften*. 7 vols. Berlin: Hildesheim.

Leibniz, G. W. 1875-1890. *Die philosophischen Schriften*. 7 vols. Berlin: Hildesheim.

Leibniz, G. W. 1989. *La naissance du calcul différentiel : 26 articles des « Acta Eruditorum »*. Paris: Vrin.

Leibniz, G. W. 1920. *The Early Mathematical Manuscripts of Leibniz*. Chicago: The Open Court Publishing Company.

Leibniz, G. W. 1923 ff. *Sämtliche Schriften und Briefe*. Berlin: Akademie-Verlag.

Leibniz, G. W. 1956. *Philosophical Papers and Letters*. Chicago: The Chicago University Press.

Leibniz, G. W. 2001. Two Papers on the Catenary Curve and Logarithmic Curve. *Fidelio*, 1:54-61.

Leibniz, G. W. 2007 ff. *Obras filosóficas y científicas*. Granada: Comares.

Leslie, J. 1821. *Geometrical Analysis and Geometry of Curve Lines*. Edinburgh: W. & C. Tait.

Nahin, P. J. 2004. *When Least is Best*. Princeton: Princeton University Press.

Salmon, G. 1879. *A Treatise on the Higher Plane Curves*. Dublin: The Dublin Trinity College.

Taton, R. 1982. Huygens vu par Leibniz. In: *Huygens et la France*. Vrin: Paris.

Yoder, J. G. 1988. *Unrolling time: Christiaan Huygens and Mathematization of Nature*. Cambridge: Cambridge University Press.

Miguel Palomo, Universidad de Sevilla, Spain
miguelpalomo@us.es

The Dialectics of Recognition in Leibniz's *Theodicy* and a Possible Route to Modern Political Thought

Ana-Maria Pascal

Abstract. This paper discusses Leibniz – the philosopher, and his contribution to the modern conceptual framework focused on the dialectics of reason. This, we shall argue, is particularly relevant within the history of socio-political ideas. The will and power of the mind to enter a dialectical dialogue with religious faith and move towards self-realisation is a symptom of a set of games of recognition that were widely played in early modernity, before the *Phenomenology of Spirit*. The 'recognition of the other' issue, which is at the core of many philosophical debates, from metaphysics to sociology, can be traced back to ancient mythology. The history of modern thought consolidates the theme through Hegel's analysis of the master-slave dialectics. We take this genealogy back to Leibniz, and read his *Theodicy* as a confrontation between reason and faith, rather than a unilateral attempt at legitimising the latter by means of the former. We tell the story as an epic, rather than a linear discourse, and follow the stages of the confrontation, which plants the seeds for the Hegelian dialectics. The issue of recognition will turn out to be at the very core of this epic, as both faith and reason need each other's acknowledgment in order to exist, and before any peace can be negotiated. The paper ends with an inquiry into the role this dialectics of recognition plays in various forms of government, and suggests ways that this might be reflected in early modern political thought, from Machiavelli to Hobbes, and Locke to Rousseau.

Keywords: Master-Slave Dialectics, God's goodness and omnipotence, Logical principles, Rapport with the other, World rationality, Rationalisation of faith, Recognition game(s), Faith and Reason, Strategy, *Theodicy*, Government, Early modern political thought.

Ana-Maria Pascal (2017) The Dialectics of Recognition in Leibniz's Theodicy and a Possible Route to Modern Political Thought. In: Pisano R, Fichant M, Bussotti P, Oliveira ARE (eds.), *The Dialogue between Sciences, Philosophy and Engineering. New Historical and Epistemological insights. Homage to Gottfried W. Leibniz 1646-1716*. College Publications, London, pp. 341-361
© 2017 College Publications Ltd | ISBN: 978-1-84890-227-5 www.collegepublications.co.uk

1 Introduction

Leibniz's contribution to a wide range of disciplines is indisputable and, perhaps, only comparable to that of Aristotle and Blaise Pascal. As a consequence, one can only agree with Fontanelle's 1740 homage in that "While antiquity made one Hercules out of many, we make many savants out of one M. Leibniz's" (Fontanelle, cit. in Garber, 2011, p. xv). The savant we are going to discuss in this chapter is Leibniz – the philosopher, and his contribution to the modern conceptual framework focused on the dialectics of reason. The will and power of the mind to enter a dialectical dialogue with religious faith and move towards self–realisation is, we shall argue, a manifestation of a whole set of games of recognition that were widely played in early and mid-modernity, before the *Phenomenology of Spirit*.

The *recognition of the other* issue, which is at the core of many philosophical debates, from metaphysical to sociological ones, can be traced back to Greek mythology. In ancient Athens, there used to be a temple dedicated to the unknown God, which Apostle Paul takes for a sign of humility in the nature of the Greek soul. In his Areopagus sermon, he confirms that

> [...] while I was passing through and examining the objects of your worship, I also found an altar with this inscription, 'TO AN UNKNOWN GOD.' Therefore what you worship in ignorance, this I proclaim to you (Acts 17:23).

The history of modern thought consolidates the theme through Hegel's analysis of the dialectics between master and slave. Indeed, Charles Taylor considers this to be the very beginning of all serious debates on the issue: "The topic of recognition is given its most influential early treatment in Hegel" (Taylor, 1994, p. 36). At the same time, the Canadian philosopher considers the need for recognition to be one of the top three "malaises of modernity" (Taylor, 1992). We shall return to this in section 5 of the paper.

First, however, we propose to take this genealogy further back to Leibniz (in the next two sections), and read his *Theodicy* as a confrontation between reason and faith, rather than a unilateral attempt at legitimising the latter by means of the former. In what follows, we shall tell the story of the Theodicy as such – as an epic, rather than a linear discourse – and follow the various stages of the confrontation therein, which plants the seeds (we shall argue) for Hegelian dialectics. The issue of recognition will turn out to be at the very core of this epic, as both faith and reason need each other's acknowledgment in order to exist, and before any peace can be negotiated.

As the epic unfolds, we highlight the strategy and the tools that reason uses (with a certain degree of cunning) in order to legitimise itself in the process of attempting to confer legitimacy to religious faith. We then proceed (in section 4) to compare and contrast this epic with the Hegelian three-stage dialectics, to see how far might Leibniz go in setting the ground for a *Phenomenology of Spirit*. This may turn out to be far enough, but not all the way through. As we shall see, while the first two stages of the master-slave dialectics can be identified in the *Theodicy*, the third one cannot. The integration of the two into something larger or higher than both (a spirit that is neither purely rational nor religious alone) will have to wait another century to occur.

After a brief conceptual interlude about the need for recognition in modern times, in section 5 we briefly explore a possible route – or fragments of a journey that would take us – from Leibniz to *early modern political thought*, from Machiavelli to Hobbes, and Locke to Rousseau. The key signpost to that route will be the dialectics of recognition, and the role this plays in both politics and the rapport between reason and faith. We shall indicate possible ways of identifying the *Theodicy* model (understood as a recognition game) in each of the main political systems, and discuss how early modernity rationalised it.

We also attempt to understand if any of these positions represents a step closer (than the *Theodicy* itself) to the Hegelian model of recognition, which is based on the master/slave dialectics. One way of pursuing this investigation is to see to what extent the early version of the dialectics of recognition (the *Theodicy* model as a confrontation, rather than a harmonious, progressive project of reason, in support of faith) can be identified in any of the forms of government theorised at the time. However, this is more of a hypothetical or interpretive journey than a categorical claim, and the reason we raise this issue is twofold – first, because of the clear connection between Leibnizian and Hegelian dialectics, and second, because of the role this (and any relationship with the other, in general) plays in any political system.

In the last section and the conclusions, we close the loop of our reflection on the issue of recognition in modern philosophy and political thought, by suggesting Leibniz's dialectics may open the way to the latter, in ways that are yet to be fully explored.

2 The *Theodicy* Plot and Main Characters

The confrontation between faith and reason takes place at the beginning of Leibniz' *Theodicy*, which is appropriately titled a "Preliminary Dissertation on the Conformity of Faith with Reason". It is worth reflecting on the fact that Leibniz dedicates the first fifty pages of his *Theodicy* to this topic, before moving on to the "Essays on the Justice of God and the Freedom of Man in the Origin of Evil". One could argue that he saw it necessary to establish the pact between faith and reason, before launching into a debate on God, evil, justice, and freedom – and that the reason for that is because that was reflective of early modernity mindset. In other words, the paradigmatic rapport between reason and faith was a fundamental preoccupation in 17th century philosophy. Furthermore, since Leibniz was not only a philosopher, but also a mathematician, physicist, lawyer, and moralist, one could indeed argue that his choice reflected a wider preoccupation with the status and role of reason in matters to do with both social and intellectual enquiry, in the period generally referred to as early modernity – namely, 16th to 18th centuries.

One thing is certain; the so-called 'conformity' of faith with reason is far from being a given at the time, or else Leibniz would not need fifty pages at the beginning of his *Theodicy* to establish it. On the contrary, it will be argued in this paper – one can assume the opposite of 'conformity' to better characterise the rapport between the two, which therefore makes the whole debate about status, recognition, and relationship absolutely necessary. There is an actual need for a debate on this rapport in an epoch historically referred to as both rationalist and profoundly religious. Hence – the perfect scenery for a whole set of recognition games, which is exactly – we shall argue – the aim not only of this preamble, but also of the *Theodicy* as a whole. The Dissertation is, indeed, organised as a staged confrontation between two well defined characters, taking turns to claim centre stage and set the rules of the game. Even the style of the Dissertation is epical and somewhat official, if not judiciary – in both its tone and the overall rigour and discipline – as a constant reminder of what (and how much) is at stake. Much more, it shall be argued, than a purely theoretical introduction.

Our first thesis is thus, that what Leibniz puts forth in his Dissertation is a debate – an actual questioning and a confrontation, rather than a mere justificatory report. In fact, what prompts Leibniz's investigation, according to his own admission in the Preface to the *Theodicy*, is a certain precedent – namely, that "one of the most gifted men of our time" (Pierre Bayle) "has applied himself with a strange predilection to call attention to all the difficulties of this subject" (Leibniz 2009, p. 62). This is certainly not a coincidence, given the influence that Bayle exercised on the Enlightenment, with his tolerance to different religions, and his general view that theology and philosophy were deeply heterogeneous. However, while "M. Bayle wishes to silence reason after having made it speak too loud: which he calls the triumph of faith", Leibniz declares himself to be "of a different opinion" (*Ivi*, p. 63) – hence, this Dissertation.

Leibniz acknowledges that "the question of the conformity of faith with reason has always been a great problem" (*Ivi*, p. 76), which preoccupied hearts and minds since antiquity, and was disputed, for instance, by Plato, Aristotle, the Stoics, early Christian theologians, and the Italian Averroists. Likewise, he says, "the question of the use of philosophy in theology was debated much amongst Christians, and difficulty was experienced over settling the limits of its use when it came to detailed consideration" (*Ivi*, p. 83). In addition, as briefly stated before, in his Preface, Leibniz announces his own stance by saying he respectfully disagrees with Bayle about the heterogeneity of faith and reason. For Bayle, reason should remain silent and bow to faith, whereas Leibniz believes that reason has an active role to play in legitimising its opponent – that is, when matters of ethics and religion are discussed (e.g. God, evil, freedom, or justice). In what follows, we propose to shadow Leibniz in his endeavours to carve just such a role for the intellect.

It is worth reflecting, first, on the significance of Leibniz' starting point – namely, Bayle's view that faith and reason are entirely distinct and separate. What does it mean that in Bayle's *Historical and Critical Dictionary* (a seminal work at the time), religion and reason appear not only as entirely separate, but also as adversaries? What else could the 'silence of reason' refer to – if not to its defeat? Leibniz does not accept this as a *sine qua non* for the triumph of faith. On the contrary, he believes reason to be an ally, with a key role to play in religious and metaphysical debates. For Leibniz, "reason is the linking together of truths" (*Ivi*, p. 73) rather than a strict opponent to either experience or faith.

The question is, however – if reason volunteers to support faith, is that a purely disinterested offer? Since the most likely result is a *rational* faith, is it not true that reason itself stands to gain from such an endeavour more than religion does? In other words, by offering its support, doesn't reason actually turn its opponent into something resembling itself? Moreover, would not *that* be a victory for rationalism, rather than for theology? Could this be the real plot in Leibniz's *Theodicy*? Is this the seed of Hegel's "cunning of reason", and Foucault's "instrument for cutting"? Such are the kind of questions that are going to guide us in the following pages, and form the background of what we have been referring to as a set of *recognition games* in modern times. Once we determine the aim and the strategies of this game in the case of the rapport between reason and faith, we shall reflect on how it might manifest itself in political theories of the time, and with what consequences.

3 Five Strategies, One Game: Round One

The whole debate on the goodness of God, the freedom of man and the origin of evil takes place in a rational context, using rational tools and techniques. Leibniz employs a series of strategies, which are officially aimed at supporting religious faith, but which may – at the same time – be meant to strengthen the role of reason itself.

One such strategy consists of trying to unify the two supreme authorities – the divine being and the moral law – into one, and give this new entity a rational foundation. For Leibniz, as we know (because he announces it in the *Preface*), the divine being itself is "Supreme Reason" and true piety does not rely on love alone; instead, it consists of "enlightened" love, where the enlightenment, of course, comes from knowledge (Leibniz 2009, pp. 51-52). In order to achieve that, Leibniz believes that "the perfecting of our understanding must accomplish the perfecting of our will"; and "when virtue is reasonable, when it is related to God, who is the supreme reason of things, it is founded on knowledge" (*Ivi*, p. 52). In other words, what Leibniz does here really is to construe both morality and metaphysics in the likeness of reason or the human intellect. Whether this amounts to a reductionist treatment or not is a separate matter; the point we would like to make is that Leibniz's aim may not be as straightforward as it is generally believed to be, and that the main character in his plot (i.e., reason) may not be as disinterested as it seems. To say that "one cannot love God without knowing his perfections, and this knowledge contains the principles of true piety" (*Ibidem*) is to declare reason to be the dominant route to God. This comes soon after declaring it the common ground between the ontological and the moral realm. It is a powerful strategy – and it works. (For a reflection on how this might be translated in terms of a recognition game, please refer to the next section).

Another strategy employed by Leibniz (or, should we say, by reason itself, as the main actor in the *Theodicy*) is to camouflage itself – and its own interest – by camouflaging one of its key traits: the negative one – doubt. Uncharacteristically for any rational process, Leibniz's rational authority does not exhibit any sceptical feature, it does not feel the need to raise or overcome any doubt. This is indeed most surprising for a rational enterprise such as this, and it begs the question whether or not it was planned this way, in order to mitigate the risk of reason showing itself too much – or in its true colour – thereby undermining its purpose of associating itself with religious faith in order to conquer it. The intellect does not openly interrogate faith because it would not suit its purpose to do so.

What Leibniz does do, however – and this is a subtler and more efficient strategy – is to turn what would otherwise be a weakness of reason into its strength, when he discusses free will and the origin of evil. Despite the fact that this constitutes one of the "famous labyrinths where our reason often goes astray" (*Ivi*, p. 53), it happens to provide a solution too, because Leibniz considers reason to be our main tool and support in confronting all difficulties. In other words, it is because we are rational human beings that we can defend ourselves from evil. Not to do so – Leibniz believes – would actually be lazy, rather than pious. This is known as the "lazy argument" ("derived from the idea of inevitable fate"), which Leibniz denounces as a pure sophism and a source of superstitious practices like fortune-telling (*Ivi*, pp. 55-56, p. 61). Instead of blind faith, which leads to such passive acceptance of necessity, Leibniz advocates an active, enlightened faith, which is based on reason.

It is unsurprising, then, that this reasoning soul acknowledges God as the supreme reason, *Intelligentia extramundana* – or *supramundana*, as Leibniz prefers to call it (*Ivi*, p. 264), rather than through any other divine features (goodness, will, power etc.). Analogously, human intelligence is "the soul of freedom" (*Ivi*, p. 303), and the world itself is the best there is, not in an ethical sense but from the point of view of its rationality: for all the evil in the world, there is always some good which more than compensates for it. All these, we might note – divine wisdom, human intelligence, and the rationality of the world – concur in establishing what could easily be interpreted as the triumph of reason, by contrast to Bayle's "triumph of faith", which Leibniz criticises (*Ivi*, p. 63).

This brings us to one of the most complex and effective strategies in *Theodicy* – the apparent canonisation of reason. Far from being some sort of illusion, this constitutes a lucid, elaborate act of conversion, which is absolutely necessary in the process of rationalisation of faith. It is, in effect, like learning to speak a foreign language – or adopting the habits of a foreigner, in order to become friends with them, and eventually win them over. Let us see how reason does this, in her rapport to religious faith. Although this is introduced, right from the beginning (in paragraph 1) of the Preliminary Dissertation, as "the linking together of truths, but especially (when it is compared with faith) of those whereto the human mind can attain naturally without being aided by the light of faith" (*Ivi*, p. 73), reason will soon claim to be 'enlightened' by it and belong to the same realm as this light.

One of the few places in the *Theodicy* where reason shows its true identity is where Leibniz distinguishes between two meanings of the term: one common or weak, of simple opinion or habit "of judging things according to the usual course of Nature", the other – strong or metaphysic, referring to "the inviolable linking together of truths" (*Ivi*, p. 88). The former type of reason would always allow views for and against it, and acknowledge a superior level of intelligibility and authority. Taken in its second sense, however, reason claims itself to be absolute – and it is this type of reason that is running the show in the *Theodicy*; this is, indeed, the level at which Leibniz pitches his arguments – the level of necessity and universality. Having first claimed that all reason can do is *uphold* the religious mysteries against objections, rather than try to prove them (*Ivi*, p. 76), the author of the Dissertation is quick to remind us, however, that

> In general, one must take care never to abandon the necessary and eternal truths for the sake of upholding the Mysteries, lest the enemies of religion seize upon such an occasion for decrying both religion and Mysteries (*Ivi*, p. 88).

Examples of mysteries are the Holy Trinity, God's miracles (such as, the creation) and the order of the universe. These, Leibniz explains, are truths *above* reason, because we cannot comprehend them though our intellect alone. We can only believe in them. But they should not be *against* reason, since the type of reason referred to is the metaphysical one, understood as "the inviolable linking of truths" (*Ibidem*, p. 88). These are the necessary, inviolable because logical types of truths.

What seems to be happening here is a transfer of legitimacy, from the eternal mysteries and logical (universal) truths onto a would-be reason, which thereby becomes as universal, inviolable and eternal as they are. Reason, in other words, places logical principles above the everyday type of human rationality (or opinion), and at the same level as the eternal truths stipulated by the religious faith. Whether this helps supporting faith itself (as it is supposed to, if we are to believe the official intention announced in the Preface), is debatable; what is certain, I think, is that it suits reason itself.

Incidentally, the strategy does not hold beyond a very restrictive point of view. It is somewhat similar to a scenario where we would be asking a patriarch to obey an army general because he never made it beyond an officer rank during his army training as a young man. The request would only make sense in a very limited context, to do with army hierarchy, which one would only be likely to encounter in a war situation. All other scenarios – in public as well as private life – would make the request untenable, or very odd. Analogously, logics and metaphysics can manifest themselves (and have a role to play) in particular contexts, but expecting them to monopolise all possible spheres of rationality is at the very least odd – and arguably unreasonable. In the real world, something resembling Hans-Georg Gadamer's "fusion of horizons" is much more likely to occur than unilateral, restrictive situations like those that require applying the strict rules of military hierarchy. And who is to say that, in a fusion of contexts and value systems, logical or metaphysical reason would still be dominant at all – let alone absolute?

That, incidentally, may be the reason why Leibniz insists on distinguishing between different meanings (or types) of reason, and argues that, in its absolute hypostasis, reason transcends the purely logical type of rationality; but he does not establish whether (or why) we can expect it to transcend it enough as to be able to establish or support (if not prove) the truth of religious faith. The author tries to hedge this by introducing another distinction – namely, that between different kinds of truth. Some truths, such as those of logic, geometry, and metaphysics, have an absolute necessity, whereas others – which are called 'positive' truths – follow from natural laws, and therefore merely have a physical or moral necessity. Leibniz suggests that nothing (not even faith) can contradict the former type (Cfr. *Ivi*, pp. 74-75), but he immediately tries to compensate for this by advancing the thesis about an would-be intrinsic agreement that allegedly exists between the truth of reason and that of faith, in virtue of their common nature – which is wisdom. Wisdom, together with goodness and justice, is the common essence that manifests itself (albeit to different degrees) at both divine and human levels. Therefore, its results "cannot be contrary to revelation" (*Ivi*, p. 76), as they come from the same source and have absolute necessity. This is a strong argument because it is profoundly rational. In addition, what it establishes is the power of reason, rather than that of religious faith. The latter seems to depend on its rapport with the former – not the other way around. A triumph, in other words, for reason – rather than faith.

4 More Strategies, *en Route* to Hegel: Round Two

Having established that what reason and faith are playing is more than a support mechanism volunteered by the former to the latter, and that it amounts to a recognition game where both players have something to gain or lose, let us compare and contrast this epical game with the Hegelian three-stage dialectics, in order to see a) who wins what, and b) how far Leibniz might go in setting the ground for a *Phenomenology of Spirit*.

We have already discussed the main endeavour in the *Theodicy* – the fact that reason is offering its support to faith, in an attempt to justify it – as a strategy used by reason, in order to triumph over it. What we are witnessing, in other words, is a projection in reverse: reason is allegedly 'helping' faith, by justifying it rationally, thereby turning faith into yet another manifestation of itself – into reason. This process of mutual recognition and justification reminds us of the Hegelian dialectics between master and slave. Reason comes very close to doing the work of a slave – namely, to faith – in order to earn recognition for itself. Let us explore this avenue in more detail, and see to what extent the rapport with the other entails a rapport to itself, in the context of the *Theodicy*.

First of all, let us look at what the master-slave relationship entails, and to what extent Leibniz's strategy amounts to a similar dialectical rapport. Most of Hegel's interpreters agree that we can distinguish three stages in it. For our purposes here, we mainly consider four such theorists: Charles Taylor (Taylor 1995), Peter Singer (Singer 1983), Alexandre Kojève (Kojève 1980), and Constantin Noica (Noica 2009). Some of them specifically name these stages, others simply describe the way in which the relationship between the two evolves up to the point that they each gain "self-consciousness". The three stages are as follows:

a) The <u>subordination</u> stage, which is explicitly assumed by one of the two characters. In the *Theodicy*, this is the moment when reason introduces its intention to serve religious faith. Kojève interprets this gesture in terms of autonomy and self-creation, as opposed to mere rapport with one another. For him, the slave "must transcend himself, "overcome" himself, as Slave" and has a predisposition towards this autonomy and self-overcoming, while "the Master is fixed in his Mastery. He cannot go beyond himself" (Kojève 1980, pp. 21-2). Could the same not be said about the "slave" in the Theodicy, doing the work for religious faith – seen as the "master"?

b) The <u>actual work</u> done on behalf of (or in service to) the other. This, in effect, has as a result the fact that one earns one's own identity and, thereby, freedom. Peter Singer interprets this stage as a conflict – not one that is out in the open, but very an important way of proving oneself and "making explicit what was already implicit" (Singer 1983), namely the self-consciousness and self-sufficiency of each of the two characters. The result of this process is a self-recognition of the one doing the work, a recognition which is due exactly to this manifestation of its own specificity (in that work). For the Theodicy, this refers to what reason does, putting itself in the service of religion – allegedly in order to serve the latter, but actually in order to earn and strengthen its own status. The legitimating process is, in fact, an argumentative one – and it results in emphasising the power of judgement itself. At this stage of the *Phenomenology*, argues Kojève, we can say that the slave creates itself (see his *In Place of Introduction*), because he acts upon the world and transforms it – which, in the *Theodicy*, corresponds to a rationalisation process.

c) The third stage in the master-slave dialectics is <u>the need for circularity (or reciprocity)</u> in the relationship with another. The problem, here, is that each of the two characters finds themselves alone with themselves, since the rapport between master and slave is a fundamentally unequal one and, as such, does not allow for genuine (truthful) mutual reciprocity or togetherness. The master's need for recognition cannot be fully fulfilled, because "recognition cannot come from those one does not recognise" (Noica 2009, p. 63); as for the slave, who is more self-aware and more 'in-charge' of its own self (through its work), more 'formed' in the sense of a *Bildung-process* (Kojève 1980, pp. 90-91), that self-awareness is still not complete without recognition from another. Each of the two needs something more, something *deeper* than what they have got – which seems to be just what the other one's got. Each of them needs both *recognition* (from another) and the comfort that *self-awareness* provides. With recognition from the slave but without a confident self-consciousness, the master keeps second-guessing himself; while the slave, although confident in its own self, lacks the confirmation that only another can provide through genuine recognition. They both need to find a deeper level than the level of their present situation, where they can reach satisfaction. That level, suggests the Romanian philosopher Constantin Noica (Noica 2009, p. 73), is the level of the spirit – where they both equally belong, and which can provide them with a sense of unity.

I wonder if we can identify this last stage in the *Theodicy* at all – this need for reciprocity, this circular or spiral-type of movement that takes us beyond the duality of faith v. reason. Is there a 'spirit' level, where faith and reason could not only be reconciled, but also enriched in a way that none of them could achieve, taken alone? Aiming for such a movement towards something else – deeper or wider – would also amount, in practice, to an effective change of behaviour: reason would not relate to faith using just rational arguments and justifications; in fact, it would not even try to justify faith anymore, because this amounts to a process of rationalisation (that is, a reduction of the other to oneself, rather than an aspiration towards something else, deeper and wider, that could comprise both). Reason does not reach this stage in the *Theodicy* – if it had, the whole process would not be a 'theodicy' anymore.

In this case, what *is* implied in the work of the slave, in Leibniz's *Theodicy* – and where does it lead to? The whole process here consists of a Preliminary Dissertation and an essay in three parts (out of which the last two are so cohesive that they could just as well form one part), plus Appendices. It is interesting to note, however, that the strong point of the whole process – the actual nexus of the *Theodicy* – lies in the Preliminary Dissertation. Let us remind ourselves what is at stake here, in these first fifty pages (less than a seventh) of the book. Officially, it is the defence of religious faith through a rational dissertation. However, suffices to look no further than the title of that dissertation – "the conformity of faith with reason" to understand that what this is intended to be is not really a service to religious faith; at best, it may be a rationalisation of faith – its adaptation to a conformity with reason. So what is really at stake could be a defence of reason itself, in an effort to be acquitted of any potential charge of opposition to faith. In other words, what reason is hoping to achieve, through this discourse, is a better position for itself (one that would be seen as being "in conformity with" faith), from which it could then progress further. What other role could such a discourse have, being as it is placed between the Preface and the actual Essay, in Leibniz's book? It would be naïve to imagine this to be a purely formal gesture – placing a discourse there for the sake of having an introduction, which in any case would be redundant after a Preface that clarifies the author's intentions. The Dissertation does more than that – it seeks to prove a point, to persuade us of something. It is a pleading aimed to produce certain effects – in this case, a consolidation of reason's position, besides (rather than in subordination of) faith. This saves reason from having to undertake a defensive effort, while in fact providing it with the opportunity to engage in an offensive. As such, although three quarters of the book (second and third parts of the Essay) appear to be building a rational defence for faith, the real "work" undertaken by reason is quite a different one, and it has quite a different purpose to the official one. We return to these two parts after a brief discussion of the Dissertation – where the real work takes place.

Let us remember that there is an essential difference between the first stage of the master-slave dialectics in Hegel, and its earlier equivalent in the *Theodicy*: faith does not present itself as a master to reason, and it does not ask the latter to obey or to be of service to it in any way. The fact that reason volunteers and attempts to do so indicates something. Here is how Leibniz explains the need to rationalise our faith in God and virtuous behaviour, by invoking a guarantor of their stability:

> The practises of virtue, as well as those of vice, may be the effect of a mere habit, one may acquire a taste for them; but when virtue is reasonable, when it is related to God, who is the supreme reason of things, it is founded on knowledge. One cannot love God without knowing his perfections, and this knowledge contains the principles of true piety (Leibniz 2009, p. 52).

The implicit assumption here – quite a presumptions one, we might add – is that knowledge is superior to both habit and feeling. Virtuous behaviour is deemed insecure, unless founded on knowledge rather than habit, while devotion to God – outright impossible without such rational foundations. Surely if we are to take this line of argumentation seriously, it would be reason itself that would triumph, rather than either virtue or religious faith, as it is claimed. That much for the 'disinterested' character of reason volunteering its services to faith!

One place where, *prima facie*, reason again seems to sincerely constrain itself is in paragraph 5 of the Dissertation, where Leibniz raises the issue of mysteries – and how we relate to them, intellectually. Here, reason does not pretend to *comprehend* them; rather, it can "only" try to "*explain* them sufficiently to justify belief in them" (Leibniz 2009, p. 76). Leibniz would like us to acknowledge this as true humility on the part of reason. But is it, really? Pretending to be able to explain something without understanding it – is that true humility? On the contrary, one could argue, it is yet another indication of reason's confident proposal that it can explain and support everything that would otherwise not lend itself to rationalisation – virtue, mysteries, faith and the love of God. In fact, some fifty paragraphs later, Leibniz admits that we should have no need "to prove the Mysteries *a priori*, or to give a reason for them; suffices us *that the thing is thus* (το ότι) even though we know not the *why* (το διότι), which God has reserved for himself." (*Ivi*, p. 104) As such, the only authority that reason can invoke here is that of a poet – as Leibniz quotes Joseph Scaliger warning us not to search for the causes of all things, not to seek to enter by force in territories covered by silence, and not to want to know what the best Teacher does not want to teach; instead, tread carefully and with timidity, and assume a wise ignorance on such matters (*Ivi*, pp. 104-105). This is one of those rare moments where Leibniz prefers to admit the limits of reason and suggests the possibility that other things – such as silence, feelings, or poetry – might be more appropriate to invoke in our rapport to mysteries and the divine.

Why, then, does the main character in the *Theodicy* (reason) insist to put itself forward most of the time, and endeavours to do the opposite of what the poet suggests? Does reason even acknowledge faith as its complete other – something entirely or mostly outside its remit? Or does it fail to recognise this degree of otherness, and instead keeps trying to reduce faith to what reason itself can justify and explain? In paragraph 29 of the Dissertation, reason seems to admit the divine reality of faith which,

> [...] when it is kindled in the soul, is something more than an opinion, and depends not upon the occasions or the motives of that have given it birth; it advances beyond the intellect, and takes possession of the will and of the heart, to make us act with zeal and joyfully as the law of God commands. Then we have no further need to think of reasons or to pause over the difficulties of argument which the mind may anticipate (*Ivi*, p. 91).

Just as above, this is another rare occasion where reason admits its limitations insofar as true faith is concerned, and bows to it, as one that may, in fact, not require its services. The little it can do – bring some explanations here and there – is but an add-on, which that divine faith kindled in the soul may welcome, but it does not *need*. Elsewhere (*Ivi*, §§ 41-43) reason plays a double role. On the one hand, it is the "advocate of faith", even willing to ignore its own convictions to protect its 'client', because it accepts both the incomprehensibility of certain religious elements, and the contradictions of others, such as the question of the evil. On the other hand, it brings its own specificity within the core of religious faith, arguing that it

> [...] triumphs over false reasons by means of sound and superior reasons that have made us embrace it; but it would not triumph if the contrary opinion had for it reasons as of faith, that is, if there were invincible and conclusive objections against faith (*Ivi*, p. 98).

Does this not amount to a triumph of demonstrative reason instead of faith? And would this not, in turn, represent a confirmation of reason, rather than faith, as supreme authority? Leibniz himself acknowledges this, albeit to a certain extent only – when he says

> [...] It is well also to observe here that what M. Bayle calls a 'triumph of faith' is in part a triumph of demonstrative reason against apparent and deceptive reasons which are improperly set against the demonstrations (*Ibidem*).

One example where this self-consciousness of reason manifests itself is the analogy between our ability to 'see' things through an *a priori* knowledge of the causes of things, and that of simply 'believing' them through faith; the latter is the superior ability that St Paul refers to, in his second letter to Corinthians, when he says that "we walk by faith, not by sight" (2 Cor. 5:7). With this analogy, Leibniz introduces the thesis of universal harmony – which we cannot wholly understand, but which we can believe because God "gives us insight into his infinite goodness […] in spite of the appearances of harshness that may repel us" (*Ivi*, p. 99). In other words, reason is supported by God's grace and, in turn, it helps justify faith. This is why, Leibniz argues, we can say "that the triumph of true reason illuminated by divine grace is at the same time the triumph of faith and love" (*Ibidem*).

St Paul is not the only religious authority quoted in support of the above rationale. A few paragraphs later, Leibniz quotes the Holy Fathers (from early Byzantine times) to suggest that reason has always been acknowledged as a useful tool in religious debates:

> […] when the Fathers entered into a discussion they did not simply reject reason. And, in disputations with the pagans, they endeavour usually to show how paganism is contrary to reason, and how the Christian religion has the better of it on that side also (*Ivi*, p. 102).

This, again, is quite a clever, albeit unexpected, strategy employed by reason – to use the already established authority of religious figures in order to strengthen its own position.

Moving on to the main Essay in three parts, we shall only refer to one or two aspects from each of these. Paragraph 44 in Part 1 is where Leibniz makes the argument about the main principles of logic as a) universal, and b) essential to our endeavour to justify religious faith. The principle of the *determinant reason*, for instance, which "states that noting ever comes to pass without there being a cause or at least a reason determining it, that is, something to give an *a priori* reason why it is existent rather than non-existent" is one, we are told, without which "we could never prove the existence of God, and we should lose an infinitude of very just and very profitable arguments whereof it is the foundation" (*Ivi*, pp. 147-8). This, in addition to the other three lines of argumentation from the Dissertation – namely, those about the conformity of reason with faith on grounds of the universal harmony, the eternal truths, and the individual reasons of things, shows that, far from being in any way humble and discrete, reason engages the most royal means when attempting to be of service to faith – and in fact, argues its own case. What we are witnessing, in fact, is a peculiar version of a master, dissimulated as slave, in order to earn self-consciousness and self-creation to complement its recognition (which is in the nature of a master). The slave, too, earns recognition by becoming a master through this dissimulated work for itself – rather than for the 'master' it alleges to serve.

Part 2 of the Essay is mostly dedicated to defending faith against Mr. Bayle's objections, which are mainly centred around the issue of free will and divine goodness. These are contrasted with the notions of necessity and predestination, on the one hand, and the existence of evil, on the other. Clarifying these would amount to overcoming several difficulties (challenges) concerning certain mysteries or other subjects of our faith. We should not forget, however, that this whole debate takes place against an entirely rational background, and that the key issues being disputed through those criticisms, and Leibniz's response to them are, in fact, the very topics that Leibniz discusses, throughout his book, in relation to how reason relates to faith. In other words, this is a discourse by reason, about reason, and aimed at strengthening reason's own status – rather than that of faith. What is actually being defended here is the former, under the pretence of defending the latter. It may well be that the average believer does not even experience these doubts, raise these kinds of rational objections, or feel the need to see them clarified; this, in fact, is irrelevant – because the very purpose in *The Theodicy* is to raise and clarify such rational objections, thereby bringing the whole issue of defending and promoting faith onto the territory of reason. Whatever happens to faith, it must happen through rationalisation. At limit, we could argue that the actual clarification of the objections is secondary; the key is that they be raised and discussed – allegedly in support of faith. It is itself that reason, in fact, promotes, when it volunteers its service to faith. First, it makes itself useful; then – indispensable.

But why is Leibniz even discussing Mr. Bayle's objections, since these assume a dualism – indeed, an opposition – between faith and reason? We believe that Leibniz willingly chooses to do so, because he knows that this very assumption (about the alleged opposition between the two) is the biggest challenge to his endeavour – so unless he addresses it head-on, it is likely to overshadow everything else he does. Indeed, we can see him trying to dismantle this assumption right from the beginning of Part 2, just as he starts addressing Bayle's objections: "M. Bayle undertakes to discomfit those who maintain that there is nothing in faith which cannot be harmonized with reason" (*Ivi*, pp. 182-183). In fact, everything that reason undertakes from now on, is an attempt to nullify this role distribution which works to its disadvantage, against the theological background of the 18th century.

One of the key strategies used by reason is to bring everything in its own court – concepts, claims, means of argumentation, and the whole arsenal of the debate. For example, Bayle's theological theses are described as purely rational ones: " 'God', he says, 'the Being eternal and necessary, infinitely good, holy, wise and powerful, possesses from all eternity a glory and a bliss that can never either increase or diminish.' This proposition of M. Bayle's is no less philosophical than theological. To say that God possesses a 'glory' when he is alone, that depends upon the meaning of the term." (*Ivi*, p. 183). In addition, if we carry on reading the paragraph, it is hard not to notice that reason – rather than theology – wins the argument, since the philosophical terms and reflections are the ones that prevail. The very "glory of God" is described as something that God acquires "only when he reveals himself to intelligent creatures; even though it be true that God thereby acquires no new good, and it is rather the rational creatures who thence derive advantage, when they apprehend aright the glory of God" (*Ibidem*). Far from being a theological argument, this is yet another example of reason using the same strategy to gain the upper hand on faith – it uses purely rational concepts and arguments to talk about an apparently religious thesis.

Another example of the same strategy is – where Leibniz comments on a particular phrase from Bayle's *Dictionary*, saying that "This proposition is also, just like the preceding one, in close conformity with that part of philosophy which is called natural theology" (*Ibidem*). As such, the phrase in question can be philosophically analysed – which is an opportunity for reason to put forward specific meanings and insist on a particular interpretation that suits its own purposes. The same happens three paragraphs further down: "This thesis is also purely philosophic, that is, recognisable by the light of natural reason." (*Ivi*, p. 185).

Another phrase, although widely accepted as a revealed truth, will also be analysed philosophically:

> This proposition is in part revealed, and should be admitted without difficulty, provided that *free will* be understood properly, according to the explanation I have given" (*Ivi*, p. 184).

There are very few places where Leibniz agrees with Bayle; on such rare occasions (as the one referred to above) he still makes the agreement conditional upon something. Usually this is something belonging to common sense, like the most literal meaning of a word. That is so because what matters is not the actual condition as such, but the fact that it is required. It represents a rational criterion or reference point – a marker, which everything else should refer to. Moreover, this is important to reason.

In Part 3 of the Essays, reason becomes even bolder in declaring philosophy as the royal route to embracing (a well-founded) faith. In paragraph 296, for example, Leibniz cites Chancellor Bacon as saying "that a little philosophy inclineth us away from God, but that depth in philosophy bringeth men's minds about to him." (*Ivi*, p. 306). The purpose of such a rational endeavour, in the context, is to acknowledge that "all (even perceptions and passions) comes to us from our inner being" rather than being imposed on us by God, hence – the free will of man.

Insofar as the way Leibniz responds to Bayle's objections is concerned, this could be described as a holistic approach. The overarching thesis is that God tolerates one individual evil or another in order to save the larger picture – the goodness and order of the universe. This principle – of the whole taking priority over each of its parts – constitutes one of the axioms of Leibniz's philosophy; using it in the context of defending faith amounts to adopting a rational criterion as absolute reference. God pursues the general good of the universe over and above individual ones, not because *He* chooses to do so (i.e., this is His will), but because this is according to *universal reason*. (We should note here that universal reason is not necessarily the same thing as God's will)

> Supreme reason constrains him [God] to permit the evil" (*Ivi*, p. 201).
> [NOT doing so – allowing individual evils in order to save the greater good – would be a] "wrong choice" [that] "would destroy his wisdom and his goodness" (*Ibidem*).

So God *must* obey supreme reason, in order to remain good and wise.

Leibniz was, of course, aware of the dangers of risking a contrast between (supreme) *reason* and divine *will*, which is why he then tries to describe them as convergent (if not one and the same). However, this requires an assumption that we agree to identify God with supreme reason: "For it is, in my judgement, the divine understanding which gives reality to the eternal verities, albeit God's will have no part therein" (*Ivi*, p. 243). This is, indeed, a judgement call, which some may be unwilling to make.

The most interesting aspect related to this line of the argumentation, however, is that a much stronger argument comes a few pages further down (see § 192), where Leibniz says that "It is

> [...] not to be wondered at that he who penetrates all things at one stroke should always strike true at the outset; and it must not be said that he succeeds without the guidance of any cognition. On the contrary, it is because his knowledge is perfect that his voluntary actions are also perfect" (*Ivi*, p. 247).

Here, the distinction between reason (or knowledge) and divine will is maintained, but Leibniz grounds his thesis about their convergence in the assumption of the perfection of both – which is much more difficult to refute, than his earlier argument about God having to obey reason in order to stay wise.

The following paragraph finds Leibniz reiterate his thesis about the independence (indeed, the apparent self-sufficiency) of supreme reason – independence, it seems, even from God, which he considers an obvious truth: "Up to now I have shown that the Will of God is not independent of the rules of Wisdom, although indeed it is a matter for surprise that one should have been constrained to argue about it, and to do battle for a truth so great and so established" (*Ivi*, p. 247). This remains the centre-point of Leibniz's *Theodicy*: everything – even God and His will – must relate to the rules of wisdom. However, given the risk that too strong a distinction between this supreme reason and the divine will would pose, Leibniz invests a lot of effort in emphasising the divine wisdom, rather than will. Whole paragraphs are dedicated to an elaborate description of the role played by divine wisdom in the creation of the world (§ 225), and to the thesis that God's actions reflect a rational order: God

> [...] cannot determine upon Adam, Peter, Judas or any individual without the existence of a reason for this determination [...]. The wise mind always acts *according to principles*; always *according to rules*, and never *according to exceptions*" (*Ivi*, p. 328).

Moreover, there is only one difference between human reason and such a supreme reason – the role that time plays in anything human, which never affects universal knowledge. What greater acknowledgement could reason ever aspire to? Indeed, for Leibniz it is "knowledge of simple intelligence (that which embraces all that is possible), wherein at last the source of things must be sought" (*Ivi*, p. 373), as he tells us in the closing sentence of the *Theodicy*. In addition, this, we believe, is the real wager here, rather than the officially declared purpose – which is, to defend faith. The latter is only the first stage of the process, and a means that leads to the second – conferring recognition to reason, instead, as the supreme entity.

To summarise, the main character in the *Theodicy* – reason – starts up as a slave, offering its services to religious faith, and ends up as a master having legitimised itself through its work. Although it may be difficult to identify an equivalent of this process in Hegel's *Phenomenology* (since the dialectics there is between a self-sufficient slave without recognition, and an insecure master acknowledged as such by their other, the slave), one could argue that there is a natural progression between the former and the latter. This would, effectively, be the story of a duality growing into a trinity, or that of a linear into a spiral movement. The story of spirit's development in Hegel may well have begun with the dance between reason and faith in Leibniz's *Theodicy*. In fact, two out of the three stages of the Hegelian master-slave dialectics (those of <u>subordination</u> and <u>work</u>, respectively) can be identified in the *Theodicy*, while the third one (<u>circularity</u>) cannot. The integration of the two into something larger or higher than both (a spirit that is neither purely rational nor religious alone) will have to wait another century to occur.

5 A Possible Route to Modern Political Thinking

5.1 The Need for Recognition – a *Malaise* of Modernity

Our underlying thesis in this paper is that what really is at stake in the *Theodicy* is a particular need for recognition – namely, that of reason, vis-à-vis religious faith. The whole process described in the two previous sections, and summarised in the last paragraph above consists of just that – a self-legitimising of reason through its work on behalf of faith, thereby overcoming its status and becoming a master itself.

Our second thesis is that this need for recognition and the ensuing dialectics was not just a passing concern of one author or another; rather, it is a constant preoccupation in modernity (from early representatives like Leibniz, to later ones like Hegel), which reflects the actual mindset of the times. We are not alone in entertaining this belief. In Charles Taylor's view, in fact, the need for recognition is a real obsession of the whole modern period – up until and including the contemporary era. Taylor discusses it both in the context of a philosophy of identity (Taylor 1992) and insofar as its socio-political implications are concerned (Taylor 1994).

In what follows, we would like to investigate the question of whether or not the early version of the dialectics of recognition (that of Leibniz) can be identified as a key thread in the context of modern political thought. By early modern political thought, we mean the period from Machiavelli to Hobbes, and Locke to Rousseau (17th and 18th centuries). We would like to reflect on the extent to which the *Theodicy* model may be seen as becoming apparent in each of the main forms of government at the time (republic, monarchy in its various guises, oligarchy, aristocracy, anarchy, and dictatorship). Whether this was in any way reflected and articulated at all in modern political thought – is a topic for another study. For now, we would simply like to establish whether the *Theodicy* model can be identified as present (in an active, rather than normative way) in any of the major governance systems. What we refer to as 'the *Theodicy* model' is a confrontation, rather than a harmonious, progressive project of one entity (e.g. reason), in support of another (e.g. faith), aimed at gaining recognition for the former. The reason we raise this issue is because the relationship with the other is at the core of any system of government. Essentially, the rapport is between the sovereign (or the governing body) and the state (or 'the people'). To what extent this can be seen in terms of a recognition game, is itself open for debate. Our assumption here is that it can be, and indeed it should be, in the case of modern theories of government. Indeed, we shall argue, the issue of ensuring mutual recognition for the two entities involved might be the key feature and what distinguishes opposite systems like absolutism and democracy. So let us first, see how exactly we can identify the *Theodicy* model in any (or all) of the major political systems, and second, indicate some possible routes for further research by looking at various modern political theorists – without going into the detail of their respective systems.

5.2 Recognition Games in Major Forms of Government

If we look at the major forms of government – republicanism, democracy, monarchy, oligarchy, anarchy, and authoritarianism – they all entail key considerations (and some strong views, in fact) on the relationship between sovereign and people. To start with, republicanism is the notion of investing sovereignty in the people, rather than in some hereditary (or wealth-based) governing elite. In a republic, power resides in *elected* individuals, and the state is governed in accordance with a constitution and a set of laws. The fact that the people are represented at the governing table means that they are *recognised* as free and equal to the governing body – in other words, there is no master and slave as such, but two mutually acknowledging groups, the representatives and the represented. We shall return to this, once we will have sketched the alternatives.

At the opposite spectrum of the political systems we find monarchy, oligarchy, aristocracy, and anarchy. There are at least three types of monarchies, according to the degree of power that the monarch has, and limits thereof. *Political absolutism*, or the "divine right of kings", was advocated by the French theologian Jacques-Benigne Bossuet in the 17th century. One such king was Louis XIV of France, at whose court Bossuet was a preacher and tutor to one of the king's children. It is for this purpose that he wrote his work, *Politics Derived from the Words of Holy Scripture*, where he sets out the principles of royal absolutism. He adopts the medieval notion of kingship, according to which kings rule because they are chosen by God; as such, they are accountable to no one except God. Kings have the right to rule by birth and usually refuse to grant constitutions; instead, they govern in autocratic style. According to this view, to question or disobey the monarch effectively meant to rebel against God.

A milder version of despotism, *enlightened absolutism*, is the notion that certain monarchs were influenced by the Enlightenment and therefore governed in a different way from absolute rulers – the key difference being an emphasis on rationality. Such enlightened monarchs would, for example, tend to allow for religious toleration, freedom of speech, and the right to hold private property; they would also protect the arts, sciences, and education. Voltaire theorised this view. However, enlightened despots – just like absolute monarchs – believed they had the right to govern by birth. So the same principle stood, in principle – disobeying them would count for an act of rebellion against God.

The mildest version of monarchy, the *constitutional one*, is the only type of monarchy that actually limits the power of the monarch. The sovereign reigns alongside a Parliament, which is a governing body the monarch does not control. Historically, this form of governance has been in place for centuries in England – where the absolute forms of monarchy were never popular, ever since the Magna Carta in 1215. Here the monarchy has been regarded as a contractual political instrument, rather than an absolute one of divine origin.

The alternative to monarchism is aristocracy – a form of government where "the best" govern, rather than those allegedly chosen by God. A distorted type of aristocracy, oligarchy, is a form of government where the power rests with a small group – who are privileged either by wealth, family, or military strength. By contrast to aristocracy (where power is exercised openly), oligarchs often prefer to remain "the power behind the throne", exerting control through economic means. Aristotle considers the term to be synonym for rule by the rich (e.g. plutocracy). However, in the 18th c., theorists in the circle of Lord Burlington discussed a different kind of oligarchy – namely, one of civic virtue, modelled on the ideal of the Roman republic (Ayres 2009, chapter 1).

The most radical view of government, anarchy, is based on the ideal of a stateless society of free people – one without rulers, because all forms of government are considered undesirable and should therefore be abolished.

The key tread in all these political systems is the relationship between government and the governed – between sovereign and the people. Unsurprisingly, the question of which one of the two is superior has been at the core of many legal and political philosophers' debates. Indeed, theorists like Hugo Grotius in the 17th century and John Austin in the 19th century discussed in detail the question of whether government is the mean and people – the purpose, or the other way around. And it is exactly the question of the *Theodicy*, asked at a socio-political level, one that entails playing a recognition game just as much as the relationship of reason and faith does.

Austin, for example, stipulates the monarch's superiority over the citizens, which in his view suffices to enforce compliance to the ruler's will, but does not dispute that the very purpose of government is to concern itself with the interests of its citizens (Austin in Morris 1959, pp. 340, p. 358). Hobbes, too, is of the view that, given the violent nature of humanity, society needs an absolutely central sovereign, whose absolute power constitutes the very 'insurance' against him acting in his own personal interests rather than those of the people.

Grotius, on the other hand, repudiates the claim that all government exists for the sake of the governed, arguing that some governments "may be established for the good of the kings, as those which are won by victory", while others yet "may have respect to the utility both of the governor and the governed; as when a people in distress places a powerful king over it to defend it" (Grotius in Morris 1959, pp. 89-90).

Rousseau (Rousseau 1998) opposes monarchy on the basis that princes promote their personal interests rather than those of the people, and advocates the merits of various kinds of aristocracy instead. For him, consent is the key ingredient for the relationship between government and the governed; each person has the ability to put the common good above self-interest, and this (rather than an artificially imposed superiority of a sovereign) is the guarantor of freedom and good government. Locke, too, criticises monarchy as incompatible with civil society, and considers that government's power must be carefully circumscribed (Locke 2011).

To summarise the major difference between all these views, which is relevant for our purpose here – namely, to identify manifestations of a recognition game (through a confrontational or a dialectical process) in the rapport in discussion, we can focus on the way the balance of purpose inclines – towards the governor or the governed (or both). While the former is privileged in certain types of monarchy, and the later – in dictatorship, republicanism and aristocracy, as well as constitutional monarchy and generally any kind of government that involves a reliance on some form of social contract, they all allow for a combination of both sets of interests – in other words, for *mutual recognition*.

6 Conclusion

What does the epic development from Leibniz's *Theodicy* mean, in the context of early modern philosophy and, more importantly, for the socio-political mindset of the time? First, it raises the issue of the relation between faith and reason (trying to justify God's existence and reconcile it with the existence of evil), at a time and a place where the average person's worldview was dominated by the religious representation of an omnipotent, omniscient divinity. Answering the question of why such a God permits evil and suffering had a certain degree of urgency at the time, which we may not fully appreciate today. Second, the way in which Leibniz chooses to do this – through a confrontation, rather than a statement – has the advantage of making the whole process ring more true (albeit even subversive, at times), and thereby more convincing than the simpler form that a syllogism would take. The personal and engaged ethos of the *Theodicy* brings the whole debate home, in a way that a logical sequence of arguments would have been unable to. Third, the various stages of the process – the way in which the relation between faith and reason develops – may open the way not only to Hegel's *Phenomenology of Spirit*, but also to a whole tradition of socio-political and legal theories, which would place the question of the rapport with the other at the core of their public ethics debates. The list of Leibniz's contributions to the modern intellectual history could continue. It is, indeed, difficult to overestimate his role in the history of human enquiry as a whole, and it is somewhat puzzling that his *Theodicy* (the only book of philosophy published during his lifetime) is yet to receive the full attention that it deserves.

References

Ayres, P. 2009. *Classical Culture and the Idea of Rome in Eighteenth-Century England*. Cambridge: The Cambridge University Press.
Garber, D. 2011. *Leibniz: Body, Substance, Monad*, Oxford: Oxford University Press.
Hegel, G. W. F. 1977. *Phenomenology of Spirit*. Translation into English by Miller A. V. Oxford: The Oxford University Press.
Hobbes, T. 2014. *Leviathan*. London: Wordsworth Editions.
Kojève, A. 1980. *Introduction to the Reading of Hegel: Lectures on the "Phenomenology of Spirit"*. Translated into English by Nichols, J. H. Ithaca and London: The Cornell University Press.
Leibniz, G. W. 2009. *Theodicy. Essays on the Nature of Goodness of God, the Freedom of Man and the Origin of Evil*. Translated into English by Huggard, E. M. New York: Cosimo Classics.
Locke, J. 2011. *The Second Treatise of Government*. Watchmaker Publishing.
Machiavelli, N. 2003. *The Prince*. Translated into English by Bull, G. London: Penguin Books.
Morris, C. (Ed.). 1959. *The Great Legal Philosophers: Selected Readings in Jurisprudence*. Philadelphia: The University of Pennsylvania Press.
Noica, C. 2009. *Povestiri despre om. Dupa o carte a lui Hegel*. Bucharest: Humanitas.
Popper, K. 1984. *The Open Society and its Enemies (vol. 2): The High Tide of Prophecy: Hegel, Marx and the Aftermath* (5th ed.). Princeton: The Princeton University Press.
Rousseau, J.-J. 1998. *The Social Contract and Discourses*. Translated into English by Cole, G. D. H. London: J. M. Dent Orion Publishing Group.
Singer, P. 1983. *Hegel*. Oxford: The Oxford University Press.
Taylor, C. 1992. *The Ethics of Authenticity*. Harvard: Harvard University Press.
Taylor, C. 1994. *Multiculturalism: Examining the Politics of Recognition*. Princeton: Princeton University Press.
Taylor, C. 1995. *Hegel*. Cambridge: The Cambridge University Press.

Ana-Maria Pascal, Regent's University London, UK
pascala@regents.ac.uk

The Equivalence of Hypotheses and Leibnizian *Vires*

Tzuchien Tho

Abstract. In Leibniz's most sustained effort at constructing a systematic physical theory, the dynamics, he patiently developed, across two decades (*c.* 1676-1700), a methodology that borrowed from Galileo and Huygens centered on the equivalence of hypothesis. This principle states that motion and rest in a given physical system are relative. This equivalence of relative motion and rest appears to contradict later remarks, made famously to Clarke in the 1710's, where Leibniz defends "true motion". This classical problem in Leibniz scholarship has provoked a wide range of different responses. I offer a different perspective on this long-standing problem by reinterpreting the causal nature of Leibnizian forces (*vires*) and its theoretical independence from the equivalence of hypotheses.

Keywords: Leibniz, Clarke, force, equivalence of hypotheses, substantivalism and relationism of space, absolute and relative motion.

1 Introduction

This article argues for the meaningfulness of the distinction between absolute and relative motion in Leibniz's work *vis-à-vis* the apparent inconsistency of this distinction with the principle of the equivalence of hypotheses.[1] Due to the constraints of article length, we cannot be comprehensive in our treatment of the dynamics here but restrain ourselves to the specific conflict between the principle of the equivalence of hypotheses and the doctrine of *vis viva* in Leibniz's dynamics. The principle of *equivalence* states that any hypothesis concerning the relative motion and rest of bodies in a physical system is equivalent and hence states of rest or motion are in an ultimate sense *relative* given the "equal" truth of various hypotheses (Leibniz 1971, 4, 247; Leibniz 1980, 131). This does not mean that *any* hypothesis will do since Leibniz offers a law governed transformation between the hypotheses that obey his theory of forces or *vires* that conserves the quantity mv^2 (the product of mass and square of speed or velocity) of physical systems. This means that although *states* of motion and rest are relative, the relation between motion and rest within a *particular* hypothesis and the invariants that govern across equal hypotheses are not themselves relative. The solution that I offer is in many

---

[1] In what follows, I shall refer to the equivalence of hypotheses as "*equivalence*".

Tzuchien Tho (2017) The Equivalence of Hypotheses and Leibnizian Vires. In: Pisano R, Fichant M, Bussotti P, Oliveira ARE (eds.), *The Dialogue between Sciences, Philosophy and Engineering. New Historical and Epistemological insights. Homage to Gottfried W. Leibniz 1646-1716.* College Publications, London, pp. 363-379
© 2017 College Publications Ltd | ISBN: 978-1-84890-227-5 www.collegepublications.co.uk

respects a traditional one that follows interpreters like Arthur and Slowik that give absolute motion an "honorific" status (Arthur 1994, 232 and Slowik 2006). That is, absolute or "true" motion in Leibniz is not *actual*. I argue that although *states* of motion and rest are ultimately relative, the distinction between absolute and relative motion can be meaningful if we take it be a claim about the inherence of Leibnizian *vires* in bodies within the larger dynamics.[2]

In what follows I demonstrate that *equivalence* is an independent thesis from the theory of *vires* in dynamics. I disassociate standard arguments about the reality of a "Leibnizian" space-time from *equivalence* by arguing that *equivalence* only governs phenomena and cannot imply any further theses about the "deep" structure of Leibnizian dynamics. Here I argue that recent attempts to demonstrate the "depth" of *equivalence* as corresponding to a dynamical rather than phenomenal level of reality actually demonstrate the opposite. As such, the true aim of *equivalence*, in Leibniz's work, only corresponds to the "surface" rather than the causal depth of corporeal phenomena. Further I argue that this "surface" nature of *equivalence* is crucial to the dynamics precisely because it provides a bridge between phenomenal effects and their dynamical causes. This bridge constrains the "depth" of dynamical causes through the limited set of possible phenomena (limited by *equivalence*). My argument here leads to two conclusions. First, insofar as *equivalence* is independent from the theory of *vis*, *equivalence* describes a limited range of phenomenal reality distinct from a level of dynamic causes. Second, *equivalence* limits the possible expressions of *vis* such as to provide a model of *vires* in physical reality. This second conclusion affirms that the causal reality of Leibnizian *vires* is expressed in phenomenal reality only if there is a physical inherence of *vis* in system of one or many bodies. Hence the distinction between absolute and relative motion is meaningful precisely because the inherence of *vis* in some particular body is a necessary condition for dynamical causation. This picture of causation will lend clarity to its structural nature but will more importantly demonstrate the meaningfulness of "true motion" in Leibniz's dynamics.

2   Forces and Vires

The concept of force, like many fundamental physical quantities (ie. energy-work, action, inertia, etc.), endured a long process of conceptualization, formalization and systematization. Both within and beyond classical mechanics, say, within the resurgence of interest in Leibniz during the period of Einstein's relativistic revolution, terminological equivocation gave rise to many opportunities to misread Leibniz's work.[3]

To establish our first footing then, we need to disentangle force from Leibnizian *vis*. Leibniz's entry into the dynamics project was greatly motivated by a critique of Cartesian laws of motion and its accompanying metaphysical foundations. The ultimate driving notion in Leibnizian dynamics was the aim to establish *entelechies*, associated with *vis viva* (living force), inherent in substances. Against the Cartesians, this means that corporeal reality cannot reduce to the

---

[2] I shall consistently refer to Leibnizian force as "*vis*", leaving the term "force" to refer more generally.

[3] Cf. Stein 1977, pp. 3-49, De Risi 2012, pp. 143-185, Reichenbach 1957 [1928].

properties of *res extensa*: size, shape and magnitude (of motion). Though this critique was present in Leibniz's earliest works on motion, the *Theoria Motus Abstracti* (1671), Leibniz's initial means of establishing this theory of *vis* in this post-Paris dynamics project was the critique of the conservation of the Cartesian "quantity of motion" (product of mass and speed) in the laws of collision. Here much of Leibniz's initial work on the dynamics took the form of indirectly arguing for a theory of inherent *vis* in bodies by showing the theoretical and empirical *inadequacy* of the Cartesian theory. Of course since the terminology for "force", "work" and "energy" would not be stabilized until the 19th century, we must proceed with caution.

In context, this dispute over "force" did have a stable reference for Leibniz. He interpreted it, like the Cartesians, as the quantity conserved in the created world. The quantity of *vis viva*, mv^2, borrowed from his mentor Huygens' work and measured in various ways by Leibniz, was argued to be the superior candidate for the "quantity conserved". This was the overarching meaning of "*force*" (*vis* or *Kraft*) shared by Leibniz and his Cartesian antagonists. Of course *vis viva* in Leibniz would extend beyond this general meaning to include more sophisticated physical demonstrations and a complex metaphysical doctrine as the dynamics matured. We can analogize the concept in Leibniz to the conservation of energy-work in 19th century classical mechanics. Through this analogy we can at least see here in rough conceptual terms that the concept of Leibnizan *vis* cannot be identified with Newtonian-classical force.

Moving from the Cartesian to the Newtonian context, we note that although Newtonian force shares a family resemblance with Leibnizian *vis* (and Cartesian quantity of motion), the difference is great enough to cast them as distant cousins. Newtonian force is defined operationally. Force is the change in the momentum of a body. As defined in Newton's second law of motion, "The change of motion is proportional to the motive force impressed; and is made in the direction of the right line which that force is impressed" (Newton 2004, 71). Leaving aside the complexity of Newton's force, we make a simple point. Leibnizian *vis* and Newtonian force address separate problems. In Newton the Cartesian "quantity of motion" undergoes a reinterpretation and is given directionality such as to be equivalent to momentum. It is clear that Leibniz also made the appropriate reinterpretation of the Cartesian "quantity of motion" and includes it, within the terminology of his dynamics, as *impetus* (scalar product of mass and speed).[4] Further, Leibniz develops the terminology of the "quantity of progress" [*progrès*], equivalent to momentum, in *Essay de Dynamique* (circa 1699-1701).[5] It is important also to note that this "quantity of progress" is conserved in the collision laws of this text. As such, Leibniz's theory of *vis viva* is not a mistaken conflation of the referent of one dispute with the Cartesians and another rivalry with the Newtonians. Instead, *vis* and momentum occupy distinct roles in Leibniz's dynamics. Of course this does not make Leibniz's dynamics equivalent to Newtonian mechanics. There are important limitations in Leibniz's dynamics given his failure to attend to force qua change of momentum. Nonetheless, the analogy of *vis* with energy-work might be a

---

[4] Cf. Leibniz 1965 [1849-1863], 4, 235; Leibniz 1980, 121.
[5] Cf. Leibniz 1971 [1849], 4, 216-217, and Spector 1975, 136.

convenient short-hand here: *vis viva* governs the distribution of extended phenomena of motion (the kinetic-potential energy translation) in a system with respect to a conserved quantity mv^2.

3   Absolute and Relative Motion

Given the appropriate distinction between Newtonian force and Leibnizian vis, we turn to grasp the difficulty involved in interpreting how *vis* is supposed to "inhere" in bodies. The problem here is of course that the distinction between Newtonian force and Leibnizian *vis* renders *vis* a structural property of a given system of bodies. With the analogical interpretation of *vis* as energy-work, we might say that *vis* is not an operational property that "acts" on particular momenta of bodies but "acts" structurally on a system of bodies across the temporal evolution of motion and momenta. Yet it is difficult to understand how *vis viva* is supposed to *inhere* in a body. My aim in this section is not to provide a direct answer to this difficult question but rather to show how attentiveness to this problem transforms our understanding of Leibnizian *vis* with respect to the larger dynamics project.

The problem of the inherence of *vis* in bodies plays the crucial role in distinguishing between absolute and relative motion. As many interpreters have noted, the assertion of an "absolute" motion appears inconsistent because Leibniz seems to understand all extensional features of motion to be relative. Leibniz himself understood the problematic nature of this issue in his assertions of the 1686 *Discours de metaphysique*:

> [M]otion is not something entirely real, and when several bodies change position among themselves, it is not possible to determine, merely from a consideration of these changes, to which body we should attribute motion or rest (Leibniz 1965, 4, p. 444; Leibniz 1980, p. 51).

A few lines further, Leibniz adds that,

> But the force or proximate cause of these changes is something more real, and there is sufficient basis to attribute it to one body more than to another [...] it is only in this way that we can know to which body the motion belongs (Leibniz 1965, 4, 444; Leibniz 1980, 51).

The stakes of this distinction between mere relational motion and the attribution of "true motion" to motion coupled with *vis* can thus be read alongside Leibniz's pronouncements to Clarke, at the very end of his life, when he argues that,

> However, I grant there is a difference between an absolute true motion of a body and a mere relative change of its situation with respect to another body. For when the immediate cause of the change is in the body, that body is truly in motion, and then the situation of other bodies, with respect to it will be changed consequently, though the cause of that change is not in them (Leibniz and Clarke 2000, 49).

To understand the stakes here, we note two points that are crucial to this absolute-true and relative-phenomenal distinction in motion. First, we should be wary of understanding Leibnizian *vis* through Newtonian force. The coupling of *vis* and motion to form "absolute true motion" does not reduce to the classical relation between force and extended motion. Leibniz's explanation of "absolute true motion" by the notion of a coupling of *vis* and motion is thus more complex than he seems to state. Without clarification on how this coupling of *vis* and motion works, it is hard to understand how the distinction between absolute and relational motion is constituted. Secondly, despite making this distinction between absolute-true and relative-phenomenal motion, it appears that Leibniz requires the concept of absolute-true motion *only* to make a metaphysical *rather* than a scientific point. This metaphysical point, a central theme in his correspondences with Clarke, but also central to his dynamics project as a whole, concerns the immanence of physical causality in bodies. We shall return to this issue in what follows.

Faced with this seeming inconsistency between absolute and relative motion, it is the appropriate place to examine the principle of *equivalence*. We turn to a statement of *equivalence*, in a central programmatic text of the dynamics, the posthumously published second part of the *Specimen Dynamicum*:

> [W]e must hold that however many bodies might be in motion, one cannot infer from the phenomena which of them really has absolute and determinate motion or rest. Rather one can attribute rest to any one of them one may choose, and yet the same phenomena will result [...] [T]he equivalence of the hypothesis is not changed even by the collision of bodies with one another, and thus, that the laws of motion must be fixed in such a way that the relative nature of motion is preserved, so that one cannot tell, on the basis of phenomena resulting from collision, where there had been rest or determinate motion in an absolute sense before the collision" (Leibniz 1971, 4, 247; Leibniz 1980, 131).

Two things are clear from Leibniz's formulation of *equivalence*. The first is an epistemological point and states that *equivalence* provides a limit to empirical knowledge. It states that we cannot infer absolute-true motion and rest from phenomena. The second is a methodological point concerning his scientific practice, an epistemic point. It states that laws of nature are invariant with respect to the variations that are fixed *across* different hypothesis. On this second point, we see an application of Galilean relativity at work. It states a non-trivial principle that laws of motion operate in the same way across inertial frames. That is, recalling both Galileo and Huygens' example of physical systems between land and a moving water vessel, what we seek is invariance. Hence, at least for Leibniz, the invariance of physical laws on land and on a moving water vessel constitutes the necessary condition for the establishment of natural law.

Reading *equivalence* in this way results in an immediate logical implication. *Equivalence* states that rest and motion vary between hypotheses but also state that laws of motion are invariant across hypotheses. Hence though *equivalence* allows us to demonstrate the laws of nature through invariance, the invariances themselves are *independent* from it.

My emphasis here on the methodological usefulness of *equivalence* aims at clarifying the fundamental grounds of the dynamics project. To understand this emphasis, it is important to note the context. Throughout the dynamics project, Leibniz never ceased to take the Cartesians as his central target. In this, the Cartesian laws of motion were framed in such a way as to provide different rules for motion (and collision) depending on the relative rest or motion of the body. This application of different laws occur in the *Principia Philosophiae* Book II §46-§52 through first determining the difference of mass and speed in two colliding bodies and then proscribing the rule that will account for the resulting motion. We know that Leibniz, drawing from Huygens, was armed with a powerful tool, through an application of Galilean relativity, to reject any such absolute (or naïve) distinctions in the motion and rest in colliding bodies (Huygens 1977, 574-597). Leibniz's general methodology in the dynamics, drawn from Huygens, was informed by the determination of invariances across variations according to arbitrary inertial frames. The result is then a unified theory of the phenomenon of motion rather than a dependence on the different cases of rest and motion in the Cartesian theory. This methodology, stated as a principle, *equivalence*, is thus primarily about the invariance of the laws of motion (and collision) within the framework established by Galilean and Huygensian relativity. Hence although we can identify an epistemological aspect of *equivalence* concerning the negative limits of empirical knowledge, the more important role of *equivalence* rightly belongs to its status as a positive epistemic principle, a method of determining invariance from variation. This methodology, for example, is seen over and over in Leibniz's different presentations for the conservation of mv^2 as *vis viva*: a quantity that Leibniz borrowed from Huygens who also defended it with the same method.

Establishing this basic formulation of *equivalence* is important to further fill out what Leibniz means when he argues for an "absolute true motion". We see, in this basic formulation, that *equivalence* means that it and the thesis of "absolute true motion" cannot govern the same domains. This explicitly means that the two can be synthesized to constitute a more general theory of motion.

4 Equivalence and Equipollence

In this section we make use of a particular interpretation of *equivalence* to clear up the Leibnizian's larger theory of motion from its Newtonian alternative. Since Jauernig's recent publications (Jauernig 2008, 2009) represent a convenient and concise opposing position on this issue, I take her statements to be symptomatic of the problem I wish to diagnose.

Jauernig's argument roughly presents two versions of Leibnizian relativism, "strong" and "weak", for making Leibniz's double commitment to relative and absolute motion consistent (Jauernig 2008, 13). Instead of fully outlining her argument, I summarize by focusing on the key step taken in her article (Jauernig 2008). The crucial step in Jauernig's argument is to show that the reduction of all motion ultimately to rectilinear motion means that a universe of only rectilinear motion only implies Galilean relativity. If this is so then this universe of rectilinear motion can also accommodate Newtonian absolute motion. This results from the fact that absolute *acceleration* is well-defined in this universe where actual motion is exclusively rectilinear:

Given *equivalence*, the velocities v_1 (of a body A) and v_2 (of a body B) are different across inertial frames (where u is the relative difference in velocity):

$$v_1(A) = v_2(B) + u \tag{1}$$

but if we take the time derivative of these velocities (where t is time):

$$(d/dt)v_1(A) = (d/dt)v_2(B) + 0 \tag{2}$$

Acceleration or $(d/dt)v$ is thus well-defined given Galilean relativity.

As such, I agree with Jauernig concerning the claim of compatibility between this Leibnizian mechanical universe and a Newtonian spacetime model but our agreement stops here. However I disagree with Jauernig precisely on how "absolute motion", identified through "absolute acceleration", implies that we interpret Leibniz through the Newtonian concept of force. It is only the framework of force as change of momentum

$$\vec{F} = m\vec{a}$$

that renders Jauernig's absolute acceleration account a convincing one. Jauernig's larger argument about compatibility of Galilean relativity and Leibnizian absolute motion falters on this conflation between Newtonian force and Leibnizian *vires*.

The key role played by the identification of absolute motion and absolute acceleration in Jauernig's argument brings much of her interpretation into focus. If one accepts a weak version of Leibniz's relativity, where we take Leibniz as only accepting Galilean relativity, then relativity applies to the phenomenal level of reality while absolutism is available to the theory through absolute acceleration. If we accept a strong version, where Leibniz's relativity corresponds to some weak form of proto-general relativity (i.e., so–called Leibnizian space-time *qua* Machian space-time with time metric), then the hierarchical distinction between mere phenomena and deep dynamic reality plays a stronger role in separating the relativity of phenomena from absolute motion at a dynamic level (Earman 1989, 30-31). The difference of "strong" and "weak" here concerns the question of which level of reality we ascribe the well-definition of absolute acceleration. If absolute acceleration is available on mere phenomenal grounds then Leibniz's relativity is "weak" since the phenomenal structure can accommodate the distinction between absolute and relative motion. Conversely, if the concept of absolute acceleration is available only with the deeper dynamic level of reality then Leibniz's relativity is "strong" because phenomenal reality cannot accommodate the distinction between absolute and relative motion, and the deeper dynamical reality has to be introduced to provide the needed distinction.

To be clear, I agree with Jauernig only concerning the claims about the *compatibility* of Leibniz's theory with absolute acceleration. I disagree with the implications resulting from this compatibility, a conflation of the concept of "force" and *vis* in Leibniz. My point in engaging Jauernig is not to point out any egregious error but rather to show how an alternative understanding of Leibnizian *vires* can significantly transform the interpretation of what is at stake in Leibniz's *equivalence* and theory of *vires*. As such since the exposition of *equivalence* was central to Jauernig's interpretation, we require a more careful look at what is at stake in *equivalence* here.

The presentation of *equivalence* that Jauernig spends time explicating comes from the "proposition 19" of the *Dynamica*:

> Proposition 19. The Law of Nature that we have established of the equipollence of hypotheses – that a Hypothesis once corresponding to the present phenomena will then always correspond to subsequent phenomena – is true not only in rectilinear motions (as we have already shown), but universally: no matter how the bodies act among themselves; but provided that the system of bodies does not communicate with others, i.e., that no external agent supervenes.[6]

What is unique about this statement of *equivalence* is, as Jauernig points out, a slightly different version which Leibniz calls the principle of the equipollence of hypotheses.[7] That is, proposition 19 states *equivalence* with special emphasis on the related problem of the determination of a hypothesis *across* the temporal evolution of a physical system. It is precise to understand this modified version of *equivalence* through the term "*equipollence*" since the idea is that the temporally evolving phenomenon is correlate to the hypothesis of a system's motion conserving the same "power" or *potentia* (hence with a further step to *equipollentia*) from an initial time t to a later time t'. We can understand this as a specification of *equivalence* principle within the context of the *Dynamica*. As such, I emphasize that this statement of *equipollence* should not be taken as the standard expression of *equivalence*.

Jauernig develops her interpretation of proposition 19 into a proof of the proposition which hinges on the reduction of all motion (including curvilinear motion) to rectilinear ones. Again, her eventual argument is that since this reduction presents a universe where motion is exclusively rectilinear, it is a universe that can accommodate Newtonian absolute motion. The only question, for Jauernig's framework, is whether this distinction occurs within phenomena or between phenomena and its dynamic cause. With respect to the proposition 19, I argue for a radically different understanding of what is at stake. First, treating proposition 19 to be a statement of the generalization of *equivalence*, I argue that it is a generalization of *equivalence* across different kinds of phenomena. Hence, proposition 19 does not fall within the scope of the distinction between relative and absolute motion in the Leibnizian sense since it is limited to phenomena. Secondly, treating proposition 19 as a problem of *equipollence* rather than *equivalence* provides a much needed insight into the relationship between Leibnizian *vis* and *equivalence*. Both of these points together demonstrate that *equivalence* and the theory of *vis* are independent but come together to determine the phenomena of motion.

On the first point, I note that Jauernig de-emphasizes the explicit logical connection between proposition 19 and the preceding propositions that allowed Leibniz to make what is the central claim in the proposition, that *equivalence* is

---

[6] Cf. Leibniz 1971, 4, 507; translated quoted from Stein 1984, 41; Jauernig provides her own translation in Jauernig 2008, 15.

[7] I shall refer to the principle of the equipollence of hypothesis in what follows as "*equipollence*". Cf. Jauernig 2008, 15-16.

established. Stein notes that the Latin is ambiguous on whether the proposition pertains more directly to the "universality" of *equipollence* or the *equipollence* itself.[8] Although the language of proposition 19 is ambiguous, the *equivalence* principle has already been demonstrated in the *Dynamica* and then reiterated just above in proposition 14 of the same section (Leibniz 1971, 4, 500). Hence it appears more likely that proposition 19 aims at a generalization or extension of the principle of *equivalence* rather than a reiteration. Emphasis then should be placed on the role played by proposition 19 in connecting a theory of *equivalence* (proposition 14) and a theory of the reduction of all motion to rectilinear ones (proposition 16), as Leibniz clearly states, in the passage that follows the statement of proposition 19 (Leibniz 1971, 4, 507). Proposition 19 states the implication drawn from preceding propositions in order to establish the equipollence of hypothesis as the "universal truth" of the "law of nature".[9]

The aim then, of proposition 19, concerns this generalization of a rule from a simple form of phenomenal motion (rectilinear motion) to a more complex form of phenomenal motion (curvilinear motion). Leibniz's claim is only that *equivalence* applies equally from simple to complex rather than make any larger claims about complex phenomena. From a methodological perspective, it states that the invariance (natural laws) determined by Leibniz's theory of motion cover more than rectilinear motion. This argument is thus a claim that extends the method of treating rectilinear motion across a range of other phenomena. In short, proposition 19 argues that if *equivalence* holds for rectilinear motions (and their collisions), and all curvilinear motion is reducible to compositions of rectilinear motions, then *equivalence* holds for all motion.

The reasonable conclusion to be drawn from proposition 19 and its role with respect to previous propositions is that it states the extension of *equivalence* from rectilinear motions to all motions appropriately decomposed to be reducible to rectilinear motion (proposition 16). With this, both of Jauernig's hypotheses, weak and strong relativism, are still applicable insofar as Newtonian absolute motion is in principle compatible with a universe of motion reducible to rectilinear ones. There is a conflict of interpretation only if we ask what Leibnizian absolute motion is. Again, absolute motion in Leibniz is the coupling of *vis* and motion. Since proposition 19

---

[8] "Non tantum in motibus rectilineis (ut hactenus ostendimus) sed et in universum vera est, quam stabilivismus Naturae Lex de aequipollentia hypothesium [....]" Leibniz 1971, 4, 507. Cf. Stein 1977, 41.

[9] Although this aspect of my interpretation draws from the difference between "aequipollere" and "aequivalere" in the context of the *Dynamica* and the dynamics project in general, I do note that in general Leibniz uses these terms rather interchangeably *in the domain of mathematics*. Hence the ongoing argument is based on the *use* of these terms within this context rather than Leibniz's general usage. With respect to the "equivalence" between an infinitely sided regular polygon and the circle, relevant here to the reduction of curvilinear motion to rectilinear ones, Leibniz notes that, "[Q]uod figura curvilinear cesenda sit *aequipollere* Polygono infinitorum laterum...." Elsewhere Leibniz states, "[S]eu latus productum polygoni infinitanguli, quod nobis curvae *aequivalet*." Emphasis added. Leibniz 1971, 5, 126; Leibniz 1971, 5, 223.

only extends *equivalence* from simple rectilinear motions to complex curvilinear motions, both phenomenal in character, it has little to do with *vis* itself.

The alternative interpretation of proposition 19, presented here, would take up two different aspects of motion. The first, above stated, concerned the scope of the principle of *equivalence* and we have stated that the reduction of complex to simple phenomena provides the extension of *equivalence* from limited phenomena (of rectilinear motion) to phenomena universally. Both simple and complex phenomena remain phenomena. The importance of the reduction is epistemic and serves to universally extend the methodology of *equivalence*. Secondly, we wanted to show how the equipollence of hypotheses indeed does imply something about Leibnizian *vis* even if it remains independent from *equivalence*. We examine this in what follows by a final look at the theory of "force" implicitly assumed by Jauernig.

The difference between *equivalence* and *equipollence* here is crucial for any reckoning of absolute and relative motion in the Leibnizian sense. *Equipollence* indeed says something different to *equivalence* of hypothesis. *Equipollence* involves time and states that the relative motions at time t will correspond to the relative motions at (later) time t'. Why is this so? The emphasis on "equipollence" or equality of power means that Leibniz emphasizes here the intensity of a "*potentia*" or the power to act. In the simplest terms, *equipollence* relies on the idea that *vis* is conserved across time t and t' regardless of choice of inertial frames. In particular, it means that the quantity mv^2 is conserved between the distribution *vis* of a system at time t and time t'. Since mv^2 was methodologically determined as the quantity of *vis* conserved in elastic rectilinear collisions, the equipollence of hypotheses naturally relies on the generalization of these motions. The specific problem addressed by proposition 19 is thus whether or not there can be expressions of *vis* that might possibly contradict the relativity of motion understood via *equivalence*. We have seen that there are not and we have done so in a way that distinguishes, at least theoretically, *equivalence* and *equipollence*.

Does *equivalence* depend on *equipollence*? *Equipollence* relies on some conserved quantity that governs the evolution of the properties of a system of bodies while *equivalence* only answers to Galilean relativity. One should emphasize here that Leibniz did indicate an understanding of the conservation of momentum but he understood this through the notion of the "quantity of progress". It is only the conservation of mv^2 that took center stage in the dynamics and assumes the role of "*potentia*" or the intensive power to act. Hence *equivalence* does not then depend on *equipollence* but vice versa because *equipollence* is an extension of *equivalence* with respect to the conservation of *vis*. *Equivalence* only establishes the relativity of motions with respect to bodies within a system at a certain given time. Proposition 19 involves the idea that the invariants determined through *equivalence* are also valid across the evolution of that physical system. What then generalizes *equipollence* across time is the conservation of a certain quantity taken together with *equivalence*. Hence what carries the truth of a hypothesis (a chosen inertial frame) at time t to t' across the temporal evolution of the system is the conservation of a certain "power" quantitatively determinable (qua measure or *aestimatio*) using *equivalence*.

Now, it is true that both momentum and energy (and thus mv^2) are conserved in motion. Yet there is a major contextual difference when the question becomes how they operate in Leibniz's dynamics. Seen through the lens of the conservation of momentum, *equivalence* or *equipollence* can be naturally understood to be compatible with absolute acceleration. The case is less clear when *equivalence* or *equipollence* is seen through the lens of the conservation of mv^2. However given the well-definition of absolute acceleration in physical systems under Galilean relativity, shown above, the conflation of absolute acceleration and Leibnizian absolute motion can only be made by conflating momentum change with *vis*. Since acceleration is change of motion, implying momentum change, rather than the motion itself, it falls outside of the domain of *equivalence*. Nothing in the static methodology of Huygens and Leibniz can account for this treatment of acceleration since all they can say with *equivalence* concerns only relative initial, terminal or average velocities with respect to arbitrary reference frames rather than the intermediate behavior of these velocities.

Within this historical context, the only direct treatment of acceleration is available only with Newtonian force where

$$\vec{F} = m\vec{a}$$

Hence, absolute acceleration is only correlated to absolute motion if we think of force in the Newtonian sense, that is, change of momentum. Thus the assumption behind Jauernig's association of absolute acceleration and absolute motion is that the "equal power" or equipollence between a system of bodies at time t and time t' is momentum rather than mv^2. The evolution of momentum (acceleration) in a system of bodies is a natural implication of Newtonian force but not Leibnizian *vis*.

In a strict sense, Leibniz possessed some idea of both the conservation of momentum and energy. Indeed, given the conservation of mv^2 and the Huygensian collision rules provided by *equivalence*, he could have easily "filled out the puzzle" and derive the conservation of *mv*.

For any two bodies A and B in collision, the quantity conserved (proportional to mv^2) before and after collision (marked by the prime) is:

$$m_A v_A^2 + m_B v_B^2 = m_A v_A^{'2} + m_B v_B^{'2} \tag{3}$$

And since Leibniz's laws of collision conform to the Huygensian reversibility of relative velocity:

$$v_A - v_B = -(v_A^{'} - v_B^{'}) \tag{4}$$

As such

$$m_A(v_A - v_A^{'})(v_A + v_A^{'}) = -m_B(v_B - v_B^{'})(v_B + v_B^{'}) \tag{5}$$

The equations here simply reduce to:

$$m_A v_A + m_B v_B = m_A v_A^{'} + m v_B^{'} \tag{6}$$

Despite the compatibility of momentum with Leibniz's conception and his recognition of such a dimension of physical reality, the "absoluteness of acceleration" does not correspond to any aspect of Leibniz's theory of absolute motion.

My analysis above has argued that a contextual interpretation allows us to distinguish *Leibniz's* theory of absolute motion with *Newton's* force through proposition 19. This is thus the second crucial point concerning the interpretation of proposition 19. What Leibniz does here is to combine *equivalence* and his theory of *vis*. Whether we accept the particular measurement of *vis* as mv^2 is rather secondary and perhaps ultimately irrelevant to our reading of proposition 19. The problem here was only that understanding *equipollence* through momentum might mislead us into conflating the conservation of *vis* with absolute acceleration. If this analysis is accurate, what we see in this proposition 19 is Leibniz's bringing together of two independent results. It states that the invariance available to any consideration of a physical system under *equivalence* continues to be invariant even in the evolution to that system of bodies in time. This invariant, for Leibniz, is *vis* qua mv^2, the "potentia" conserved across time constituting an "equipollence" of hypothesis that relies both on *equivalence* and the quantity of *vis* qua mv^2.

5   Leibnizian Absolute Motion

As we have analyzed above, the compatibility of *equivalence* and Newtonian absolute acceleration cannot serve as a criterion for Leibniz's own theory of absolute motion. In the attempt to do so, we find quite the opposite. Understanding the nature of absolute acceleration and relative motion within *equivalence* demonstrated that the generalization of *equivalence* was a generalization of methodology. The analysis does not remove us from the idea that empirical phenomenon is inadequate to supply knowledge of the inherence of force in a body.

The relation between *equivalence* and the theory of *vis* can be neatly reduced to a methodological relationship. Leibniz often repeats, against the Cartesians, that the nature of *vis* requires metaphysical judgment and cannot be reduced to extensional features like size, shape and motion. At the same time, we know that Leibniz's commitment to the quantity mv^2 as the measure of *vis* was an entrenched one, a conclusion he reached (at least by 1678) long before he dared to exclaim a "new science of dynamics" based on *vis* (in a letter to Bodenhausen in 1689) (Leibniz 1994, 69-166; Leibniz 1965, 4, 469). Leibniz's view was that mere empirical measurements could never produce a theory of motion adequate to the status of natural law. Yet the principle of *equivalence* provided the means to establish invariance from variation. At the same time, *equipollence*, the idea of the conservation of a certain "power" within physical systems (in temporal evolution) cannot be justified from mere empirical grounds. As theoretically separate results, we propose that the theory of *vis* and *equivalence* are synthesized insofar as *equivalence* determines the form of physical phenomena and the theory of *vis* determines the invariants determined through this form. What does this then imply about Leibniz's distinction between absolute and relative motion?

What is at stake here is that the same principle that allows us to determine the invariance that measures *vis*, *equivalence*, is also the one that forbids any identification of *vis* with a particular body. With our examination of the relation between *equivalence* and *equipollence*, we are provided an alternative formulation of this problem. That is, if *vis* is only expressed in physical systems through *equivalence* and *equipollence* then *vis* can belong to any body (or any distribution among bodies) in a given physical system. This interpretation, while removing our inquiry concerning absolute motion from identification with direct physical quantities, also provides a significant constraint on what absolute motion means.

Leibnizian absolute motion implies that the causal *vis* of a motion inheres in a body such that, as Leibniz explains,

> [...] the immediate cause of the change is in the body, that body is truly in motion, and then the situation of other bodies, with respect to it will be changed consequently, though the cause of that change is not in them (Leibniz 2000, 49).

The passage associates "true" motion to the inherence of cause in a body. Though not explicit here, we can systematically understand this inherence of cause in body through the dynamics as the inherence of *vis* in body. As we discussed above, the idea is that the inherence of *vis* can be understood as a certain intensity of the power to move in a body or system of bodies.

We have invoked the mechanical interpretation of intensity above. It is certainly an intuitive empirical model for the dynamics. Yet this intensity appears to challenge the theory of structural causality that I have claimed. Although I will not venture into a larger account of structural causation, the present analysis of absolute motion is aided by a defense of my overarching interpretation against the reduction of Leibniz's dynamics to mechanics.

To summarize, I have argued that the independence of the theory of *vis* and *equivalence* means that while *equivalence* governs the range of possible phenomena in corporeal motion, it does not, by itself, provide a theory of *vis*. Although the theory of *vis* is independent from *equivalence*, *equivalence* nonetheless allows us the methodological apparatus to determine the effects of *vis* through the determination of invariance through variations. Thinking of the quantity of *vis* through the shorthand of classical work-energy, a bridge is made between the causality of *vis* and its expression in extended phenomena. That is, physical systems are not only governed by *equivalence* but also by *equipollence*. A physical system at time t is equipollent with respect to *vis* in that system at the later time t'. Through this, the theory of *vis* remains a systematic concept that governs the particular values taken up by particular bodies within the framework of *equivalence*. The causality of *vis* is thus structural. Hence although we can comfortably treat Leibnizian dynamics through an empirical model of intensity, the causal nature of *vis* does not simply or exclusively reduce to this model.

This interpretation can help clarify what follows from the passage quoted above:

> It is true that, exactly speaking, there is not any one body that is perfectly and entirely at rest, but we frame an abstract notion of rest by considering the thing mathematically. Thus have I left nothing unanswered of what has been advanced for the absolute reality of space (Leibniz 2000, 49).

Recall that Leibniz's problem here is not Newtonian force or the change of momentum but only the "change of place". However why would the abstract notion of rest (reiterating *equivalence* principle) be invoked by Leibniz to clarify his notion of absolute motion and rejection of absolute space? It appears that such an abstract notion of rest works against the meaningfulness of the absolute-relative distinction.

If we think of the inherence of *vis* in a body as intensity, then the distinction between absolute and relative motion is meaningful irrespective of the capacity to determine motion and rest absolutely. Although the principle *equivalence* forbids any concrete information about the actual inherence of *vis* in any particular body, it is nonetheless the case that *any* hypothesis requires that we distinguish, for that inertial frame, relative motion and rest. Hence any hypothesis admits to a distinction between intensity and the expression of that intensity in extended motion. At the same time however no hypothesis admits to a distinction between absolute and relative space.

Although *equivalence* and the theory of *vis* are independent, they are both relevant to actual phenomena. As such *equivalence* constrains how *vis* is extensionally expressed. This means that any mechanical theory of intensity is subject to the constraints of phenomena and hence hypothetical in nature. Thus understanding the inherence of *vis* through intensity is a helpful illustration but is ultimately limited precisely because it is a mechanical model. Intensity here is only a meaningful concept when it is understood as an indirectly measured magnitude of motion. As Leibniz remarks to Papin in the 14 December 1686 letter,

> Hence if you attribute true motion to some body as true intrinsic denomination [*denominatione vera intrinseca*], I attribute to it also a true action of change that I measure equally through intension or promptitude as through extension and duration (Leibniz cited in Ranea 1989, 62).[10]

The conditional expressed in this passage is clear. The hypothesized true motion is tantamount to true action of change (cause). However, intensity and extensity are equivalent expressions of this this "true intrinsic denomination". Intensity is not taken as a more absolute determination of "true action of change". Hence the intensity-extensity mechanical model is thus only an indirect measure of effects rather than causes themselves. Any equipollence relationship can only relate extended measure to magnitudes of intensity. Intensity, by itself, cannot be adequate to causality. As such, the distinction between absolute and relative motion, understood within the framework of *equivalence*, is a meaningful one even if intensity can only be hypothetically attributed to a particular body. This is consistent with what Leibniz later says in the 1692 *Animadversiones in partem generalem Principiorum Cartesianorum*. When speaking about *equivalence*, he remarks that,

> For even in astronomy one can explain the same phenomena with different hypotheses and so it will always be possible to attribute *real movement* to one or another body that changes in place [*viciniam*] or position [*situm*] with respect to each other. Hence one could arbitrarily

---

[10] Author's translation. Leibniz cited in Ranea 1989, 62.

consider one of the two [bodies] as being at rest or in movement on a line and with given speed and one can then geometrically define the move or rest that one must attribute to the other [bodies] [...] (Leibniz 1965, 4, 369).[11]

Notice here again that the invoking of equivalence is a methodological one. Real movement is hypothesized to determine the properties of the other bodies in the physical system "geometrically". Hence despite the ultimate relative nature of motion, the distinction between absolute and relative motion is a meaningful one that serves the purposes of the dynamics *in practice* far more than the general metaphysical thesis of relative motion.

In this analysis of the notion of intensity, we also see that the theory of absolute motion is a necessary condition for the dynamics insofar as the inherence of vis in bodies is a necessary condition for the laws of motion (invariants) that Leibniz sets out to determine. The mechanical model of intensity-extensity played out in examples of pendulum, colliding bodies and the like recurring throughout Leibniz's work provide a natural way of understanding this. But the causal reality of *vis* does not reduce to mere mechanics. Hence although the causal reality of *vires* and the effective reality of motion constitute hierarchically distinct levels, without the inherence of *vires* in physical reality, the variations resulting from *equivalence* would result merely the trivial exchange of velocities if not for invariance grounded by the conservation of *vis viva* or mv^2.

The distinction then, between absolute and relative motion, is meaningful even if we take the distinction as hypothetical in phenomena. That is, regardless of variation across hypotheses, the distinction holds in any particular hypothesis. More importantly, through *equipollence*, some invariant *potentia* is conserved across hypotheses and this fact relies on the theory of the inherence of *vires* as a necessary condition for the translation of cause to effect. As such, a structural interpretation of the causality of *vis* allows us to construct a bridge between the absolute-relative distinction of motion and the metaphysical thesis of the inherence of *vis* in bodies.

6   Concluding Remarks

The problem that we sought to resolve in this article was the seeming contradiction between the principle of the equivalence of hypotheses and the distinction between absolute and relative motion. If all hypotheses concerning the relative states of rest and motion for bodies within a physical systems, given that they respect relative velocities, are true, then the distinction between absolute and relative motion appears to be meaningless. We examined Leibniz's explicit explanation for how both of these principles can be held together. Leibniz defined absoluteness of motion through the *inherence* of vis in a body while motion itself is relative to a chosen inertial frame. In this way, the determination of the absoluteness of extended motion relies on a realm of non-phenomenal and unextended *vires*.

This solution is however very difficult to understand without a broader systematic grasp of the dynamics. That is, it is not clear how extended motion could

---

[11] Author's translation. Emphasis added.

be relative while forces absolute. By examining the many misunderstandings that could result from this difficult problem, we have outlined the idea that the action of *vis* is structural insofar as it acts on a physical system taken as a whole rather on specific bodies and the evolution of their motion in time. At the same time, we examined how the intrinsic determination of motion is indifferent to notions of intensity and extendedness. Both expressions of dynamical causation equally satisfy the intrinsic determination of motion. As such, there is can be no corresponding physical mechanism which accounts for the inherence of *vis* in a particular body.

A possible solution for understanding absolute motion is found in understanding the synthesis of the two theoretically independent aspects of Leibniz's dynamics. Firstly, the equivalence of hypotheses provides a theory of inertial motion which delimits the range of locomotive phenomena. Through this principle, Leibniz argues for the reduction of all motion to a composition of rectilinear motion. Secondly, a theory of vires provides the determination for the invariant around which the variables involved in motion are structured. The invariant mv^2 thus provides a principle for the action of a physical system beyond the mere account of the relative velocities of a system. It is in this sense that motion can be absolute. The states of rest and motion (and everything in between) for individual bodies remain irreducibly relative (to other bodies in the system) but what is absolute is their systematic invariant action. This means that *vis* inheres in bodies, but only as a property of a system of bodies.

The solution proposed here for the meaningfulness of Leibniz's distinction between absolute and relative motion provides another way of understanding the distinction as "honorific", as some commentators have argued. This means that for any system of bodies, some arbitrary body can be chosen to represent the quantity mv^2 in its locomotive effects. Of course, some other body in that same system can be chosen for the same purposes. What is absolute would then not be the body that is moving but rather the measure of *vis* that is expressed through its movement. Due to the independence of the equivalence of hypotheses and the theory of vires, the inherence of *vis* in a physical system is not something that mere *equivalence* could determine. With a contextual understanding of the dynamics as a whole, we are thus in a position to provide a deeper understanding of the principle of the equivalence of hypotheses and grasp its consistency with a distinction between relative and absolute motion.

Acknowledgements

I wish to thank the Berlin-Brandenburg Academy of Sciences and Humanities as well as the Institute for Research in the Humanities (University of Bucharest) for their formal support for the research involved in this article. In particular, I wish to thank Eberhard Knobloch, Vincenzo Di Risi, Ed Slowik, Richard T.W. Arthur and Colin Mcquillan for comments on various previous drafts of this paper. I also wish to thank the anonymous referees for their helpful comments in the process of blind review.

References

Arthur, R. T. W. 1994. Space and Relativity in Newton and Leibniz. *British Journal for the Philosophy of Science* 45:219-240.

De Risi, V. 2012. Leibniz on Relativity: The Debate between Hans Reichenbach and Dietrich Mahnke on Leibniz's Theory of Motion and Time. In *New Essays in Leibniz Reception*. Ralf Krömer and Yannik Chin-Drian (Eds.). Basel: Springer Verlag, pp. 143-185.

Huygens, C. 1977. *The Motion of Colliding Bodies*. Richard J. Blackwell (trans.). *Isis* 68(4):574-597.

Jauernig, A. 2008. Leibniz on Motion and the Equivalence of Hypotheses. *The Leibniz Review* 18:1-40.

Jauernig, A. 2009. Leibniz on Motion- reply to Edward Slowik. *The Leibniz Review* 19:139-147.

Leibniz, G. W. 1923–. *Sämtliche Schriften und Briefe*. Deutsche Akademie der Wissenschaften (ed). Darmstadt and Berlin: Akademie Verlag.

Leibniz, G. W. 1965 [1875–1890]. *Die Philosophischen Scriften von Gottfried Wilhelm Leibniz*. C.I. Gerhardt (Ed.). Hildesheim: Olms.

Leibniz, G. W. 1971 [1849–1863]. *Leibnizens Mathematische Scriften*. C.I. Gerhardt (Ed.). Hildesheim: Olms.

Leibniz, G. W. 1980. *G.W. Leibniz Philosophical Essays*. Roger Ariew and Daniel Garber (Eds.). Indianapolis: Hackett Publishing Company.

Leibniz, G. W. 1994. De Corporum Concursu. In *La réforme de la dynamique*. Michel Fichant (Ed.). Paris: Vrin, pp. 69-166.

Leibniz, G. W., Clark, S. 2000. *Correspondence*. Roger Ariew (Ed. and Introduction). Indianapolis: Hackett Publishing Company.

Newton, I. 2004. *Philosophical Writings*. Andrew Janiak (Ed). Cambridge: The Cambridge University Press.

Ranea, A. G. 1989. The *a priori* Method and the *actio* Concept Revised. *Studia Leibnitiana* 21(1):42-68.

Reichenbach, H. 1957 [1928]. *Philosophie der Raum-Zeit-Lehre*. Berlin: De Gruyter; *The Philosophy of Space and Time*. Translation by Maria Reichenbach and John Freund. New York: Dover Publications.

Slowik, E. 2006. The 'dynamics' of Leibnizian relationism: Reference frames and force in Leibniz's plenum. *Studies in History and Philosophy of Modern Physics* 37:617-634.

Stein, H. 1977. Some Philosophical Prehistory of General Relativity. In: John Earman, Clark N. Glymour, John J. Stachel (Eds.). *Foundations of Space-Time Theories*. Minneapolis: The University of Minnesota Press, pp 3-49.

Spector, M. 1975. Leibniz vs. the Cartesians on Motion and Force. *Studia Leibnitiana* 7(1):135-144.

Tzuchien Tho, University of Milan, Italy
tzuchien.tho@gmail.com

Leibniz's Defence of Heliocentrism

Friedel Weinert

Abstract. This paper discusses Leibniz's view and defence of heliocentrism, which was one of the main achievements of the Scientific Revolution (1543-1687). As Leibniz was a defender of a strictly mechanistic worldview, it seems natural to assume that he accepted Copernican heliocentrism and its completion by figures like Kepler, Descartes and Newton without reservation. However, the fact that Leibniz speaks of the Copernican theory as a *hypothesis* (or plausible assumption) suggests that he had several reservations regarding heliocentrism. On a first approach Leibniz employed two of his most cherished principles to defend the Copernican hypothesis against the proponents of geocentrism: these were the principle of the relativity of motion and the principle of the equivalence of hypotheses. A closer analysis reveals, however, that Leibniz also appeals to dynamic causes of planetary motions, and these constitute a much stronger support for heliocentrism than his two philosophical principles alone.

Keywords: Copernican Revolution, Dynamics, Equivalence of hypotheses, Leibniz's heliocentrism, Relativity of motion.

1   Introduction

Like many of Newton's contemporaries – R. Boyle (1627-91); P.L. Moreau de Maupertius (1698-1759) – G. W. Leibniz (1646-1716) was a believer in the mechanical worldview: a clockwork universe. I. Newton (1643-1727) still believed that the clockwork universe needed occasional adjustments – corrective interferences – on the part of the 'Artificer', making God a supreme engineer. But both Boyle and Leibniz saw the need for occasional repair work as a diminution of the power of the supernatural agent. Once the clockwork was set in motion by its divine author, the universal laws took over and kept it in reliable order. Leibniz holds that nature is subject to the laws of nature:

> The natural forces of bodies are all subject to 'mechanical laws'
> (Leibniz 1715-1716, Fifth Paper, Section 124, pp. 237-38).

Friedel Weinert (2017) Leibniz's Defence of Heliocentrism. In: Pisano R, Fichant M, Bussotti P, Oliveira ARE (eds.), *The Dialogue between Sciences, Philosophy and Engineering. New Historical and Epistemological insights. Homage to Gottfried W. Leibniz 1646-1716.* College Publications, London, pp. 381-402
© 2017 College Publications Ltd | ISBN: 978-1-84890-227-5 www.collegepublications.co.uk

I consider it sufficient that the mechanism of the world is built with such wisdom that these wonderful things depend upon the progression of the machine itself [...] (Leibniz 1698, p. 499).[1]

It is therefore natural to assume that Leibniz accepted Copernican heliocentrism, and its completion by figures like J. Kepler (1571-1630), R. Descartes (1596-1650) and I. Newton, without reservation. With the discovery and mathematical formulation of his three laws of planetary motion, Kepler had placed heliocentrism on a more secure footing than Copernicus's original version. In his *Tentamen de motuum coelestium* (1689), Leibniz showed himself to be fully aware of Kepler's achievements and proposed a mechanical theory to 'explain the causes of celestial motions'. In an assessment of the Leibnizian defence of heliocentrism it is therefore important to be aware of the development of the Copernican hypothesis from its originator through Kepler and Galileo to Newton.

On a first approach it looks, however, as if Leibniz's endorsement of heliocentrism was hedged by several reservations (cf. Finocchiaro 2005, Chapt. 5; Bertoloni Meli 1988). 1) On the *scientific* level, Leibniz continues to speak of heliocentrism in the Copernican manner as a hypothesis, even though the term had acquired a negative connotation at the hands of figures like Kepler and Newton. 2) On the *political* level, Leibniz shared Galileo's concern that the anti-Copernican censure of the Catholic Church could do serious harm to scientific progress and threaten its recent achievements. 3) On the *philosophical* level, Leibniz nevertheless offered a guarded defence of heliocentrism. He declared it more 'intelligible' than its rival – Ptolemaic geocentrism. He employed two of his most cherished philosophical principles to defend heliocentrism against the proponents of geocentrism. These were the principles of the relativity of motion and of the equivalence of hypotheses. Closer analysis shows, however, that Leibniz also advanced dynamic reasons, which offered stronger support in favour of Copernicanism than his two philosophical principles alone.

The purpose of this paper is to analyse how these considerations led Leibniz to a balanced approach to this still hotly debated topic of his day. Leibniz's defence cleverly steers its way between support for heliocentrism and avoidance of its condemnation by the Censors. It will be helpful to offer the reader a brief summary of Copernicus's achievements and to describe the changes in connotation, which the term 'hypothesis' underwent during the course of the 16th and 17th centuries: from an educated conjecture for Copernicus to a 'gratuitous' fiction for Newton. These changes in connotation reflect the development of heliocentrism during the course of the 16th and 17th centuries.

---

[1] The fact that Leibniz also held that the principles of mechanics were metaphysical principles does not change his commitment to the mechanical universe, since the mechanical laws can be derived from the metaphysical principles: 'everything happens mechanically in nature but the principles of mechanics are metaphysical' (Leibniz 1690a, p. 245; Leibniz 1695, p. 441; Leibniz 1710-6, p. 399; Antognazza 2003; Garber 1995).

Homage to Gottfried W. Leibniz 1646-1716

1.1 A Summary of Heliocentrism

For readers unfamiliar with the astronomical theory of Nicholas Copernicus (1473-1543), it may be useful to remind them of some of his achievements. In a departure from a long tradition, which had its roots in Greek thought, Copernicus proposed a *heliocentric* – sun-centred – view of the universe, in opposition to the established *geocentric* – Earth-centred -view, whose chief proponents were Aristotle (354-322 BC) and Ptolemy (100-175 AD). Copernicus made the Earth a planet, which orbited the central (mean) sun. To place the sun at the centre of the then known universe was not in itself an original idea. The Greek astronomer Aristarchus of Samos (çirca 310-230 BC) had already constructed a heliocentric world system, which made the Earth rotate daily on its own axis and annually around the sun. The diurnal rotation of the Earth was proposed by several thinkers throughout the ages (Herakleides, Buridan, Oresme, Nicolas of Cusa). But no technical details of Aristarchus's system have survived so that Copernicus became the first known astronomer to construct a coherent, mathematical system of planetary motion from a heliocentric perspective. In the Greek geocentric tradition all the planets and their motions were treated separately but Copernicus's aim was to derive all the observational data of the planets' orbits from the assumption of a moving Earth. Thus Copernicus became the first astronomer to propose a detailed account of the astronomical consequences of the Earth's motion, as part of a planetary system (Kuhn 1957, pp. 142-144; Weinert 2009, § 3.1). It is important to observe that although Copernicus reports his own observations of the sky, these observations do not reach beyond the discoveries of his Greek predecessors. He does not discover *new* facts about the planets. It is equally important to realize that Copernicus still adheres to much of the Greek tradition in his mathematical techniques. Like his illustrious Greek predecessors, Aristotle and Ptolemy, he uses geometry to describe the motions of the then known 6 planets. Most importantly, Copernicus does not abandon the fundamental Greek idea that all celestial objects must move in circles around a central body, since the circle was considered to be the most perfect geometric figure. Perfection and harmony, to the Greek mind, characterized the heavens.

Given these few rudimentary facts about the Copernican system, especially his profound indebtedness to the Greek tradition, the obvious question, which many historians of science have asked, is whether the Copernican model constitutes a scientific revolution. The epithet 'Copernican revolution' is sometimes bestowed on the whole period from the publication of the Copernican treatise (1543) to the publication of Newton's *Principia* (1687) and sometimes on the Copernican theory itself. In the present context the question is only whether the Copernican theory itself is revolutionary, since there is little disagreement that the period from Copernicus to Newton constitutes indeed a scientific revolution. When Leibniz took up his defence of heliocentrism, he had the whole development of heliocentrism in mind. Answers to this question help to understand the whole extent of the Copernican revolution. Historical judgements on this question have therefore varied widely. The historian of science de Solla Price saw in Copernicus's book 'little more than a reshuffled version of [Ptolemy's] *Almagest*' (de Solla Price 1962, p. 215). Arthur Koestler also detected little originality in Copernicus, characterizing him as a 'stuffy pedant', but also recognized in him a 'crystallizer of thought' (Koestler 1964, pp. 205, 113). E. Rosen found that 'Copernicus did not foment a "Copernican Revolution"' (Rosen 1984, pp. 132-3), whilst for A. C. Crombie (Crombie 1961, p. 168) the Copernican Revolution consisted in the link Copernicus established between the diurnal and annual revolution of the Earth and the motion of the planets. J. H. Randall (Randall 1962, pp. 308-15) was more willing to grant Copernicus the title of a scientific revolutionary, whilst H. Blumenberg (Blumenberg 1955, 1965) acknowledged Copernicus above all as an intellectual reformer. Similarly, for O. Gingerich, Copernicus was a 'sensitive visionary who precipitated a scientific revolution' (Gingerich 1993, p. 201).

Thomas S. Kuhn is best known for his book *The Structure of Scientific Revolutions* (1962 [^{2}1970]), which characterizes numerous brief episodes in the history of science, including Copernican heliocentrism, as 'revolutionary' periods. But Kuhn's most elaborate exploration of a scientific revolution is provided by the masterly analysis in his earlier book *The Copernican Revolution* (1957). In this book, Kuhn describes Copernicus as a *precursor* of a scientific revolution. His book *De Revolutionibus* (Copernicus 1543) is a 'revolution-making rather than a revolutionary text' (Kuhn 1957, p. 183).

Kuhn's most careful investigation of a scientific revolution is to be found in his analysis of the early history of astronomy from the Greeks to Newton. In the book *The Copernican Revolution* (1957) Kuhn goes beyond the assessments of de Solla Price and Koestler and agrees with O. Gingerich that Copernicus is best described as a precursor of a scientific revolution. Unlike Rosen he sees in Copernicus's book *De Revolutionibus* (Copernicus 1543) a 'revolution-making rather than a revolutionary text' (Kuhn 1957, p. 183). The Copernican system has aesthetic advantages, he holds, since it derives from the principle of a moving Earth a natural explanation of one of the gross planetary irregularities in Greek astronomy: the apparent retrograde (westward) motion of planets becomes a matter of the perspective of an Earth-bound observer who assesses the motion of planets around the sun against the background of the fixed stars (Fig. 2). Although Copernicus abides by the Greek notion of uniform circular motion, he departs from Ptolemy by adopting a simple 'distance-period' relationship to assess the relative distances of the planets from the sun. The rule states that the further a planet is away from the sun, the longer is its orbital period. But Copernicus produced no decisive evidence, which could demonstrate that a Copernican hypothesis is more probable than a geocentric hypothesis. However, Kuhn's tone changes in *Structure*, where he states, in many passages, that the replacement of Aristotelian-Ptolemaic geocentrism by Copernican heliocentrism is a paragon of a scientific revolution. Copernicus is discussed in the same breath as Newton, Lavoisier and Einstein and is hailed as the originator of a new paradigm (Kuhn 1962 [^{2}1970], p. 6, p. 66, p. 92, p. 116, p. 180, p. 200).

In order to appreciate the problem situation, which Leibniz faced, the question arises whether the Copernican model of 1543 was a scientific revolution or a precursor to a scientific revolution – two aspects of Kuhn's assessment of the situation. The answer to this question depends on the criteria adopted but, crucially, the criteria themselves must be adequate for a historical judgement of a particular episode, like the Copernican heliocentric model. To appreciate the reasons why the Copernican hypothesis does not amount to a scientific revolution, it will be helpful to add some further historical material regarding the Copernican model. It has already been mentioned that Copernicus's commitment to circular orbits and geometry marks a significant element of *continuity* between his work and that of his

Greek predecessors. But there is also a significant element of *discontinuity*, hinted at by Crombie, which has not been sufficiently emphasized in the literature. Copernicus becomes the first astronomer to successfully treat the planets and the sun as a coherent *system*. As a cosmologist, Aristotle had provided a qualitative model of the whole cosmos, consisting of two spheres. The *supralunary* sphere extended from the moon to the 'fixed' stars and was characterized by harmony, immutability, perfection and symmetry. As it was the realm of planetary orbits, the only possible trajectory for planets was the circle, since the circle was the most perfect geometric figure. The *sublunary* sphere extended from the (central) Earth to the moon and was characterized by imperfection, flux and change. As an astronomer, Aristotle had proposed a concentric model of planetary motion, according to which the planets were carried around the central Earth on homocentric shells, which consisted of a fifth element, called the ether. This planetary model was bound to be a failure because it could not account for the 'appearances': as the Greeks knew, planets do not keep the same distance from the 'centre' and consequently astronomers observe a change in brightness.

Ptolemy, the mathematical astronomer, accepted Aristotle's cosmological principles – especially the centrality of a stationary Earth – but, for computational reasons, treated each planet separately and in isolation from each other. Furthermore, Ptolemy introduced a number of geometric devices – in particular epicycles, eccentrics and equants (see Figs. 1-3) – to bring the geocentric model in closer agreement with the 'appearances' (i.e., the known observable planetary orbits).

Copernicus departs from the mathematical treatment of individual celestial objects. Instead he binds the planets into a coherent system, with the sun at the 'centre', such that the removal or displacement of one element would disrupt the entire system. Such a commitment imposes an important constraint on the model, which became a permanent feature of astronomical model-building.

> And so, having laid down the movements which I attribute to the Earth farther on in the work, I finally discovered by the help of long and numerous observations that if the movements of the other wandering stars are correlated with the circular movement of the Earth, and if the movements are computed in accordance with the revolution of each planet, not only do all their phenomena follow from that but also this correlation binds together so closely the order and magnitudes of all the planets and of their spheres or orbital circles and the heavens themselves that nothing can be shifted around in any part of them without disrupting the remaining parts and the universe as a whole (Copernicus 1543 [1995], p. 6).

The conception of the coherence of planetary phenomena obliges the Copernicans to build a model of the planetary system, which must accommodate all the known empirical data. They were not altogether successful but the balance of successes and failures of the Copernican system provides useful indicators as to the criteria of scientific revolutions. Given these main lines of continuity and discontinuity it may be best to characterize Copernicus's work as a Copernican *turn*: a change in perspective but not a revolution, in line with Kuhn's original 1957 verdict.

Let us briefly consider why the original Copernican position falls short of a scientific revolution and review some of the main reasons why many historians of science tend to withhold the status of a scientific revolution from the Copernican heliocentric model.

- Although Copernicus was a 'realist' regarding the physical distribution and order of the planets in the sky, he accepts the 'equipollence of hypotheses', a philosophical device which can already be found in Ptolemy's *Almagest*. It is also employed by Leibniz in his assessment of the Copernican turn. This device encourages the acceptance of different geometric techniques, which are regarded as equivalent for the purpose of describing planetary motions. Two devices – one based on the eccentric circle (Fig. 1) and the other based on the epicyclic circle (Fig. 2) – were of particular importance for the geometric modelling of the apparent motions of the planets, as seen from the assumption of a stationary or rotating Earth. Note that a planetary system, of the Ptolemaic or Copernican flavour, faces two observational anomalies, which it must explain. One is the apparent non-uniform motion of the planets around the 'centre', i.e. the known fact that the planets do not travel around the centre at a uniform speed, contrary to both the Ptolemaic and Copernican assumptions of the uniform circular motion of the planets around either the central Earth or the central sun. Ptolemy solved this problem by the employment of the eccentric circle (Fig. 1). Despite his heliocentric hypothesis, Copernicus also still required small epicycles and eccentrics for his geometric constructions because he did not abandon the Greek ideal of circular motion.

The second anomaly was the apparent retrograde motion of planets, as seen from the Earth. All planets move from west to east around the 'centre' but at certain periods they seem to reverse their motion (Fig. 2) and appear to move temporarily from west to east, as measured against the 'fixed' stars. In the Copernican model the retrogression is satisfactorily explained through the distance-period relationship. The Earth, being closer to the sun than, say, Mars overtakes Mars in its annual journey around the sun, which creates the impression that Mars temporarily retrogrades. Leibniz cites this natural explanation as one of the advantages of the Copernican system. Ptolemy solved the problem by the device of epicyclical motion (Fig. 2).

Copernicus regards the employment of both eccentric and epicyclic circles as equivalent techniques – both can be used to model planetary motion (Copernicus 1543 [1995], Bk. III, §20, Bk. IV, §4).

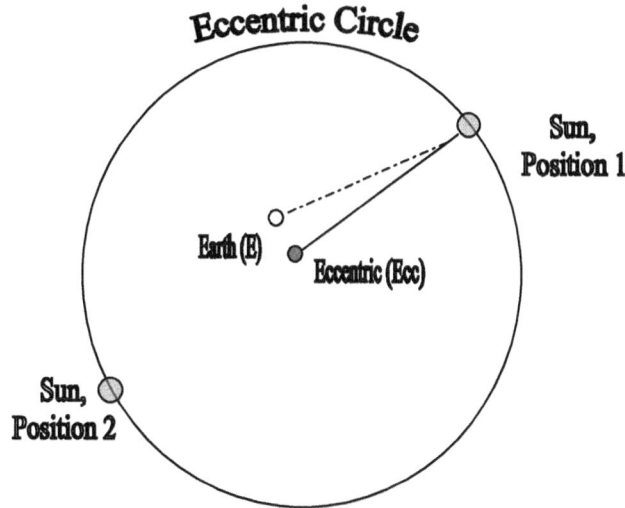

Fig. 1. *Eccentric Motion*. Explanation of apparent non-uniform motion on the assumption of uniform motion. The sun moves uniformly around point (*Ecc*). Seen from the Earth (*E*), however, the uniform motion looks non-uniform. At positon 1 the Sun appears furthest away from the Earth (apogee), while at position 2, it appears at its closest approach to the Earth (perigee). Source: author's own drawing.

The Copernican indifference towards different geometric techniques shows that he is content with the Greek ideal of 'saving the appearances'. He is satisfied that these different kinds of 'motion' reproduce the appearance of planetary orbits as obtained from observational data. But Copernicus is not concerned with the further question whether either of these different geometric techniques may be a better way of modelling the kinematics of planetary motions. Of these different models he says: 'I could not really say which one is right' (Copernicus 1543 [1995], Bk. III, § 20; cf. § 15). Nor is he concerned with establishing whether these geometric devices can be regarded as a physical explanation of the apparent motion of the planets. Kepler later complained that his predecessors had sought the 'equipollence of their hypotheses with the Ptolemaic system' (Kepler 1618-1621 [1995], Bk. IV, Pt. II, § 5). Kepler went on to investigate 'physical' causes of planetary motion – a process during which he abandoned many of the ideas still important to Copernicus.

As we shall see the equivalence of hypotheses is one of the central pillars of Leibniz's philosophy, which he deploys in his defence of the Copernican hypothesis. But Leibniz stood in the tradition of an evolved Keplerian version

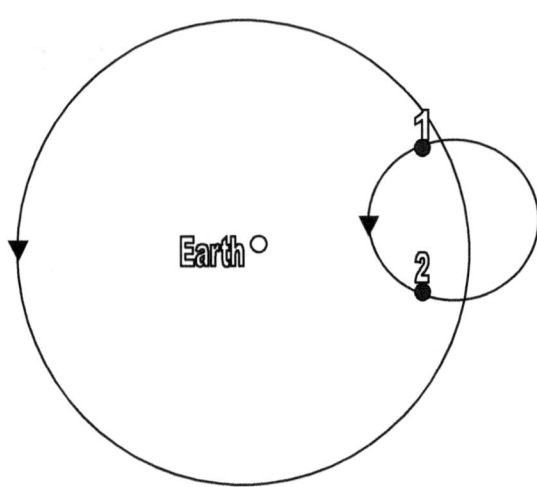

Fig. 2. *Epicyclic Motion.* Retrograde motion, as seen from the Earth, occurs, when the planet moves from P_1 to P_2 on its epicycle. Source: author's own drawing.

of Copernicanism. Like Kepler, he also appeals to dynamic notions to argue his case in favour of the Copernican 'hypothesis'.

• Copernicus was still committed to the Greek ideal of circular motion for planetary orbits. To be precise, Copernicus believed that planets were carried on spheres, which themselves performed circular motion around a centre (Barker 1990, 2002). The title of his book refers to the 'revolutions of heavenly spheres'. He tells his readers that

> [...] the movement of the celestial bodies is circular. For the motion of a sphere is to turn in a circle [...] (Copernicus 1543 [1995], Book I, §4; cf. Book VI, §§ 1-2).[2]

In early Greek geocentric astronomy the centre coincided with the position of the Earth. But a simple homocentric model of planetary motion, according to which the planets orbit the central Earth on concentric rings, fails to match the observations. Planets move at varying speeds and distances from the central body and sometimes seem to go into retrograde motion. Retrograde motion is

---

[2] The original title of Copernicus's book was changed from *De Revolutionibus orbium mundi* to *De Revolutionibus orbium caelestium* in order to avoid the disapproval of the Church.

the apparent periodic westward deviation of planets, as seen from Earth, from their normal eastward motion. Various devices were introduced to cope with this difficulty. (Fig. 2) In order to improve the accuracy of his geocentric model even further, in particular with respect to retrograde motion, Ptolemy introduced a new device: the *equant* (Fig. 3), which was meant to explain more precisely the retrograde motion of the planets. Copernicus strongly objected to the use of the equant because it violated the ideal of uniform circular motion. Although Copernicus puts the mean sun at the centre of his heliocentric model, he admits only circular motion, which forces him to apply minor epicycles to improve the 'fit' between his model and the apparent motion of the planets.

- Copernicus lacks dynamic concepts like inertia and gravity, which were needed to advance towards a physical explanation of planetary orbits. Copernicus possessed no modern concept of lawful physical behaviour, no notion of laws of science as quantified functional relationships between various physical parameters. The lack of these tools meant that Copernicus had to content himself with the geometry of kinematic relationships, like his Greek predecessors. When Kepler broke with the presupposition of circular motion, abandoned the idea of 'celestial spheres' and replaced geometry with mathematical analysis, which permitted him to establish the three laws of planetary motion, he went a significant step beyond the Copernican model of heliocentrism. In particular, Kepler began to think of the physical causes of planetary motion and thus introduced dynamic considerations. For these reasons, Kepler is regarded as the true revolutionary in astronomy. Leibniz, too, stood in the Keplerian tradition since he invented, independently of Newton, the differential calculus and proposed a vortex theory (1689) to account for the Keplerian non-circular orbits of the planets.

- Modern defenders of the computational equivalence of the geocentric and heliocentric models could add a further argument to their case by considering the explanation of the seasons on the two models. On the geocentric view the seasons are a result of a tilt of the eccentric, ecliptic circle by 23.5° with respect to the plane of the stationary Earth. The tilt of the ecliptic circle explains the sun's variation in latitude in different locations around the globe. The explanation is more cumbersome on the Copernican model. Copernicus naturally stipulates that the Earth is tilted at the same degree with respect to the solar plane (Copernicus 1543 [1995], Bk. I, §2, §11). But Copernicus introduces a third motion to the Earth, which he calls the 'deflexion of the axis of the moving Earth.' This movement can be visualized as a wobble in the Earth's axis in its orbit around the sun. The third motion (in addition to the daily and annual motion) has the function of explaining the change of seasons. This 'deflexion' is necessitated by the Copernican assumption that planets are not free-moving in space but are attached to spheres, which serve as their orbital vehicles. This means that the Earth's axis shifts its orientation in the

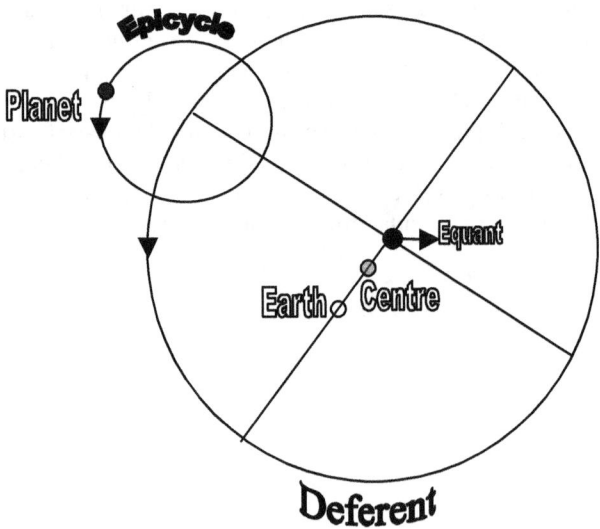

Fig. 3. *The Equant.* Explanation of retrograde motion with a new geometric device, the equant (see Copernicus, 1543 [1995], Bk. III, §15-6; Ptolemy 1984, §IX.6; Andersen, Barker and Chen 2006, Chapter 6.3). This representation is supposed to be a closer fit of the model to the data than the elementary homocentric model. From the point of view of the equant, the motion of the planet on the epicycle would appear uniform. Further flexibility is introduced by letting the Earth either sit at the Centre of the deferent or off-centre, as indicated in the diagram. Source: author's own drawing.

annual orbit around the sun. As Kepler abandons the spheres, on which the planets are carried in the Copernican model, he is able to dispense with the third motion of the Earth. The axis of inclination remains constant with respect to the plane of the orbit around the sun. The Keplerian model of free-moving planets and a constant tilt of the Earth's axis with respect to the ecliptic are sufficient to explain the seasons.

Leibniz was aware of the developments in astronomy from the original Copernican to the later Keplerian, Cartesian and Newtonian versions of the theory. Leibniz was equally aware of the philosophical discussions surrounding the epistemological status of the heliocentric 'hypothesis': was it to be understood as a mere calculating device, in the instrumentalist fashion, or as a realist claim about the planetary system? Leibniz did not follow Kepler and Copernicus's only 'pupil' Rheticus in embracing a realist reading of the Copernican hypothesis. He did not simply ignore the instrumentalist reading, which, as reflected in Leibniz's writings, was still a viable option in his own time. As we shall see, the principle of the equivalence of hypotheses does not offer sufficient ground to declare the heliocentric hypothesis the clear winner in the dispute with the geocentric hypothesis. A realist or instrumentalist reading of the Copernican model depended on an understanding of the notion of 'hypothesis' in the 16th and 17th century. In order to explain Leibniz's defence of the Copernican model, we need to understand the career of the notion of 'hypothesis' and its changing connotation during the crucial period from 1543-1687, a period which is nowadays dubbed the 'Scientific Revolution'.

1.2   On Hypotheses

Newton is famous for his statement: 'Hypotheses non fingo.' This Latin phrase can be rendered alternatively as 'I do not feign hypotheses'; 'I do not make use of fictions'; 'I do not use false propositions or premises or explanations.'[3] Historians of science have identified several senses in which Newton uses the word 'hypothesis.' Sometimes he meant a plausible though not provable conception. In his later years he came to regard a hypothesis as a gratuitous fiction (Koyré 1965, pp. 36-7).

> That which cannot be derived from phenomena is called a hypothesis and these do not belong to experimental philosophy (quoted in Dijksterhuis 1956, p. 537; see also Burtt 1924 [1980], pp. 215-220).

Newton was not the first scientist to worry about the term 'hypothesis'; the worry reaches back to the Greeks. The concern is about whether the geometric constructions, with which the Greeks attempted to explain the appearances, i.e. the observable behaviour of the planets, including the retrograde and non-uniform motion around the centre, have to be regarded as fiction or reality. This uncertainty

---

[3] Koyré 1965, p. 35; Dijksterhuis 1956, p. 541; Crombie distinguishes 3 senses of 'hypotheses': improvised propositions, heuristic aids, illegitimate fictions (Crombie 1994, II, p. 1071).

about astronomical hypotheses is reflected in the attitudes of Aristotle and Ptolemy. Aristotle adopted a much more realist attitude than Ptolemy since he considered that his homocentric spheres, which carried the planets around their circular orbits, actually existed in nature. But his homocentric model cannot be correct because it fails to account for the 'appearances' – the variation in brightness of the planets and their temporary retrogression. In order to account for the observations, Ptolemy introduced his geometric devices (epicycles, eccentrics, equants, which made sense of the observations) but at the price of abandoning Aristotelian realism. Ptolemy adopted an instrumentalist attitude towards his geometric devices as useful fictions, which made sense of the observations, but he did not expect his geometric models to properly represent the celestial phenomena (Ptolemy 1948, pp. 600-601).

The contrast between instrumentalism and realism shaped the discussion of astronomers well beyond the death of Copernicus in 1543 (cf. Donahue 1975; Westman 1975a). Duhem (Duhem 1908, Chapter 4) holds the view that this contrast – and the desire to overcome it – lay at the root of Copernicus's reform of astronomy.[4] Copernicus and his pupil Rheticus had corresponded about the usefulness of hypotheses in astronomy with the Lutheran theologian Andreas Osiander (1498-1552). Osiander wrote a Preface to *De Revolutionibus* in which he also adopted an instrumentalist interpretation of the Copernican hypothesis in order to protect this book from a 'realist' misinterpretation at the hand of a hostile clergy (cf. Wrightsman 1975). Copernicus and his pupil considered that certain astronomical hypotheses were more probable than others. More probability accrued to the heliocentric hypothesis than to the geocentric hypothesis, in view of the observations. Acceptable hypotheses in astronomy had to explain all the observable phenomena. They had to explain the phenomena in a coherent way. The Ptolemaic hypothesis, says Rheticus, does not suffice to establish the harmony of celestial phenomena (Rheticus 1540, p. 132; see also *Correspondence* reprinted in Rosen 1959, pp. 31-2; 1984, pp. 125-126, pp. 193-194, pp. 198-205). Kepler later agreed that the Copernican hypothesis enjoyed more probability than the Ptolemaic hypothesis. The notion of hypothesis had great repercussions throughout the next 140 years. The ambiguity of the term as reflected in Newton's views on hypotheses in science, invited opposing interpretations of the Copernican model. In his *Dialogue Concerning The Two Chief World Systems* (1632), Galileo epitomizes the ambivalent status of hypotheses in the 16th and 17th century. The Preface states that his spokesman, Salviati, will defend the Copernican system but only as a purely mathematical hypothesis. But as the dialogue unfolds, Salviati is drawn towards probability arguments. Eventually he adopts the Copernican position that the acceptance of the dual motion of the Earth as a physical assumption leads to a more coherent explanation of the appearances. Note that these probability arguments invoke belief in a model, because its physical assumptions are more probable. It is not believable, says Kepler, that the 'fixed stars move at incalculable speed', whilst the Earth stands still. It is more probable that the apparent daily rotation of the fixed stars is an effect of the rotating Earth (Kepler 1618-1621 [1995], Pt. II, § 5).[5] The

---

[4] An alternative, more technical reason is that Copernicus was disturbed by Ptolemy's equant and wished to return to truly circular motion (cf. Wilson 1975).

[5] Kepler's probability argument states that we should attach more plausibility to the heliocentric view because the evidence - the apparent motion of the 'fixed' stars in a 24-hour rhythm about the Earth – is more probable on the view that the Earth

Copernican hypotheses are more like conjectures than useful fictions. They have a much closer association with the phenomena than Newton would later accept. They form, as Rheticus tells us, the basis of inferences.

By contrast, labelling hypotheses as 'useful fictions' in astronomy had, according to Osiander, certain advantages. It reassured Copernicus's adversaries that his heliocentric model did not force them to abandon their cherished geocentric beliefs. Cardinal Bellarmine reminded Galileo that Copernicus had always spoken *hypothetically*: it is possible to use the motion of the Earth as a mathematical device to render the calculations more economic, since fewer epicycles and eccentrics are needed. However to affirm the centrality of the sun as a physical hypothesis is in conflict with the Scriptures.[6]

In order to avoid a clash between the Church and heliocentrism, Osiander inserts his Preface in an attempt to present the Copernican hypotheses as mere calculating devices. They have the license to be false or replaceable as long as 'they reproduce exactly the phenomena of the motions' (Osiander 1541). By the time Newton appeared on the scene, and in the wake of Kepler's work, hypotheses had lost their appeal. Newton declared that the laws of motion are deduced from 'Phenomena and made general by Induction', and this is the 'highest evidence that a proposition can have in this Philosophy' (quoted in Koyré 1965, pp. 36-37; Dijksterhuis 1956, pp. 544, 546-547). Phenomena are (reliable) observational or experimental data, from which are derived laws or axioms. Newton rejects any explanation of natural phenomena, which appeals to 'metaphysical' hypotheses, for which no evidence can be cited. It was against this historical backdrop that Leibniz took up his defence in support of heliocentrism (1689).

Copernicus still used the term 'hypothesis' quite freely in his *De Revolutionibus*. The Greeks, he says, reserved the term 'hypothesis' for 'principles and assumptions.' Throughout his work Copernicus calls the motion of the Earth a 'hypothesis'. He does not mean it in Newton's pejorative sense. Copernicus is a realist about the motion of the Earth and the spatial distribution of the planets around the sun. But Copernicus is not a realist about the geometric devices, which he and his Greek predecessors employed to 'save the phenomena'. As mentioned

---

rotates on its own axis. These probability arguments can be supported by a consideration of the angular velocities involved under the two scenarios. Under some simplifying assumptions, the angular velocity of the rotating Earth for an observer at the equator is *464 m/s = 1670 km/h*. The geocentric view, by contrast, has to assume an angular velocity of the 'fixed' stars about the stationary Earth. A calculation produces a value of *5.45×10^6 m/s = 1.96×10^7 km/h*. It is such an enormous rotational velocity of the stars – 19.6 million kilometres per hour, compared to 1670 km per hour for the Earth at the equator – which the Copernicans consider implausible on mechanical grounds. By comparison, the orbital velocity of the Earth around the sun is 30km/s and the velocity of the solar system around the galactic centre is 225km/s. The evidence – the apparent rotation of the sphere of fixed stars – is more likely on account of heliocentrism than on account of geocentrism (see Weinert 2010).

[6] See Koestler 1959, p. 454; similar statements, reflecting Osiander's instrumentalist attitude, are found in Kuhn 1957, pp. 191, p. 194 and in Crombie 1994, I, pp. 599-600.

above, Copernicus accepts the 'equipollence of hypotheses', as had his Greek predecessors. Thus he declares that the apparent irregular motion of the planets can either be accounted for by the use of an eccentric circle, i.e. one whose 'centre is not the centre of the sun' or through an epicycle on a homocentric circle' (a deferent) (Copernicus 1543 [1995], p. 151). 'Accordingly, it is not easy to determine which of them exists in the heavens' (*Ivi*, p. 154). By contrast Kepler aims to practice astronomy without the use of 'hypotheses', since he is interested in the physical causes of planetary motion.

> It can easily do without the useless furniture of fictitious circles and spheres. But there is such a great need of imagining the true figures, in which the routes of the planets are arranged, that we are impoverishing Astronomy and that the big job to be worked on by the true astronomer is to demonstrate from observations what figures the planetary orbits possess; and to devise such hypotheses, or physical principles, as can be used to demonstrate the figures which are in accord with the deductions made from observations (Kepler 1618-1621 [1995], p. 124).

Kepler goes beyond Copernicus by moving from kinematics to dynamics. This distinction between a kinematic description and a dynamic explanation of planetary motion becomes important in the defence of Copernicanism. When Leibniz took up this challenge, he employed two of his most fundamental principles – the relativity of motion and the equivalence of hypotheses. Strictly speaking, these two principles will not deliver a decisive defence of Copernicanism since by the verdict of these two principles astronomical phenomena could be explained equivalently by the geocentric or the heliocentric hypothesis. As indicated above, the seasons can either be explained by the assumption that the path of the sun around the central Earth is inclined by 23.5° with respect to the plane of the Earth or by the assumption of an orbiting Earth whose axis of rotation is inclined by the same angle with respect to the stars. In the face of the equivalence of hypotheses Leibniz appeals to other criteria - like simplicity, intelligibility, probability - to argue in favour of an elimination of geocentrism. Above all, however, it is his attempt at a dynamic explanation of Keplerian elliptical orbits, which renders his case in favour of the Copernican hypothesis much more solid than a mere reliance on his kinematic principles.[7]

---

[7] It is interesting to note that this ambivalent attitude towards hypothesis survived the consolidation of Newtonian heliocentrism. L. Boltzmann (1844-1906), for instance, calls hypotheses 'arbitrary pictures' (Boltzmann 1905 [1974], p. 161), whilst E. Mach (1838-1916) regards hypotheses as thought constructs, which aid the economy of thought (Mach 1883 [^{9}1933] p. 468). Duhem himself concludes his survey of astronomy (1908) with a general plea for instrumentalism in science.

2 Leibniz's Defence of Heliocentrism

2.1 Galilean Relativity

Galileo introduced into physics the principle of the relativity of motion. According to the principle of relativity, the kinetic motion of an object can be described from either a stationary or a moving reference frame. As long as the motion is inertial (either at rest or moving at constant velocity) both views are equivalent. They must lead to the same numerical results. It is a matter of choice, which system we regard as the frame at rest and the frame in motion, respectively. This makes no difference to the physics of the situation. Galileo offers a famous thought experiment to demonstrate the equivalence of inertial systems. In a cabin below the deck of a large ship observe the behaviour of 'flies', other 'small winged creatures' and 'fish in a bowl'. At first the ship is at rest. When the first set of observations is completed, let the ship proceed with uniform speed. The observations will reveal no difference in the behaviour of the creatures (Galileo 1632 [1953], pp. 199-201).

Leibniz was obviously aware of the Galilean relativity principle, since he employs a Galilean-type thought experiment against absolute motion. The principles of the relativity of motion and the equivalence of hypotheses are closely related. In fact, the (Galilean) relativity of motion implies the equivalence of hypotheses, as Leibniz observed in his *Specimen Dynamicum* (Leibniz 1695, p. 445; italics in original):

> Therefore we must hold that if any numbers of bodies are in motion, we cannot determine from the phenomena which of them are in absolute determinate motion or rest; rest can be attributed to any one of them you may choose, and yet the same phenomena will be produced. It follows therefore (Descartes did not notice this) that the *equivalence of hypotheses is not changed by the impact of bodies upon each other* and that such rules of motion must be set up that the relative nature of motion is saved […].

Leibniz applies this reasoning to Copernicus. But it raises the question whether one of the hypotheses may be said to be more probable than its rival. Leibniz deploys the criterion of simplicity:

> That is to say, if the given phenomena appear the same, whatever may be the true hypothesis or however we may ascribe motion or rest to them, the same result will be produced in the unknown or the resulting phenomena, even with respect to the action of bodies upon each other. This conforms to our experience; we feel the same pain whether our hand strikes a stone which is at rest […] or the stones strikes our hand at rest with the same velocity. Meanwhile we speak as the situation demands, in whatever way provides the more fitting and simpler explanation of the phenomena, just as we make use of the motion of a *primum* mobile in the study of spheres and must use the Copernican

hypothesis in planetary theory (Leibniz 1695, pp. 445-6; italics in original).

Nevertheless, from the point of view of (Galilean) relativity it makes no difference whether we adopt a geocentric or a heliocentric view (de Solla Price 1962, p. 198; Rosen 1984, pp. 183-184).[8] We can follow Ptolemy: regard the Earth as a stationary frame and the sun as a moving frame. Or we can follow Copernicus: regard the Earth as a moving frame and the sun as a stationary frame. According to the principle of relativity our choice makes no difference to the physics of the situation. As Leibniz says, so it appears to be. The Earth turns on its own axis once in a 24-hour rhythm to give us day and night. If the sun turned around the stationary Earth once in a 24-hour rhythm it would give us day and night. The seasons result from either a tilted plane of the sun around the Earth or a tilted axis of the Earth around the sun. However there is more to a description of the solar system than mere kinematics. From a strictly kinematic point of view, the models are equivalent. The kinematic point of view is only concerned with pure motion, without regard to its causes (Dijksterhuis 1956, Pt. I, § 83; Pt. IV, § 18, Pt. IV, II. C). This is the Ptolemaic and Copernican perspective. But there is also the question of dynamics: What causes the planetary bodies to move? Consider a slightly amended version of Leibniz's example: the encounter of a hand with a wall. Whether we regard the hand or the wall as being at rest, we experience the same pain. Physics informs us that both can be regarded as reference frames, either at rest or in motion. The kinematics will be the same. But experience also tells us that the hand is more likely to move than the wall. The dynamic situation is no longer equivalent. The body causes the hand to move. The wall has no cause of motion. Kepler was preoccupied with the question of physical causes. He suspected that energetic rays from the sun drove the Earth around its elliptical orbit. When a planet shows its 'friendly face' to the sun, its magnetic lines attract it. When a planet shows its 'unfriendly face' to the sun, its magnetic lines repulse it. The game of attraction and repulsion constrains the planet to its orbital motion around the sun (Kepler 1618-1621 [1995], Pt. II, § 93). As Newton later showed, this dynamic explanation was mistaken. Nevertheless, Kepler advanced dynamic arguments in favour of the orbital motion of the Earth. Once Newton had shown why the planets stay in their elliptical orbits around the sun, the heliocentric model gave a better representation of physical reality than the geocentric model. Newton improved the mathematical structure of the model. He provided a dynamic explanation of planetary orbits in a heliocentric model: it is the vectorial result of the first law of motion (inertia) and gravitational attraction.

---

[8] Galilean relativity only applies to inertial motion. Einstein's general principle of relativity applies to both inertial and non-inertial motion. In General relativity it is possible to distinguish inertial from accelerated motion by observing the geometry of world-lines. (Thanks to an anonymous referee who suggested a clarification on this point.)

2.2 Some Dynamical Considerations

It seems at first that the central message of his essay 'On Copernicanism and the Relativity of Motion' (1689) is the equivalence of hypotheses. Leibniz affirms once again that 'motion is not something absolute, but consists in a relation' (Leibniz 1689a, p. 91). Hence

> [...] an astronomer makes no greater mistake by explaining the theory of the planets in accordance with the Tychonic hypothesis than he would make by using the Copernican hypothesis in teaching spherical astronomy and explaining day and night, thereby burdening the student with unnecessary difficulties (Leibniz 1689a, p. 91).[9]

Adopting a criterion of simplicity, the Copernican account is the 'truest theory', that is, the most intelligible one, since it avoids the 'perplexities' (like retrograde motion) with which other theories are burdened. In a letter of 1688, Leibniz calls the Copernican hypothesis 'confirmed by [...] many arguments drawn from new discoveries (quoted in Bertoloni Meli 1988, p. 21).

> The truth of a hypothesis is nothing but its intelligibility (Leibniz 1689a, p. 91; cf. Lodge 2003).

For if the truth of a hypothesis lies in its intelligibility and the Copernican hypothesis has 'greater intelligibility' than the geocentric hypothesis,

> [...] there would be no more distinction between those who prefer the Copernican system as the hypothesis more in agreement with the intellect, and those who defend it as the truth. For the nature of the matter is that the two claims are identical; nor should one look for a greater or a different truth here. And since it is permissible to present the Copernican system as the simpler hypothesis, it would also be permissible to teach it as the truth in this particular sense (Leibniz 1689a, p. 92).

On this understanding astronomers need not hold back 'by the fear of censure.' By adopting this approach, Leibniz hopes to 'free Rome and Italy from the slander that great and beautiful truths are there suppressed' (Leibniz 1689a, p. 93). But is Leibniz not adopting Osiander's instrumentalist attitude? After all, Ptolemy already believed that his theory of eccentrics, epicycles and equants provided the simplest devices to account for the phenomena, without believing in their reality. Yet Ptolemy added that the hypotheses should only be as simple as to allow them to save the phenomena as accurately as possible (Duhem 1908, Chapter 3). Osiander, too,

---

[9] Tycho Brahe (1546-1601) presented a compromise system between heliocentrism and geocentrism. On Tycho's model the Earth remains the centre of the 'universe' with the moon and the sun orbiting it, but the other planets orbiting the sun.

recommends the Copernican hypothesis for its simplicity and greater intelligibility, without admitting any correspondence to reality.

A closer reading, however, reveals that Leibniz goes well beyond Ptolemy's and Osiander's instrumentalism. In his *Specimen Dynamicum* (1689) Leibniz concludes his remarks on Copernicanism with an observation, which will allow him to express a preference for one of the competing hypotheses, not offered by the relativity of motion and the principle of simplicity.

> But whenever the equipollence of hypotheses is involved, every factor contributing to the phenomena must be included (Leibniz 1695, p. 450).

Equally, 'truth is found not so much in phenomena as in their causes' (Leibniz 1695, p. 446). As we shall see now these additional factors are the dynamics of planetary motions. The equivalence of hypotheses shows that all hypotheses are equally possible but dynamics demonstrates that not all hypotheses are equally probable. Thus, it is more probable that the appearance of certain celestial motions, like the 24-hour rotation of the fixed stars, is the result of the Earth's motion on its own axis.

The criteria of intelligibility and simplicity cannot be decisive in moving from an instrumentalist to a realist position, for Leibniz concedes that 'the Ptolemaic account is the truest one in spherical astronomy.' The simplest hypothesis is not necessarily the true one – contrary to what Leibniz writes to Huygens in September 1694 (Leibniz 1690b, p. 308) – for Ptolemy also appealed to the criterion of simplicity, whilst adopting an instrumentalist position regarding his geometric devices. There must be other reasons to prefer the Copernican hypothesis. Leibniz writes to Huygens in June 1694 that 'other than extension and its variations, which are purely geometric things, we must acknowledge something higher, namely force' (Leibniz 1690b, p. 308). In other words, there are additional criteria to judge astronomical hypotheses, in particular dynamical reasons.

Although Leibniz was aware of Newton's theory of gravitation, he was dissatisfied with Newton's account and attempted his own mechanical explanation of planetary motions by way of a vortex theory. He hints at it in his paper on Copernicanism (1689): Kepler's laws, he claims, can be given a physical explanation 'by means of a vortex around the sun' (Leibniz 1689a, p. 93). He divides planetary trajectories into two components: a) harmonic circulation and b) paracentric motion.

> And thus we may consider a planet to be moved by a two-fold motion, or composed from the harmonic circulation of the orbit of its carrying fluid, and from a paracentric motion, as if of a certain weight or an attraction, that is an impulse towards the sun, or the primary planet.
> [...] the paracentric motion of the planets is required to be explained, arising from the force of the circulation on a planet being made to change orbit, and composed from the attraction between itself and the sun (Leibniz 1689b, §§ 8-9).

The sun is compared to a magnet, as in Kepler's dynamical explanations, but in order to account for the stability of the elliptical orbits, in agreement with Kepler's laws, Leibniz stipulates that paracentric motion consists of an attracting and a receding part, with respect to the sun (Leibniz 1689b, §27). During its harmonic circulation the planet at first falls towards the sun and is then repelled by the magnetic force. With his attempted solution Leibniz stands more in the tradition of Kepler and Descartes than in the footsteps of Newton, whose solution appeared at the same time. But it is clear that Leibniz is at pains to formulate his causal theory of planetary motion in such a way that it is in accordance with Kepler's laws (Leibniz 1690b, pp. 309-11; 1689b). By contrast both Galileo and Descartes ignored Kepler's laws (see Schmaltz 2015).

By accepting Kepler's laws of planetary motion, Leibniz accepts the Copernican system as a reality, since Kepler's laws take the sun as a focal point of planetary orbits. He also dispenses with epicycles and eccentrics, in line with Kepler's stipulation. The relativity of motion only applies to kinematics, proved through geometrical demonstrations' (Leibniz 1689a, p. 92). But Kepler's laws require a physical explanation. Whilst Leibniz attempts such a dynamic explanation in terms of Cartesian vortex theory, the Newtonian way is to derive these laws from the laws of motion. Leibniz's vortex theory suffered the same fate as its Cartesian cousin. Irrespective of the inadequacy of the vortex theory, from the dynamical point of view Leibniz's defence of the Copernican hypothesis looks considerably stronger than Osiander's instrumentalist approach. For the equivalence of hypotheses does not bestow a greater degree of credibility on the rival hypothesis; and the simplicity of a hypothesis is relative to background knowledge. In theory, Aristotle's homocentric circles are simpler than Ptolemy's geometric devices but not in view of the observable appearance of planetary motion.

Dynamic causes, however, allow a much stronger defence of the Copernican hypothesis, since they can be coupled with probability considerations. Some causes are more probable than others: the hand is more likely to move than the wall. Leibniz even claims that the 'magnetic properties', which planets like the Earth, Jupiter and Saturn 'exert' on their respective moons, as well as the natural explanation of retrograde motion, constitute confirmation of the Copernican view (Leibniz 1689a, p. 93). He could have derived even stronger evidence from Galileo's observations of the phases of Venus or the bounded elongation of inferior planets.

3   Concluding Remarks

It looks at first as if Leibniz offered only a guarded defence of the Copernican hypothesis: a) the relativity of motion and the equivalence of hypotheses do not allow him a stronger stance; b) he wants to protect the reputation of the Catholic Church, which would only allow an instrumentalist interpretation (as expressed by Osiander and Bellarmine); c) nevertheless, the Copernican hypothesis can be regarded as more intelligible than the Ptolemaic one.

The greater intelligibility of the Copernican hypothesis is revealed, despite the equivalence of hypotheses, by its greater simplicity. It should be added that it is only simpler than Ptolemy in its Keplerian form for it dispenses with Copernican epicycles and spheres. It is the Keplerian version of heliocentrism, which Leibniz defends. It is also revealed by the dynamic reasons – the vortex theory – at which Leibniz hints in his paper on Copernicanism and which he spells out in greater detail in his letter to Huygens (1690) and his *Tentamen* (1689). Although the vortex explanation does not succeed, it reveals that Leibniz was eager to provide a causal theory which respected Kepler's laws. The sun does not exert a magnetic attraction on the planet but a gravitational force. What matters, however, is that Leibniz goes beyond kinematics towards dynamics. It is the dynamic reasons, which ultimately provide the best defence of heliocentrism.

References

Andersen H., Barker P., Chen X. 2006. *The Cognitive Structure of Scientific Revolutions*. Cambridge: The Cambridge University Press.
Antognazza, M. R. 2003. Leibniz and the post-Copernican universe. Koyré revised. *Studies in History and Philosophy of Science* 34:309-327.
Barker, P. 1990. Copernicus, The Orbs, and the Equant. *Synthese* 83:317-323.
Barker, P. 2002. Constructing Copernicus. *Perspective on Science* 10:208-227.
Bertoloni Meli, D. 1988. Leibniz on the Censorship of the Copernican System. *Studia Leibnitiana* XX(1):19-42.
Blumenberg, H. 1955. Der Kopernikanische Umsturz und die Weltstellung des Menschen. *Studium Generale* 8:637-648.
Blumenberg, H. 1965. *Die kopernikansiche Wende*. Frankfurt a./M.: Suhrkamp.
Boltzmann, L. 1905 [1974]. On Statistical Mechanics. In: McGuinness. B. (ed). *Theoretical Physics and Philosophical Problems, Selected Writings*. Dordretcht: Reidel, pp. 159-172.
Burtt, E. A. 1924 [1980]. *The Metaphysical Foundations of Modern Science*. London: Routledge & Kegal Paul.
Copernicus, N. 1543 [1995]. *On the Revolutions of Heavenly Spheres*. Amherst, New York: Prometheus Books.
Crombie, A. C. 1961. *Augustus to Galileo*. London: Mercury.
Crombie, A. C. 1994. *Styles of Scientific Reasoning in the European Tradition*. London: Duckworth.
De Solla Price, D. J. 1962. Contra-Copernicus, in *Critical Problems in the History of Science*, edited by M. Clagett. Madison: The University of Wisconsin Press, pp. 197-218.
Dijsterhuis, E. J. 1961. *The Mechanization of the World Picture*. Oxford: Clarendon.
Donahue, W. H. 1975. The Solid Planetary Spheres in Post-Copernican Natural Philosophy. In Westman 1975b, pp. 244-275.
Duhem, P. 1908. *To Save the Phenomena*. Chicago–London: The Chicago University Press.
Finocchiaro, M. A. 2005. *Retrying Galileo, 1633-1992*. Berkeley: The University of California Press.
Galilei, G. 1632 [1953]. *Dialogue on the Great World Systems*. G. de Santillana (ed.). Chicago: The Chicago University Press.
Garber, D. 1995. Leibniz – physics and philosophy. In: Jolley N (ed). *The Cambridge Companion to Leibniz*. New York–Cambridge: The Cambridge University Press, pp. 270-352.
Gingerich, O. 1993. *The Eye of Heaven*: Ptolemy, Copernicus, Kepler. New York: The American Institute of Physics.

Kepler, J. 1618-1621 [1995]. *Epitome of Copernican Astronomy. Epitome of Copernican Astronomy & Harmonies of the World.* Amherst–New York: Prometheus Books, pp. 165-245.

Koestler, A. 1964. *The Sleepwalkers.* Harmondsworth: Penguin Books.

Koyré, A. 1965. *Newtonian Studies.* Chicago: The Chicago University Press.

Kuhn, T. S. 1959. *The Essential Tension.* Reprinted in Kuhn 1977, pp. 225-239.

Kuhn, T. S. 1977. *The Essential Tension.* Chicago: The Chicago University Press.

Kuhn, T.S. 1962 [^{2}1970]. *The Structure of Scientific Revolutions.* Chicago: The Chicago University Press.

Leibniz, G. W. 1689a. On Copernicanism and the Relativity of Motion. In Leibniz 1989, pp. 90-94.

Leibniz, G. W. 1689b. Tentamen de motuum coelestium causis (An attempt to explain the causes of celestial motions). In G. W. Leibniz, *Mathematische Schriften*, herausgegeben von C. I. Gerhardt, Bd. VI (Die Mathematischen Abhandlungen). Hildesheim: Olms 1971, pp. 144-161.

Leibniz, G. W. 1690a. On the Nature of Bodies and the Laws of Motion. In Leibniz 1989, pp. 245-250.

Leibniz, G. W. 1690b. Planetary Theory, from a Letter to Huygens. In Leibniz 1989, pp. 309-12.

Leibniz, G. W. 1695. Specimen Dynamicum. In Loemker ^{2}1970, pp. 435-452.

Leibniz, G. W. 1698. On Nature Itself or on the Inherent Force and Action of Created Things. In Loemker ^{2}1970, pp. 498-507.

Leibniz, G. W. 1710-1716. Against Barbaric Physics. In Leibniz 1989, pp. 312-320.

Leibniz, G. W. 1715-1716. Correspondence with Clarke. In: Parkinson G. H. R. (Ed.). *Philosophical Writings.* London: Dent & Sons, pp. 205-38.

Leibniz, G. W. 1989. *Philosophical Essays.* Edited and translated by Roger Ariew and Daniel Garber. Indianapolis & Cambridge: Hackett Publishing Company.

Lodge, P. 2003. Leibniz on Relativity and the Motion of Bodies. *Philosophical Topics* 31:277-308.

Loemker, L. E. (Ed.) ^{2}1970. *Gottfried Wilhelm Leibniz – Philosophical Papers and Letters.* Dordrecht: Reidel.

Mach, E. 1883 [^{9}1933]. *Die Mechanik.* Darmstadt: Wissenschaftliche Buchgesellschaft (1976) [English Translation: *The Science of Mechanics.* Open Court Publishing 1915].

Miyake, T. 2015. Underdetermination and decomposition in Kepler's *Astronomia Nova. Studies in History and Philosophy of Science* 50:20-27.

Osiander, A. 1541. Letter to Copernicus. In Rosen, 1984, pp. 193-194.

Ptolemy, C. 1984. *Ptolemy's Almagest.* Toomer G. J. (Ed.). London: Duckworth.

Randall, J. R. 1962. *Career of Philosophy.* I: From the Middle Ages to the Enlightenment. New York: The Columbia University Press.

Rheticus, J. 1540. *Narratio Prima.* In Rosen 1959, pp. 107-96.

Rosen, E. (Ed.) 1959. *Three Copernican Treatises.* New York: Mineola.

Rosen, E. 1984. *Copernicus and the Scientific Revolution.* Malabar: Robert E. Krieger Publishing Company.

Schmaltz, T. M. 2015. Galileo and Descartes on Copernicanism and the cause of the tides. *Studies in History and Philosophy of Science* 51:70-81.

Weinert, F. 2007. Why Copernicus was not a Scientific Revolutionary. *Central and Eastern European Review* 1:1-22.

Weinert, F. 2009. *Copernicus, Darwin & Freud*: Revolutions in the History and Philosophy of Science. Chichester: Wiley Blackwell.

Weinert, F. 2010. The Role of Probability Arguments in the History of Science. *Studies in History and Philosophy of Science* 41:95-104.

Westman, R. S. 1975a. Three Responses to the Copernican Theory. In Westman 1975b, pp. 285-345.

Westman, R. S. (Ed.) 1975b. *The Copernican Achievement*. Berkeley–Los Angeles/London: The University of California Press.

Wilson, C. A. 1975. Rheticus, Ravetz and the 'Necessity' of Copernicus's Innovation. In Westman 1975b, pp.17-39.

Wrightsman, B. 1975. Andreas Osiander's Contribution to the Copernican Achievement. In Westman 1975b, pp. 213-243.

Friedel Weinert, Bradford University, U.K.
f.weinert@bradford.ac.uk

Index

A
a priori, 123, 130, 134, 135
absolute space, 7
absolute true motion, 366, 367, 368
Acta Eruditorum, 333–337, 339
action at a distance, 1, 4–6, 49, 85, 86
ad absurdum, 127, 129, 132–135
Adams, R. M., 169, 170, 173, 174, 176
aggregate, 171
Aiton, E. J., 88, 89
algebra, 21, 26-28, 30, 32, 228, 236, 237, 244, 245, 248, 309–312, 314–323, 325, 328
algorithm, 132
analysis situs, 21, 37, 45, 336
analysis, 227–250
analytic, 123, 131
analytical geometry, 289, 291, 293
analytical mechanics, 301, 302
anarchy, 358, 359
Anaximander, 240
Anderson, A., 112
Anselm, 134
Antibarbarus Physicus, 1, 3, 6, 9
antinomies, 124
Archimedes, 133
    Archimedean point, 242
Aristarchus of Samos, 383
aristocracy, 358, 359, 360
Aristotle, 129, 230, 240, 241, 245, 247, 248, 291, 383, 385, 391, 392, 399
arithmetic, 21–30, 32, 33, 35–37, 39, 40, 43-45, 231, 238, 309, 310, 313, 314, 318, 322, 321, 322, 324, 326
    arithmetic mean 164
Arndt H. W., 311
Ars combinatoria, 309-312, 314, 317, 319–323, 328
astronomy, 290
atoms, 133
Attfield, R., 4, 19
Ausdehnungslehre, 181, 182, 188–191
Austin, J., 359
Axiomatics, 309, 310, 319

B
Bacon, F., 291, 293
Barrow I., vii, 290
Basic Law V, 44,
Bayle, P., 344, 346, 352–355
Belaval, Y., 318
Bellarmine, R., 393, 399
Bergson H., 163, 266, 278, 279
Berkeley G., 169, 175, 268, 269
Berlioz D., 326
Bernoulli, J., vii, 152, 153, 161, 331–342
Bernstein R., 264
Bertoloni Meli, 2, 19, 58, 89
best of possible worlds, 207, 208
Beth, E. W., 133
bijection, 39, 41, 42
binary system, 32
Blanché R., 309
Bodenhausen, 374
body, 171–175, 242, 243
Bonner, A. 270
Bos, H. J. M. 239, 240
Boyle, R., 381
Bradley F. H., 269, 270
Brans-Dicke equation, 223
Breton Ph., 254, 262
Browder F., 257
Brown, R., 270, 273, 274–277
Brown, G., 2, 4, 5, 8, 19
Bühler, M., 322
Burke, C., 259
Burkhardt, H., 313
Busch, C., 88

Index

Bussotti, P., 49, 51, 52, 53, 56–58, 65, 66, 70, 71, 79, 81, 89

C

calculating machine, 29, 305, 306
calculus, 22, 23, 121, 126, 129, 231, 235, 236, 239, 240
Cantor, G., 39, 269
Caracteristica universalis, 309
Carnot S., 126
Carnot, L., 126, 127
Caroline of Ansbach, 4, 13, 14, 19
Cartan, E., 183, 188, 191–195, 198
Cartesian laws of motion, 364, 368
Cartesianism, 228, 231, 236, 237, 240, 242, 243
Cassirer, E., 126, 128, 136, 288, 313
categories, 136
Catenary, 331-342
Catholic Church, 10
cause, 2, 3, 123, 124, 125, 128, 131
celestial motion, 382, 398
centrifugal force, 57, 63, 65, 80
centripetal force, 57, 78,
characteristica universalis (universal characteristic), 21, 27, 46
Chareix, Fabien, 277
chemistry, 133
Child J. M., vii
Chomsky, N. 243
Church, A., 126, 127
circle, 390, 394, 399
Clarke, S., 1, 3, 4, 6, 7, 13–19
classical, 121-124, 126 128-31, 133, 233, 241, 363, 366, 367
clockwork universe, 381
Cockcroft, J., 277
coexistence, 36, 42
Cohen, I. B., 15, 19, 20
cohomology, 196, 197
combinatorial, 236–238, 245, 249
common good, 360
complete individual concept, 230, 232, 235
conate to recede, 81

conatus explosivus, 80–83, 85
Conatus, 336
Concept of Force, 299, 301, 303
conceptualism, 230, 248,
Conring, H., 312
consent, 360
conservation principle, 296, 297
constraints, 121, 122, 132-136
contingent, 122, 124, 125, 128–131, 133-135
continuum, 172, 174, 236, 241, 244
convergence 145, 150, 153, 154, 155, 157, 161
Copernicus, N., 382-396
    Copernican model, 383, 384, 386, 389–392
    Copernican revolution, 383, 384
    Copernican turn, 385, 386
    Copernicanism, 5, 10, 11
corporeal substance, 170, 171, 174, 176
cosmological Λ-term, 220
cosmological, 52, 53, 79, 84,
Cotes, Roger., 3, 5, 17,
Couturat, L., 22, 23, 45, 46, 122, 123, 133, 136, 229, 241, 261, 288, 311, 312, 315, 316, 319, 320, 323
Cover, J. A., 172
curvilinear line, 237, 241, 242, 246
Cusanus, N., 126, 134

D

Dascal, M., 124
Davis, M., 260
De Causa Gravitatis, 3, 58, 80, 82,
de l'Hôpital, G., 114
De Luca A, 254
De Mora-Charles M., 318
De quadratura arithmetica circuli ellipseos et hyperbolae cujus corollarium est trigonometria sine tabulis 141, 142, 144, 147, 148, 149, 151, 154, 157, 158, 161
De Ruggiero, G., 269

De vera proportione circuli ad quadratum circumscriptum in numeris rationalibus expressa 142, 145, 153
De Volder., 10, 176
Dedekind, R., 267
deductive, 126
definition, 231, 232, 240, 249
  of number, 28
Democritus, 12
Des Bosses., 10
Descartes and Fermat methods, 289
Descartes R., 4, 26, 27, 47, 55, 71, 81, 97, 117, 237, 242, 243, 245, 287–292, 304, 330, 337, 382, 395, 398
Dewey J., 253, 257, 264
Dhombres J., 56
dialectics, 341–361
dialogues on metaphysics and on religion, 14, 19
difference, 236, 239, 240, 243, 244, 246,
differential and integral calculus, 287, 292, 296, 306
differential forms, 183, 191–194, 196–198, 202, 203
Dinostratus, 329
discourse on Metaphysics, 173, 175
distance-period relationship, 384, 386
doubly negated statements (DNP), 121, 126–130, 133-135
Drago, A., 121, 125–127, 132, 134
Drapeau-Contim, F., 324
Duchesneau, F., 71
Duhem, P., 392, 394, 397
Dummett, M., 126, 129, 132,
dynamic, 382, 399
dynamica, 3, 369, 370, 371
dynamical, 50–54, 56–58, 63, 70, 84, 88
dynamics reformation, 300
dynamics, 1, 3, 169–175, 49, 50, 51, 55, 70, 394, 396, 398

E

eccentric, 385–387, 390, 392-394, 397, 398
edition of complete works and letters, viii
Edwards, P., 133
Einstein, A., 126, 275, 364
Einstein's equations, 221
Elastic Bodies, 294, 299, 302, 303, 306
electrodynamics, 207, 208, 215, 223
elite governing, 358
energy, 364, 365, 366, 372, 373, 375
Engineering Sciences, 287, 293, 305, 306
Enlightenment, 287, 344, 345, 359
Enriques, F., 122, 132
entity (pl.), 29, 35, 43
epicycle, 385, 386, 388-390, 392–394, 397, 399
Epistola ad V. Cl. Christianum Wolfium um Grandi 142
equality, 29, 30, 39
equant, 385, 389, 390, 392, 397
equivalence (of hypotheses), 391–399
et defensio sententiae Autoris de veris Naturae Legibus contra Cartesianos, 3
Euclid, 27, 36, 227, 238, 240
Euler, L. 164
existence of evil, 343, 344, 346, 352, 353, 355, 360
expansion of elementary function 155–158
explanation, 124, 130, 135

F

Fabri, H., vii, 110
Fermat, P., 289, 290
Fermats little theorem, 322
Ferraro, G. 141–167
Fichant, M., 126
financial mathematics, vii
force, 1–4, 9, 14, 15, 19, 20, 169-176, 364, 365, 366, 367, 369, 372, 373, 374, 375
formal manipulation 142, 143

foundations of arithmetic, 21, 24, 36, 39, 40, 44
foundations of logic, 228, 229
foundations of mathematics, 228, 234, 239, 245, 248
foundations of physics, 183, 195, 202
foundations of science, 125, 126
freedom, 241
Frege G., 228, 261
function, 228, 239, 243
functions recursive, 127, 128
fundamental physics, 210, 215, 223
fusion of horizons, 347

G

Gadamer, H.-G., 347
Gale G., 280
Galilei, G., 4
Galileo, G., 238, 246, 333, 363, 367, 382, 392–395, 399, 397
Gamow G., 277
Garber, D., 342
gauge theory, 183, 195, 196, 198
general relativity, 369
geocentric, 383–385, 388–399
geocentrism, 382, 384, 393, 394, 397
geometric calculus, 182, 189, 191, 194
Geometrical Curves, 329, 330
geometrical, 50, 53, 59–64, 67, 68, 70, 84
geometry, 21, 22, 25–27, 29, 32, 33, 36, 37, 40, 43, 45, 126, 227, 228, 237, 238, 239, 245, 383, 384, 389, 396
Gerhardt, K. I., 21–28, 30, 36–47
Gerhardt, C. L., 50, 61, 64,
Gibbs, W., 265, 266, 269, 271, 277
Gillispie C.C., 302
God, 4–8, 10, 12–18, 134
Gods omniscience and omnipotence, 360
Gödel, K., 127, 247, 248
goodness of God, 345
government, 343, 357–360
Grandi, G. 141, 142, 162
Grandis series 161

Granger G. G., 316
Grassmann, H., 181–183, 188–191, 198
GrattanGuinness I., 264, 265
gravitation, 1-6, 8, 11, 13, 15–18
gravity, 57, 58, 66, 67, 70, 71, 74–88, 223
Grégoire de St. Vincent, vii
Gregory, J., 93, 96–105, 108
Grize, J-B., 126
Grosholz E., 322
Grotius, H., 359, 360
Grzegorczyk, A., 127

H

Haldane J.B.S., 253, 279, 280
Hall., A. R., 2, 20
Hand, M., 131
harmonic circulation, 60, 62, 65–67, 80, 86, 87, 398
harmonic series 151
Hartz, G. A., 169–178
Hegel, G. W. F., 342, 343, 345, 348, 351, 357, 360
Heidegger, M., 130
Heims S,.J., 257
heliocentric, 383, 384, 386, 389–399
heliocentrism, 382–384, 386, 388-390, 393, 394, 397–399
Hilbert D., 261
history of the force concept, 299, 301, 303
Hobbes, T. 8, 231, 338, 343, 357, 359
Hofmann, J. E., 96, 99, 279
homogeneity (homogeneous), 21, 25, 28, 29, 31, 33–37, 44
Honoratus Fabri, 76, 77, 79
Horn, L., 126
Hughes, R.I.G., 122
Hume D., 39 133
Humes principle, 39, 44
Husserl E., 253, 257, 265, 311
Huygens, C., vii, 49, 56, 57, 93, 93, 94, 96–101, 103–105, 108, 117, 235, 279, 290, 293, 294, 363, 365, 367, 368, 373
Huygens, Joaquim Jungius, 332-338

hypotheses, 2, 4, 6, 7, 9, 11, 12, 18, 382, 384, 386-388, 391-399
Hypothesis Physica Nova, 58, 62, 70–77

I

Idealism, 169–171, 173–177
identity, 11–13, 18, 25, 26, 30, 36-38, 42, 230–232, 245
Ignace Gaston Pardies, 331
Illustratio Tentaminis de Motuum Coelestium Causis, 77, 80, 83, 86
impetus, 365
impossibility 93, 96-108, 110,113–115, 117
incommensurability, 128, 129
inconsistency, 176, 177
indeterminate coefficients 159, 161
indeterminism, 133
infinitangular polygon, 240, 241
infinite analysis, 228, 235, 239, 246
infinite series 100, 101, 116–118, 239, 245, 249
infinitely small, viii
infinitesimal calculus, 292, 299, 300, 304
infinitesimal geometry viii, 94, 104
infinitesimal, 235, 239, 240, 248, 249
infinity, 10
influx, 6
inherence (conceptual containment; praedicatum inest subjecto), 228, 230, 235,
instrumentalism, 392, 397
insurance mathematics, viii
Intellectus, 126
intelligibility, 1, 2, 4–13, 16–18, 394, 397, 398, 399
interest accruing in the meantime, vii
intuitionist, 121, 126, 127, 129–131

J

James W, 264
Janiak, A., 2, 18–20
justice, viii

K

Kac, M., 274
Kahn, P., 259
Kant, I., 39, 41, 122, 169, 175
Kepler, J., 51, 57, 80, 81, 83, 86, 87, 289–291, 382, 387-380, 391–394, 396, 398, 399
Keplers laws, 288
kinematical, 50–54, 56, 58, 59, 84, 88
kinematics, 49, 59, 387, 394, 396 398, 399
Kleene, S. C., 127
Kline, M., viii
Knobloch, E., 51, 93, 94, 142, 154, 228, 237, 238, 246, 321
Knorre, M., 99
Koch, H. von, 273, 275
Kojève, Al., 348
Kolmogorov, A., 126,127
Koyré, A., 58, 62,
Kuhn, T. S., 4, 383–385, 393

L

L'Hôpital, 337
L'Hospital Marquis de, vii
labyrinth, 122, 236, 241, 249
Lagrange, J. L., 301–303
Lavoisier ,A. L., 127
law of continuity, 235, 236, 238
Lebesgue H., 267, 268, 271, 273, 275, 277
Leeuwenhoek A. Van, 271, 272,
Leibniz, G. W., 1–20, 21–47, 49–52, 54–88, 93–120, 121–135, 141-164, 207–224, 228-250, 257–305, 360, 396–399
Leibnizs criterion 154
Leibnizs Law, 172
Leibnizs mathematical career, vii
Leibnizian Dynamics, 296, 298,
Leibnizian Metaphysics, 288, 295, 296
Lenzen, W., 326
Levey, S. 236, 244, 246
levity, 66, 67, 75

limit, 239, 240, 244, 247
Lindgren, N. A., 260
lineation, 243, 244
Lobachevsky, N. I., 127, 128
Locke, J., 5, 8, 343, 357, 360
logic, 21-26, 28, 29, 37, 40, 44, 45, 309–328
logica mathematica, 309, 310, 313, 315–316, 320, 323, 328
logical calculus, 312, 315, 328
logical principle, 347
long division 144
Longo G. O., 275
Louis XIV, 289
Ludewig J., 314, 316
Lully R., 260

M

Machiavelli, N., 343, 357
Mahnke, D., vii
Malebranche, N.,
Mandelbrot, B., 272, 275, 276
Markov, A. A., 121, 122, 126, 132, 133, 135, 136
Masani, P. R., 257
master-slave, 342, 348, 349, 351, 357
materially equivalent, 174
Mates, B. 228, 230, 235, 247
Mathematical logic, 309–317, 319, 328
mathematical, 121, 122, 124, 136
mathematics, 21, 24, 25, 27, 32, 33, 37, 40–42, 44, 45, 309–328
Mathesis Universalis, 27, 309–320, 324–326
matter, 230, 235
Maupertuis, P. L., 267
Maxwell C., 260, 264, 268
Maxwell's equations, 216, 223
McCracken, C. J., 15, 20
McCulloch W., 263
McTaggart J., 269
Mechanical curves, 331, 332
mechanical worldview, 391
mechanics, 126, 133, 169, 175

mechanism, 5
mechanization of geometry and geometrization of mechanics, 289
Melmnaas, P.-E., 126
mereology, 309
Mersenne, 333
metaphysics, 25, 26, 37, 45
Michel-Pajus, A., 322
Mill, J.S., 39
mind, 233, 234, 235, 237, 241, 242–246, 247-249
miracles, 9, 20
Mittelstrass, J., 313
modal, 134
monad, 169–172, 174, 176, 177
monadology, 179, 181, 183–185, 187, 201, 207
monads, 10
monarchy, 357, 358, 360
Monk, R., 261
Montagnini, L., 253–286
Moreau de Maupertius, P. L., 381
Morrison, P., 133
motion, 3 4, 14, 15
Mugnai, M. 228–230, 232, 235, 247

N

natural philosophy, 2-4
necessary, 122, 124, 128–131, 133, 135
Neumann, J. Von, 262, 263
New Essays on Human Understanding, 7, 8
Newton, I., viii, 1-6, 8, 10-20, 43, 45, 49, 51, 53, 56–58, 62, 65, 70, 85, 86, 156, 157, 241, 279, 287–290, 292, 231, 305, 306, 339, 365, 366, 367, 368, 369, 370, 371, 381-384, 389, 391–394, 396, 398
Nicholas of Cusa, 232
Noica, C., 348–350,
nominalism, 230, 241
Nyce J. M., 259

O

Occams razor, 230
occasionalism, 3, 6, 8 14, 15
occult qualities, 5, 12, 17
    of contradiction, 122–124, 127–130
    of excluded middle, 127, 128
    of Markov, 121, 122, 126, 132
    of motion, 393, 399
    of nature, 391
    of Peirce, 132
    of science, 389
    of sufficient reason, 121, 124, 128, 133, 134
oligarchy, 357-359
Oliveira, A. R. E., 287–306
Olms, G., 88
optics, 5
organization of a theory 121, 129, 133
osculation, 242–244
Osiander, A., 392, 393, 397, 399
outward impression of the circulation paracentric motion, 63, 87

P

Panza, M. 142
Papin, 376
paracentric motion, 398
Parkinson, G. H. R. 229, 231
parsimony, 9
Pascal, B., vii
Pascal, A.-M. 341–361
Peano, G., 44, 273, 275
Peirce, C. S., 132, 264, 266, 270
Pelletier, A., 313
perfection, 10-13, 17
Perrin, J., 273–275
Petzold, C., 305
phenomena, 169–175, 385, 391-395, 397, 398
Philosophy of mathematics and logic, 288–293, 301, 305
Philosophy of science, 125
physical-structural model, 51, 53, 56, 63, 71, 85–87

physical, 389, 394, 396
physics, 2–4, 9, 15, 17, 19, 133
Piro, F., 124
Pisano R., 49, 50–52, 56, 71, 79, 89, 304
Pitts, W., 261
planetary, 49–64, 68–71, 74–76, 79, 80, 82–86
Plato, 12, 230, 241
political thought, 343, 357
Popper, K., 125, 264
possible worlds, 1, 2, 13, 17
Prawitz D., 126
predicate, 121–123, 127–134, 228, 230–235, 245, 246, 249
Princess of Wales, 4
Principia Mathematica, 3, 5, 15, 19
Principle architectonic 131
principle of contradiction, 25–27, 36, 38, 40
principle of sufficient reason, 13, 17, 18, 231, 237, 247, 248
Principles of Nature and Grace, 10
probability (of arguments), 392
problem-based, 130
Proclus, 27
Pruss, A. B., 121, 122, 135
Ptolemaic, 382, 384, 386, 387 392, 396, 398, 399
Ptolemy, C., 383–386, 389–392, 397, 399

Q

Quadratura arithmetica communis sectionum conicarum 141, 142, 144, 147–149, 151, 154, 157, 158, 161
quadrature of the circle 93, 94, 96, 98–104, 106, 117
quadrature, 237, 240
quality, 28, 42
quantifier, 129, 132
quantity, 41
quantum mechanics, 133
quatenus, 234
Quine, W. V. O., 123, 245

R

Rabouin D., 314
radial velocity, 70
rapport with the other, 348, 360
ratio, 122, 126
ratio, recta (straight line), 28, 29, 34–36
rationalism of faith, 342–348, 358–360
rationalist philosophy, 289
rationality of the world, 346
real, 171, 174
realism, 171, 172, 175, 176, 229, 247–249, 392
recognition game, 343–345, 348, 358–360
rectilinear motion, 368, 370, 371
rectilinear motions, 371
reduction, 9
reference frame, 395, 396
relation of faith and reason, 342–344, 350, 354, 360
relation, 228–235, 238, 247, 249
relational space, 183, 188
relativity of motion, 382, 384, 394, 395–399
Remond, 339
republicanism, 358, 360
res extensa, 364
Rescher, N., 11, 20, 123, 126, 228, 231, 232, 235, 239, 241
retrograde motion, 386, 388–390, 397, 399
Rheticus, G. J. 391–393
Rosenblith, W., 257
Rossi, P., 258
Rousseau, J.-J., 343, 357, 360
Royal Society, 3
Royce J., 253, 264, 266–271, 273
Ruin H., 124
Russell, B., 4, 20, 23, 229, 233, 236, 239, 249, 253, 257, 261, 262, 265, 270, 273, 288, 311, 315, 316
Rutherford, D., 169–174, 176

S

saltus (leap), 236, 238, 239, 241
Sasaki, C., 316
saving the appearances, 387
scalar-tensor theories of gravity, 221–223
Schmidt, J. A., 313, 314
Schneider, M., 312, 313, 316, 318, 320
scholasticism, 3
scholastics, 290, 291
Scholtz, L., viii
Schuster, J. A., 71
Science, 125, 126, 128, 136
Scientific Revolution, 306, 383–386, 391
scientific theories, 121, 125–133, 136
seasons, 390, 391, 394, 396,
Segal, J., 276
self-awareness, 349, 350
self-consciousness, 349, 350, 352
sequence, 134, 236, 239, 240, 248
Serfati, M., 321
Serres, M., 313
Shannon, C.E., 262, 263, 276–278
similarity, 21, 23, 25, 27–33, 35, 38, 45
simple substance, 10
simplicity, 9, 10, 18, 394, 395, 397–399
Singer, P., 348, 349
situs, 231, 235,
Sleigh, R. Jr., 121, 122, 130, 131, 134
Smith, J.E.H., 272
Smith, G. E., 19, 20
Smoluchowski, M., 275
social contract, 360
solar attraction, 87, 88
solar system, 57, 79, 82, 83, 85, 88
solicitation of gravity or levity, 88
solidarity, viii
sovereign, 358–360
special relativity, 126
Specimen Dynamicum, 3, 51, 171, 172
Speciosa generalis, 310, 312, 319
Spencer, M., 268
Spinoza, B., 13, 15, 16, 235, 241, 264, 265, 272, 279
spirit, 343, 350, 357
St Paul, 352, 353
Stapp, H.P., 280

Stevenson, G. P., 20
Stifel, 232
strategy, 342, 345–348, 353, 354
Stratton, J.A., 262
Suárez, F., 27, 47
subject (hypokeimenon), 230
subordination, 349, 350, 357
substance, 169–176
sum of a series 141–143, 146, 147, 164
Supplementum geometriae practicae sese ad problemata transcendentia extendens, ope novae methodi generalissimae per series infinitas 142, 159–161
supreme reason, 345, 346, 351, 355, 356
Swift, J., 271
Swoyer, C. 229, 237
symmetries, 133
synthesis, 228, 229, 239
synthetic, 123

T

Tabacchi, M. E., 254
tangent, 235, 237–240
Taylor, C., 342, 348, 357
Tennant, N., 125, 131
Tentamen de motuum coelestium causis, 3, 57, 79, 82, 83, 85, 88
Termini, S., 254
the lazy argument, 346
the Meditations, 242
The Search for Truth, 19
Theaetetus, 233, 234
Theocentrism 290
Theodicy, 4, 360, 361
Theology, 126
theory of numbers, 289
thermodynamics, 126, 127
Thomas Aquinas, 232, 234, 245,
to apeiron, 240, 247
transcendentals, 237, 245
transverse velocity, 62, 65
Trendelenburg, F.A., 47

triumph of faith over reason, 344, 346, 352, 353
triumph of reason over faith, 346
Troelstra, A., 126, 131
Truth, 122–124, 128–130, 133, 135, 346–349, 353, 355, 356, 397, 398
Tschirnhaus, E. W. von, 319, 332
Turing, A., 126, 127, 263
Turing machines, 303
Turing-Churchs thesis, 126, 127

U

unity, 25, 28, 29, 31–38, 40, 41, 43, 45
universal algebra, 312, 316
Universal Mathematics, 1, 4, 309, 311, 312, 318, 319, 326, 328
unknown God, 342

V

vacuum, 9, 10
van Dalen, D., 126, 131
variety or variance, 179, 183, 184, 188, 192, 200
Varignon, P., 162, 163
Venezia A., 125
virtue v. vice, 342, 348, 351
vis viva, 296
vortex theory 389, 398, 399
vortex, 63, 71, 74, 79, 82, 85–87
vortices, 79, 83, 84

W

Wallis, J., 318, 326
Walton, E., 262
Weaver, W., 262
Weierstrass, K., viii
Weigel, E., 35, 36, 47
Westfall, R. S., 175, 288, 290, 296
Weyl, H., 311
Whitehead, A. N., 261, 270
whole greater than its part , 37, 38
Whole-parts relationship, 311, 324–327
Wiener, N., 253, 254, 258–281
Wiesner, J., 257

Wiggins, D., 124, 131, 135
Wildes, K.L., 262
Wilson, C., 176, 177
wisdom, 346, 348, 355, 356
Wolff, C., 1, 11–13, 19, 122, 123
work and energy history, 287, 291, 306
work, 348, 349–351, 353, 357
World 124, 125, 135, 136

Y
Yakira, E. 324

www.ingramcontent.com/pod-product-compliance
Lightning Source LLC
Chambersburg PA
CBHW050830230426
43667CB00012B/1947